Floral Treasures of Shimla Hills, India: Conservation and Discovery

ABOUT THE AUTHORS

Dr. Suman Rawat is working as Assistant Professor (Botany) in Himachal Pradesh University. She has obtained her M.Sc. M.Phil, Ph.D (BOTANY) from Himachal Pradesh University and awarded Gold Medal for her work in M.Phil. She has published several research papers and books in the field of ethnobotany. She has 10 years of experience in teaching PG and UG classes. Right now, she has a keen interest in 'Ethnobotany' and 'Biodiversity Conservation' and wants to excel in this field.

Dr. S.K. Sood was the former Dean and Professor in the Department of Biosciences, Himachal Pradesh University, Shimla. He obtained his M.Sc. (1974) and M.Phil. (1976) from Himachal Pradesh University and Ph.D. (1981) from Nagarjuna University, Guntur, specializing in Plant Systematics and Embryology. For his post-doctoral, he underwent specialized training in Ethnobotany. He had a brilliant academic career (triple gold medalist) followed by over 35 years of teaching and conducting research in Botany at postgraduate as well as research levels. Dr. Sood has published several research papers in various national and international journals He has attended and participated in many seminars, workshops and conferences. He is also associated with many educational and research bodies in India and abroad. A profile writer, his books 'Ethnobotany of Cold Desert Tribes of Lahoul-Spiti (N.W. Himalaya)'; 'Ethnobotany of Rewalsar Himalaya'; 'Ethnic Aphrodisiac Plants'; 'Ethnic Indian Plants in Cure of Diabetes'; 'Sacred and Magico-religious Plants of India'; 'Ethnic Plants of India Used in Cancer Cure – A Compendium'; Orchidaceae and Mankind': 'Honey Plants of the North-West Himalaya'; Edible Roots and Underground Stems of Ethnic India'; 'Fibre and Fodder Resources of India'; 'Root Drug Plants of Ethnic India'; 'Ethnobotanical Studies on Trees, Shrubs and Climbers of Himalaya'; 'Dye, Masticatory and Fumitory Resources of India'; 'Flavouring and Fragrant Resources of India'; 'Healing Herbs: Traditional Medications for Wounds, Sores and Bones'; 'Handbook on Wild Edible Plants'; 'Ethnic Healing Herbs for Cold, Flu and Lungs Ailments'; 'Indian Herbs for perfumery and Beauty Care'; 'Ethnobotany of Himalayan Region' were widely acclaimed. His current major interests are in the theory and practice of systematics, flora of North-West Himalaya, ethnic tribes and conservation research.

Floral Treasures of Shimla Hills, India: Conservation and Discovery

Dr SUMAN RAWAT
M.Sc., M.Phil., Ph.D, (Gold Medalist)
ASSISTANT PROFESSOR

Prof. S. K. SOOD
(M.Sc., M.Phil., Ph.D., F.I.B.S., F.E.B.S.),
FORMER CHAIRMAN & DEAN OF LIFE SCIENCES

DEPARTMENT OF BIOSCIENCES
HIMACHAL PRADESH UNIVERSITY
SHIMLA-171005
INDIA

SCIENTIFIC®
PUBLISHERS
New Delhi • Jodhpur

SCIENTIFIC® PUBLISHERS
New Delhi • Jodhpur

Published by:

Jodhpur
5 A, New Pali Road, P.O. Box 91
Jodhpur - 342 001 Rajasthan (India)
E-mail: info@scientificpub.com
Website:
www.scientificpubonline.com

New Delhi
4806/24, Ansari Road, Daryaganj,
New Delhi - 110 002 (India)

Print 2025

ISBN: 978-81-962464-7-1
eISBN: 978-81-962464-8-8

Printed in India

प्रो० (डॉ.) सिकन्दर कुमार
Prf. (Dr.) Sikander Kumar

सत्यमेव जयते
आज़ादी का अमृत महोत्सव

सांसद राज्यसभा
व पूर्व कुलपति हिमाचल प्रदेश
विश्वविद्यालय शिमला
Member of Parliament
Rajya Sabha & Former Vice
Chancellor H.P.U. Shimla

क्रमांक MPRS/SML-179

दिनांक 13/9/2023

Foreword

In my role as the Vice-Chancellor of Himachal Pradesh University, I had the privilege of immersing myself in the captivating tapestry of Shimla, where I discovered a harmonious blend of natural beauty, rich culture, and an astonishing diversity of plants. Shimla, renowned worldwide for its architectural gems from the British colonial era, boasts a unique landscape that bears the imprints of its historical past. These heritage sites, largely shaped by the colonial rulers, hold within them a wealth of plant biodiversity.

Within the precincts of Shimla's heritage sites, one can encounter a fascinating array of herbs, shrubs, and trees, each possessing ornamental, medicinal, or economic value, drawn from both indigenous and exotic sources. Regrettably, the exuberance of plant life within these heritage locations faces a perilous decline, owing to the detrimental impact of human activities.

In their meticulously researched book, the authors dedicated five years to explore the heritage sites of Shimla, compiling a comprehensive repository of information about plants of significant importance. Moreover, the book not only delves into the botanical aspects but also offers a profound insight into the history, culture, and architectural heritage of Shimla.

Remarkably, the book unfolds the narrative of numerous plant species that thrived during the colonial period, further enriching our understanding of the region's history. It leaves no stone unturned in its vivid depiction of Shimla's culture, natural beauty, and plant diversity, serving as a powerful call to action for the conservation of this splendid heritage.

With genuine enthusiasm, I extend my heartfelt congratulations to the authors. This book stands as an invaluable resource, catering to a diverse audience of entrepreneurs, scientists, conservationists, geneticists, explorers, and researchers. Its pages are a testament to the enduring legacy of Shimla, beckoning us to safeguard and cherish this unique treasure.

(Sikander Kumar)

4, फिरोजशाह रोड, नई दिल्ली – 110001
पता – गांव : ढगोह, डाकघर व उपतहसील : कांगू, जिला : हमीरपुर (हि०प्र०)–177040
कार्यालय : दीपकमल भवन, कामना नगर, चक्कर, शिमला (हि०प्र०)–177040
E-mail : sikanderkumar.2010@rediffmail.com & Office E-mail : mediacell.bjp.hp@gmail.com
Mo : 9418025437, Office 0177-2831892 (R) 0177-2830024

PREFACE

Plants and flowers have been planted all over the world since time immemorial. This leads the attention of humanity toward the development of planned horticulture. The history of organized horticulture in India dates back to the Ramayana and Mahabharata period or even earlier. Many beautiful trees like the sacred fig, banyan tree, fire of the forest, basil, etc. are mentioned in the 'Ashtadhyayi,' an ancient piece of literature by the great Sanskrit scholar Panini. Even older literature described the importance of gardens and plants used in the Indus Valley Civilization (2500–1750 BCE). Early evidence suggests that the Aryans who invaded India were responsible for the introduction and naturalization of exotic plants in Indian territory.' Bio-aesthetics planning refers to the conscious planning of flora and fauna with the aim of beautifying the country. This includes the plantation of ornamental flowering trees in city streets, parks, public places, and compounds of houses of towns and villages. As, 'Shimla' is known worldwide for its unique architectural British-era heritage buildings and landscape. Most of these heritage sites are rich in plant biodiversity mainly introduced by the colonial rulers. The precincts of heritage sites in Shimla hills are rich in genetically diverse populations of herbs, shrubs, and trees of ornamental, medicinal and of other economic value, derived from indigenous and exotic resources. Due to such a diverse richness of plant diversity, it is of paramount importance to identify these sites as ex-situ centres of conservation. Apart from this, these cited plants play an important role in the development of plantation culture all over the world. Unfortunately, most of the plant wealth at these heritage sites is decreasing alarmingly due to degradation caused by anthropogenic activity. The authors have thoroughly explored these heritage sites for five years and have provided detailed information on '202' plants of various importance, which is a very good database for geneticists, conservationists, researchers, and ethnobotanists. The history, architecture and uniqueness of 121 heritage buildings, 5 cemeteries and 8 socio-cultural sites are beautifully illustrated with photographs. The textural work is arranged according to the IPNI (International Plant Names Index) and expounded with the help of tables, figures and photographs. It is hoped that our work will inspire researchers to conduct such studies to validate traditional knowledge.

ACKNOWLEDGEMENT

Well-deserved appreciation is due to Prof. Sikander Kumar (Former Vice-Chancellor, Himachal Pradesh University), Prof. Anand Sagar (Former Dean of Life-Sciences), Dr. Dheeraj Singh Rawat (Chairman, Biosciences), Dr. Meena Kumari, Dr. Suresh Kumar for their encouragement and well wishes.

Furthermore, we would like to express our gratitude towards Dr. Sunil Kumar Dhatwalia for his cooperation and critical suggestions. The authors are grateful to their family members for their support and encouragement during the work.

At last, I am thankful to Almighty God who has given me the best thoughts and strength to complete this work

ABBREVIATIONS

amsl	Above mean sea level
Alc.	Alcoholic
Alk.	Alkaline
Aq.	Aqueous
C	Carbon
C.	Central
Ca	Calcium
cm	Centimeter
D	dextro
E.	East
EL	Ethnobotany Laboratory
Eng.	English
et al.	(et alia) and other authors
Etc	(et cetera) and so on
Extr. (s)	Extract (s)
Fig. (s)	Figure (s)
Fl.	Flowering
Fr.	Fruiting
HCN	Hydrogen Cyanide
H.P.	Himachal Pradesh
i.e.	(id est) that is
L	Laevo
M	Metre
mg	Milligram
mg	Magnesium
M.P.	Madhya Pradesh
ml	Milliliter
N.	North
(-)	Negative
N.W.	North West
O	Oxygen
OAc	Acetoxy
OH	Hydroxy
OMe	Oxymethyl
%	Per Cent
P	Phosphorus
(+)ve	Positive
(-)	Negative
Ph	Phenyl
Pl. (s)	Plate (s)
S.	South
Sans.	Sanskrit
Sp	Specie
Spp	Species
Sq m	Square meter
Sq km	Square kilometer
St.	Saint
Syn.	Synonym
S.W.	South West
TIV	Total Importance Value
U.P.	Uttar Pradesh
Var.	Variety
Vern.	Vernacular
viz.	Vide licet (namely)
Vet.	Veterinary
Vit.	Vitamin
Vol. (s)	Volume (s)
W.	West
&	And
°	Degree

LIST OF ILLUSTRATIONS

Figures:

COLOUR PHOTOGRAPHS

PHOTOGRAPHS OF HERITAGE SITES OF SHIMLA

HERITAGE SITES OF SHIMLA HILLS

A. Aira Holme

B. Alderton

C. Allan Bank

D. Bantony

E. Bishop Cotton School (B.C.S)

F. Sadar Circuit House

G. Chapslee

H. Christ Church Complex

A. Clarkes Hotel

B. Cleremont

C. Conny Cottage

D. Crow-Brough

E. . Curzon House

F. . Dalziel House

G. Eddleston

H. Ellerslie Building

A. Foss Well

B. Glen Uquard

C. Gorton Castle

D. Hainault

E. Holcombe

F. Holly Lodge

G. Inverarm

H. . Kali Bari Temple

A. Knockdrin

B. Manor Villa

C. Mythe

D. Pari Mahal

E. Post office, Chhota Shimla

F. Primrose Hill Cottage

G. Railway Board B[illegible]

H. Railway Station, Summer Hill

A. Vice-regal Lodge Complex

B. Winter Field

C. Wood Villa

D. Annandale Ground

E. Naldehra Golf Course

F. Nuns Graveyard St. Bede's at Nav Bahar

Cemetery

PLANT LIST OF COLOURED PHOTOGRAPHS	
1. *Abutilon* X *hybridum*	27. *Hydrangea macrophylla*
2. *Abutilon megapotamicum*	28. *Hypericum oblongifolium*
3. *Acer palmatum*	29. *Indigofera dosua*
4. *Aesculus indica*	30. *Jasminum humile*
5. *Aloysia citriodora*	31. *Kerria japonica*
6. *Araucaria columnaris*	32. *Liriodendron tulipifera*
7. *Berberis lycium*	33. *Magnolia grandiflora*
8. *Buddleja asiatica*	34. *Magnolia obovata*
9. *Callistemon lanceolatus*	35. *Magnolia X soulangeana*
10. *Camellia japonica*	36. *Nerium oleander*
11. *Campsis radicans*	37. *Padus cornuta*
12. *Cerasus cerasoides*	38. *Paulownia tomentosa*
13. *Cestrum elegans*	39. *Platanus orientalis*
14. *Clematis buchananiana*	40. *Pyrus pashia*
15. *Clematis montana*	41. *Rhododendron arboreum*
16. *Colquhounia coccinea*	42. *Rhododendron simsii var. vittatum*
17. *Diospyros lotus*	43. *Robinia pseudoacacia*
18. *Drepanostachyum falcatum*	44. *Rosa moschata*
19. *Erythrina arborescens*	45. *Salvia leucantha*
20. *Euonymus fimbriatus*	46. *Salvia splendens*
21. *Euonymus tingens*	47. *Solanum laxum*
22. *Fuchsia magellanica*	48. *Spiraea cantoniensis*
23. *Ginkgo biloba*	49. *Tropaeolum majus*
24. *Hibiscus rosa-sinensis*	50. *Wisteria sinensis*
25. *Hibiscus syriacus*	51. *Yucca aloifolia*
26. *Himalrandia tetrasperma*	

ETHNOBOTANICAL INFORMATION OF PLANTS AT HERITAGE SITES OF SHIMLA

A. *Abutilon* x *hybridum*

B. *Abutilon megapotamicum*

C. *Acer palmatum*

D. *Aesculus indica*

E. *Aloysia citriodora*

F. *Araucaria columnaris*

G. *Berberis lycium*

H. *Buddleja asiatica*

A. *Callistemon lanceolatus*

B. *Camellia japonica*

C. *Campsis radicans*

D. *Cerasus cerasoides*

E. *Cestrum elegans*

F. *Clematis buchananiana*

G. *Clematis montana*

H. *Colquh*

A. *Diospyros lotus*

B. *Drepanostachyum falcatum*

C. . *Erythrina arborescens*

D. *Euonymus fimbriatus*

E. *Euonymus tingens*

F. *Fuchsia magellanica*

G. *Ginkgo biloba*

H. *Hibiscus rosa-sinensis*

A. *Hibiscus syriacus*

B. *Himalrandia tetrasperma*

C. *Hydrangea macrophylla*

D. *Hypericum oblongifolium*

E. . *Indigofera dosua*

F. *Jasminum humile*

G. *Kerria japonica*

H. *Liriodendron tulipifera*

A. *Magnolia grandiflora*

B. *Magnolia obovata*

C. *Magnolia* X *soulangeana*; flowers

D. *Nerium oleander*

E. *Padus cornuta*

F. *Paulownia tomentosa*

G. *Platanus orientalis*

H. *Pyrus pashia*

A. *Rhododendron arboreum*

B. *Rhododendron simsii* var. *vittatum*

C. *Robinia pseudoacacia*

D. *Rosa moschata*

E. *Salvia leucantha*

F. *Salvia splendens*

G. *Solanum laxum*

H. *Spiraea cantoniensis*

A. Tropaeolum majus

B. Wisteria sinensis

C.Yucca aloifolia

CONTENTS

Chapter 1
INTRODUCTION

PLANT CULTIVATION AND BIO-AESTHETIC PLANNING

India (Lat. 8°-04'N to 37°-06'N and Long. 68°-07'E to 97°-25'E) with its botanical canopy comprising over 15,000 species out of the total 4,22,000 flowering plants reported from the world ranks as one of the 17 mega-diverse countries in the world (Govaerts, 2001; Mc Neely *et al.*, 1990). Its diversity is unmatched due to the 16 different agroclimatic zones, 10 vegetative zones, and 15 biotic provinces (Perumal Samy & Gopalakrishnakone, 2007). The country has four out of thirty-four global biodiversity hot spots, which is an indicator of high degree of endemism of the species (Goyal & Arora, 2009). About 168 domesticated crop species and 336 species of their wild relatives originated and/ or developed diversity in India (Arora & Nayar,1984; Zeven & de wet, 1982). It has been estimated that about 6,000 species belonging to over 141 genera of 47 families (about 33% of the flowering plants) are endemic to India (Hajra & Mudgal, 1997). The flora of India shows close affinity with the flora of Indo-Malayan and Indo-Chinese region. Based on the distribution of flora, the Indian sub-continent has been sub-divided into various botanical or floristic provinces by Clarkes (1898), Hooker (1906), Chatterjee (1939) and Razi (1955). According to Nayar (1977) 35% of Indian flora has South-East Asian and Malayan, 8% Temperate, 1% Streppe, 2% African, 5% Mediterranean Iranian elements and 18% adventive weeds and naturalized aliens. In fact, the ten dominant families in Indian flora include Orchidaceae, Fabaceae, Poaceae, Rubiaceae, Euphorbiaceae, Acanthaceae, Asteraceae, Cyperaceae, Lamiaceae and Urticaceae (Hajra & Mudgal, 1997). There are over 550 native or tribal communities under 227 ethnic groups occupying about 18.74% of the total area under varied geographical and climatic zones in India (Singh & Arora, 1978). Harmony with nature is inherent in the culture and way of life of tribal societies. There is growing recognition that biodiversity is a global asset of tremendous value to present and future generations. Unfortunately, plant diversity is facing tremendous threat mainly because of unsustainable harvesting for their multifarious utilization and habitat degradation. Each day, a minimum of 140 plant species/ biotypes are becoming endangered and some 30,000 are condemned to extinction every year (Paroda, 1996). According to IUCN, if the current trends continue unchecked, 25% of the world's species may be lost by 2050 AD. The loss will immensely effect the present and future generations as the species lost today may have food, medicinal and industrial value presently not known to mankind (Chowdhery & Murti, 2000).

Since time immemorial all over the world, plants and flowers have engaged the attention of humanity leading to the development of planned gardening. The history of organized gardening in India dates back to the periods of Ramayana and Mahabharata or even earlier. There is a mention of 'Ashokvana', the garden having predominantly Ashoka (*Saraca indica*) trees, in which Sita (wife of Lord Rama) was held captive. Several beautiful trees like Sacred fig, Banyan tree, Flame of Forest etc. find their mention in Ashtadhyayi, an ancient (6th Century BC) piece of literature by the great Sanskrit scholar Panini (Paliwal, 2009). The importance of gardens and plants for the people in ancient India has also been evident from the literature on Indus Valley Civilization (2500-1750 BC). Regarding introduction and naturalization of foreign plants to Indian region, Chatterjee (1947) stated that it dates back to the early Aryans who invaded India from countries in the N.W. Eurasia. However, the term 'bioaesthetic planning' was coined by Professor Lancelot Hogben and it refers to conscious planning of the flora and fauna with the object of beautifying the country. Since the planning embraces both the plant and animal sciences, it may further be defined as planned ecology of living beings from the artistic and aesthetic points of view. It includes the plantation of ornamental flowering trees along city roads, in parks, public places and compounds of houses both in towns and villages, and development of national parks for the preservation of beautiful, non-carnivorous animals, and the creation of bird's sanctuaries (Randhawa, 1947). Bio-aesthetic planning, of course, embraces landscape gardening as well; and runs hand in hand with town planning. Of late, the world over, people have begun thinking of the future needs of the community and have realized the value of bio-aesthetic planning.

The State of Himachal Pradesh with a geographical extent of 55,673 sq kms nestling under the lesser Himalayas between latitude 30°22'-33°12'N and longitude 75°47'-79°46'E, in general and its summer capital - Shimla in particular is known worldwide due its unique architectural British-era heritage buildings that dot the cityscape. Most of these heritage sites are rich in plant biodiversity introduced mainly by colonial rulers, and thus constitute the precisely delineated areas of outstanding universal value from the point of view of genetic diversity, conservation and natural beauty. Unfortunately, much of the plant wealth in courtyards of these built heritage structures, cemeteries and socio-cultural spaces is depleting alarmingly due to degradation caused by anthropogenic activities and wheels of progress, and needs to be documented for devising any comprehensive conservation strategy before it is too late.

ORNAMENTAL AND CONSERVATIONAL ASPECTS

The term 'Ornamental Plants' or 'Ornamental' is used to describe those species primarily cultivated for their aesthetically – pleasing characteristics such as form, bark, leaves, flowers, fruits or some combination of thereof (Davidson & Miller, 1990), Nature has given a wealth of wild flowers and ornamental plants, unfortunately many of them have been destroyed to such an extent that several have become extinct and survival of many is endangered due to exploitation by human beings (Arora, 1993). Of late, identification and exploration of wild ornamental species is one of the new areas of research and accounts wide spectrum of uses in environmental management (Reddy *et al.,* 2012). During the past, a number of books, monographs and treatises containing wealth of information on ornamental horticulture, horticultural crops, landscape gardening, ornamental germplasm and bioaesthetic planning have appeared. In this regard, important contributions include works such as *Indian Trees. An Account of Trees, Shrubs, Woody Climbers Bamboos and Palms Indigenous or Commonly Cultivated in British Indian Empire* (Brandis, 1921), *Flowering Trees and Shrubs in India* (Cowen, 1950), *The Book of Garden Flowers* (Phillips, 1954), *Manual of Cultivated Plants* (Bailey, 1957), *Flowering Trees of India* (Randhawa, 1957), *Flowering Trees of the World (for Tropics and Warm Climate)* (Menninger, 1962), *Flowering Shrubs* (Pal & Krishnamurthi, 1967), *A Field Guide to the Flowers of Europe* (Polunin, 1969), *Flowering Vines of the World: An Encyclopedia of Climbing Plants* (Menninger, 1970), *Beautiful Climbers of India* (Pal, 1972), *Ornamental Horticulture in India* (Randhawa, 1973), *A Dictionary of the Flowering Plants in India* (Santapau & Henry, 1973), *Some Exotic Ornamentals of Kashmir* (Singh & Misri, 1974), *Exotica, Pictorial Cyclopedia of Exotic Plants from Tropical and Near-Tropic Regions* (Graf, 1980), *Tropica, Color Cyclopedia of Exotic Plants and Trees for Warm Region Horticulture - in Cool Climate the Summer Garden or Sheltered Indoor* (Graf, 1981), *The Illustrated Encyclopedia of Flowering Plants* (Krisa, 1982), *The Dictionary of Indoor Plants in Colour* (Hay *et al.*, 1983), *Exotic Trees and Shrubs of Kashmir* (Singh & Kachroo, 1983), *Field Guide to the Common Trees of India* (Bole & Vaghani, 1986), *Shrubs for Tropical and Subtropical Gardens* (Oakman, 1990), *Tropical Garden Plants in Colour* (Bose & Chowdhury, 1991), *The New RHS Dictionary of Gardening* (Huxley, 1992), *The American Horticultural Society A-Z Encyclopedia of Garden Plants* (Brickell *et al.*, 1997), *Conservatory and*

Indoor Plants (*Vols I & II*) (Phillips & Rix, 1998), *Encyclopedia of Garden Plants* (Booth, 1999), *Some Beautiful Indian Climbers and Shrubs* (Bor & Raizada, 1999), *Trees of India (Medicinal, Commercial, Religious and Ornamental): A Colour Atlas* (Warman, 1999), *Illustrated Encyclopedia: Garden Plants* (Berry & Bradley, 2000), HIMFLORDAT: *A Computerized Database of Ornamental Plants for Himachal Pradesh* (Thakur *et al.* 2003), *RHS A-Z Encyclopedia of Garden Plants* (Brickell, 2003), *The Complete Encyclopedia of Trees and Shrubs: Descriptions, Cultivation Requirements, Pruning, Planting* (Wasson, 2003), *Tropical Flowering Plants* (Llamas, 2003), *Garden Plants in Colour: House Plants* (Bose *et al.*, 2004), *Advances in Ornamental Horticulture, 6 Vols* (Bhattacharjee, 2006), *Ornamental Plants and Flowers of Tropical Mexico* (Trapp, 2006), *Flowers and Gardens in India-A Manual For Beginners* (Temple Wright, 2007), *Flowers of Sahyadri* (Ingalhalikar, 2007), *Trees and Shrubs for Landscape Use* (Sharma *et al.*, 2007), *Dirr's Hardy Trees and Shrubs-An Illustrated Encyclopedia* (Dirr, 2009), *Wild Ornamental Bulbous Plants of Himachal Pradesh for Landscaping* and *Wild Ornamentals of Himachal Pradesh* (Thakur *et al.*, 2009, 2010) and *Trees of Vidarbha –A Field Guide to Flowering Trees* (Dhore, *et al.*, 2011).

Likewise, a number of publications describing conservational practices of tribal of Purulia and their relevance in eco-development programmes (Pal & Mudgal, 1985), strategies for sustainable utilization of plants resources by the tribals of Ton Valley, Western Himalaya (Rana *et al.*, 1996), role of ethnobotany in development and conservation of resources (Maheshwari, 1995), plant conservation practices in Maharashtra during 17th Century (Kulkarni & Kumbhojkar, 2000), sacred conservation practices at species level through tree worship in ancient India (Sane & Ghate, 2006) and plant diversity conservation in academic institutions: an efficient approach for conserving biodiversity across ecological regions in India (Nautiyal, 2010) have also been done from Indian perspective. Undeniably, the study of various aspects of the interactions between humans and local ecosystems/landscapes is a big concern all over the world, as it provides the basis for designing and developing strategies for ensuring the future sustainability of landscapes (Mander & Jongman, 1998; Schmidt-Vogt, 1998; Roberts *et al.*, 2002).

STATUS OF FLORISTIC DIVERSITY IN INDIA

From the Indian perspective, the earliest credit for scientific beginning of plant exploration work goes to Heinrich Van Rheede-the Governor of Dutch possession in Malabar in 1667, who made the first real attempt to explore systematically the plant wealth of India. His monumental work *Hortus Indicus Malabaricus* written and published in latin (12 Vols) from Amsterdam, is the oldest comprehensive printed book on the natural plant wealth and still regarded as a valuable contribution to floristics in India. This book attracted the attention of numerous workers to the richness of flora of India. Thereafter, account of Indian plants in the form of lists was included in the works entitled *Flora Indica* (Burman, 1768) and *Flora Malabarica* (Burman, 1769), followed by other valuable contributions viz., *Plants of Coast of Coromandel* (1795-1819), *Hortus Bengalensis* (1814) and *Flora Indica* (1832) by William Roxburgh, *Prodromus Florae Peninsulae Indiae Orientalis* (Wight & Walker-Arnolt, 1834), *Icones Plantarum Indiae Orientalis* (Wight, 1838-1853); *Illustrations of Botany and Other Branches of the Natural History of the Himalayan Mountains and Flora of Cashmere* (Royle, 1833-1840), *A Catalogue of the Plants Growing in Bombay and its Vicinity* (Graham, 1839), *Plantae Rarios Quas in Indae Orientali Collegit Victor Jacquemont* (Cambessedes, 1844), *Plants of Canara and Neighbouring Areas of Maharashtra and Malabar* (Hohenacker, 1849), *Notulae und Plantas Asiaticas* (Griffith, 1854), *Flora Indica* (Hooker & Thomson, 1855), *Catalogue of the Indigenous Flowering Plants of the Bombay Presidency* (Dalzell, 1858), *The Bombay Flora* (Dalzell & Gibson, 1861), *Handbook of Indian Flora* (Drury, 1864), *Report on the Vegetation of the Andaman Islands* (Kurz, 1870), *Forest Flora of North-West and Central India* (Brandis, 1874), and *A Sketch of the Vegetation of the Nicobar Islands* (Kurz, 1876). During the closing quarter of the 19th Century, publication of Hooker's *Flora of British India* in 7 Vols during 1872-1897 provided necessary impetus for all the later provincial or local Indian floras published till date. Since then, many comprehensive floristic studies on Indian flora had been published for different regions by various workers like Edgeworth (1867), Stewart, (1868), Brandis (1874, 1906), Gamble (1898), Coventry (1901), Cooke (1901-1908), Kanjilal (1901, 1911, 1933), Collett (1902), Prain (1903), Duthie (1903-1929), Rama Rao (1914), Fyson (1915, 1932), Gamble & Fischer (1915-1936), Bamber (1916), Osmaston (1927), Blatter & Hallberg (1918), Parker (1918), Haines (1921-1925), Parkinson (1923), Kanjilal & Gupta (1928), Kanjilal *et al.* (1934-1940), Santapau (1953, 1958, 1962), Bor (1960), Maheshwari (1963, 1966), Champion & Seth (1968), Ramaswamy & Razi (1973), Saldanha & Nicolson (1976), Singh & Kachroo (1976), Srivastava (1976), Babu (1977), Kachroo *et al.* (1977), Oommachan (1977), Singh & Tiagi (1979), Rao & Razi (1974), Deb (1981, 1983), Manilal & Sivarajan (1982), Rao & Verma (1982), Dhar & Kachroo (1983), Chowdhery & Wadhwa (1984), Saldanha (1984), Haridasan & Rao (1985), Rao (1985-1986), Joseph (1986, 1987), Manilal (1988), Henry *et al.*, (1989), Roy *et al.* (1992), Aswal & Mehrotra (1994), Saxena & Brahmam (1994-1996), Deorani & Naithani (1995), Chowdhery (1998), Gaur (1999), Bedi (2000), Hynniewta *et al.* (2000), Kumar (2001), Sharma (2002), and Pullaiah (2004). In recent years *Dictionary of Angiosperm Families and their Genera* (Sambamurty, 2002) and *Flora of Barmati* (Bhagat *et al.*, 2009) have been also published.

STATUS OF FLORISTIC DIVERSITY IN N.W. HIMALAYAS

Since the early 19th Century, efforts of numerous botanical explorers, mountaineers, surveyors, engineers and foresters contributed significantly towards our present knowledge of the vegetation and flora of N.W. Himalayas (Strachey, 1852; Duthie, 1906; Stewart, 1917; Osmoston, 1927; Gupta, 1957, 1962; Rau, 1974, 1975; Rao, 1978). Extremely rich and varied flora of the region, several kinds of vegetational complexes, and scenic grandeur are the reasons for attracting the attention of workers from far off countries. Though Hardwick (1798; cited in Burkill, 1965) seems to be the first botanist who collected plants from the Himalayan region but the real credit for most comprehensive collections from the region (Bashahr, Kinnaur, Cashmere Himalayas) goes to Royle (1833-1840) who gave an illustrated account of species collected during these expeditions in *Illustrations of Botany and Other Branches of the Natural History of the Himalayan Mountains and Flora of Cashmere* around 1830's. In the mid 19th Century, Hooker & Thomson (1855) collected about 2,000 plants from N.W. Himalayas. Duthie (1882) documented plants of W. Himalayas from Kashmir to Kumaon region. Hooker's *Flora of British India*, published during the closing quarter of the 19th Century, described the spermatophytes of the country including the Himalayas. Collett's (1902) *Flora Simlensis* described 1,326 species of flowering plants from Shimla and adjoining areas. Though nomenclaturally outdated, this document still holds its sway as a major work for identification of plants of the region. Based upon extensive survey of the Kumaon region including the adjoining areas of Garhwal between 1913-1926, Osmaston published the *Forest Flora of Kumaon* in 1927 whereas Kanjilal & Gupta (1928) compiled *Forest Flora of the Chakrata, Dehra Dun and Saharanpur Forest Divisions*. Pennel (1943) carried out in depth studies on the Scrophulariaceae of the W. Himalayas. Singh & Kachroo (1976) documented *Forest Flora of Srinagar and Plants of Neighbourhood.* In 1977, Kachroo *et al.* compiled *Flora of Ladakh* and C. R. Babu documented and described *Herbaceous Flora of Dehradun*. The work on *Flora of Mussoorie* has been that of Raizada & Saxena (1978). *Flowers of Himalaya* by Polunin & Stainton (1984, 1987) with 689 colour photographs and 316 line-drawings and description of over 1,500 plant species has been considered one of the most useful books for studying the floristic diversity of Nepal and parts of India falling in the W. Himalayas. Work on critical taxonomic treatment of 401 weed species represented by 223 native and 178 introduced species has been published by Kaul (1986). Based on exsiccata housed at BSD, BSIP, CAL, CDRI, DD, LWG and RRL, Chowdhery & Wadhwa (1984) brought out 3 Vols of the *Flora of Himachal Pradesh, Analysis*, documenting 3,424 species of flowering plants under 1,137 genera belonging to 194 families. Present status of the range in diversity of plants occurring in hill districts of Garhwal division has been given by Gupta (1989). Notably, Maithani *et al.* (1991) provided important details for the identification of some important Himalayan shrubs. Useful information pertaining to 985 species belonging to 353 genera under 79 families of gymnosperms and angiosperms has been published by Aswal & Mehrotra (1994) in their book entitled *Flora of Lahaul Spiti (A Cold Desert in North-West Himalaya*). An extensive account of the *Forest Flora of Pir Pinjal* has been presented by Singh & Kachroo (1994). A concise field guide to the *Flowering Plants of the Himalayas* has been contributed by Dang (1998). On the basis of CAMP workshop, Ved & Tandon (1998) red-listed 42 medicinal plants for the states of H.P. and Jammu & Kashmir. Gaur (1999) compiled *Flora of District Garhwal, North-West Himalaya* with ethnobotanical notes. A comprehensive floristic account of 930 plant species belonging to 504 genera under 126 families has been dealt by Dhaliwal & Sharma (1999) in their useful contribution on *Flora of Kullu District.* Singh & Rawat (2000) studied *Flora of Great Himalayan National Park.* Important publication *Flora of Cold Desert of Western Himalaya, Vol.1 (Monocotyledons)* by Murti (2001) provided detailed floristic information of 347 species belonging to 103 monocotyledonous genera under 16 families. A comprehensive account of *Flora of Tons Valley* enlisting over 761 species from the region has been compiled by Rana *et al.* (2003). In recent years, Kaur & Sharma (2004) exhaustively enumerated floristic account of 911 plant species belonging to 544 genera under 139 families of higher plants in their book on *Flora of Sirmaur* (*Himachal Pradesh*) and Sood *et al.* (2006) gave a beautiful account of 290 species under 221 genera of the *Honey Plants of the N.W. Himalayas*. Another important publication on Trans-Himalaya (Chaurasia *et al.*, 2007) provided in depth study on the ethnobotany and plants of the region. Not only that, contributions of Aslam *et al.* (2010), Jaryan *et al.* (2012) and Sekar (2012) also enriched our understanding on alien plants and invasiveness of alien ornamental plants of Indian Himalayan region.

STATUS OF FLORISTICS DIVERSITY IN HIMACHAL PRADESH

The earliest attempt to initiate systematic floristic in H.P. goes to William Moorecroft who collected plants from the areas of Kangra, Kullu, Lahaul and Spiti in 1819. Countess of Dalhousie had been credited for pioneering collection of 600 plants in 1829 from Shimla region. Some other plant explorers of the 19th Century who made collections from various localities of H.P. (Bashahr, Kangra, Kinnaur,

Kullu, Lahul-Spiti, Mandi, Shimla, Sirmour) include V. Jacquemont, J.F. Royle, R. Inglis, M.P. Edgeworth, G.T. Vigne, R.S. Simpson, W.H. Parish, W. Grant, H. Collett, J.S. Gamble, D. Brandis, J. F. Duthie and G.A. Gammie (*see* Moorcroft & Trebeck, 1841). During the 20th Century, notable collections from the State have been those of R.N. Parker, C. E. Parkinson, W. Koelz, R.S. Troup, N.L. Bor, M.B. Raizada, M.A. Rau, U.C. Bhattacharya, M. Sharma and B.S. Aswal. Apart from these, the other contributions in the form of books such as *The Himalayan District of North West Provinces of India* (Atkinson, 1882), *Flora of British India* (Hooker, 1872-1897), *Flora Simlensis* (Collett, 1902), *Flora of Bashahr Himalaya* (Nair,1977), *Flora of Himachal Pradesh*, Vols I-III (Chowdhery & Wadhawa, 1984), *Concise Flowers of* the *Himalaya* (Polunin & Stainton, 1987), *Flowers of the Himalaya-A Supplement* (Stainton, 1988), *Flora of Lahaul-Spiti-A Cold Desert in North West Himalaya* (Aswal & Mehrotra, 1994), *Flora of Kullu District* (Dhaliwal & Sharma, 1999), *Flora of Great Himalayan National Park, Himachal Pradesh* (Singh & Rawat, 2000), *Flora of Chamba District, Himachal Pradesh* (Singh & Sharma, 2006) as well as a large number of research publications more particularly from the different parts of H.P., namely *Chamba* (Gupta, 1961, 1964, 1971), *H.P.* (Sarin & Chopra, 1985; Chauhan, 1983, 1989a, b, c, 1990, 1999, 2003; Singh & Aswal 1992; Sharma & Sood, 1997; Singh, 1999; Badola, 2001), *Kangra Valley* (Ahluwalia, 1952; Uniyal & Chauhan, 1971, 1973; Kapur, 1985), *Kullu* (Rastogi, 1960; Uniyal & Chauhan,1972; Dobriyal *et al.*,1997), *Lahul-Spiti* (Rau, 1961; Sarin, 1967; Uniyal *et al.*, 1973; Arora *et al.*, 1980; Chauhan, 1989a, b, c; Chaurasia *et al.*, 2001), *Mandi* (Singh, 1993) and *Una* (Chauhan, 1974) added to the wealth of literature on floristics of the region. Furthermore, publications of Chauhan (1995: Economically useful plant resources of H.P.), Sood *et al.*, (2001: Comprehensive ethnobotanical account of cold desert tribes of Lahaul-Spiti -129 ethnobotanically important plants belonging to 92 genera and 44 families), Kala (2003: Trans –Himalayan area -220 medicinal plants) and Sood & Thakur (2004: Rewalsar Himalaya- 262 plant species belonging to 82 families and 220 genera) also enriched our knowledge on the floristic diversity and medicinal and aromatic plants of the State.

STATUS OF ETHNOBOTANICAL DIVERSITY

Since the publication of the first book on the ethnobotany, *An Introduction to Ethnobotany* by Faulks (1958), the scope, concept and literature on the subject have been mushrooming all over the world (Schultes, 1983, 1986; Martin, 1972; Schultes & Raffauf, 1990, Taylor, 1990; Redford & Mansaur, 1997; Neuwinger, 2000; Payee, 2000; Martin *et al.*, 2001; Seitel, 2001; Grotenhermen & Russo,

2002). Regarding Indian scenario, *Glimpses of Indian Ethnobotany* (Jain, 1981) is the first book dealing with the ethnobotany whereas the first bibliography of ethnobotany was published in the early eighties (Jain *et al.*, 1984). By 1995, as many as 420 research papers and 14 books on ethnobotany had been published covering ethnobotanical investigations of about 250 ethnic groups (Rao, 1996). In recent times, considerable work has been done in India which is evident from two published comprehensive dictionaries (Jain, 1991; Jain & Srivastava, 1999) and analysis presented by Jain & Srivastava (2001), Khan (2005) and Goel & Tripathy (2009) as well as a number of useful treatises on the subject (Maheshwari, 1992; Joshi, 1995; Arora & Pandey, 1996; Rama Rao & Henry, 1996; Kaushik, 1998; Pandey, 1998; Singh & Pandey, 1998; Dagar & Dagar, 1999; Mudgal *et al.*, 1999; Singh & Kumar, 2000a, Sood *et al.,* 2001, 2009a, b, 2010, 2013; Singh *et al.*, 2003a, b; Sood & Thakur, 2004; Trivedi, 2004, 2009, 2010; Sood & Kaushal, 2008; Sood & Sudershna, 2008) etc. Despite the occurrence of a large number of ethnic communities inhabiting different regions, the available ethnobotanical information in the context of H.P. is less fortunate (Goel & Tripathi, 2009). As many as 60 publications including 8 books have been published so far, and the data documented for some of its districts pertain to *Chamba* (Uniyal *et al.*, 2006), *Hamirpur* (Prakash *et al.*, 2003; Rawat *et al.*, 2009), *Kangra* (Uniyal & Chauhan, 1973b; Kapur, 1993; Sharma & Chauhan, 2000; Singh & Kumar, 2000b), *Kinnaur* (Negi & Subramani, 2002), *Kullu* (Singh 1990; Jain & Puri, 1994; Singh *et al.*, 1998; Uniyal, 2003; Sharma *et al.*, 2004, 2005; Negi & Subramani, 2006), *Lahaul-Spiti* (Koelz, 1979; Duhoon *et al.,* 1996; Chauhan & Lakhanpal, 2000; Kala & Manjrekar, 2000; Sood *et al.*, 2001; Chandrasekar & Srivastava, 2003, 2005; Kala, 2003; Srivastava & Chandrasekar, 2004; Singh & Chauhan, 2005; Chaurasia *et al.*, 2007), *Mandi* (Gaur & Singh, 1993; Singh, 1996; Sood & Thakur, 2004; Rajasekaran & Singh, 2009; Sood *et al.*, 2009b), *Shimla* (Chauhan, 1996; Sharma *et al.*, 2009; Verma *et al.*, 2012), *Solan* (Verma & Chauhan, 2006, 2007), *Sirmaur* (Chauhan & Chauhan, 1985; Sharma, 2002; Kumar, 2005; Thakur, 2011) and the entire *State* (Sarin *et al.,* 1963; Sarin & Chopra, 1985; Chauhan, 1988, 1989; Gupta, 1988; Sagwal & Gupta, 1988; Singh & Aswal, 1992; Singh, 1993; Lal *et al.*, 1996; Sharma & Chandel, 1996; Sharma & Rana, 1999; Singh & Kumar, 2000a; Badola & Pal, 2002; Negi & Bhalla, 2002; Mishra & Gokhale, 2003; Sethi, 2003; Ved *et al.*, 2003; Aggarwal *et al.*, 2004; Rana, 2004; Bhalla & Bhalla 2006; Sood *et al.*, 2012a, 2013).

It is evident from the above foregoing that the literature abounds in information on floristics and time tested ethnobotanical practices of this hilly region but it is rather unfortunate that no work has so far exclusively been done to document the rich plant diversity having the elements of both indigenous and exotic flora in courtyards of heritage buildings, cemeteries and socio-cultural spaces in Shimla Hills of H.P. Against this backdrop, the present work has been envisaged with an aim to explore and enlist under-exploited diversity in the courtyards of heritage buildings, cemeteries and socio-cultural spaces to serve as valuable clues for social forestry and horticultural endeavours, therapeutic purposes, supplementary foods and sustainable management of species as well as their habitats.

Chapter 2

HISTORY, CULTURE AND ENUMERATION OF DATA

SHIMLA -'QUEEN OF HILLS': A GENERAL PERSPECTIVE

Geographical Extent

Shimla-'Queen of Hills' lies in the north-western ranges of the Himalayas between 31°06 N latitude - 77°13 E longitude. The town with an average altitude of 2,397.59m above mean sea level spreads over an area of 25 km^2. It was planned and developed by Britishers on seven hills, namely Elysium Hill (2,257m, North-Western Shimla), Summer Hill (2,070m, Western Shimla), Jakhoo Hill (2,449m, Central Shimla), Observatory Hill (1,983m, Western Shimla), Inverarm Hill (2,150m, Western Shimla), Bantony (2,105m, Central Shimla) and Prospect Hill (2,175m, Western Shimla). The highest point is the Jakhu Hill (2,454m). The place is bounded by Mandi and Kullu in the North, Kinnaur in the East, Solan district in the West, State of Uttarakhand in South-East and Sirmour district in the South. The place - a unique combination of hills, spurs and valleys, stretches nearly 9.2 km from east to west. The lovely hill station is famous for its Colonial buildings (institutional buildings, bungalows, churches, socio-cultural spaces, hotels, cemeteries, schools, hospitals) styled in Tudor Bethan and Neo-Gothic architecture dating 19th Century. The department of Town and Country Planning has identified more than 100 heritage buildings in Shimla town. The place is approximately 115 km from Chandigarh, the nearest major city, and 365 km from New Delhi, the national capital.

Etymology and Historical Perspective

The city is named after the local Hindu temple of goddess Shyamala Devi, which is considered an incarnation of the Goddess Kali. Various tribes such as Koilis, Halis, Dagis, Dhaugris, Dasa, Khasas, Kinnars and Kirats were known to inhabit the region during pre-historic era. During the Vedic period, several small republics, 'Janapada' existed here which were later conquered by Gupta Empire (Anonymous, 2008). The region was captured by Gurkhas from Rajput principalities in the beginning of 19th Century. For rendering invaluable services to the British empire during Anglo-Gurkha war, the land in the neighbourhood of the area new comprising Shimla was rewarded to the Maharaja of Patiala in 1815. It is said that the first person who brought Shimla to international map was a British officer, who, when moving Gurkha troops from Sabathu to Kotegarh in about 1816, passed through the place and was impressed by its cool climate. In 1819, Lt. Ross (Assistant Political Agent in the Hill States) probably built the first British cottage of wood and thatch in Shimla. By 1831, Shimla had about sixty permanent houses and a bazaar. Some of the social institutions through a central authority promoted the Municipal Committee in 1851. The Committee was responsible for establishment of the Town Hall with a library, Gaiety Theatre, and Police Station. A new road named Grand Hindustan-Tibet road, 93 km in length, passing through Dharampur, Solan and Kiaree Ghat was built in 1861. Subsequently, Shimla was declared the summer capital of the British Raj in India in 1864. Not only that, Shimla was also notified as the capital of the undivided state of Punjab in 1871. Lord Lytton, Viceroy 1876 -1880 also made efforts to plan for a Viceregal Lodge. The electrical lighting system was introduced in 1903. The famous Kalka-Shimla railway line was commissioned in 1906. With the re-organisation of territory into Punjab, Haryana and H.P., Shimla became the capital of H.P. in 1966. Over the past three decades some of the most elegant structures of the British era, Kennedy House (the first permanent structure built in 1822), Peterhoff, Armsdell, Bentick Castle, B.C.S. (part), Chelsea, Snowdown Hospital, Walker Hospital, Western Command (part), General Post Office (part), Ripon Hospital and Auckland School had been gutted in recurring fires. In 2008, the Shimla railway line was granted UNESCO World Heritage Status.

Land and People

Shimla urban agglomeration comprises a population of 1,71,817, out of which 94,797 are males and 77,020 are females. The majority of population consists of natives of H.P. Pahari, Hindi and Punjabi are the main languages. The literacy rate is 94.14 percent. The natives of Shimla have been influenced by the Tibetan lifestyle, which is evident in the way they live and dress up. Locals mostly use 'Pahari' for communicating with each other. The major religion is Hinduism (93.5%), followed by Sikhism (2.5%) and small minorities of Christians, Tibetan Buddhists and Muslims (Anonymous, 2012). The life style is simple, down to earth and uncomplicated. By and large, people are very friendly and hospitable. The houses are built in Indo-Tibetan style with slanted roofs and slates. The economy of Shimla is mainly dependent upon the revenue from travel and tourism industry and the agriculture and horticulture produce in the region. The place is also well known for 'Sunni' embroidery, traditional jewellery works, wood crafts, woolen garments, Himachali caps, metal works, local paintings, various jams, candies, pickles, squashes and dried fruits. The place is also the hub of top reputed boarding and day schools in India. This strong educational background dates back to the time of Colonial rule when a number of missionary schools, Bishop Cotton School (B.C.S.) (Asia's oldest public school), Auckland House, Convent of Jesus and Mary, and Loreto Convent, were established here.

Fairs and Festivals

Fairs and festivals are an essential part of culture in Shimla. These predominantly are: Ice Skating Carnival (winters), Rhyali festival (rainy season), Baisakhi (mid-April), Summer festival (May), Sipi fair (May), Dussehra (end of Navaratri festival), Diwali, Bhoj fair (November), Christmas (December) and Lohri (January). All these are celebrated with much joy, fervor and enthusiasm.

Faith and Worship

People are god fearing. Various temples like Tara Devi, Dhingu Mata, Sankat Mochan, Shodash Ganpati Temple, Kali Bari, Shimleshwar Mahadev, Jakhu and Kamna Devi for Hindus; Christ Church and St. Michael's Cathedral for Christians; Gurudwaras (Gurudwara Sahib Shri Guru Singh Sabha, Gurudwara Sahib Sanjauli, Gurudwara Sahib Boileauganj) and a few mosques exist here.

Vegetation (Flora & Fauna)

The main forests in and around the city are that of pine, deodar, oak and rhododendron. As many 1,326 species of flowering plants grow wild in the area (Collett, 1921). Faunal elements comprise langoors, monkeys, barking deer, cheer pheasants, pine marten, porcupine, gorals, leopard, kakkar, woodpecker, monal etc.

Climate

The climate in Shimla is predominantly cool during winters and moderately warm during summers. The temperature ranges from -4°C to 31°C over the year. The average temperature during summers is between 19°C and 28°C, and between -1°C and 10°C in winter. The average total annual precipitation is 1520 mm. It usually snows during winters.

PLAN AND COLLECTION OF DATA

The Shimla is ethnobotanically and phyto-geographically a very diverse terrain and also rich in the enormous wealth of built heritage inherited over the years during the British regime (**Table I**). Most of these heritage sites are rich in plant biodiversity introduced mainly by colonial rulers. For collecting first-hand information on the plants at heritage sites in Shimla hills of H.P. (**Table I**), intensive and extensive surveys were re-conducted from 2008-2013. The field surveys were planned in such a way so as to collect the plant species either in flowering or fruiting stage. The desired information on ethnobotanical aspects was also collected through interviews with knowledgeable people (traditional practitioners, caretakers, managers, gardeners, elderly persons of the community, foresters etc.), as per the methodology outlined by Jain (1987). Ethnobotanical lore was considered valid if at least three informants made similar comments about the uses. Besides, scrutiny of floras, monographic and revisionary works, travelogues, historical documents, record of Municipal Corporation, D.C. Office and Town and Country Planning Department Shimla, ethnologies and collections in musea and herbaria for notes on uses were also undertaken as per the standard ethnobotanical practices (Schultes, 1962).

TABLE I

Heritage Buildings, Cemeteries and Socio-Cultural Spaces of Shimla

Sr. No.	Name	Location
	A. Heritage Buildings	
1.	Aira Holme	Kasumpti
2.	Alderton	Chaura Maidan
3.	Allan Bank	Nigam Bihar
4.	Armsdell	Boileauganj
5.	Auckland House	Lakkar Bazaar Bus Stand
6.	Ava Lodge	Chaura Maidan
7.	Band Stand (Tourism Restaurant)	Ridge
8.	Bantony	Near G.P.O. Shimla
9.	Barnes Court (Raj Bhawan)	Chhota Shimla
10.	Belvedere	Near Circular Road
11.	Bemloe Cottages	Bemloe
12.	Benmore Estate	Chhota Shimla
13.	Bentick Castles (Grand Hotel)	Near Kali Bari Temple
14.	Bishop Cotton School	B.C.S. Stop
15.	Burj	Chaura Maidan

Sr. No.	Name	Location
16.	Carton House (Sanitarium)	Chaura Maidan
17.	Catholic Church	Near Deputy Commissioner's Office and Cantonment Area.
18.	Catholic Club	Near Fingask
19.	Cedar Circuit House	Oak Over
20.	Chalet Day School	Near Oak Over
21.	Chapslee	Longwood
22.	Christ Church Complex, Ridge	Ridge
23.	Clarkes Hotel	Near H.P. Tourism Lift
24.	Cleremont (Labour Beureu Office)	Near Vidhan Sabha
25.	Conny Cot (Telegraph Office Building - Telephone Exchange)	The Mall near PNB
26.	Conny Cottage (G. P. O Building)	Near Scandal Point
27.	Convent of Jesus and Marry	Nav Bahar
28.	Corner House	Sanjauli
29.	Council Chamber (Vidhan Sabha)	Near Victory Tunnel
30.	Courteen House	Near Boileauganj
31.	Craig Garden	Chhota Shimla
32.	Crow-Brough	Near Railway Station Shimla
33.	Curzon House	Boileauganj
34.	Dalziel House (State Bank of India)	Near Kali Bari Temple
35.	Delphine Lodge	Chhota Shimla
36.	Dimple Lodge	Chhota Shimla
37.	Dixie	Chhota Shimla
38.	Eagle Mount	Nav Bahar
39.	Earnstone	Nigam Bihar
40.	Eddleston	Nigam Bihar
41.	Ellerslie	Chhota Shimla
42.	Elysium Spur	Longwood
43.	Emm Villa	Chhota Shimla
44.	Firwood Cottage	Near Tara Hall
45.	Foss Well	Chhota Shimla
46.	Gaiety Theatre	The Mall Road
47.	Gaston Hall (Old D.C. Office)	Near C.T.O. Shimla
48.	Glen Uquard	About ½ km from Vidhan Sabha
49.	Gorton Castle	Near Vidhan Sabha
50.	Green Gate Longwood	½ Km from Longwood
51.	Grindlay's Bank	The Mall Road

Sr. No.	Name	Location
52.	Hainault	Benmore
53.	Holcombe	Chaura Maidan
54.	Holly Lodge	Jakhu Hill
55.	Inverarm (State Museum)	Chaura Maidan
56.	Jakhu Temple	Jakhu Hill
57.	Jammu Castle	Kasumpti
58.	Kali Bari Temple	Near S.B.I. Main Branch, Shimla
59.	Kennedy House	Near Vidhan Sabha
60.	Knockdrin	Near Vidhan Sabha
61.	Lady Reading Hospital	Oak Over
62.	Manor Villa	Summerhill
63.	Morvyn (I.T.I. Shimla)	1 Km from Summerhill
64.	Mythe	Tara Hall
65.	North Bank	Near Kali Bari Temple
66.	North Stoneham (Govt. Boy'Sanjauli)	Sanjauli
67.	Oak Lodge	Near Vidhan Sabha
68.	Oak Over	Oak Over
69.	Observatory House	Boileauganj
70.	Pari Mahal	Kasumpti
71.	Parkash Niwas	Near S.B.I. Main Branch, Shimla
72.	Peterhoff	Chaura Maidan
73.	Police Station Sadar, Boileauganj	Boileauganj
74.	Police Station Sadar, Sabji Mandi, Shimla	Near Sabji Mandi, Lower Bazaar, Shimla
75.	Police Station, Chhota Shimla	Chhota Shimla
76.	Post Office, Chaura Maidan	Chaura Maidan
77.	Post office, Chhota Shimla	Chhota Shimla
78.	Post office, Summerhill	Summerhill
79.	Primrose Hill Cottage	Chaura Maidan
80.	Race View	Near Vidhan Sabha
81.	Railway Board Building	Near A.G. Office
82.	Railway Station, Shimla	Below Cart Road, Shimla
83.	Railway Station, Summerhill	Summerhill
84.	Ravenswood	High Court
85.	Retreat	Chaura Maidan
86.	Ripon Hospital	Near Local Bus Stand
87.	Rothney Castle	Jakhu Hill
88.	Shimleshwar Mahadev (Shiv Mandir)	Near The Mall Road

Sr. No.	Name	Location
89.	Sidhowal Lodge	Near Ridge
90.	Snowdown	Next to Correstophen Building on way to Sanjauli from Lakkar Bazaar
91.	Snow View	Near Lakkar Bazaar Bus Stand
92.	South Gate	Chhota Shimla
93.	Spring Field	Chhota Shimla
94.	Squire's Hall	Summerhill
95.	St. Andrew's Church	Near G.P.O. Shimla
96.	St. Bede's College	Nav Bahar
97.	St. Edward's School	Adjoining Hotel Himland, Cart Road, Shimla
98.	St. Mark	Near SBI Main Branch, Shimla
99.	St. Thomas School	Near Western Command, Shimla
100.	State Library Ridge	Ridge
101.	Stirling Castle	1.5 Km from Long Wood
102.	Strawberry Hill	Near Chhota Shimla
103.	Tara Hall	Upper Kaithu, Cart Road, Shimla
104.	Tendril Cottage (Cecil Hotel)	Chaura Maidan
105.	The Whole Range of Buildings Starting from Gainda Mull up to Northern Railway Booking Agency, Mall Road	The Mall Road
106.	Thistle Bank	Near S.B.I. Main Branch, Shimla
107.	Torney House	Chhota Shimla
108.	Torrentium	Chhota Shimla
109.	Town Hall	Ridge
110.	U.S. Club	Jakhu Hill
111.	Vice Regal Lodge Complex	Near Boileauganj
112.	Walker Hospital	Below G.C. Boy's College, Sanjauli
113.	Walsingham (D.C. Residence)	Jakhu Hill
114.	Western Command Complex	Parallel to The Mall
115.	Winter Field	Near Old Bus Stand
116.	Wood Bank	Near Railway Station, Shimla
117.	Wood Villa	Benmore
118.	Y.M.C.A.	Ridge
119.	Y.W.C.A.	Near G.P.O. Shimla
120.	Yarrows	Chaura Maidan
121.	Yates Place	Chhota Shimla

Sr. No.	Name	Location
B. Cemeteries		
1.	Boileauganj Barrier Cemetery	Boileauganj Barrier
2.	Kanlog Cemetery	Kanlog
3.	Nav Bahar Cemetery (Nuns Graveyard below St. Bede's College)	Nav Bahar
4.	Oak Over Cemetery	Oak Over
5.	Cemetery Sanjauli	Near Dhalli
C. Socio-Cultural Spaces		
1.	Annandale	Annandale
2.	Golf Course, Naldehra	Naldehra
3.	Ladies Park (Rani Jhansi Park)	Mall
4.	Ridge	Ridge
5.	Rothney Castle Ground	Jakhu Hill
6.	Scandal Point	Mall
7.	Ice Skating Rink	Near Lakkar Bazaar Bus Stand
8.	Town Hall Square	Mall

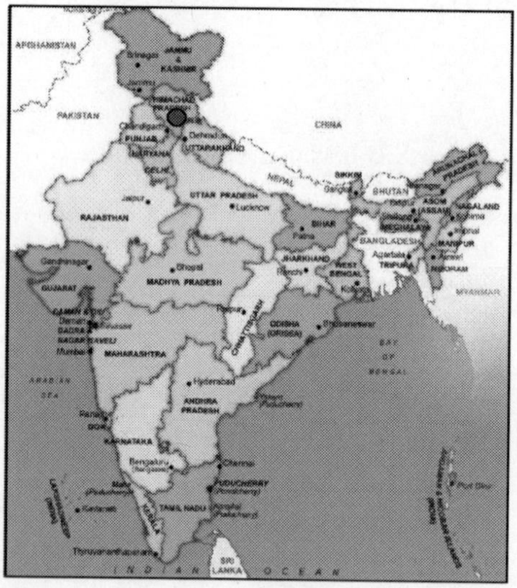

Fig. 1. Map Showing the Location of H.P. in India

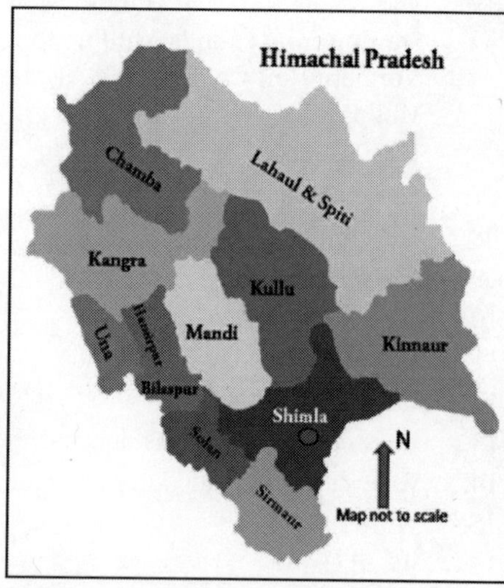

Fig. 2. Map Showing the Various Districts of H.P.

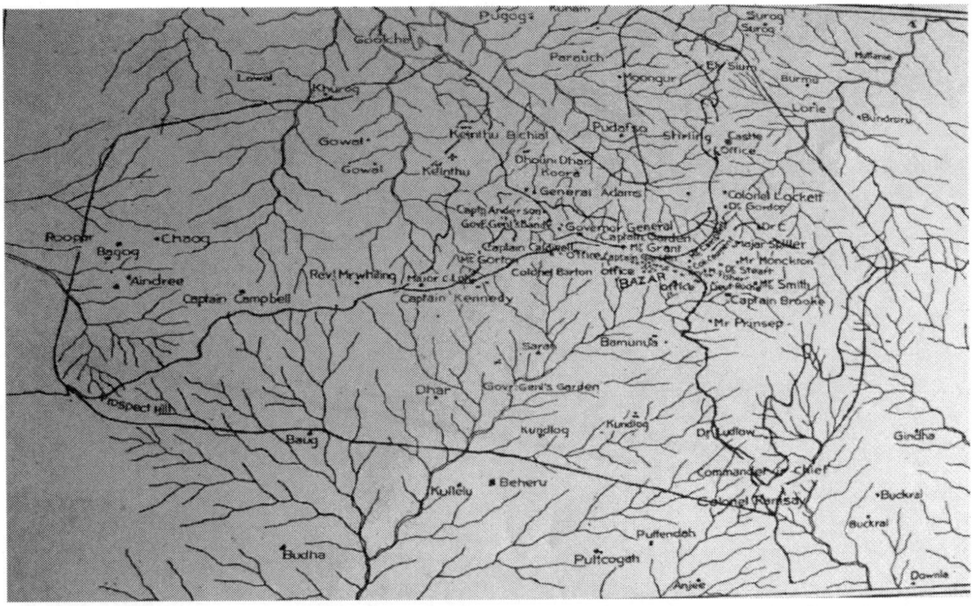

Fig. 3. Old Map of Shimla (1832)

The collected plant specimens were dried, preserved and mounted as per the known herborizing practices outlined by Jain & Rao (1977). Botanical identification of the selected species was first done with the help of regional floras (Chauhan, 1999; Chowdhery & Wadhwa, 1984; Collett, 1902; Dhiman, 1976; Hooker, 1872-1897; Nair, 1977; Polunin & Stainton, 1984; Stainton, 1988), dictionaries and illustrated encyclopedia of flowering plants (Bailey, 1957; Berry & Bradley, 2000; Brandis, 1921; Brickell, 1990; Cowen, 1950; Graf, 1980, 1981; Hay *et al.*, 1983; Krisa, 1982; Perry & Hay, 1982; Randhawa, 1960) and later carefully matched with the authenticated specimens at the herbaria of Botanical Survey of India, Northern Circle and Forest Research Institute, Dehradun. Nomenclature of these taxa was confirmed from Bennett (1987), Quattrocchi (2000) and International Plant Names Index (www.ipni.org.). Excepting for the minor modifications to conform to the present day circumscription, families and genera are delimited mainly after '*The Flora of British India*' (Hooker, 1872-1897). Economic valuation of all the presently recorded ethnobotanical species was also carried out to calculate the total importance values (TIV) as per detailed methodology outlined by Belal & Springuel (1996). The nativity of the species has been determined on the basis of Matthew (1969), Maheshwari & Paul (1975), Nayar (1977), Hajra & Das (1982), Sharma (1984), Saxena (1991), Pandey & Parmar (1994), Reddy *et al.* (2000), Reddy & Raju (2002), Reddy & Reddy (2004), Murthy *et al.* (2007),

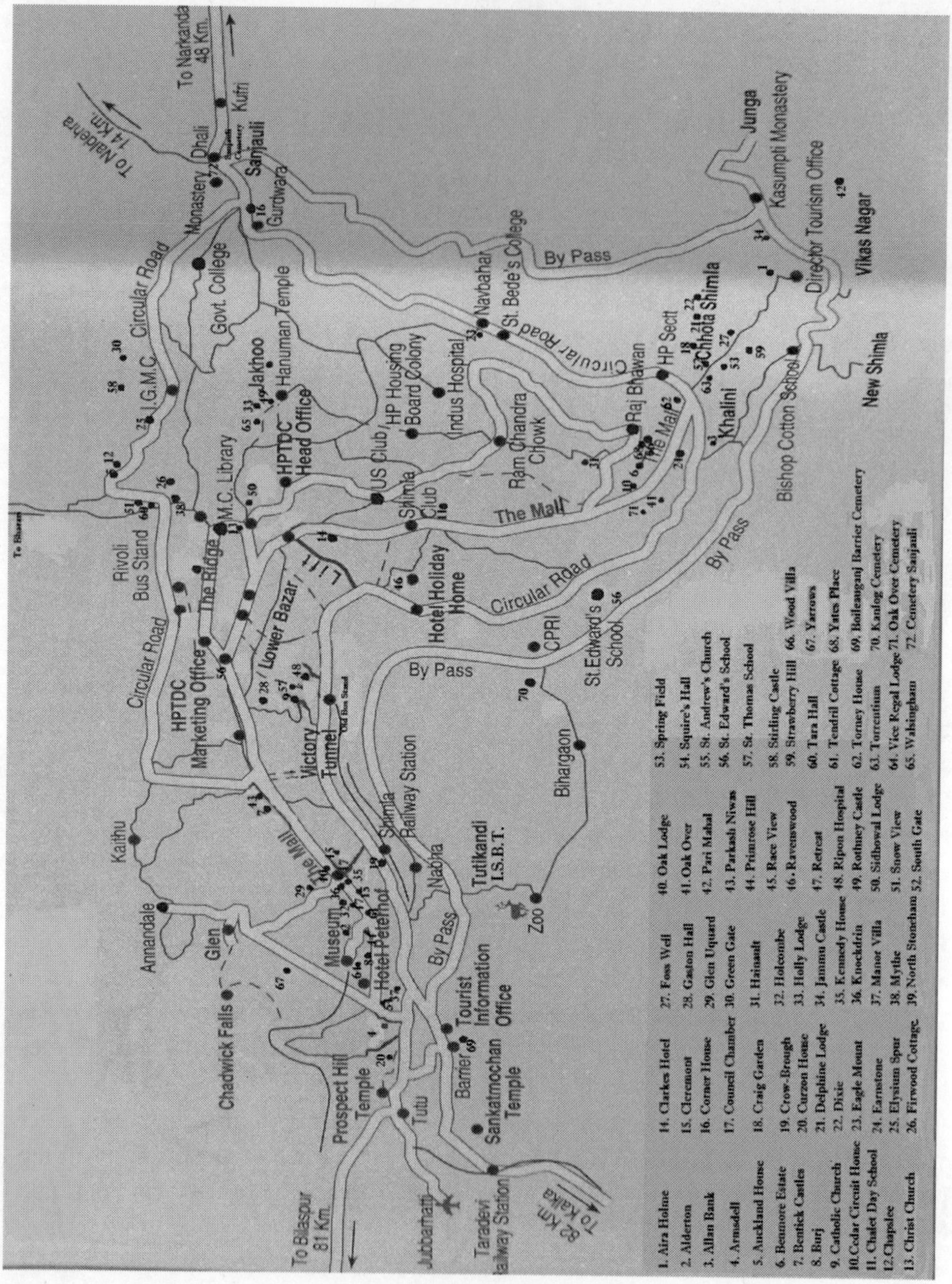

Fig. 4. Map of Shimla Showing Location of Some of the Important Heritage Buildings, Cemeteries and Socio-Cultural Spaces

Negi & Hajra (2007), Reddy (2008), Aslam *et al*., (2010), Jaryan *et al*., (2012) and Seker (2012). For future reference and studies voucher specimens of afore-mentioned taxa (HPU-EL-SKS-5401-5745) have been deposited in the Ethnobotanical herbarium of Department of Biosciences, Himachal Pradesh University, Shimla. Besides, colour photographs of the plant species in their natural habitat were also clicked.

Systematic enumeration of the plants is in alphabetical order of their botanical names, synonym of the species followed by their respective families. Besides this, information on English, Hindi and Sanskrit Names, distribution, description, reproductive cycle, material examined, propagation notes, part/s used, folk use/s, active constituents, and uses in literature has also been furnished and presented, if available. The data gathered were screened with the help of available literature (Ambasta, 1986; Anonymous, 1948-1976; Bailey, 1957; Chopra *et al*., 1956, 1969; Graf, 1980, 1981; Jain, 1991; Kirtikar & Basu, 1984; Maheshwari & Singh, 1984; Singh, 1988; Singh *et al*., 1987) besides many other books and articles published in different journals.

TABLE II
Flowering and Fruiting Season of the Plant Species Collected from the Study Area

Name of the Species	Flowering & Fruiting Season
Abies pindrow	April-May
Abutilon X *hybridum*	May-August
Abutilon megapotamicum	April-September
Acacia dealbata	February-May
Acer negundo	April-June
Acer oblongum	February-April
Acer palmatum	May-July
Actinidia chinensis	February-November
Aesculus indica	May-September
Afrocarpus gracilior	April-October.
Agave americana	March-May
Aloysia citriodora	August-September.
Amygdalus mira	March-April
Araucaria bidwillii	June-September
Araucaria columnaris	August-October
Araucaria cunninghamii	June-October
Arundo donax	September-October
Asparagus aethiopicus	April-June

Name of the Species	Flowering & Fruiting Season
Asparagus densiflorus	May-June
Asparagus racemosus	April-August
Asparagus setaceus	May-August
Asparagus virgatus	April-June
Berberis aristata	May-September
Berberis lycium	April-July
Bougainvillea glabra	April-October
Buddleja asiatica	February-June
Buddleja crispa	March-May
Buddleja davidii	April-September
Buxus wallichiana	April-November
Callistemon lanceolatus	May-September
Camellia japonica	November-April
Campsis radicans	August-September
Cedrus deodara	June-July
Celtis australis	April-October
Cerasus cerasoides	November-May
Cestrum elegans	October-December
Cestrum nocturnum	August-October
Cestrum parqui	November-January
Chamaerops humilis	February-March
Citrus limon	November-February
Citrus maxima	November-March
Clematis buchananiana	July-October
Clematis montana	April-June
Colquhounia coccinea	August-October
Coriaria nepalensis	March-June
Cornus capitata	May-June
Cornus macrophylla	May-August
Cotoneaster bacillaris	May-October
Cryptomeria japonica	March-October
Cunninghamia lanceolata	March-October
Cupressus cashmeriana	May-August
Cupressus torulosa	February-April
Cycas revoluta	February-May
Daphne papyracea	April-October
Datura stramonium	June-October
Debregeasia saeneb	April-August
Desmodium elegans	August-November

Name of the Species	Flowering & Fruiting Season
Desmodium multiflorum	June-October
Diospyros kaki	April - July
Diospyros lotus	April-August
Drepanostachyum falcatum	May-November
Duranta erecta	April-August
Elaeagnus umbellata	May-October
Eriobotrya japonica	July-September
Erythrina arborescens	August-October
Eucalyptus globulus	February-October
Euonymus fimbriatus	April-June
Euonymus japonicus	May-August
Euonymus lucidus	April-August
Euonymus tingens	April-August
Fagus sylvatica	April-October
Ficus auriculata	April-August
Ficus benjamina	February-June
Ficus elastica	April-July
Ficus neriifolia	April-November
Ficus palmata	April-September
Ficus religiosa	March- October
Ficus sarmentosa	March-July
Ficus subincisa	June-September
Forsythia X *intermedia*	February-April
Forsythia ovata	March-April
Fuchsia X *hybrida*	March-November
Fuchsia magellanica	April -November
Ginkgo biloba	February-March
Girardinia diversifolia	July-August
Grevillea robusta	March-July
Hedera canariensis	August-November
Hedera colchica	September-November
Hedera helix	September-November
Hedera nepalensis	September-November
Hibiscus rosa-sinensis	February-October
Hibiscus syriacus	July-September
Himalrandia tetrasperma	July-October
Hydrangea macrophylla	May-July
Hypericum hookerianum	July-October
Hypericum oblongifolium	February-October

Name of the Species	Flowering & Fruiting Season
Hypericum patulum	August-September
Ilex dipyrena	April-October
Indigofera dosua	April-July
Inula cappa	August-February
Ipomoea purpurea	June-August
Jacaranda acutifolia	March-June
Jasminum grandiflorum	April-September
Jasminum humile	May-September
Jasminum mesnyi	March-May
Jasminum multiflorum	November-April
Juglans regia	April-October
Juniperus chinensis	April-October
Juniperus horizontalis	March-September
Juniperus procera	March-September
Juniperus pseudosabina	March-August
Juniperus squamata	April-October
Kerria japonica	March-May
Lagerstroemia indica	June-October
Leea asiatica	July-August
Ligustrum lucidum	July-November
Ligustrum X *vicaryi*	May-July
Liriodendron tulipifera	June-August
Lonicera japonica	April-August
Lonicera quinquelocularis	April-September
Lyonia ovalifolia	April-May
Magnolia grandiflora	April-August
Magnolia obovata	January-April
Magnolia X *soulangeana*	January-March
Malus pumila	May-October
Malvaviscus arboreus var. *mexicanus*	Major part of the year.
Morus alba	March-May
Morus nigra	March-June
Morus serrata	April-May
Musa paradisiaca	Throughout the year
Myrica esculenta	November-May
Myrsine africana	April-October
Myrtus communis	June-October
Nerium oleander	April-November
Opuntia dillenii	April-June

Name of the Species	Flowering & Fruiting Season
Padus cornuta	April-May
Parthenocissus semicordata	July-November
Paulownia tomentosa	May-September
Picea smithiana	April-May
Pinus roxburghii	March-May
Pinus wallichiana	April-June
Pistacia khinjuk	March-May
Platanus orientalis	April-July
Populus ciliata	April-July
Populus nigra	April-July
Prinsepia utilis	June-October
Prunus amygdalus	February-September
Prunus armeniaca	February-August
Prunus domestica	April-August
Prunus laurocerasus	April-September
Prunus persica	February-May
Prunus salicina	January-June
Punica granatum	May-September
Pyracantha crenulata	March-June
Pyrus communis	June-August
Pyrus pashia	May-August
Pyrus pyrifolia	March-June
Quercus floribunda	April-October
Quercus incana	April-October
Quercus oblongata	April-October
Quercus robur	April-October
Rhamnus persica	April-October
Rhamnus virgatus	April-June
Rhapis excelsa	May-July
Rhododendron arboreum	March-May
Rhododendron simsii var. *vittatum*	January-March
Ricinus communis	March-May
Robinia pseudoacacia	April-May
Rosa alba	June-August
Rosa banksiae	April-June
Rosa chinensis	June-September
Rosa damascena	April-July
Rosa jacquemontii	June-September
Rosa moschata	April-October

Name of the Species	Flowering & Fruiting Season
Rosa multiflora	June- July
Rubia cordifolia	July-October
Rubus biflorus	June-August
Rubus ellipticus	January-May
Rubus macilentus	April-September
Rubus niveus	March-May
Rubus paniculatus	April-August
Salix alba	March-May
Salix babylonica	February-May
Salix pycnostachya var. *oxycarpa*	May-September
Salvia leucantha	Late summer or autumn
Salvia splendens	Summer onwards
Sambucus nigra	March-August
Sarcococca pruniformis	May-October
Schefflera arboricola	August-September
Skimmia laureola	April-May
Smilax aspera	September-November
Solanum laxum	June-October
Solanum pseudocapsicum	June-October
Solanum virginianum	April-September
Spartium junceum	May-September
Spiraea canescens	May-October
Spiraea cantoniensis	April-July
Taxodium distichum	April-October
Taxus baccata	May -November
Thuja occidentalis	April-October
Thuja orientalis	July-October
Thujopsis dolabrata	April-October
Trachelospermum jasminoides	July-October
Tropaeolum majus	May-November
Tsuga dumosa	April-November
Ulmus wallichiana	March-May
Viburnum cotinifolium	April-September
Viburnum mullaha	May-July
Wisteria sinensis	April-May
Xanthium strumarium	April-November
Yucca aloifolia	July-August
Zanthoxylum armatum	April-June

Chapter 3

HERITAGE SITES OF SHIMLA HILLS

AIRA HOLME

Features:

Year of Construction : 1870.

Location : Kasumpti.

Extent : About 6,257 sq m.

Building Material : Lime-Surkhi (Mixture of Burnt Clay, Sand, Lime), Lath, Sand Stones, Bricks.

Building Design : Colonial Architecture.

Past Owner : Sir Jogendra Singh, then Home Minister of Patiala.

Past Use : Residential house.

Present Owners : B.S. Malhan and Billy Malhan.

Present Use : As a residential house.

Miscellaneous Aspects : This was originally named 'Morefield Grange'. In the early 20th century, this was acquired by Sir Jogendra Singh, Home Minister of Patiala and renamed it as 'Aira Holme' after one of his 'jagir' (manor) lands.

Plants in Courtyard : *Cedrus deodara*, *Chamaerops humilis*, *Cupressus torulosa*, *Daphne papyracea*, *Euonymus japonicus*, *Hedera helix*, *Hibiscus syriacus*, *Hydrangea macrophylla*, *Hypericum patulum*, *Jasminum mesnyi*, *Malus pumila*, *Parthenocissus semicordata*, *Prunus armeniaca*, *Quercus oblongata*, *Rhododendron arboreum*, *Robinia pseudoacacia*, *Rosa banksiae*, *Rosa chinensis*, *Sarcococca pruniformis*, *Solanum laxum*.

ALDERTON

Features:

Year of Construction : 1894.

Location : Near Chaura Maidan.

Extent : About 2,128 sq m.

Building Material : Dressed Stones, Lath, Plaster, Mud, Wooden Posts (Devdaar).

Building Design : Colonial Architecture.

Past Owner : J. Elston, the then Director of Alliance Bank of Shimla.

Past Use : Residential house.

Present Owner : Indian Army.

Present Use : Official residence of Lt. Gen Ragesh Kochhar, HQ ARTRAC.

Miscellaneous Aspects : In 1920, British Empire acquired this as residence for its officers. Alderton was a part of Annandale View Estate.

Plants in Courtyard : *Cedrus deodara*, *Cerasus cerasoides*, *Chamaerops humilis*, *Cryptomeria japonica*, *Euonymus japonicus*, *Girardinia diversifolia*, *Hedera helix*, *Hibiscus syriacus*, *Hydrangea macrophylla*, *Hypericum patulum*, *Jasminum mesnyi*, *Parthenocissus semicordata*, *Pistacia khinjuk*, *Quercus oblongata*, *Rhododendron arboreum*, *Rosa banksiae*, *Rosa chinensis*, *Sarcococca pruniformis*.

ALLAN BANK

Features:

Year of Construction : 1840.

Location : Near Nigam Bihar.

Extent : About 2,148 sq m.

Building Material : Dhajji, Lath, Sand Stones, Plaster (Lime, Sand), Iron Sheet.

Building Design : Pent Roof Style with Dormers.

Past Owners : Three British sisters.

Past Use : Nunnery.

Present Owner : H.P. Government.

Present Use : Residential house for IAS officers.

Plants in Courtyard : *Cedrus deodara*, *Girardinia diversifolia*, *Hibiscus syriacus*, *Hydrangea macrophylla*, *Malus pumila*, *Rosa banksiae*, *Rosa chinensis*, *Solanum laxum*.

ARMSDELL

Features:

Year of Construction : 1888.

Location : Near Boileauganj.

Extent : About 5,124 sq m.

Building Material : Sand Stones, Mud, Lath, Plaster (Dhajji), Galvanised Iron Sheet.

Building Design : Roofing in 'Nainital Pattern' and Art-Deco Style of Architecture.

Past Use : Lady Reading used it as a hostel for English nurses.

Present Owner : H.P. Government.

Present Use : Residential house of Chief Justice of H.P.

Plants in Courtyard : *Acacia dealbata, Araucaria columnaris, Arundo donax, Asparagus setaceus, Callistemon lanceolatus, Cedrus deodara, Celtis australis, Chamaerops humilis, Citrus maxima, Clematis buchananiana, Cupressus torulosa, Drepanostachyum falcatum, Elaeagnus umbellata, Eriobotrya japonica, Magnolia grandiflora, Quercus oblongata, Robinia pseudoacacia, Rosa chinensis.*

AUCKLAND HOUSE

Features:

Year of Construction : 1920-1921.

Location : Near Lakkar Bazaar Bus Stand.

Extent : About 6,752 sq m.

Building Material : Bricks, Lime, Iron Sheet.

Building Design : Mixed Colonial and Western Style of Architecture.

Past Owner : Governor-General, Lord Auckland.

Past Use : Used as residential house for himself and his sisters.

Present Owner : Managed by Diocese of Amritsar, Church of North India (CNI).

Present Use : School for girls from from Nursery to Class XII.

Miscellaneous Aspects : The original house at this site was built by Dr Blake who was in service of the East India Company. This was then taken by the Governor-General, Lord Auckland (1835 - 1842) for his use. It was here, in

1838, that Auckland used to receive embassies from both the Hill Chiefs and Punjab's Maharaja Ranjit Singh. While retaining residence, Auckland then sold the remaining portion to Dr Ransford. Later, Auckland's successors the Lords Ellenborough and Hardinge occupied it. The school was founded in 1866 and was initially called Punjab Girls School. At the instance of the wife of Bishop Cotton, the school began with 32 girls, most of them being Europeans. The school authorities bought the Auckland House (originally the residence of Lord Auckland, Governor-General) in 1868 and shifted there on November 6, 1869. Unfortunately, the building was damaged in an earthquake in 1905. Later, the entire structure was pulled down and rebuilt by January 1921. From 1908 - 1952 the St. Hilda's Society, Lahore, provided the staff for school. Chief among them was Miss Pearce who was the Principal from 1915 - 1936.

Plants in Courtyard : *Aesculus indica, Agave americana, Callistemon lanceolatus, Cedrus deodara, Chamaerops humilis, Hedera helix, Hibiscus syriacus, Quercus oblongata, Robinia pseudoacacia, Rosa chinensis, Thuja orientalis, Wisteria sinensis.*

AVA LODGE

Features:

Year of Construction : 1920-1921.

Location : Chaura Maidan.

Extent : 2,950 sq m.

Building Material : Bricks, Lath, Plaster, Tin, Wood.

Building Design : Colonial Architecture.

Past Owner : H.W. Emerson.

Past Use : Official residence of a Member of Viceroys Council.

Present Owner : H.P. University.

Present Use : It is now University Institute of Legal Studies.

Plants in Courtyard : *Aesculus indica, Buddleja crispa, Cedrus deodara, Chamaerops humilis, Girardinia diversifolia, Hibiscus syriacus, Hydrangea macrophylla, Quercus oblongata, Rosa chinensis, Sarcococca pruniformis, Thuja orientalis.*

BAND STAND (TOURISM RESTAURANT)

Features:

Year of Construction : 1920.

Location : Ridge.

Extent : About 625 sq m.

Building Material : Slates, Wooden Post, Sand Stones, Lath, Plaster.

Building Design : Circular Base having Conical Roof.

Past Owner : Kanwar Jiwan Dass of Jubbalpur.

Past Uses : Resting place and later military bands used to play music here.

Present Owner : Managed by State Tourism Department.

Present Use : Restaurant.

Miscellaneous Aspects : The upper portion houses 'Ashiana' restaurant whereas the lower ones known as 'Goofa'.

Plants in Courtyard : *Lagerstroemia indica*, *Platanus orientalis*, *Populus ciliata*, *Thuja orientalis*.

BANTONY BUILDING

Features:

Year of Construction : 1880.

Location : Near G.P.O. Shimla.

Extent : About 2,856 sq m.

Building Material : Cast-Iron Railing from Nahan Foundry, Stones, Mud, Lath and Plaster (Dhajji).

Building Design : Hybrid Architecture of English Country House and the Local Sirmouri Baroque Design.

Past Owner : Maharaja of Sirmour.

Past Uses : Summer palace of the Maharaja of Sirmour and later it was used as temporary office for the British army.

Present Owners : M/S Lehnu Mal Thakur Das, Atul Thakur Das, Anuj Thakur, Ramkrishan Thakur, Shimla.

Present Uses : Presently, the above property is being repaired by the owners after being got vacated from D.I.G, Police, H.P. Government.

Miscellaneous Aspects : Bantony was bought by the Maharaja of Sirmaur from its British owner Captain Gordon to build his Ridge-top summer house in 1880. He in turn sold it to Maharaja of Darbhanga (Bihar) from whom it was bought by the present owners.

Plants in Courtyard : *Berberis aristata, Buddleja crispa, Buddleja davidii, Cedrus deodara, Chamaerops humilis, Clematis buchananiana, Ficus palmata, Rosa chinensis, Rubia cordifolia.*

BARNES COURT (RAJ BHAWAN)

Features:

Year of Construction : 1879-1886.

Location : Near Chhota Shimla.

Extent : About 9,647 sq m.

Cost of Construction : ₹ 3,02,257/-.

Building Material : Sand Stones, Mud, Lath, Wooden Posts, Plaster (Dhajji), Teak Wood.

Building Design : Tudor Style.

Past Owner : British Empire.

Past Uses : It was occupied by Sir Edward Barnes in 1832. After a break from 1849 - 1864, it was again the residence of various British Commanders-in- Chief, General Napier, General Gomm, General Anson, General Campbell and General Rose.

Present Owner : H.P. Government.

Present Use : Governor House.

Miscellaneous Aspects : Lord Edward Barnes, the then Commander built it in the year 1828 for ₹ 14,500/- with slate roofing on the top. The new structure was built during 1879-1886. After reorganization, when Shimla was allotted to H.P., it was converted into a State Guest House. In late seventies, it was converted into a State Guest House-cum-Tourist Bungalow. In August 1972, Shimla Agreement was signed here by Smt Indira Gandhi and Mr Zulfikar Ali Bhutto of Pakistan. In 1994, a fire escape stair case and a lift were incorporated in the building.

Plants in Courtyard : *Abutilon megapotamicum, Acacia dealbata, Araucaria columnaris, Asparagus densiflorus, Callistemon lanceolatus, Cedrus deodara, Celtis australis, Chamaerops humilis, Cunninghamia lanceolata, Elaeagnus umbellata, Euonymus japonicus, Ficus palmata, Forsythia* X *intermedia, Hibiscus syriacus, Hydrangea macrophylla, Jasminum grandiflorum, Jasminum mesnyi, Juniperus horizontalis, Juniperus procera,*

Juniperus squamata, Ligustrum X *vicaryi, Pinus wallichiana, Quercus floribunda, Salix alba, Thuja orientalis.*

BELVEDERE

Features:

Year of Construction : Around 1870.

Location : Near Circular Road.

Extent : About 995 sq m.

Building Material : Sand Stones, Bricks, Pebble- Work, Lath, Plaster (Sand, Lime).

Building Design : Colonial Architecture.

Past Uses : To start with, this building was the residence of Lieutinant Governor of Punjab followed by Mr Jacob (Art Dealer after whom the world's sixth largest diamond, the Jacob Diamond, was named). Since 1915, this building has been under the occupation of Seventh-day Adventist Church, Shimla who then used it as sanitarium and hospital.

Present Owner : Managed by Diocese of Amritsar, Church of North India (CNI).

Present Use : Junior Section of Auckland House School.

Plant in Courtyard : *Aesculus indica.*

BEMLOE COTTAGES

Features:

Year of Construction : 1918-1922.

Location : Bemloe.

Extent : About 25,678 sq m.

Building Material : Burnt Bricks set in Lime, Iron Sheet, Stones.

Building Design : Colonial Architecture with Pent Roof.

Past Owner : British Empire.

Past Use : Houses of European officers.

Present Owner : H.P. Government.

Present Uses : Residential house of IAS, HAS, Army Officers.

Plants in Courtyard : *Abutilon* X *hybridum, Acacia dealbata, Aesculus indica, Agave americana, Amygdalus mira, Araucaria columnaris, Asparagus setaceus, Asparagus virgatus, Berberis aristata, Berberis lycium, Callistemon lanceolatus, Campsis radicans, Cedrus deodara, Celtis australis,*

Cerasus cerasoides, Cestrum nocturnum, Chamaerops humilis, Citrus limon, Clematis montana, Cornus macrophylla, Cupressus torulosa, Datura stramonium, Debregeasia saeneb, Desmodium multiflorum, Drepanostachyum falcatum, Duranta erecta, Euonymus japonicus, Ficus palmata, Ficus subincisa, Grevillea robusta, Hedera helix, Hedera nepalensis, Hibiscus rosa-sinensis, Hibiscus syriacus, Hydrangea macrophylla, Jasminum humile, Jasminum mesnyi, Juniperus chinensis, Juniperus horizontalis, Juniperus procera, Juniperus squamata, Lagerstroemia indica, Leea asiatica, Lonicera japonica, Lonicera quinquelocularis, Opuntia dillenii, Padus cornuta, Pinus wallichiana, Pistacia khinjuk, Populus ciliata, Prinsepia utilis, Prunus armeniaca, Prunus domestica, Prunus persica, Prunus salicina, Malus pumila, Quercus floribunda, Ricinus communis, Robinia pseudoacacia, Rosa alba, Rosa damascena, Rosa moschata, Solanum laxum, Tropaeolum majus, Yucca aloifolia.

BENMORE ESTATE

Features:

Year of Construction : 1850.

Location : Chhota Shimla.

Extent : About 3,756 sq m.

Building Material : Sand Stones, Mud, Lath, Wooden Posts, Plaster.

Building Design : Colonial Architecture.

Past Owner : British Empire.

Past Uses : It was a residential house of Lord William Hay, Superintendent of Hill States till 1862. This passed into the hands of Major Innes in 1863 who later sold it to Felix von Goldstein, a leading estate agent of Shimla.

Present Owners : Government Property; H.P. Government and Haryana Government.

Present Use : Guest houses (Haryana Circuit House).

Miscellaneous Aspects : Several improvements were made to the property by Goldstein and as the property of his 'Shimla Rink Company, Limited' this became a focus of town's social life. In 1885, this was purchased by the Punjab Government who placed their offices in the main house while its Secretariat was built at Chotta Shimla. Even after that, Benmore retained its role in housing various offices - and for several years the office of the Superintending Engineer, Hydro-Electric Circle was located here.

Plants in Courtyard : *Aesculus indica, Cedrus deodara, Chamaerops humilis, Cryptomeria japonica, Hibiscus syriacus, Hydrangea macrophylla, Jasminum humile, Jasminum mesnyi, Nerium oleander, Picea smithiana, Prinsepia utilis, Sarcococca pruniformis, Schefflera arboricola, Solanum laxum, Thuja orientalis.*

BENTICK CASTLE (GRAND HOTEL)

Features:

Year of Construction : 1829.

Location : Near Kali Bari Temple.

Extent : About 20, 670 sq m.

Building Material : Burnt Bricks, Sand Stones, Mud, Lath, Wood, Plaster, Dhajji.

Building Design : Colonial Architecture.

Past Owner : British Empire.

Past Uses : Residence of Governor-General Lord William Bentinck and various British Commanders-in chief of the Indian Army.

Present Owner : Central Government.

Present Use : Guest House for Central Government employees.

Miscellaneous Aspects : This building was named 'Bentinck Castle' after its first occupant Lord William Bentinck. Subsequently, the house became the residence of various British Commanders-in-chief of the Indian Army-Sir Henery Fane, Sir Jasper Nicholls and Sir Hugh Gough. Later, it became property of the famous Sir Henry Lawrence-brother of the Viceroy, John Lawrence. The Shimla Bank became its next owner and subsequently the property was acquired by Chevalar Peliti the Viceregal Confectioner in 1892 for ₹ two lakh and converted it to the famous Peliti's Grand Hotel. A great portion of the hotel was damaged by fire in 1922. There is a natural source of water within the premises of Grand Hotel.

Plants in Courtyard : *Agave americana, Araucaria columnaris, Asparagus aethiopicus, Asparagus racemosus, Callistemon lanceolatus, Camellia japonica, Cedrus deodara, Celtis australis, Chamaerops humilis, Drepanostachyum falcatum, Euonymus japonicus, Ficus palmata, Fuchsia* X *hybrida, Girardinia diversifolia, Grevillea robusta, Hedera helix, Hibiscus rosa-sinensis, Hibiscus syriacus, Hydrangea macrophylla, Jasminum humile, Jasminum mesnyi, Jasminum multiflorum, Juniperus chinensis, Juniperus horizontalis, Lonicera japonica, Magnolia grandiflora, Nerium oleander,*

Prinsepia utilis, Pyrus pyrifolia, Quercus oblongata, Rhapis excelsa, Robinia pseudoacacia, Rosa chinensis, Rosa damascena, Salvia leucantha, Salvia splendens, Sambucus nigra, Sarcococca pruniformis, Schefflera arboricola, Smilax aspera, Thuja occidentalis, Yucca aloifolia.

BISHOP COTTON SCHOOL (B.C.S.)

Features:

Year of Construction : 1866-68.

Location : B.C.S Station.

Extent : About 7,00,000 sq m.

Building Material : Sand Stones, Mud, Lath, Wood (Teak, Devdaar), Plaster, Dhajji, Iron Sheet.

Building Design : Mixture of Colonial and Gothic Architecture.

Past Owner : British Empire.

Past Use : School.

Present Owner : Managed by Diocese of Amritsar, Church of North India (CNI).

Present Use : School for Boys from Class I - XII.

Miscellaneous Aspects : Bishop Cotton School was founded by the Most Revd. G.E.L. Cotton, Bishop of Calcutta and Metropolitan of India in 1859 as a part of the offertories held in a service at Calcutta. In 1863, the School was initially established at Jutogh, and on September 26, 1866 the foundation stone was laid at this site. As a mark of respect and to perpetuate the memory of its founder, the school was named Bishop Cotton School in 1867. On September 29, 1868, the school moved from Jutogh and occupied the new building. B.C.S was furnished by Mr Crawford Campbell of the Corps of Engineers. It is one of the oldest boarding schools in Asia. In 1905, the school was damaged in the fire and was rebuilt and occupied in July 1907.

Plants in Courtyard : *Acacia dealbata, Aesculus indica, Araucaria columnaris, Callistemon lanceolatus, Cedrus deodara, Celtis australis, Cerasus cerasoides, Chamaerops humilis, Cupressus torulosa, Drepanostachyum falcatum, Duranta erecta, Ficus neriifolia, Ficus palmata, Forsythia* X *intermedia, Girardinia diversifolia, Grevillea robusta, Hibiscus syriacus, Hydrangea macrophylla, Hypericum oblongifolium, Jacaranda acutifolia, Jasminum mesnyi, Juniperus squamata, Morus alba, Paulownia tomentosa, Pinus roxburghii, Platanus orientalis, Populus ciliata, Prunus armeniaca, Prunus persica, Pyrus pashia, Robinia pseudoacacia, Rosa banksiae, Rosa* *chinensis, Salvia splendens, Solanum laxum, Spartium junceum, Zanthoxylum armatum.*

BURJ

Features:

Year of Construction : 1876.
Location : Chaura Maidan Road.
Extent : About 627 sq m.
Building Material : Stones, Dhajji, Wood, Mud.
Building Design : Colonial Architecture with Pent Roof and Gable Ends.
Past Owner : British Empire.
Past Uses : Initially, this was used to receive the Princes from India when they came to call on the Viceroy at Peterhoff. Later, this became the residence of the Surgeon to the Viceroy. When Peterhoff served as the residence of Himachal's State Governor, this was used as the residence of his Secretary.
Present Owner : H.P. Government.
Present Use : Government residence for HAS officer/s.
Plants in Courtyard : *Cedrus deodara, Eriobotrya japonica, Ficus palmata, Pinus wallichiana, Quercus oblongata, Thuja orientalis.*

CARTON HOUSE (SANITARIUM)

Features:

Year of Construction : 1903-1905.
Location : Chaura Maidan Road.
Extent : About 5,258 sq m.
Building Material : Bricks, Dhajji, Lath, Plaster.
Building Design : Colonial Architecture.
Past Uses : It was originally a guest house till Dr H. C. Menkel shifted the'Shimla Sanitarium and Hospital' along with the Chapel of the Seventh Day Adventists from Belvedere in the Lakkar Bazaar area to these premises in the 1940s.
Present Owner : Managed by Medical Trust India of Seventh Day Adventist.
Present Use : Private Hospital.

Miscellaneous Aspect : It was modified by Dr I.R. Bazliel in 1949 into a new building.

Plants in Courtyard : *Callistemon lanceolatus, Cedrus deodara, Celtis australis, Cerasus cerasoides, Citrus limon, Euonymus fimbriatus, Euonymus tingens, Juglans regia, Magnolia grandiflora, Pinus wallichiana, Prunus domestica, Prunus persica, Pyrus pashia, Quercus incana, Quercus oblongata.*

CATHOLIC CHURCH

Features:

Year of Construction : 1885.

Location : Near Deputy Commissioner's Office and Cantonment Area.

Extent : About 2,578 sq m.

Cost of Construction : ₹ 80,000/-.

Building Material : Grey Stones, Plaster (Lime, Sand).

Building Design : Mixture of Gothic Style of Architecture of an English Renaissance.

Past Owner : During British period it was managed by Shimla-Agra Archi Diocese.

Past Use : Church.

Present Owner : Managed by Shimla Chandigarh Diocese.

Present Use : Church.

Miscellaneous Aspects : The site was purchased in 1850 for just ₹ 1600/-. Mr William Edward exempted the site from ground rent as it was a religious place. Later, Lord Ripon showed interest and spent most of total expense from his own pocket for its furnishing brought from Germany.

Plants in Courtyard : *Aesculus indica, Cedrus deodara, Cerasus cerasoides, Lagerstroemia indica, Populus ciliata.*

CATHOLIC CLUB

Features:

Year of Construction : 19th Century.

Location : Near Fingask.

Extent : About 2,878 sq m.

Building Material : Bricks, Lath, Plaster (Lime, Sand), Dhajji.

Building Design : Colonial Architecture.

Past Owner : Catholic Club.

Past Use : Used as a club by the members of St. Michaels Cathedral Catholic Church for various activities such as playing cards, parties and badminton.

Present Owner : Catholic Club.

Present Use : Used as residential houses by Catholic families and others.

Plants in Courtyard : *Buddleja crispa, Cedrus deodara, Clematis buchananiana, Drepanostachyum falcatum, Girardinia diversifolia, Hibiscus syriacus, Hydrangea macrophylla, Juniperus pseudosabina, Malus pumila, Quercus oblongata, Rosa banksiae, Rosa chinensis, Rubus ellipticus, Salvia splendens, Sambucus nigra, Sarcococca pruniformis, Solanum laxum.*

CEDAR CIRCUIT HOUSE

Features:

Year of Construction : 19th Century.

Location : Near Oak Over.

Extent : About 3,859 sq m.

Building Material : Stones, Mud, Lath, Wooden Posts, Plaster (Dhajji).

Building Design : Swiss Bavarian Style and Alpine Architecture.

Past Owner : Maharaja of Patiala.

Past Use : Guest House of Maharaja of Patiala.

Present Owner : Punjab Government.

Present Use : Used as Circuit Rest House for class one officers of Punjab Government.

Plants in Courtyard : *Aesculus indica, Chamaerops humilis, Cryptomeria japonica, Daphne papyracea, Euonymus japonicus, Euonymus tingens, Ficus palmata, Hibiscus syriacus, Hydrangea macrophylla, Prunus persica, Quercus oblongata, Robinia pseudoacacia, Sarcococca pruniformis, Thuja orientalis.*

CHALET DAY SCHOOL

Features:

Year of Construction : 1909.
Location : Near Oak Over (Chief Minister Residence).
Extent : About 1,879 sq m
Building Material : Stones, Lath, Wood, Plaster (Dhajji).
Building Design : Swiss Alpine Style.
Past Owner : British Empire.
Past Use : It was once a part of the former 'United Services Club'.
Present Owner : Managed by Evangelical Church of India.
Present Use : Co-educational School from Nursery to Class X.
Miscellaneous Aspects : Old Chalet was built in 1890. In 1909, the old Chalet was demolished and replaced by a large modern building in which club dances were held on Tuesdays and Fridays. During 2nd World War, the Chalet was used as an office and afterwards it housed the junior wing of the Loreto Convent. In 1982, it was taken over by the ECI Chalet Days school with D. F. Das as its founder principal. The school, a recognised unaided co-educational Christian school, operates under the guidance of Rt Rev Dr Ezra Sargunnam, Bishop, President, Evangelical Church of India, through the management board of ECI Chalet Day School.
Plants in Courtyard : *Acacia dealbata*, *Araucaria columnaris*, *Berberis lycium*, *Buddleja crispa*, *Cedrus deodara*, *Clematis buchananiana*, *Ficus palmata*, *Girardinia diversifolia*, *Rosa banksiae*, *Rosa chinensis*.

CHAPSLEE

Features:

Year of Construction : Between 1828-1835.
Location : Longwood.
Extent : About 4,980 sq m.
Building Material : Dressed Stones, Mud, Lath, Plaster.
Building Design : Colonial Architecture.
Past Owners : Lord Auckland acquired this property in 1836 to house his private and Military Secretaries, 'Aides De Camps' (ADC's) and named it as 'Secretary Lodge'. Later, General Peter Innes of the Bengal Army purchased it in 1848 and changed its name to Chapslee. General Innes

sold the house in 1870. Thereafter, it changed hands every few years and served as the residence of some famous persons, including Sir Courtenay, Ilbert, General Pemberton, General Sir C.E. Nairne, Surgeon-General Bradshaw, Surgeon-General Cleghorn and so on. Eventually, it was purchased by Sir Arthur Milford Ker, Manager of the Alliance Bank and improved it beyond all recognition in 1896. After the demise of Sir Arthur Ker, the property was purchased by Raja Charanjit Singh of Kapurthala as his summer residence and refurnished the interiors including electrification and plumbing.

Past Use : Residential house.

Present Owner : Kanwar Ratan Jit Singh, the grandson of the Late Raja Charan Jit Singh of Kapurthala.

Present Uses : Residential house and Junior school.

Miscellaneous Aspects : It was built by Dr Blake, a surgeon in the service of the East India Company. Presently, the school is being managed by Mr Singh and his wife Mrs Purnooti Singh.

Plants in Courtyard : *Cedrus deodara*, *Cunninghamia lanceolata*, *Cupressus torulosa*, *Girardinia diversifolia*, *Hedera helix*, *Hibiscus syriacus*, *Hydrangea macrophylla*, *Jasminum mesnyi*, *Prinsepia utilis*, *Pyracantha crenulata*, *Rosa alba*, *Rosa banksiae*, *Rosa moschata*, *Rosa multiflora*, *Rubus niveus*, *Sambucus nigra*, *Sarcococca pruniformis*, *Solanum laxum*, *Spiraea cantoniensis*, *Thuja orientalis*, *Tsuga dumosa*, *Wisteria sinensis*.

CHRIST CHURCH COMPLEX, RIDGE

Features:

Year of Construction : 1844.

Location : Ridge.

Extent : About 3,487 sq m.

Cost of Construction : ₹ 89,000/-.

Building Material : Stones, Lath, Plaster (Lime, Sand).

Building Design : Neo-Gothic Style; designed by Col. J.T. Boileau in 1844.

Past Owner : British Empire.

Past Use : Church.

Present Owner : Amritsar Diocese Trust Association, Church of North India.

Present Use : Chruch.

Miscellaneous Aspects : It is second oldest Christ Church in North India and its foundation was laid on September 9, 1844 on a portion of the Ballyhack Estate owned by Colonel Gouch. The Church was taken over by Government as public building in 1856. The famous clock was put up by Col. Dumbletodn in 1860, and the porch was added in 1873. The Pipe-Organ of Christ Church, biggest in the Indian subcontinent, was erected in September 1899. The Church contains five fine stained glass windows.

Plants in Courtyard : *Cedrus deodara, Cupressus torulosa, Buddleja davidii, Hydrangea macrophylla, Robinia pseudoacacia, Solanum laxum.*

CLARKES HOTEL

Features:

Year of Construction : 1898.

Location : Near H.P. Tourism Lift.

Extent : About 5,678 sq m.

Building Material : Sand Stones, Mud, Lath, Plaster (Lime, Sand).

Building Design : Neo-Tudor Structure.

Past Owner : Mr Ernest Clarkes.

Past Use : Hotel.

Present Owner : P.R.S. Oberoi (son of Mr Mohan Singh Oberoi).

Present Use : Hotel.

Miscellaneous Aspects : In the history of Indian hotel industry, the story of Clarkes began in 1920s when Mr Ernest Clarke arrived as Manager of the Cecil. Both Mr Clarke and his wife Gertrude had a keen eye for observation. They took a great liking to the honesty of a hardworking, young gentleman named Mr Mohan Singh Oberoi, who was working at the Cecil that time. Wanting to take a break from their busy schedule, Mr Clarke and his wife decided to hand over the responsibility of managing Hotel Carlton - now renamed as Clarkes to this impressive young man. During their six months absence, Mr Oberoi doubled up the occupancy to eighty percent which gave them enough reason to offer the hotel - on a decided amount to Mr Oberoi as they wanted to return to England. After continuous hard work for five years, on August 14, 1934, Rai Bahadur Mohan Singh Oberoi became the sole and absolute owner of Hotel Carlton, Shimla.

Plants in Courtyard : *Aesculus indica, Buddleja asiatica, Chamaerops humilis, Cycas revoluta, Hedera helix, Hibiscus syriacus, Hydrangea* *macrophylla, Jasminum mesnyi, Ligustum lucidum, Quercus incana, Solanum laxum*

CLEREMONT (LABOUR BUREAU OFFICE)

Features:

Year of Construction : 1927-1928.

Location : Near Vidhan Sabha.

Extent : About 2,375 sq m.

Building Material : Stones set in Mud, Dhajji, Lath, Plaster, Wooden Posts.

Building Design : Colonial Architecture.

Past Owner : British Empire.

Past Use : It housed offices of the Inspector General, Imperial Service Troops and the Inspector General of Explosives in India.

Present Owner : Government of India.

Present Use : Labour Bureau Office.

Miscellaneous Aspects : At various times in the past this building also housed the offices of Central Board of Revenue, the Director-General of Posts and Telegraphs, the Ex-Services Association, India and the Department of Industries and Labour.

Plants in Courtyard : *Cedrus deodara, Celtis australis.*

CONNY COT (TELEGRAPH OFFICE BUILDING - TELEPHONE EXCHANGE)

Features:

Year of Construction : 1921.

Location : The Mall near PNB.

Extent : About 2,434 sq m.

Cost of Construction : ₹ 9,00,000/-.

Building Material : Ashlar Worked Stones, Bricks, Lath, Plaster.

Building Design : Swiss Bavarian Style.

Past Owner : British Empire.

Past Use : Library (Conny Cot).

Present Owner : H.P. Government.

Present Uses : Telephone Exchange and Telegraph Department, Satellite Communication Centre, Internet Cafe, BSNL Office.

Miscellaneous Aspects : At this site once stood the old Station library in a house called Conny Cot. Telegraph Office was shifted here in 1913-1914. In the original plan, the entire structure was designed in 'Ashlar worked stone' (the stones measured and dressed to be placed in pre-fix position). Due to the economy required during the Ist World War, only the lower section of the building was done in this style and the upper portion was completed in brick work. The design of this building is seismic resistant. A small corner stone on lower section of the building carries an inscription in latin (*Molem Aedificii Mvlti Construxerunt: Rationem Exegit* I. Begg) and means "Many men erected the Stones work of this building. The work was directed by J. Begg". In 1922, one of the first automatic exchanges (Pre-2000 type Strowger exchange) in the S. E. Asia was installed. The prominent clock facing towards both sides of the Mall with a weight of 90 kg is a peculiar feature of this unique building.

Plants in Courtyard : *Aesculus indica*, *Berberis lycium*, *Celtis australis*, *Chamaerops humilis*, *Cupressus torulosa*, *Desmodium elegans*, *Euonymus japonicus*, *Ficus palmata*, *Hibiscus syriacus*, *Lonicera japonica*, *Prinsepia utilis*, *Rosa alba*, *Spartium junceum*, *Thuja orientalis*.

CONNY COTTAGE (G. P. O. BUILDING)

Features:

Year of Construction : 1882.

Location : Near Scandal Point.

Extent : 1,810 sq m.

Cost of Construction : ₹ 1,12,500/- in the beginning; later additions and alteration to the extent of ₹ 1,16,731/- was carried out resulting in a total expenditure of ₹ 2,29,231/-.

Building Material : Six Huge Iron Pillars, Lath, Dhajji, Plaster, Wooden Frames.

Building Design : Tudor Architecture.

Past Uses : Prior to acquisition for Post Office, the building had shops of European tailors like Enjalbat & Co. and Contts & Co.

Present Owner : Central Government.

Present Use : Used as General Post Office (Main Branch).

Miscellaneous Aspects : General Post Office, Shimla is the oldest post office of Northern India. It was purchased from Mr Peterson, Manager of the Shimla Bank in 1883 and used since as post office. Mr F. Dalton was its first Post Master and Mr L.G. Pigott was the last. From Indian perspective, Mr A.K. Hazari was its first Post Master in 1947. Due to a fire engulfing its first floor on September 21, 1972, most of the old records housed in it were burnt which later was rebuilt with its architectural features. In 1992, it was declared as one of the heritage post office buildings in the country.

Plants in Courtyard : *Aesculus indica*, *Buddleja crispa*, *Hedera helix*.

CONVENT OF JESUS AND MARY

Features:

Year of Construction : 1864.

Location : Nav Bahar.

Extent : About 7,578 sq m.

Building Material : Stones, Lath, Plaster (Sand, Lime), Dhajji, Wood (Devdaar, Kail).

Building Design : Western Style.

Past Owner : Arch Bishop of Agra.

Past Uses : It was a military orphanage and boarding school for European and Anglo-Indian Girls with Mother St. Lewis Gonzaga as its first superior. In 1869, the school was enlarged to accommodate 155 pupils.

Present Owner : Congregation or Society of Jesus and Mary.

Present Use : School for girls from Nursery to Class XII.

Miscellaneous Aspects : Due to a devastating fire in April 1946, most of Chelsea was destroyed resulting in loss of property worth ₹ 10.00 lacs. With the generous support of the Archbishop of Delhi-Shimla, the spontaneous response of the civil and military authorities and help rendered by convents and ex-pupils, the school was re-opened within a fortnight of the disaster.

Plants in Courtyard : *Araucaria columnaris*, *Campsis radicans*, *Cedrus deodara*, *Hedera canariensis*, *Hedera helix*, *Hibiscus syriacus*, *Hydrangea macrophylla*, *Ficus elastica*, *Forsythia* X *intermedia*, *Jasminum mesnyi*, *Juniperus squamata*, *Quercus oblongata*, *Rosa banksiae*, *Rosa chinensis*, *Salvia splendens*, *Spartium junceum*, *Solanum laxum*, *Thujopsis dolabrata*.

CORNER HOUSE

Features:

Year of Construction : 19th Century.

Location : Sanjauli.

Extent : About 2,250 sq m.

Building Material : Dressed Sand Stones, Lath, Plaster (Lime, Sand).

Building Design : Lower Portion Constructed in Gothic Style and the Upper one in Colonial Style.

Past Use : Occupied by Municipal Engineer of Shimla.

Present Owner : H.P. Government.

Present Use : Residence of M.C. Commissioner, Shimla.

Plants in Courtyard : *Acacia dealbata*, *Berberis aristata*, *Buddleja crispa*, *Cedrus deodara*, *Eriobotrya japonica*, *Girardinia diversifolia*, *Hibiscus syriacus*, *Hydrangea macrophylla*, *Magnolia* X *soulangeana*, *Ligustrum lucidum*, *Prinsepia utilis*, *Quercus oblongata*, *Robinia pseudoacacia*, *Rosa chinensis*, *Sarcococca pruniformis*.

COUNCIL CHAMBER (VIDHAN SABHA)

Features:

Year of Construction : 1923-1924.

Location : Near Victory tunnel.

Extent : About 2,678 sq m.

Cost of Construction : ₹ 10,00,000/- (designed by Mr W. George of Delhi; built by Sir Claude Hill).

Building Material : White Stones and Baked Brick in Lime, Cement.

Past Owner : British Empire.

Past Use : Legislative Assembly Chamber of India.

Present Owner : H.P. Government.

Present Use : Vidhan Sabha - Assembly of Legislators.

Miscellaneous Aspects : Chamber was opened by the Viceroy, Lord Reading on August 20, 1925. Today, this serves as the Vidhan Sabha - Assembly of Legislators - of Himachal Pradesh who held their first session here on October 1, 1963.

Plants in Courtyard : *Buddleja crispa*, *Cedrus deodara*, *Cupressus torulosa*, *Euonymus japonicus*, *Hedera helix*, *Jasminum mesnyi*, *Juniperus procera*, *Juniperus squamata*, *Quercus oblongata*.

COURTEEN HOUSE

Features:

Year of Construction : Late 19th Century.
Location : Near Boileauganj.
Extent : About 1,440 sq m.
Building Material : Lath, Dhajji, Plaster (Lime, Sand), Sand Stones.
Building Design : Colonial Architecture.
Past Owner : British Empire.
Past Use : Office of the Military Secretary to the Viceroy.
Present Owner : Central Government.
Present Use : Staff quarters for Indian Institute of Advance Study.
Plants in Courtyard : *Berberis aristata, Buddleja crispa, Cedrus deodara, Hedera helix, Hibiscus syriacus, Malus pumila, Prinsepia utilis, Prunus armeniaca, Prunus persica, Quercus oblongata, Rosa chinensis.*

CRAIG GARDEN

Features:

Year of Construction : 1892-95.
Location : Chhota Shimla.
Extent : About 2,356 sq m.
Building Material : Rough Cast Walls, Stones, Lime, Plaster.
Building Design : Colonial Architecture.
Past Owner : British Empire.
Past Use : Residence for officers of Punjab Government secretariat.
Present Owner : H.P. Government.
Present Use : Office of H.P. Bal Kalyan Parishad.
Plants in Courtyard : *Aesculus indica, Cedrus deodara, Celtis australis, Cornus capitata, Elaeagnus umbellata, Ficus sarmentosa, Lonicera quinquelocularis, Morus serrata, Myrsine africana, Populus ciliata, Prinsepia utilis, Pyrus pashia, Sarcococca pruniformis, Viburnum mullaha.*

CROW-BROUGH

Features:

Year of Construction : 1921.

Location : Near Railway Station Shimla.

Extent : About 2,350 sq m.

Building Material : Dressed-Stones, Brick Masonry, Wooden Strap-Work.

Building Design : Mixed Gothic and Colonial Architecture with Pent Roof and Gables.

Past Owner : British Empire.

Past Use : As rest house.

Present Owner : Railway Board.

Present Use : Railway Board Rest House.

Plants in Courtyard : *Cedrus deodara*, *Hibiscus syriacus*, *Lonicera japonica*, *Prunus armeniaca*, *Prunus domestica*, *Prunus persica*, *Pyrus communis.*

CURZON HOUSE

Features:

Year of Construction : 1903.

Location : Near Boileauganj.

Extent : About 1,670 sq m.

Building Material : Stones, Bricks, Lath, Plaster, Wooden Posts.

Building Design : Tudor Style of Architecture.

Past Owner : Lord Curzon.

Past Use : Residential house.

Present Owner : H.P. Government.

Present Use : H.P. State Judicial Academy.

Miscellaneous Aspects : The original house was named after Lord Lansdowne. Eventually, it was damaged in a fire in 1888 and rebuilt by Lord Curzon.

Plants in Courtyard : *Afrocarpus gracilior*, *Araucaria columnaris*, *Asparagus aethiopicus*, *Asparagus densiflorus*, *Berberis lycium*, *Cedrus deodara*, *Chamaerops humilis*, *Euonymus japonicus*, *Hedera helix*, *Hydrangea* *macrophylla, Magnolia grandiflora, Rubus ellipticus, Schefflera arboricola, Quercus oblongata, Thuja orientalis.*

DALZIEL HOUSE (STATE BANK OF INDIA BUILDING)

Features:

Year of Construction : 1924.
Location : Near Kali Bari temple.
Extent : About 3,945 sq m.
Building Material : Pebble-Dash Walls, Stones, Lath, Plaster.
Building Design : Neo - Tudor Style.
Past Owner : Captain H.B. Dalzell.
Past Use : Residential house.
Present Owner : S.B.I.
Present Use : S.B.I. Main Branch, Shimla.
Miscellaneous Aspects : This property first was owned by the Bank of Bengal which in 1924, became the Imperial Bank of India.
Plants in Courtyard : *Buxus wallichiana, Cedrus deodara, Celtis australis, Cupressus torulosa, Hedera helix, Juglans regia, Morus serrata, Rosa banksiae, Wisteria sinensis.*

DELPHINE LODGE

Features:

Year of Construction : 1901-1904.
Location : Below Chhota Shimla.
Extent : About 2,489 sq m.
Building Material : Sand Stones, Mud, Lath, Plaster (Lime, Sand).
Building Design : Pent Roof with Western Style of Architecture.
Past Owners : British officers.
Past Use : Residential house.
Present Owners : Champeshwar Lall Sood, Himesh Sood, Sarveshwar Lal, Dhruv Sarveshwar Lal.
Present Use : Used as residential house.
Plants in Courtyard : *Acer oblongum, Cedrus deodara, Desmodium multiflorum, Euonymus japonicus, Juglans regia.*

DIMPLE LODGE

Features:

Year of Construction : 19th Century.
Location : Chhota Shimla.
Extent : About 4,256 sq m.
Building Material : Dhajji, Sand Stones, Lath, Plaster (Sand, Lime).
Building Design : Pent Roof with Western Style of Architecture.
Past Owners : British officers.
Past Use : Residential house.
Present Owners : Harpreet Singh, Gurpreet Singh, Smt Gur Amarjit Kaur, Smt Jasjeet Kaur, Smt Manjeet Kaur, Major General Gurdayal Singh.
Present Use : Residential house.
Plants in Courtyard : *Amygdalus mira*, *Berberis aristata*, *Cedrus deodara*, *Cerasus cerasoides*, *Chamaerops humilis*, *Hypericum oblongifolium*, *Lonicera japonica*, *Malus pumila*, *Prunus armeniaca*, *Prunus domestica*, *Prunus persica*, *Pyrus pashia*, *Rosa moschata*.

DIXIE

Features:

Year of Construction : 19th Century.
Location : Chhota Shimla.
Extent : About 2,376 sq m.
Building Material : Dhajji, Mud, Wooden Posts, Sand Stones.
Building Design : Pent Roof Style with Dhajji Walls.
Past Owner : Lady Dixie.
Past Use : Residential house.
Present Owner : Contractor Vijay Sood.
Present Use : Not put to use as it is in neglected state.
Plants in Courtyard : *Cedrus deodara*, *Inula cappa*, *Prunus salicina*, *Quercus oblongata*, *Rubus paniculatus*, *Skimmia laureola*, *Solanum virginianum*, *Spiraea canescens*.

EAGLE MOUNT

Features:

Year of Construction : 19th Century.

Location : Nav Bahar.

Extent : About 2,178 sq m.

Building Material : Dhajji, Sand Stones, Lath, Plaster, Wooden Posts.

Building Design : Tudor Style of Architecture.

Past Uses : Catholic orphanage and residence of the Archbishop of Shimla.

Present Owner : St. Bede's College.

Present Use : Management of Girls Hostel.

Plants in Courtyard : *Agave americana, Cedrus deodara, Euonymus tingens, Hibiscus syriacus, Ilex dipyrena, Inula cappa, Juglans regia, Lyonia ovalifolia, Pistacia khinjuk, Prinsepia utilis, Quercus floribunda, Quercus oblongata, Rhododendron arboreum, Rosa banksiae, Solanum laxum, Spartium junceum.*

EARNSTONE

Features:

Year of Construction : 19th Century.

Location : Nigam Bihar.

Extent : About 2,245 sq m.

Building Material : Stones set in Mud while some sections employ Dhajji Walls, Lath, Plaster.

Building Design : Colonial Architecture.

Past Use : This was purchased by the Government in 1919-1920 for its use as an official residence.

Present Owner : Central Government.

Present Use : Residential houses for the employees of Central Government.

Plants in Courtyard : *Cedrus deodara, Hibiscus syriacus.*

EDDLESTON

Features:

Year of Construction : 19th Century.

Location : Near Nigam Bihar.

Extent : About 2,750 sq m.

Building Material : Stones set in Mud, Dhajji Walls, Lath, Plaster.

Building Design : Colonial Architecture.

Past Use : Occupied by British officers.

Present Owner : Central Government.

Present Use : Residential houses for the employees of Central Government.

Miscellaneous Aspects : In 1919-1920, the property was acquired by the Government for use as an official residence and the same function it is serving till to date.

Plants in Courtyard : *Cedrus deodara*, *Cupressus torulosa*, *Prunus domestica*, *Prunus persica*, *Pyrus pashia*, *Rosa chinensis*, *Thuja occidentalis*.

ELLERSLIE BUILDING

Features:

Year of Construction : 1902 (Present building).

Location : Chhota Shimla.

Extent : About 57,500 sq m.

Cost of Construction : ₹ 3,47,824/-.

Building Material : Dressed Stones, Dhajji, Bricks, Concrete, Iron, Lime Mortar Cementing.

Building Design : Tudor Style of Architecture.

Past Owners : Captain John Pengree, and Lieutenant H.H. Pengree.

Past Uses : To start with 'Ellerslie' was the residence of Colonel Keith Young who served as Judge Advocate General of the last India Company's army in Bengal. Later, the property was sold to the Punjab Government in 1881 for housing the Punjab's Military Department.

Present Owner : H.P. Government.

Present Use : H.P. Secretariat.

Miscellaneous Aspects : The work on the present building began on a design by Lieutenant Colonel H.F.S. Abbott in 1899. This was taken over by the Government of H.P. in 1967 to house its Secretariat.

Plants in Courtyard : *Cedrus deodara, Hibiscus syriacus, Hydrangea macrophylla, Pinus wallichiana, Quercus oblongata.*

ELYSIUM SPUR (RAJKIYA KANYA MAHA VIDHYALAYA)

Features:

Year of Construction : 1834.

Location : Longwood.

Extent : About 3,456 sq m.

Building Material : Lath, Dhajji, Sand Stones, Plaster, Wooden Frames.

Building Design : Tudor Style.

Past Owner : M/S Hussain Bux and Company.

Past Use : Hotel.

Present Owner : H.P. Government.

Present Use : Government College for Girls.

Miscellaneous Aspects : The word Elysium, the 'blessed fields' of Greek mythology was given to this area to compliment Lord Auckland's sisters, Fanny and Emily Eden who had accompanied their brother to India when he became Governor-General between 1835 and 1842. In 1952, a private college-the Sanatam Dharam College was opened here and was given its present name in 1977.

Plants in Courtyard : *Agave americana, Buxus wallichiana, Forsythia* X *intermedia, Hibiscus syriacus, Pistacia khinjuk, Prunus armeniaca, Robinia pseudoacaia, Rosa damascena, Sambucus nigra, Solanum laxum, Spartium junceum.*

EMM VILLA

Features:

Year of Construction : 19th Century.

Location : Chhota Shimla.

Extent : About 2,127 sq m.

Building Material : Dhajji, Lath and Plaster as well as Stones set in Mud.

Building Design : Pent Roof and with Lean to Roof Style.
Past Owner : British Empire.
Past Use : British officer's residence.
Present Owners : Private property of Lakhanpals Family (Surender Lakhanpal, Sudesh Lakhanpal, Naresh Lakhanpal, Indu Lakhanpal).
Present Use : None, as it is in bad state.
Plants in Courtyard : *Buddleja crispa*, *Cedrus deodara*, *Girardinia diversifolia*, *Pinus wallichiana*, *Quercus oblongata.*

FIRWOOD COTTAGE

Features:

Year of Construction : 19th Century.
Location : Near Tara Hall.
Extent : About 6,250 sq m.
Building Material : Bricks, Glasses, Stones, Wooden Frames.
Building Design : Pent Roof Style.
Past Use : Residential house.
Present Owner : Shri B.D. Sharma.
Present Use : Property given to five families on lease for residential purposes.
Miscellaneous Aspect : It was once occupied by a British lady who used to play Piano in Gaiety Theater for Viceroy.
Plants in Courtyard : *Aesculus indica*, *Cedrus deodara*, *Ficus palmata*, *Girardinia diversifolia*, *Populus ciliata*, *Quercus oblongata*, *Robinia pseudoacacia.*

FOSS WELL

Features:

Year of Construction : 19th Century.
Location : Chhota Shimla.
Extent : About 2,560 sq m.
Building Material : Dhajji, Mud, Sand Stones, Wood (Devdaar), Lath, Plaster.
Building Design : Pent Roof Style with Dhajji Wall and Dormer Windows.
Past Use : As residential house.

Present Owner : Shri Karam Singh Barar.

Present Use : Residential house.

Plants in Courtyard : *Cedrus deodara, Drepanostachyum falcatum, Girardinia diversifolia, Hibiscus syriacus, Opuntia dillenii, Parthenocissus semicordata, Pinus wallichiana, Prunus amygdalus, Prunus armeniaca, Prunus domestica, Prunus persica, Sarcococca pruniformis, Smilax aspera, Solanum laxum, Spiraea cantoniensis.*

GAIETY THEATRE BUILDING

Features:

Year of Construction : 1887.

Location : Mall Road.

Extent : About 2,250 sq m.

Building Material : Dressed Stones, Slates, Plaster, Lath, Teak, Devdaar Wood.

Building Design : Gothic Style.

Past Owner : British Empire.

Past Uses : Used as theatre for entertainment and socialization.

Present Owner : Under the Deptt. of Language, Art and Cultural.

Present Uses : Used as theatre and exhibition place.

Miscellaneous Aspects : The architect Henry Irwin who designed the Viceregal Lodge, also designed the theatre building for Anglo-Indian Society. 'Time will Tell' was the first play performed here. However, first Indian play by local artist was performed in 1893.

Plant in Courtyard : *Cupressus torulosa.*

GASTON HALL (OLD D.C. OFFICE, SHIMLA BUILDING)

Features:

Year of Construction : 19th Century.

Location : Near C.T.O. Shimla.

Extent : About 2,278 sq m.

Building Material : Heavy Stones, Plaster, Lath.

Building Design : England's Norman Baronial Castles Style.

Past Owner : British Empire.

Past Use : D.C. office.

Present Owner : H.P. Government.

Present Uses : Offices of District Superintendent of Police, D.C and District Attorney.

Miscellaneous Aspects : It was also known as Rosna Hall.

Plants in Courtyard : *Cedrus deodara*, *Celtis australis*, *Cryptomeria japonica*, *Cupressus torulosa*, *Robinia pseudoacacia*.

GLEN UQUARD

Features:

Year of Construction : 1893.

Location : About half km from Vidhan Sabha.

Extent : About 1,978 sq m.

Building Material : Lath, Stones, Dhajji, Plaster, Wood (Devdaar).

Building Design : Pent Roof Style.

Past Owner : Mr P.N. Kaul (1937-1951).

Past Use : Residential house.

Present Owner : Mr Ajay Goel.

Present Use : As Holiday Resort.

Miscellaneous Aspects : Today, it is known as Shimla British resort. This cottage was built by Mr Henery Irwin (chief architect) for his use. He owned this cottage for around 41 years, i.e. from 1893 - 1934. Afterwards this cottage was owned by Nawab of Karachi (1934 - 1937). After independence, it also remained the property of Nehru family for long time. Now, it had been purchased by Mr Ajay Goel who converted it into a holiday resort. The walls are decorated with sketches of J.M.W. Warner.

Plants in Courtyard : *Araucaria cunninghamii*, *Cedrus deodara*, *Girardinia diversifolia*, *Hibiscus syriacus*, *Hydrangea macrophylla*, *Jasminum mesnyi*, *Ligustrum* X *vicaryi*, *Picea smithiana*, *Pinus wallichiana*, *Quercus oblongata*, *Rhododendron arboreum*, *Rosa banksiae*, *Solanum laxum*.

GORTON CASTLE BUILDING

Features:

Year of Construction : 1901-1904.

Location : Near Vidhan Sabha.

Extent : About 17,688 sq m.

Cost of Construction : ₹ 14,02,258/-.

Building Material : Dressed Stones, Iron Sheets, Lath, Plaster.

Building Design : Neo-Gothic Style.

Past Owner : British Empire.

Past Uses : It was Civil Secretariat of the Imperial Government of India and housed legislative, land, education, health and finance departments.

Present Owner : H.P. Government.

Present Use : Accountant General Office.

Miscellaneous Aspects : This site originally owned by Mr Gorton, a bureaucrat in 1840. This house went into hands of many owners, and finally in 1890, purchased by Mr James Walker of the Alliance Bank of Shimla for ₹ 80,000/- with a desire to open a hospital for Europeans. But Lord Ripon, then Viceroy wanted to use it as a Civil Secretariat. So it was purchased by Government for ₹ 1,20,000/-. A new building was built where the actual Gorton Castle stood for ₹ 14,02,258/-. The building was designed by the famous S.S Jacob K.C.I.E. with detailed design and construction carried under supervision of Major H.F Chesney R.E. Executive Engineer.

Plants in Courtyard : *Aesculus indica, Agave americana, Asparagus aethiopicus, Buddleja asiatica, Buddleja crispa, Buddleja davidii, Cedrus deodara, Celtis australis, Cupressus torulosa, Debregeasia saeneb, Euonymus japonicus, Euonymus tingens, Grevillea robusta, Hedera helix, Hibiscus syriacus, Jasminum mesyni, Juniperus procera, Juniperus pseudosabina, Morus serrata, Nerium oleander, Pyrus pashia, Quercus incana, Quercus oblongata, Taxodium distichum, Thuja orientalis.*

GREEN GATE LONGWOOD

Features:

Year of Construction : First decade of the 20th Century.

Location : 1/2 Km from Longwood.

Extent : About 2,488 sq m.

Building Material : Sand Stones, Lath, Dhajji, Plaster (Sand, Lime).

Building Design : Pent Roof Style.

Past Use : Residence of the Medical Officer till this began housing the Assistant Commissioner of the Shimla, Municipal Corporation.

Present Owner : H.P. Government.

Present Use : Government residence for HAS officers.

Plants in Courtyard : *Cedrus deodara, Girardinia diversifolia, Hibiscus syriacus, Inula cappa, Ipomoea purpurea, Pinus wallichiana, Rhamnus virgatus.*

GRINDLAY'S BANK BUILDING

Features:

Year of Construction : 1920.
Location : Mall Road.
Extent : About 1,280 sq m.
Building Material : Stones, Plaster (Sand, Lime).
Building Design : Gothic Style of Architecture.
Past Use : As a bank.
Present Owner : ICICI Bank.
Present Use : Presently used as a ICICI Bank since 2002.
Miscellaneous Aspects : The building has had a chequered history. Fifty years ago, the Mercantile Bank of India sold the building to Grindlay's Bank Ltd. To start with Standard Chartered Bank was opened in it followed later by ANZ Grindlay's Bank.
Plants in Courtyard : *Cupressus torulosa, Rhapis excelsa, Rosa chinensis.*

HAINAULT

Features:

Year of Construction : 1891-1892.
Location : Benmore.
Extent : About 6,568 sq m.
Building Material : Lath, Plaster (Lime, Sand), Sand Stones.
Building Design : Colonial Architecture.
Past Owner : Raja of the erstwhile princely state of Jubbal.
Past Use : Residential House.
Present Owner : Hainault Educational Society, Shimla, since 1986.
Present Use : Co-educational School from Nursery to Class XII.
Plants in Courtyard : *Afrocarpus gracilior, Araucaria columnaris, Camellia japonica, Cedrus deodara, Chamaerops humilis, Daphne papyracea, Euonymus japonicas, Hibiscus syriacus, Hydrangea macrophylla,*

Lagerstroemia indica, Magnolia X soulangeana, Malus pumila, Rosa alba, Rosa banksiae, Rosa chinensis, Salvia splendens, Solanum laxum.

HOLCOMBE

Features:

Year of Construction : 19th Century.

Location : Near Chaura Maidan.

Extent : About 1,980 sq m.

Building Material : Sand Stones, Lath, Plaster, Dhajji, Wood (Teak, Devdaar).

Building Design : Mixed Colonial and Dhajji Style with Pent Roof and Bay Windows.

Past Owners : It was the property of Mr J. E. O'Conor, Director-General of Statistics to the Government of India. Holcombe afterwards became the property of Mr Everard C. Cotes, a well-known Anglo-Indian Journalist, and here Mrs Cotes (Sara Jeannette Duncan of literary fame) wrote more than one of her delightful books.

Past Use : Residential house.

Present Owner : Indian Army.

Present Use : Defense purposes.

Miscellaneous Aspects : Other name for this building is Jim's Lodge.

Plants in Courtyard : *Cedrus deodara, Euonymus japonicus, Girardinia diversifolia, Hibiscus syriacus, Hydrangea macrophylla, Rosa banksiae, Rosa chinensis.*

HOLLY LODGE

Features:

Year of Construction : 19th Century.

Location : Jakhu Hills.

Extent : About 2,480 sq m.

Building Material : Lath, Dhajji, Plaster, Sand Stones, Wood (Devdaar).

Building Design : Colonial Architecture.

Past Owner : Major S.B. Goad.

Past Uses : At the instance of Sophia Anne Cotton, wife of G.E.L. Cotton, Bishop of Calcutta and Metropolitan of India, a girl school was established here in 1866. Eventually, it moved to its present premises at Auckland House in 1868. Holly Lodge then passed into the hands of Raja Sir Padam Singh of Bushehr (Bushair) who was one of the leading hill chiefs.

Present Owner : Raja Vir Bhadara Singh (Chief Minister, H.P.)

Present Use : Residential house.

Plants in Courtyard : *Agave americana, Araucaria columnaris, Asparagus densiflorus, Berberis lycium, Buddleja davidii, Campsis radicans, Cedrus deodara, Cryptomeria japonica, Drepanostachyum falcatum, Euonymus japonicus, Euonymus tingens, Fuchsia X hybrida, Hedera colchica, Hedera helix, Hydrangea macrophylla, Ilex dipyrena, Juniperus squamata, Magnolia grandiflora, Nerium oleander, Prinsepia utilis, Quercus incana, Rosa chinensis, Rosa multiflora, Salvia leucantha, Salvia splendens, Schefflera arboricola, Spartium junceum, Thuja orientalis, Thujopsis dolabrata, Tropaeolum majus.*

INVERARM (STATE MUSEUM)

Features:

Year of Construction : 19th Century.

Location : Near Chaura Maidan.

Extent : About 5,348 sq m.

Building Material : Stones set in Mud, Lath, Plaster, Dhajji.

Building Design : Colonial Architecture.

Past Owner : General Innes.

Past Uses : Inverarm' belonged to General Innes in the 1860s – who also owned several other properties in Shimla. This then became the property of the Raja of Sirmaur and was later purchased by the Government. In the last quarter of 19th Century, the house was totally rebuilt and was often allotted to Members of the Viceroy's Council. Its occupants included Lord William Beresford, General Sir Edwin Collen, Sir Edward Law, Sir Sankaran Nair, Sir Mohamed Shafi and Sir Mahomed Habibullah.

Present Owner : H.P. Government.

Present Use : State Museum.

Plants in Courtyard : *Aesculus indica, Araucaria columnaris, Cedrus deodara, Euonymus japonicus, Girardinia diversifolia, Hibiscus syriacus, Ligustrum lucidum, Pyrus communis, Quercus oblongata, Schefflera arboricola.*

JAKHU TEMPLE

Features:

Location : Jakhu Hills.

Extent : About 3000 sq m.

Building Material : Stones, Devdaar Wood, Lath, Mud.

Building Design : Sikhara Style.

Past Use : Temple.

Present Owner : H.P. Government.

Present Use : Temple of Lord Hanuman ji.

Miscellaneous Aspects : Earliest reference to this one finds in the diary of Captain Alexander Gerard who stopped here to find a temple in August 1817. As per the legendary story Lord Hanuman stopped here to rest while searching for 'Sanjivini' required to cure younger brother of Lord Rama, Laxman who laid mortally wounded on a battlefield in Lanka. It is also believed that this was also the site of a local deity, Jakhra from where the sacred wooden ridge beams of many local temples were brought before being taken to their intended positions elsewhere in the area.

Plants in Courtyard : *Cedrus deodara*, *Rosa chinensis.*

JAMMU CASTLE

Features:

Year of Construction : 1905.

Location : Kasumpti.

Extent : About 1,690 sq m.

Building Material : Lath, Burnt Bricks, Plaster, Wooden Posts (Devdaar).

Building Design : Colonial Architecture.

Past Owner : Raja of Keonthal.

Past Uses : It was once the residence of Raja of princely State of Keonthal. When the State of H.P. came into being, this then became office and Court House of the district of Mahasu.

Present Owner : H.P. Government.

Present Use : Office of H.P. Public Works Department, Electrical Wing.

Plants in Courtyard : *Cedrus deodara*, *Girardinia diversifolia*, *Lonicera japonica*, *Quercus oblongata*, *Schefflera arboricola*, *Solanum laxum.*

KALI BARI TEMPLE

Features:

Year of Construction : 1823.

Location : Near S.B.I. Main Branch, Shimla.

Extent : About 1,500 sq m.

Building Material : Stones, Lath, Plaster.

Building Design : Sikhara Style.

Past Owner : This was established by Ram Charan Bhramachari, a Bengali Brahmin.

Past Use : Temple.

Present Owner : Temple Committee.

Present Use : Temple.

Miscellaneous Aspects : The statue of Goddess Kali is made of black marble and flanked by folk carving in Stones – which are considered to have belonged to an older temple that existed once on the slopes of Jakhu hills.

Plants in Courtyard : *Cedrus deodara*, *Hydrangea macrophylla*, *Thuja orientalis*.

KENNEDY HOUSE

Features:

Year of Construction : 1822.

Location : Near Vidhan Sabha.

Extent : About 1,480 sq m.

Building Material : Cement, Bricks, Stones.

Building Design : Colonial Architecture.

Past Owner : Charles Pratt Keneddy.

Past Use : Residential house.

Present Owner : Government of India.

Present Use : Office of Central Public Works Department.

Miscellaneous Aspects : It was the first 'pucca' house in Shimla. Residence of Charles Pratt Keneddy (Political Officer of the State). Later, it was occupied by Lord Amherst, the Viceroy of India with his family from August 1823 - March 1828. The Commander - in - Chief, Viscount Combermere also used this house as his headquarter from 1825-1832. Afterwards, this was acquired by Major S.B. Goad. Eventually, it was lost in a fire in 1975.

Plants in Courtyard : *Araucaria columnaris, Celtis australis, Juniperus squamata, Pyrus pashia, Thuja orientalis, Yucca aloifolia.*

KNOCKDRIN

Features:

Year of Construction : 1868.

Location : Near Vidhan Sabha.

Extent : About 2,456 sq m.

Building Material : Lath, Dhajji, Plaster, Sand Stones, Wooden Posts (Devdaar).

Building Design : Colonial Architecture.

Past Owner : British Empire.

Past Uses : Residential house to start with but later used as army mess for British soldiers.

Present Owner : Indian Army.

Present Use : Officers Mess.

Miscellaneous Aspects : This army mess so called as 'Will Hall' was built in 1862 by General Butler. Later, it was modified, enlarged and rechristened as 'Knockdrin' in 1868 by Captain Levinge. In 1906, Knockdrin became the residence of the Foreign Secretary of Government of India. Finally, it was converted into officers mess on June 25, 1985.

Plants in Courtyard : *Araucaria columnaris, Buxus wallichiana, Cedrus deodara, Celtis australis, Chamaerops humilis, Ficus elastica, Grevillea robusta, Lonicera japonica, Magnolia grandiflora, Parthenocissus semicordata, Platanus orientalis, Prunus domestica, Quercus oblongata, Schefflera arboricola, Solanum laxum, Yucca aloifolia.*

LADY READING HOSPITAL

Features:

Year of Construction : 1922-1924.

Location : Near Oak Over.

Extent : About 2,880 sq m.

Building Material : Lath, Dhajji, Sand Stones, Plaster.

Building Design : Western Style.

Past Owner : British Empire.

Past Use : Hospital.

Present Owner : H.P. Government.

Present Use : Hospital.

Miscellaneous Aspects : To start with the female wards of the Ripon hospital, Shimla were converted into a separate hospital, 'The Lady Dufferin Hospital', and were put under the charge of a qualified Lady doctor to commemorate Queen Victoria's Diamond jubilee. The name of hospital was later changed into the Lady Reading Hospital. The hospital, finally came into being on April 25, 1924, as part of the philanthropic activities undertaken by Lady Reading during the British rule; the largest donation of ₹ 2,62,000/- being from the Raja of Nabha. The hospital was re-christened as Kamla Nehru Hospital in September, 1981.

Plants in Courtyard : *Araucaria cunninghamii*, *Buddleja crispa*, *Cedrus deodara*.

MANOR VILLA (RAJ KUMARI AMRIT KAUR'S BUILDING)

Features:

Year of Construction : 1886-1888.

Location : Summerhill.

Extent : About 1,948 sq m.

Building Material : Bricks, Lath, Plaster.

Building Design : Colonial Architecture.

Past Owner : Raj Kumari Amrit Kaur.

Past Use : Residence of Raj Kumari Amrit Kaur (1889 – 1964), daughter of Raja Harnam Singh of Kapurthala.

Present Owner : Government.

Present Use : Guest House of All India Institute of Medical Science (AIIMS).

Miscellaneous Aspect : During India's struggle for freedom, the father of Nation Mahatma Gandhi often stayed here while visiting Shimla.

Plants in Courtyard : *Buddleja asiatica*, *Buddleja crispa*, *Cedrus deodara*, *Hibiscus syriacus*, *Quercus oblongata*, *Salvia leucantha*, *Solanum laxum*.

MORVYN (I.T.I. SHIMLA)

Features:

Year of Construction : 1920-1921.

Location : 1 km from Summerhill.

Extent : About 4,180 sq m.

Building Material : Sand Stones, Lath, Plaster, Bricks.

Building Design : Colonial Architecture.

Past Owner : British Empire.

Past Use : British Officers' residence.

Present Owner : H.P. Government.

Present Use : I.T.I. Shimla.

Plants in Courtyard : *Cedrus deodara, Euonymus tingens, Ficus palmata, Girardinia diversifolia, Hedera nepalensis, Hibiscus syriacus, Pyrus pashia, Quercus oblongata, Solanum virginianum, Spartium junceum, Viburnum cotinifolium, Viburnum mullaha.*

MYTHE

Features:

Year of Construction : 1901-1903.

Location : Near Tara Hall.

Extent : About 7,456 sq m.

Building Material : Lath, Dhajji, Mud, Plaster, Wood (Devdaar), Sand Stones.

Building Design : Colonial Architecture with Dhajji Walls.

Past Owners : It was the property of Mr Whitley Stokes, the legal officer, during whose occupation it was known as 'Laurel Hall'. Later, it was occupied by Lt. General George Leslie and General Sir Edmund Elles.

Past Use : Residential house.

Present Owners : Family Members of Late Shri Mela Ram Sood.

Present Use : Residential house.

Plants in Courtyard : *Cedrus deodara, Chamaerops humilis, Cupressus torulosa, Drepanostachyum falcatum, Hedera nepalensis, Malus pumila, Rosa banksiae, Rosa chinensis, Solanum pseudocapsicum, Thuja orientalis.*

NORTH BANK

Features:

Year of Construction : 19th Century.

Location : Below Kali Bari temple.

Extent : About 420 sq m.

Building Material : Lath, Plaster, Stones, Wooden Posts.

Building Design : Pent Roof Style with Colonial Features.

Past Owners : Mr C.S. Bayley (agent of Governor General for Central India), Sir Edward Buck.

Past Use : Residential house.

Present Owner : Half of the house is private property of Rani Jayanti Devi (Rani of Koti), and the other half with two families.

Present Use : Residential purposes.

Miscellaneous Aspects : It was also head office of the Countess of Dufferin's Fund. The house was also famous for staying of Rudyard Kipling (writer, Nobel prize winner) with Sir Edward Buck.

Plant in Courtyard : *Cedrus deodara*.

NORTH STONEHAM (GOVT. BOY'S DEGREE COLLEGE, SANJAULI)

Features:

Year of Construction : 1904-1905.

Location : Sanjauli.

Extent : About 6,258 sq m.

Building Material : Stones, Lath, Plaster.

Building Design : Colonial Architecture.

Past Owners : Mr & Mrs Tytler.

Past Uses : It was 'Asiatic Christian Orphanage' to start with and formally declared open by Viceroy, Lord Mayo in June 1871 and renamed as 'The Mayo Institute and the Himalayan Christain Orphanage'- popularly known as the 'Mayo Industrial School'. On the death of her husband, Mrs Tytler passed on the institution to a committee of townspeople constituted for the purpose with the objective to provide education and industrial training to European and Eurasian orphaned girls.

Present Owner : H.P. Government.

Present Use : Govt. Boy's Degree College.

Miscellaneous Aspects : It was established by Mrs Harriet Tytler.

Plants in Courtyard : *Cedrus deodara, Chamaerops humilis, Hibiscus syriacus, Rosa chinensis, Thuja orientalis.*

OAK LODGE

Features:

Year of Construction : 1918-1919.

Location : Near Vidhan Sabha.

Extent : About 2,158 sq m.

Building Material : Burnt Bricks, Plaster.

Building Design : Colonial Architecture.

Present Owner : Indian Army.

Present Use : Officer's Transit Mess (Sashastra Seema Bal).

Plants in Courtyard : *Berberis aristata, Buddleja crispa, Cedrus deodara, Prinsepia utilis, Quercus oblongata, Robinia pseudoacacia, Rubus ellipticus.*

OAK OVER

Features:

Year of Construction : 19th Century.

Location : Oak Over.

Extent : About 3,980 sq m

Building Material : Lath, Plaster, Stones, Iron Sheets.

Building Design : Colonial Architecture with Pentagonal Plan.

Past Owner : Maharaja of Patiala.

Past Use : Summer residence of Maharaja of Patiala.

Present Owner : H.P. Government.

Present Use : Official residence of the Chief Minister of H.P.

Miscellaneous Aspect : The house came into hands of Indian Government after first world war.

Plants in Courtyard : *Cedrus deodara, Chamaerops humilis, Cryptomeria japonica, Cunninghamia lanceolata, Cupressus torulosa, Drepanostachyum falcatum, Eriobotrya japonica, Euonymus japonica, Hedera helix, Hibiscus syriacus, Hydrangea macrophylla, Jasminum mesyni, Juniperus squamata, Ligustrum* X *vicaryi, Lonicera japonica, Prinsepia utilis,*

Quercus oblongata, Rhapis excelsa, Rosa chinensis, Schefflera arboricola, Thuja orientalis, Thujopsis dolabrata.

OBSERVATORY HOUSE

Features:

Year of Construction : 1844.

Location : Near Boileauganj.

Extent : About 2,268 sq m.

Building Material : Dressed Stones, Lath, Plaster, Wooden Frames.

Building Design : Pent Roof Style.

Past Owner : Col. J.T. Boileaue.

Past Uses : This was used as an observatory equipped with magnetic devices. Later, it was used as a residence for the Private Secretary to the Viceroy.

Present Owner : Indian Institute of Advance Study (IAAS), Govt. of India.

Present Use : Guest House.

Miscellaneous Aspect : Name of observatory hill owes to the name of house.

Plants in Courtyard : *Cedrus deodara, Ficus palmata, Girardinia diversifolia, Hedera helix, Pyrus pashia, Quercus oblongata.*

PARI MAHAL

Features:

Year of Construction : 1892-1896.

Location : Kasumpti.

Extent : About 4,280 sq m

Building Material : Sand Stones, Lath, Plaster, Mud, Dhajji.

Building Design : Colonial Architecture.

Past Owner : Rulers of Jubbal.

Past Use : Palace for Rulers of Jubbal.

Present Owner : H.P. Government.

Present Use : Administrative wing of the State Health and Family Planning Welfare Training Centre.

Miscellaneous Aspect : This was also the residence of the noted sitar maestro, Ustad Vilayat Khan.

Plants in Courtyard : *Agave americana, Araucaria columnaris, Cedrus deodara, Hedera helix, Hibiscus syriacus, Ficus palmata, Populus ciliata, Prinsepia utilis, Pyrus communis, Pyrus pashia, Quercus oblongata, Robinia pseudoacacia, Rosa chinensis.*

PARKASH NIWAS

Features:

Year of Construction : 19th Century.

Location : Near S.B.I. Main Branch, Shimla.

Extent : About 1,098 sq m.

Building Material : Dressed Stones, Brick, Lath, Plaster and Wooden Framing in the local Dhajji Style.

Building Design : Colonial Architecture

Past Owner : British Empire.

Past Use : Type Writing School.

Present Owner : Tarun Chatwal.

Present Use : Hotel.

Miscellaneous Aspect : This house was once a part of the old Dalzell (Dalziel) estate.

Plants in Courtyard : *Cedrus deodara, Hedera helix.*

PETERHOFF

Features:

Year of Construction : 1991 (New Building).

Location : Chaura Maidan.

Extent : About 4,480 sq m.

Building Material : Bricks, Cement, Stones.

Building Design : Colonial Architecture.

Past Owners : Initially, General Inn was its owner. In 1880, he sold it to the King of Sirmour and then it came into hands of government in 1880.

Past Uses : The old Peterhoff building in Shimla housed at least seven Viceroys and Governor Generals of India during the British Raj. After India's independence, the building served as the Punjab High Court. It was in this building the trial of Nathuram Godse, who assassinated Mahatma

Gandhi, took place in 1948-1949. Later, it was also used as the Governor's Residence and popularly called as Raj Bhavan.

Present Owner : State Tourism Department.

Present Use : Used as a hotel by State Tourism Department.

Miscellaneous Aspects : Its first occupant was James Bruce, 8th Earl of Elgin who moved into the building in 1863. Peterhoff was gutted in a devastating fire on January 12, 1981. It was restructured as a luxury hotel in 1991.

Plants in Courtyard : *Aesculus indica*, *Araucaria columnaris*, *Cedrus deodara*, *Celtis australis*, *Cupressus torulosa*, *Debregeasia saeneb*, *Drepanostachyum falcatum*, *Ficus neriifolia*, *Ficus palmata*, *Hibiscus syriacus*, *Hydrangea macrophylla*, *Juniperus pseudosabina*, *Nerium oleander*, *Pyrus communis*, *Pyrus pashia*, *Spartium junceum*, *Thuja orientalis*.

POLICE STATION SADAR, BOILEAUGANJ

Features:

Year of Construction : 1914.

Location : Boileauganj.

Extent : About 1,280 sq m.

Cost of Construction : ₹ 13,978/-.

Building Material : Lath, Dressed Stones, Dhajji, Plaster, Wooden Frames.

Building Design : Tudor Style of Architecture.

Past Owner : British Empire.

Past Use : Police Station.

Present Owner : H.P. Government.

Present Use : Police Station.

Plant in Courtyard : *Cedrus deodara*.

POLICE STATION SADAR, SABJI MANDI, SHIMLA

Features:

Year of Construction : 1832.

Location : Near Sabji Mandi, Lower Bazaar, Shimla.

Extent : About 6,678 sq m.

Building Material : Dressed Stones, Lath, Plaster, Mud, Tin.

Building Design : Pent Roof Style.

Past Owner : British Empire.

Past Use : Police Station.
Present Owner : H.P. Government.
Present Use : Police Station.
Miscellaneous Aspect : Old record available since 1886.
Plants in Courtyard : *Debregeasia saeneb*, *Euonymus japonicus Robinia pseudoacacia*, *Thuja orientalis*.

POLICE STATION, CHHOTA SHIMLA

Features:

Year of Construction : 19th Century.
Location : Chhota Shimla.
Extent : About 1,678 sq m.
Building Material : Sand Stones, Lath, Plaster, Dhajji, Tin.
Building Design : Neo-Tudor Style.
Past Owner : British Empire.
Past Uses : Office of Post and Telegraph.
Present Owner : H.P. Government.
Present Use : Police Station.
Miscellaneous Aspect : The earliest 'First Information Report' was registered in this station on May 23,1905.
Plants in Courtyard : *Cedrus deodara*, *Prunus armeniaca*, *Quercus oblongata*, *Thuja orientalis*.

POST OFFICE, CHAURA MAIDAN

Features:

Year of Construction : 1910.
Location : Chaura Maidan.
Extent : 396.82 sq m.
Cost of Construction : ₹ 24,400/-.
Building Material : Half-round Dressed Stones, Pebble-dash Walls, Wood.
Building Design : Neo-Tudor Style.
Past Owner : British Empire.
Past Use : Sub - Post office.
Present Owner : Central Government.

Present Use : Sub - Post Office.
Plants in Courtyard : *Cedrus deodara*, *Quercus oblongata*, *Thuja orientalis*.

POST OFFICE, CHHOTA SHIMLA

Features:

Year of Construction : 1924-1925.
Location : Chhota Shimla.
Extent : 1315 sq m.
Cost of Construction : ₹ 37,790/-.
Building Material : Lath, Plaster, Wooden Frames, Dressed Stones.
Building Design : Colonial Architecture.
Past Owner : British Empire.
Past Use : Sub-Post Office.
Present Owner : Central Government.
Present Use : Sub-Post office.
Plants in Courtyard : *Quercus oblongata*, *Robinia pseudoacacia*.

POST OFFICE, SUMMERHILL

Features:

Year of Construction : 1923-1924.
Location : Summerhill.
Extent : 177.09 sq m.
Cost of Construction : ₹ 41359/-.
Building Material : Pebble-Dash Walls, Dressed Stones, Lath, Plaster.
Building Design : Pent Roof Style with Gable Ends.
Past Owner : British Empire.
Past Use : Sub-Post office.
Present Owner : Central Government.
Present Use : Sub-Post Office.
Plant in Courtyard : *Quercus oblongata*.

PRIMROSE HILL COTTAGE

Features:

Year of Construction : 1886.

Location : Chaura Maidan.

Extent : About 2,398 sq m.

Building Material : Bricks, Lath, Plaster, Wooden Posts.

Building Design : Colonial Architecture.

Present Owner : Rajeev Madan.

Past Owner : Mr Bazalgetter.

Past Uses : It was residence of Mr Bazalgetter (Political agent of Punjab Hills State). Later, it was used as summer residence of Governor of Myanmar and Mr Simpson (Indian Meteorological Services).

Present Use : Hotel.

Miscellaneous Aspects : It was also known as Khud Cottage or Madan Kunj.

Plants in Courtyard : *Aesculus indica, Cedrus deodara, Cornus macrophylla, Forsythia* X *intermedia, Hydrangea macrophylla, Quercus oblongata, Rosa banksiae.*

RACE VIEW

Features:

Year of Construction : 1918-1919.

Location : Near Vidhan Sabha.

Extent : About 1,180 sq m

Building Material : Stones set in Mud, Dhajji, Lath, Plaster.

Building Design : Mixture of Colonial and Indian Architecture.

Past Owner : British Empire.

Past Uses : Offices of Superintending Engineers, Shimla East and West Divisions.

Present Owner : Government of India.

Present Use : Quarters of C.P.W.D.

Miscellaneous Aspects : This building probably derived its name from its view of Annandale – which was the site of gymkhanas, fancy dress parties, polo, cricket and football matches and horse races.

Plants in Courtyard : *Cedrus deodara*, *Debregeasia saeneb*, *Pyrus pashia.*

RAILWAY BOARD BUILDING

Features:

Year of Construction : 1896-1897.

Location : Near A.G. Office.

Extent : About 2,280 sq m.

Cost of Construction : ₹ 7,31,551/-.

Building Material : Iron, Steel, Burnt Bricks.

Building Design : British Colonial Style.

Past Owner : Major H.B. Goad.

Past Use : Public Work Secretariat.

Present Owners : Central and H.P. Government.

Present Uses : Offices of Excise, Income Tax, Central Bureau of Investigation, Public Works Department Office etc.

Miscellaneous Aspects : This building was built by M/S Richardson and Cruddas, a Bombay based firm for Major H.B. Goad after demolishing two small houses, Herbert House and Lowville. This was designed structurally to be fire and earth quake resistant and was not affected during fire which broke on February 10, 2001. Besides Railway Board Building, the famous Raffles Hotel in Singapore is only such building in South East Asia. Total height of the building is 25m including three basements.

Plants in Courtyard : *Cedrus deodara*, *Quercus oblongata.*

RAILWAY STATION, SHIMLA

Features:

Year of Construction : 1903.

Location : Below, Cart Road, Shimla.

Extent : About 2,256 sq m.

Building Material : Sand Stones, Mud, Lath, Plaster, Dhajji.

Building Design : Colonial Architecture.

Past Owner : British Empire.

Past Use : Railway Station.

Present Owner : H.P. Government.

Present Use : Railway Station.

Miscellaneous Aspects : Through its length of 96.5 km it runs at places 2133m above the sea level through 103 tunnels and 800 bridges. It took 3 years and investment of one million pounds sterling to complete and considered to be one of the most spectacular narrow gauge lines in the world. This stretch of railway was inaugurated on November 9, 1903.

Plants in Courtyard : *Cedrus deodara, Ficus palmata, Hibiscus syriacus, Pinus wallichiana, Quercus oblongata.*

RAILWAY STATION, SUMMERHILL

Features:

Year of Construction : 1902-1903.

Location : Summerhill.

Extent : About 1,480 sq m.

Building Material : Lath, Mud, Dhajji, Plaster, Wooden Posts.

Building Design : Colonial Architecture.

Past Owner : British Empire.

Past Use : Railway Station.

Present Owner : H.P. Government.

Present Use : Railway Station.

Miscellaneous Aspect : Neale's token instrument of signalling is still followed to check accidents.

Plants in Courtyard : *Aesculus indica, Cedrus deodara, Chamaerops humilis, Parthenocissus semicordata, Populus ciliata, Prinsepia utilis.*

RAVENSWOOD

Features:

Year of Construction : 1867.

Location : High Court.

Extent : About 1,580 sq m.

Building Material : Lath, Plaster, Stones, Bricks.

Past Owners : To start with, it was the property of Mr F. Lushmgton, financial Secretary of the Government of India who lived here during his term of office. Later, this exchanged hands with Mr O'Meara, Raja of Faridkot, Colonel Cecil Beadon, Colonel R. Home and General Sir Henry Brackenbury.

Past Use : Residential house.

Present Owner : H.P. Government.

Present Use : H.P. High Court.

Plants in Courtyard : *Cedrus deodara, Celtis australis, Chamaerops humilis, Cupressus torulosa, Daphne papyracea, Hedera helix, Hibiscus syriacus, Hydrangea macrophylla, Jasminum mesnyi, Quercus incana, Robinia pseudoacacia, Rosa banksiae, Rosa chinensis, Sarcococca pruniformis.*

RETREAT

Features:

Year of Construction : 1905.

Location : Chaura Maidan.

Extent : About 4,648 sq m.

Cost of Construction : ₹ 1,17,977/-.

Building Material : Sand Stones, Lath, Plaster, Dhajji, Wood (Teak, Devdaar).

Building Design : Colonial Architecture.

Past Use : Residence of Viceroy's Executive Council Member.

Present Owner : Indian Army.

Present Use : Official residence of Lieutenant General.

Miscellaneous Aspect : During Indian freedom movement some well known personalities, Nehru, Azad, Vallabh Pant stayed here.

Plants in Courtyard : *Aesculus indica, Callistemon lanceolatus, Cedrus deodara, Chamaerops humilis, Euonymus tingens, Grevillea robusta, Magnolia grandiflora, Myrtus communis, Parthenocissus semicordata, Platanus orientalis, Populus ciliata, Prinsepia utilis, Prunus domestica, Punica granatum, Robinia pseudoacaia, Salix babylonica.*

RIPON HOSPITAL

Features:

Year of Construction : 1882.

Location : Near Local Bus Stand.

Extent : About 4,345 sq m.

Building Material : Sand Stones, Lath, Plaster, Dhajji, Wooden Posts (Devdaar).

Building Design : Alpine Villa Style.
Past Owner : British Empire.
Past Use : Hospital.
Present Owner : H.P. Government.
Present Use : Hospital.
Miscellaneous Aspects : The architectural plan was prepared by Henry Irwin for this site that had originally two houses, 'The Briars' and 'Glen Cottage' which were gutted in 1881. The foundation stone was laid by Viceroy of India, the Marquis of Ripon in 1882 and the hospital was thrown open for patients on May 14, 1885 by Lord Dufferin.
Plants in Courtyard : *Aesculus indica, Cedrus deodara, Cestrum elegans, Chamaerops humilis, Debregeasia saeneb, Pyrus pashia, Sambucus nigra.*

ROTHNEY CASTLE

Features:

Year of Construction : 1838.
Location : Jakhu Hills.
Extent : About 17, 509 sq m.
Building Material : Lath, Mud, Stones, Plaster, Iron, Wooden Posts.
Building Design : Colonial Architecture.
Past Owners : Colonel Rothney, followed by Mr A.H. Mathews, Mr P. Mitchell, Mr A.O. Hume.
Past Use : Residential house.
Present Owner : Lala Chuna Mull.
Present Use : No use.
Miscellaneous Aspects : This was built by Colonel Rothney in 1838. In 1843, the house passed into hands of Dr Carte who established the 'Shimla Bank Corporation' – which later shifted to the site of Grand Hotel. Indian National Congress was founded here by A.O. Hume.
Plants in Courtyard : *Cedrus deodara, Chamaerops humilis, Cryptomeria japonica, Drepanostachyum falcatum, Euonymus tingens, Hedera helix, Hedera nepalensis, Hydrangea macrophylla, Prinsepia utilis.*

SHIMLESHWAR MAHADEV (SHIV MANDIR)

Features:

Year of Construction : 19th Century.

Location : Near Mall Road.

Extent : About 350 sq m.

Building Material : Burnt Bricks, Cement.

Building Design : Shikara Type.

Past Owner : Shri Madan Giri.

Past Use : Temple.

Present Owner : Temple Committee.

Present Use : Shiva Temple.

Miscellaneous Aspect : The temple of Lord Shiva is famous as Shimleshwar Mahadev temple.

Plants in Courtyard : None.

SIDHOWAL LODGE

Features:

Year of Construction : 1820-1826.

Location : Near Ridge.

Extent : About 2,358 sq m.

Building Material : Mud, Lath, Plaster, Dhajji, Wooden Posts (Teak).

Building Design : Colonial Architecture.

Past Owner : Lord Amherst (Ist Governor General).

Past Use : Residential house.

Present Owners : Sidhowal family.

Present Use : Residential house.

Miscellaneous Aspects : This was initially called 'Ballyhack' and later, renamed first as Christ Church Lodge and subsequently Sidhowal Lodge in the early part of 20th Century after being acquired by Sidhowal family.

Plants in Courtyard : *Buddleja davidii*, *Cedrus deodara*, *Cestrum parqui*, *Cotoneaster bacillaris*, *Cupressus torulosa*, *Debregeasia saeneb*, *Hibiscus syriacus*, *Hydrangea macrophylla*, *Prinsepia utilis*, *Prunus armeniaca*, *Spartium junceum*, *Tropaeolum majus*.

SNOWDOWN

Features:

Year of Construction : 1872.

Location : Next to Correstophen building on way to Sanjauli from Lakkar Bazaar.

Extent : About 14,679 sq m.

Building Material : Stones, Cement, Bricks, Wooden Posts.

Building Design : Gothic to Mock Tudor.

Past Owner : General Peter Innes.

Past Use : Used as a small dispensary.

Present Owner : H.P. Government.

Present Uses : Hospital and Indira Gandhi Medical College.

Miscellaneous Aspects : Considering the cold zone where the building was structured, the first British owner decided to name it Snowdown. The place which houses the grandiose and stately Indira Gandhi Medical College today, was a small dispensary in 1873-1874 and owned by General Peter Innes. Lord Roberts purchased the building from Innes in 1873 and materially improved the property after getting it vacated. The then Commander-in-Chief General Sir Donald Martin Stewart who lived in Woodville shifted to Snowdown in 1881. The place remained the official residence of the Commander-in-Chief of the Indian Army up to December 31, 1947. After partition, the Government of India purchased the building for ₹ 79, 817/- and declared it to be official residence of the Commander-in-Chief in India. The reorganisation of Punjab in 1966 brought Shimla to the fold of H.P. and a State Medical College was set up in 1966 at Snowdown. The fire on the night of December 31, 1971 gulped the old heritage building of Snowdown. Later, it was rebuilt into a huge complex and was renamed as Indira Gandhi Medical College in 1984.

Plants in Courtyard : *Cedrus deodara, Celtis australis, Desmodium elegans, Desmodium multifloum, Euonymus tingens, Hedera nepalensis, Hibiscus syriacus, Hydrangea macrophylla, Ilex dipyrena, Quercus incana, Quercus oblongata, Robinia pseudoacacia.*

SNOW VIEW

Features:

Year of Construction : 19th Century.

Location : Near Lakkar Bazaar Bus Stand.

Extent : About 1,760 sq m.

Building Material : Lath, Plaster, Dhajji, Mud, Sand Stones, Wooden Posts.

Building Design : Colonial Architecture.

Past Owner : Major H.B. Goad. It was one of thirty properties of Major Goad.

Past Use : Rented to various British officers.

Present Owner : A.K. Soni.

Present Uses : Residential house and car parking area also available.

Miscellaneous Aspect : Name important due to distant view of snow covered hills from the site.

Plants in Courtyard : *Cedrus deodara*, *Ficus palmata*, *Populus ciliata*, *Robinia pseudoacacia.*

SOUTH GATE

Features:

Year of Construction : 19th Century.

Location : Chhota Shimla.

Extent : About 1,348 sq m.

Building Material : Bricks, Cement.

Building Design : Colonial Architecture.

Past Owner : Residence of Col. Khosla of the Indian Army Medical Services.

Past Use : Residential house.

Present Owner : Susheel Kapur (Sheel Jewellers).

Present Use : Residential house.

Plants in Courtyard : *Cedrus deodara*, *Euonymus japonicus*, *Thujopsis dolabrata*, *Wisteria sinensis.*

SPRING FIELD

Features:

Year of Construction : 19th Century.

Location : Near Chhota Shimla.

Extent : About 4,440 sq m.

Building Material : Lath, Dhajji, Plaster, Sand Stones.

Building Design : Colonial Architecture.

Past Owner : Mohammed Yahya Khan of Lahore.

Past Use : Residential house.

Present Owner : Kunwar Vijay Singh.

Present Use : It is a hotel taken on lease by a unit of Usha-Lexus Hotels and Resort.

Miscellaneous Aspect : After 1947, the property passed into hands of the former Raja of Sheikhupura.

Plants in Courtyard : *Asparagus aethiopicus*, *Asparagus virgatus*, *Cestrum elegans*, *Eriobotrya japonica*, *Euonymus japonicus*, *Jasminum mesnyi*, *Juniperus squamata*, *Lagerstroemia indica*, *Ligustrum* X *vicaryi*, *Myrtus communis*, *Nerium oleander*, *Platanus orientalis*, *Rosa banksiae*, *Salix alba*, *Solanum laxum*, *Solanum pseudocapsicum*.

SQUIRE'S HALL

Features:

Year of Construction : Late 19th Century.

Location : Near Summerhill.

Extent : About 2,878 sq m.

Building Material : Bricks, Stones, Cement, Concrete.

Building Design : Colonial Architecture.

Past Owner : British Empire.

Past Use : It was occupied by the members of Viceregal Staff.

Present Owner : Indian Institute of Advance Study.

Present Use : Residence for the Director of Advance Study.

Plants in Courtyard : *Cedrus deodara*, *Euonymus japonicus*, *Hibiscus syriacus*, *Hydrangea macrophylla*, *Jasminum mesnyi*, *Kerria japonica*, *Lonicera japonica*, *Quercus oblongata*, *Wisteria sinensis*, *Yucca aloifolia*.

ST. ANDREW'S CHURCH

Features:

Year of Construction : 1914.

Location : Near G.P.O. Shimla.

Extent : About 1,876 sq m.

Building Material : Baked Bricks, Plaster.

Building Design : Gothic Style.

Past Use : Church.

Present Owner : H.P. Government.

Present Use : Evening College and Library H.P. University.

Miscellaneous Aspects : Formerly it was the Union Church built by Mr W.H. Carey in 1869. The church held 240 people and was built on the Constantia estate in Shimla. In 1910, Rav James Black made efforts for rebuilding the old Union Church. St Andrew's or the Scots Kirk was built on the site of an earlier Union Church. This was designed by architect John Begg. The cost of its construction was met almost entirely by private subscription. The dedication service was held on May 8, 1915 by the senior Presidency Chaplain of Bengal, the Rev. DN Gillan assisted by Rev. James Black, the first minister of St Andrew's Kirk at Shimla. The building had now been converted into an evening college and a part of it also used as a public library. It is also called as Gandhi Bhawan. It was started in 1962 by the Punjab University to cater educational needs of employed persons.

Plants in Courtyard : *Cedrus deodara*, *Cupressus torulosa*, *Hibiscus syriacus*, *Pyrus pashia*, *Quercus incana*, *Thuja orientalis*.

ST. BEDE'S COLLEGE

Features:

Year of Construction : 1904.

Location : Nav Bahar.

Extent : About 6,978 sq m.

Building Material : Lath, Plaster, Heavy Stones, Cement.

Building Design : Gothic Style of Architecture.

Past Owner : British Empire.

Past Use : School.

Present Owner : Congregation of Jesus and Mary.

Present Use : College for Girls.

Miscellaneous Aspects : Founded in 1904 by Rev. Mother St. Clare for the training of teachers. College forms an integral part of the Chelsea estate and imparting education to students for B.A. Degree since 1947. The components of a general college were added to it after independence, and its affiliation was transferred to H.P. University in 1970.

Plants in Courtyard : *Euonymus japonicus, Hedera colchica, Juniperus squamata, Lagerstroemia indica, Nerium oleander, Platanus orientalis, Solanum laxum, Solanum pseudocapsicum, Thujopsis dolabrata.*

ST. EDWARD'S SCHOOL

Features:

Year of Construction : 1932 (New Building).

Location : Adjoining Hotel Himland, Cart Road, Shimla.

Extent : About 8,234 sq m.

Building Material : Bricks, Stones, Lath, Plaster.

Building Design : Colonial Architecture.

Past Owners : Irish Christian Brothers.

Past Use : Boys School.

Present Owner : Managed by Shimla Chandigarh Educational Society under the Bishop of Shimla Chandigarh Diocese.

Present Use : School for Boys upto Class XII.

Miscellaneous Aspects : It was at the initiative of the first Archbishop of the Shimla Diocese, Rev. A.E.J. Kenealy that work for building the school began in March 1925. In 1984, the Christian Brothers handed over the School to the Shimla Chandigarh Diocese.

Plants in Courtyard : *Abies pindrow, Buddleja asiatica, Callistemon lanceolatus, Cedrus deodara, Celtis australis, Cestrum elegans, Chamaerops humilis, Cryptomeria japonica, Cupressus cashmeriana, Cunninghamia lanceolata, Eriobotrya japonica, Eucalyptus globulus, Euonymus japonicus, Ficus elastica, Fuchsia magellanica, Grevillea robusta, Hedera canariensis, Hibiscus syriacus, Hydrangea macrophylla, Jacaranda acutifolia, Juniperus procera, Juniperus squamata, Lagerstroemia indica, Ligustrum lucidum, Ligustrum* X *vicaryi, Paulownia tomentosa, Platanus orientalis, Populus ciliata, Robinia pseudoacacia, Rosa chinensis, Salvia splendens, Schefflera arboricola, Thujopsis dolabrata.*

ST. MARK

Features:

Year of Construction : 1875.

Location : Near S.B.I. Main Branch, Shimla.

Extent : About 678 sq m.

Building Material : Lath, Plaster, Dressed Stones, Bricks.

Building Design : Colonial Architecture.

Past Uses : Home Office and then as the Y.M.C.A. The Union Church also occupied a portion of the building.

Present Owner : Private Property.

Present Use : Residential house of Justice Deepak Gupta.

Plants in Courtyard : *Cedrus deodara, Fuchsia magellanica, Hedera colchica, Magnolia grandiflora, Solanum laxum.*

ST. THOMAS CHURCH

Features:

Year of Construction : 1912.

Location : Near Western Command Shimla.

Extent : About 1,956 sq m.

Building Material : Half-Round Dressed Stones, Cement, Lath.

Building Design : Gothic Style of Architecture.

Past Owner : Church Missionary Society.

Past Use : Church.

Present Owner : Managed by Society of St. Thomas School under Diocese of Amritser, (CNI).

Present Use : Co - educational School from Nursery - Class XIII.

Miscellaneous Aspects : The foundation of St. Thomas' was laid by the then Viceroy of India, H.E. The Rt. Hon. Lord Hardinge. The Church was consecrated by Bishop French on August 9, 1885. Services at the Church were held both in English and Urdu. The Viceroy, Lord Dufferin also used to be a part of the congregation. This church was closed for public worship in December 1947 after its amalgamation with the Christ Church.

Plants in Courtyard : *Cedrus deodara*, *Celtis australis*.

STATE LIBRARY, RIDGE

Features:

Year of Construction : 1860.
Location : Ridge.
Extent : About 1,230 sq m.
Building Material : Dressed Stones, Lath, Plaster.
Building Design : Tudor Style of Architecture.
Past Owner : Municipal Committee, Shimla.
Past Use : Library.
Present Owner : H.P. Government.
Present Use : State Library.
Miscellaneous Aspect : The library was started in 1934.
Plants in Courtyard : *Cedrus deodara*, *Robinia pseudoacacia*.

STIRLING CASTLE

Features:

Year of Construction : 1830.
Location : 1.5 Km from Long Wood.
Extent : About 3,750 sq m.
Building Material : Rough Cast Walls, Lath, Plaster, Dhajji.
Building Design : Colonial Architecture.
Past Uses : Used earlier for residential purposes by Sir W.W. Hunter, Sir Courtney Ilbert, Maharaja of Gwalior and Maharaja of Nabha.
Present Owner : Lehnu Mal Ram Kishen.
Present Use : Residential house.
Miscellaneous Aspects : In 1962, under the influence of Jawahar Lal Nehru, Maharaja of Nabha sold the Castle for contributing to Save the Children Fund (STCF) for a paltry sum of ₹ 50,000/- rupees. The Shimla STCF made the Stirling Castle a hostel for Tibetian children who had fled from Tibet after the Chinese incursion. The present owners purchased it from STCF in May 1970 and have been living here since then.
Plants in Courtyard : *Berberis lycium*, *Cedrus deodara*, *Clematis montana*, *Prunus armeniaca*, *Robinia pseudoacacia*, *Rosa moschata*.

STRAWBERRY HILL

Features:

Year of Construction : 19th Century.

Location : Near Chhota Shimla.

Extent : About 3,178 sq m.

Building Material : Bricks, Lath, Plaster, Stones.

Building Design : Colonial Architecture.

Past Owners : Colonel Tapp; Raja Sir Daljit Singh of Kapurthala.

Past Use : Residential house.

Present Owner : Rani Paramjit Singh of Kapurthala.

Present Use : Residential house.

Miscellaneous Aspects : The house was expanded by Colonel Churchill who served as Military Secretary to Edward Barnes, the Commander in Chief (1832-1833). Governor-General, Lord Dalhousie also lived here in 1849. After several other owners that included the German Consul -General Count Quadt, this was eventually purchased by Raja Sir Daljit Singh of Kapurthala and extensively renovated it.

Plants in Courtyard : *Araucaria columnaris*, *Berberis aristata*, *Cedrus deodara*, *Cerasus cerasoides*, *Daphne papyracea*, *Debregeasia saeneb*, *Drepanostachyum falcatum*, *Eucalyptus globulus*, *Hibiscus syriacus*, *Hydrangea macrophylla*, *Ligustrum lucidum*, *Malus pumila*, *Prunus armeniaca*, *Pyrus pashia*, *Robinia pseudoacacia*, *Rosa banksiae*, *Rosa chinensis*, *Salix pycnostachya* var. *oxycarpa*, *Sarcococca pruniformis*, *Spiraea cantoniensis*, *Thuja orientalis*.

TARA HALL (DARBANGA HOUSE)

Features:

Year of Construction : 1930.

Location : Upper Kaithu, Cart Road, Shimla.

Extent : About 10,256 sq m.

Building Material : Bricks, Stones, Mud, Lath, Wooden Posts, Plaster (Dhajji).

Building Design : Tudor Style.

Past Owner : Major S.B. Goad.

Past Uses : It was once a residential house of Major S.B. Goad. Later, it was purchased by Colonel J. Robertson President, Shimla

Municipality for providing the site for a Convent that was established in 1895 by the Loreto Order of Ireland.

Present Owner : Property of Loreto Education Society.

Present Use : Girls School from Nursery - Class XII.

Miscellaneous Aspects : Loreto Convent, Tara Hall, Shimla is one of 15 Loreto schools in India. The buildings were rebuilt in 1930. An adjoining property, Wheatfield and later, 'Darbanga' House was also purchased by the Convent in the 1960s and integrated with the school. From 1995 onwards for 13 years the school was under the management of the Sacred Heart Sisters, but in 2008 the Loreto Sisters took over the management again. In 2010 the School changed over to the CBSE Board and introduced Classes for 11 and 12.

Plants in Courtyard : *Cupressus cashmeriana, Hibiscus syriacus, Hydrangea macrophylla, Lonicera japonica, Pyrus communis, Rosa banksiae, Rosa chinensis, Spiraea cantoniensis, Solanum laxum, Thuja orientalis.*

TENDRIL COTTAGE (CECIL HOTEL)

Features:

Year of Construction : 1877.

Location : Chaura Maidan.

Extent : About 8,758 sq m.

Building Material : Bricks, Stones, Mud, Lath, Wooden Posts, Plaster (Dhajji).

Building Design : Mixture of Colonial Architecture and Gothic Style.

Past Owners : It was owned first by Mr David Colyear who later sold it to Mr H.R. Cooke, an Assistant Secretary in the foreign office. In 1902 it was purchased by Mr R. Hotz, a well known photographer of Shimla, for ₹ 45000/-. Later, Mrs Hotz gifted it to Mr J. Faletti (a caterer of Viceroy).

Past Use : Used initially as residential house and then as hotel.

Present Owner : Mr P.R.S. Oberoi (son of Mr Mohan Singh Oberoi).

Present Use : Hotel.

Miscellaneous Aspects : In 1946, Hotel Cecil became famous when national leaders of freedom movement like Pandit Jawahar Lal Nehru, Sardar Patel, Abdul Kalam Azad, M.A. Jinnan and others came to Shimla to attend the Shimla Conference. In the recent past, Hotel Cecil (Tendril Cottage) had been renovated on modern lines.

Plants in Courtyard : *Asparagus aethiopicus, Chamaerops humilis, Cupressus torulosa, Debregeasia saeneb, Euonymus japonicus, Ficus auriculata, Ficus palmata, Hibiscus syriacus, Juniperus squamata, Opuntia dillenii, Pinus roxburghii, Pyrus pashia, Salix babylonica, Solanum laxum, Solanum pseudocapsicum, Thuja orientalis.*

THE WHOLE RANGE OF BUILDINGS STARTING FROM GAINDA MULL UP TO NORTHERN RAILWAY BOOKING AGENCY, MALL ROAD

Features:

Year of Construction : 19th Century.

Location : Mall Road.

Extent : About 4,270 sq m.

Building Material : Western Style of Architecture having Pent Roof with Gable Ends.

Building Design : Moc-Tudor Architecture.

Past Owners : Britishers (M/S Barrett, M/S Hamilton, Mr R.B. MacDonald, Mr O'meara (Dentist of Napolean), M/S Cotton & Morris C. Crayden, M/S Bourne, Shepherd, Craddock, De Russet, M/S A. Ramsey Phelps, W. Coutts, M/S Watson & Summers, M/S Campbell, Francis & Phelps.

Past Use : Mostly as shops for various purposes, some as hotels.

Present Owner : Chander Kumar Baljees, Maya Devi, Susheel Sood, Brij Kumar Manchand, Sham Sunder, Satish Jain, Munshi Ram, Nanak Chand, Krishan Chand Sunderlal, State Toursium Department, etc.

Present Use : Used as shops (Jagsons Furniture, Asia Book House, Showrooms (Tommy Hilfiger, Reliance, Madame, Pume, Reebok), Fire Brigade, Sathe's shop, Baljees Sweets, Gainda Mull, Kashmir and Himachal emporium, Jewellery shops, and various other small shops on rent.

Plants in Courtyard : *Cedrus deodara, Cestrum elegans, Cupressus torulosa, Euonymus japonicus, Hedera helix, Hydrangea macrophylla, Nerium oleander, Prinsepia utilis, Thuja orientalis.*

THISTLE BANK

Features:

Year of Construction : Early 20th Century.
Location : Near S.B.I. Main Branch, Shimla.
Extent : About 2,234 sq m.
Building Material : Brick, Plaster, GCI Sheeting.
Building Design : Pent Roof Style.
Present Owner : Sh. Yoginder Pal Sood.
Present Use : Residential house.
Plant in Courtyard : *Cedrus deodara.*

TORNEY HOUSE

Features:

Year of Construction : 19th Century.
Location : Near Chhota Shimla.
Extent : About 1,870 sq m.
Building Material : Lath, Plaster, Dhajji, Wooden Posts.
Building Design : Colonial Architecture.
Past Use : Residential house of various British officers.
Present Owner : H.P. Government.
Present Use : Office of the Divisional Commissioner.
Plants in Courtyard : *Buxus wallichiana*, *Cedrus deodara*, *Robinia pseudoacacia.*

TORRENTIUM

Features:

Year of Construction : 19th Century.
Location : Near Chhota Shimla.
Extent : About 3,350 sq m.
Building Material : Lath, Plaster, Stones, Bricks.
Building Design : Anglo-Indian Architecture.
Past Owner : British Empire.
Past Use : Ayrcliff High School for Girls (established in 1888) and then, a wing of Bishop Cotton School.
Present Owner : Central Government.

Present Use : Central School for Tibetians.

Miscellaneous Aspects : This house draws its name from Major General R. Torrens, Adjutant-General of the Army. On October 27, 1983, the original building was gutted in a fire.

Plants in Courtyard : *Cedrus deodara*, *Malus pumila*, *Pyrus communis*.

TOWN HALL

Features:

Year of Construction : 1860.

Location : Ridge.

Extent : About 1,240 sq m.

Building Material : Stones, Timber, Slanting Slate Roofing.

Building Design : Colonial Architecture.

Past Owner : British Empire.

Past Use : Municipal Committee office.

Present Owner : Nagar Nigam Shimla.

Present Use : Municipal Corporation Office.

Plants in Courtyard : *Euonymus japonicus*, *Juniperus pseudosabina*, *Ligustrum* X *vicaryi*, *Rosa chinensis*, *Solanum laxum*, *Thuja orientalis*.

U.S. CLUB

Features:

Year of Construction : 1930.

Location : Jakhu Hills.

Extent : About 7,580 sq m.

Building Material : Dressed Stones, Plaster, Timber Frames.

Building Design : Tudor Style.

Past Owner : British Empire.

Past Use : It used to be houses for England civil officers. Besides accommodating at least 70 members, it had an honour of hosting several Viceroys, Secretary Generals and top British Empire officials.

Present Owner : H.P. Government.

Present Use : State Public Work and Irrigation-cum-Public Health Departments.

Plants in Courtyard : *Buxus wallichiana, Cedrus deodara, Jasminum grandiflorum, Jasminum mesyni, Pistacia khinjuk, Platanus orientalis, Populus nigra, Prinsepia utilis, Prunus armeniaca, Robinia pseudoacacia.*

VICEREGAL LODGE COMPLEX

Features:

Year of Construction : 1884-1888.

Location : Near Boileauganj.

Extent : About 15,37,500 sq m.

Building Material : Blue-Grey Stones, Tiled Pitch Roofing, Teak, Walnut, Deodar Wood.

Building Design : Neo–Gothic Style and Scottish Baronial Architecture.

Past Owner : British Empire.

Past Use : Occupied by British Viceroys.

Present Owner : Ministry of Human Resource Development, Government of India.

Present Use : Indian Institute of Advanced Studies.

Miscellaneous Aspects : It was the first building with electric lighting in Shimla. The site was selected by Lord Lytton, the British Viceroy of India during 1876-1880. The building was constructed during the Viceroyalty of Lord Dufferin during 1884-1888. The Architect Henry Irwin of the building was assisted by Executive Engineers F.B. Hebbest and L.M. St. Clair. Lawns of Advance Study were laid by Lord Curzon. Various Viceroys Lord Elgin, Lord Minto, Lord Curzon, Lord Wavell etc. occupied this building. Rare Tulip tree in Advance Study was planted by Marquis of Landsdowne in 1888, still adorns the eastern corner of lawn. The building has unique firefighting system which is still working. This building was also the venue of Shimla Conference in 1945 and partitioning of India in 1947. After Independence, the building was renamed as Rashtra-Pati Niwas. As it was occupied for only fifteen days in a year, Dr Sarvepalli Radha Krishanan decided to hand over this building in 1965 to the Ministry of Education to use it as the Indian Institute of Advance Study.

Plants in Courtyard : *Abies pindrow, Abutilon* X *hybridum, Abutilon megapotamicum, Acacia dealbata, Acer negundo, Acer palmatum, Actinidia chinensis, Aesculus indica, Afrocarpus gracilior, Agave americana, Aloysia citriodora, Araucaria bidwillii, Araucaria columnaris, Araucaria* *cunninghamii, Arundo donax, Asparagus aethiopicus, Asparagus densiflorus, Asparagus setaceus, Asparagus racemosus, Asparagus virgatus, Buddleja asiatica, Buddleja crispa, Buddleja davidii, Camellia japonica, Campsis radicans, Cedrus deodara, Chamaerops humilis, Clematis buchananiana, Clematis montana, Colquhounia coccinea, Cornus macrophylla, Cunninghamia lanceolata, Cupressus torulosa, Cycas revoluta, Daphne papyracea, Debregeasia saeneb, Drepanostachyum falcatum, Euonymus fimbriatus, Euonymus japonicus, Euonymus lucidus, Ficus sarmentosa, Forsythia* X *intermedia, Forsythia ovata, Fuchsia* X *hybrida, Fuchsia magellanica, Ginkgo biloba, Girardinia diversifolia, Hedera canariensis, Hedera colchica, Hedera helix, Hibiscus syriacus, Hydrangea macrophylla, Hypericum hookerianum, Hypericum oblongifolium, Inula cappa, Ipomoea purpurea, Jasminum grandiflorum, Jasminum mesyni, Juniperus procera, Juniperus squamata, Kerria japonica, Lagerstroemia indica, Ligustrum* X *vicaryi, Liriodendron tulipifera, Lonicera japonica, Lyonia ovalifolia, Magnolia grandiflora, Magnolia obovata, Morus alba, Morus nigra, Morus serrata, Myrsine africana, Nerium oleander, Parthenocissus semicordata, Picea smithiana, Pinus wallichiana, Platanus orientalis, Populus ciliata, Prinsepia utilis, Prunus laurocerasus, Pyrus pashia, Pyrus pyrifolia, Quercus floribunda, Quercus oblongata, Quercus robur, Rhapis excelsa, Rhododendron arboreum, Rhododendron simsii* var. *vittatum, Robinia pseudoacacia, Rosa alba, Rosa banksiae, Rosa chinensis, Rosa jacquemontii, Rosa multiflora, Rubus ellipticus, Salvia leucantha, Salvia splendens, Sarcococca pruniformis, Smilax aspera, Solanum laxum, Spartium junceum, Taxus baccata, Thuja orientalis, Thujopsis dolabrata, Trachelospermum jasminoides, Tropaeolum majus, Ulmus wallichiana, Viburnum cotinifolium, Viburnum mullaha, Wisteria sinensis, Yucca aloifolia.*

WALKER HOSPITAL

Features:

Year of Construction	:	1902 (Old building).
Location	:	Below G. C. Boy's College, Sanjauli.
Extent	:	About 14,678 sq m.
Building Material	:	Bricks, Cement, Sand, Stones.
Past Owner	:	Sir James Walker.
Past Use	:	Hospital.
Present Owner	:	Indian Army.

Present Use : Military Hospital till December 22, 1998.

Miscellaneous Aspects : Walker Hospital came into being in 1902 by a banker, Sir James Walker who originally planned to open it in the Gorton Castle which he had bought for ₹ 80,000/-. After much discussion and persuasion, the building was acquired for the Civil Secretariat of the Imperial Government of India, and Sir Walker was given an alternate site where the hospital was constructed for European patients. The hospital was finally converted into a military hospital in 1954. But due to a devastating fire on December 22, 1998, the building was reduced into ashes.

Plants in Courtyard : *Cedrus deodara, Quercus oblongata, Robinia pseudoacacia.*

WALSINGHAM (D.C. RESIDENCE)

Features:

Year of Construction : Mid 19th Century.

Location : Jakhu Hills.

Extent : About 1,278 sq m.

Building Material : Stones, Lath, Plaster.

Building Design : Colonial Architecture with Dormer Windows.

Past Owner : British Empire.

Past Use : D.C. Residence.

Present Owner : H.P. Government.

Present Use : D.C. Residence.

Plants in Courtyard : *Aesculus indica, Campsis radicans, Cedrus deodara, Euonymus japonicus, Grevillea robusta, Hedera helix, Rosa alba, Spartium junceum, Thuja orientalis.*

WESTERN COMMAND COMPLEX

Features:

Year of Construction : 1860.

Location : Parallel to The Mall, Shimla.

Extent : About 28,780 sq m.

Building Material : Ironwork Frames, Magnum Opus, Lath, Plaster, Wooden Frames.

Building Design : Mixture of Colonial Architecture and Gothic Style.

Past Owner : F. Dalton.

Past Uses : The property originally belonged to F. Dalton, the Post Master of Shimla, who was killed in 1857 in Delhi after which the government acquired the property and got main building constructed between September 1882 and March 1885. This place was the Summer headquarter of the British Indian Army. The ARTRAC shifted here on October 1, 1991.

Present Owner : Indian Army.

Present Use : Various official works of Army.

Miscellaneous Aspects : Unique in its design due to earthquake resistant and its ironwork frames.

Plants in Courtyard : *Aesculus indica*, *Cedrus deodara*, *Celtis australis*, *Cupressus torulosa*, *Euonymus japonicus*, *Forsythia* X *intermedia*, *Hedera canariensis*, *Hedera helix*, *Jacaranda acutifolia*, *Picea smithiana*, *Platanus orientalis*, *Robinia pseudoacacia*, *Rosa alba*, *Salix alba*, *Salix babylonica*, *Spartium junceum*, *Thuja orientalis*.

WINTER FIELD

Features:

Year of Construction : Late 19th Century.

Location : Near Old Bus Stand.

Extent : About 1,578 sq m.

Building Material : Lath, Dhajji, Plaster, Sand Stones.

Building Design : Tudor Fashion.

Past Use : Office of Public Works Department.

Present Owner : H.P. Government.

Present Use : Office of H.P. Public Works Department.

Miscellaneous Aspects : This has long served as the office of the Public Works Department and was rebuilt on the present lines after a fire destroyed the building in the late 19th Century.

Plants in Courtyard : *Cedrus deodara*, *Prunus armeniaca*.

WOOD BANK

Features:

Year of Construction : 1920.

Location : Near Shimla Railway Station.

Extent : About 1,856 sq m.

Building Material : Lath, Plaster, Bricks, Tin, Wooden Frames.

Building Design : Colonial Architecture.

Past Use : Rest House.

Present Owner : Indian Railway.

Present Use : Used presently as officers rest house.

Miscellaneous Aspects : It is also known as 'Kunj Bihar'. The campus has a 245 years old *Cedrus deodara* tree.

Plants in Courtyard : *Cedrus deodara, Euonymus japonicus, Hibiscus syriacus, Hydrangea macrophylla, Morus serrata.*

WOOD VILLA

Features:

Year of Construction : 1881 (Old);1938 (New).

Location : Near Benmore Estate.

Extent : About 3,890 sq m.

Building Material : Lath, Pebble-Dash Walls, Plaster, Stones.

Building Design : Mixture of Colonial and Gothic Architecture.

Past Owner : Once, this was residential house of General Sir William Rose Mansfield. In the year 1881, the Woodville was bought by Sir James Walker and passed it to the Alliance Bank of Shimla for its use as the manager's residence. In 1926, the property was bought by Maharaja Sir Bhagvat Singh of Gondal State in Western India for his daughter, Princess Laila. In 1938, the old Victorian building was demolished by her husband Raja Rana Sir Bhagat Chand for constructing the present one.

Past Use : Residential owner.

Present Owner : Kr. Uday Singh.

Present Use : Hotel.

Plants in Courtyard : *Aesculus indica, Aloysia citriodora, Campsis radicans, Cedrus deodara, Cestrum elegans, Drepanostachyum falcatum, Hedera helix, Hedera nepalensis Hibiscus syriacus, Hydrangea macrophylla, Ligustrum* X *vicaryi, Magnolia grandiflora, Myrtus communis, Picea smithiana, Pinus wallichiana, Pyrus pyrifolia, Quercus robur, Rosa chinensis, Rubus niveus, Solanum laxum, Solanum pseudocapsicum, Thuja orientalis, Wisteria sinensis, Yucca aloifolia.*

Y.M.C.A. BUILDING

Features:

Year of Construction : 1920-1930.

Location : Near Ridge.

Extent : 2,200 sq m.

Building Material : Bricks, Sand Stones, Lath, Plaster.

Building Design : Victorian Style.

Past Use : Y.M.C.A.

Present Owner : Y.M.C.A. of Delhi.

Present Use : Used as guest house.

Miscellaneous Aspects : Earlier it was briefly housed in St. Marks', near the State Bank of India (S.B.I.). It was founded by Sir George Williams in 1907.

Plants in Courtyard : *Cedrus deodara*, *Prunus armeniaca.*

Y.W.C.A. BUILDING

Features:

Year of Construction : 1907.

Location : Near G.P.O. Shimla.

Extent : About 1,870 sq m.

Building Material : Lath, Plaster, Sand Stones, Dhajji.

Building Design : Western Style of Architecture.

Past Use : Residence of Viceroy of India, Earl Minto and Lady Minto.

Present Owner : Managed by Y.W.C.A. of India (Delhi).

Present Uses : Guest house and part-time school for vocational courses.

Plants in Courtyard : *Cedrus deodara*, *Celtis australis*, *Euonymus japonicus*, *Hedera helix*, *Hibiscus syriacus*, *Juniperus squamata*, *Lonicera japonica*, *Nerium oleander*, *Quercus oblongata*, *Robinia pseudoacacia.*

YARROWS

Features:

Year of Construction : Late 19th Century.
Location : Near Chaura Maidan.
Extent : About 3,870 sq m.
Building Material : Dressed Stones, Bricks, Slates, Lath, Plaster, Steel Pipes.
Building Design : Colonial Architecture.
Past Owner : British Empire.
Past Use : Office of Finance Department.
Present Owner : Indian Government.
Present Use : Presently houses the office of National Academy of Audits and Accounts.
Miscellaneous Aspect : Once occupied by Muhammed Ali Jinnah, founder of Pakistan.
Plants in Courtyard : *Cedrus deodara, Lonicera japonica, Populus ciliata, Quercus oblongata, Rosa banksiae, Rosa chinensis.*

YATES PLACE

Features:

Year of Construction : 19th Century.
Location : Near Chhota Shimla.
Extent : About 1,456 sq m.
Building Material : Wooden Frames, Brickbats, Stones, Mud, Lime or Cement, Dhajji, Lath, Plaster.
Building Design : Pent Roof Style and Roofs having Gable Ends.
Past Owner : Major H.B. Goad.
Past Uses : Used by British officers for various purposes.
Present Owner : H.P. Government.
Present Use : Official residence of Education Minister of H.P.
Plants in Courtyard : *Cedrus deodara, Celtis australis, Erythrina arborescens, Euonymus tingens, Hedera nepalensis, Hibiscus syriacus, Hydrangea macrophylla, Hypericum oblongifolium, Populus nigra, Prunus armeniaca, Pyrus communis, Robinia pseudoacacia.*

SHIMLA CEMETERIES

BOILEAUGANJ BARRIER CEMETERY

Features:

Year of Use : Mid 20th Century.

Location : Barrier, Boileauganj.

Extent : About 17,500 sq m.

Past Use : Used as cemetery by Muslim community.

Present Use : Still used as cemetery.

Miscellaneous Aspect : A mosque located inside the cemetery.

Plants in Courtyard : *Acacia dealbata*, *Acer oblongum*, *Aesculus indica*, *Agave americana*, *Berberis aristata*, *Berberis lycium*, *Buddleja crispa*, *Cedrus deodara*, *Chamaerops humilis*, *Clematis buchananiana*, *Coriaria nepalensis*, *Cornus capitata*, *Daphne papyracea*, *Debregeasia saeneb*, *Desmodium multifloum*, *Drepanostachyum falcatum*, *Elaeagnus umbellata*, *Eucalyptus globulus*, *Ficus auriculata*, *Ficus palmata*, *Ficus subincisa*, *Girardinia diversifolia*, *Hedera nepalensis*, *Hibiscus syriacus*, *Hypericum oblongifolium*, *Indigofera dosua*, *Inula cappa*, *Ipomoea purpurea*, *Jasminum mesyni*, *Lonicera quinquelocularis*, *Morus alba*, *Musa paradisiaca*, *Myrica esculenta*, *Myrsine africana*, *Opuntia dillenii, Parthenocissus semicordata*, *Pinus roxburghii, Pinus wallichiana*, *Pistacia khinjuk*, *Prinsepia utilis*, *Punica granatum*, *Pyracantha crenulata*, *Pyrus pashia*, *Quercus oblongata*, *Rhamnus persica*, *Rhamnus virgatus*, *Rhododendron arboreum*, *Ricinus communis*, *Rubia cordifolia*, *Rubus ellipticus*, *Sarcococca pruniformis*, *Viburnum cotinifolium*, *Viburnum mullaha*, *Xanthium strumarium*, *Yucca aloifolia*.

KANLOG CEMETERY

Features:

Year of Use : Around 1850.

Location : Kanlog.

Extent : About 30,000 sq m.

Past Use : Used as cemetery by Christian community.

Present Use : This was closed for burial in the 1920s as no space was left in it.

Miscellaneous Aspects : Earliest grave in this burial ground for the town's Christian community was of Joseph Anderson. The cemetery had been created in phases, as a result the older one lies above the road.

Plants in Courtyard : *Acacia dealbata, Aesculus indica, Agave americana, Arundo donax, Berberis aristata, Cedrus deodara, Celtis australis, Cerasus cerasoides, Chamaerops humilis, Clematis montana, Colquhounia coccinea, Coriaria nepalensis, Cornus macrophylla, Cotoneaster bacillaris, Cunninghamia lanceolata, Cupressus torulosa, Datura stramonium, Debregeasia saeneb, Desmodium elegans, Drepanostachyum falcatum, Erythrina arborescens, Fagus sylvatica, Ficus auriculata, Ficus palmata, Ficus sarmentosa, Girardinia diversifolia, Himalrandia tetrasperma, Hypericum patulum, Ilex dipyrena, Indigofera dosua, Inula cappa, Leea asiatica, Lyonia ovalifolia, Morus alba, Myrsine africana, Padus cornuta, Parthenocissus semicordata, Pinus wallichiana, Pistacia khinjuk, Populus ciliata, Pyrus pashia, Quercus incana, Quercus oblongata, Rhamnus persica, Rhamnus virgatus, Rhododendron arboreum, Robinia pseudoacacia, Rosa moschata, Rubus biflorus, Rubus ellipticus, Rubus mecilentus, Rubus niveus, Rubus paniculatus, Sarcococca pruniformis, Skimmia laureola, Solanum virginianum, Smilax aspera, Spiraea canescens, Ulmus wallichiana, Viburnum mullaha, Viburnum stellulatum.*

NAV BAHAR CEMETERY (NUNS GRAVEYARD BELOW ST. BEDE'S COLLEGE)

Features:

Year of use : 1870.

Location : Near Nav Bahar.

Extent : About 2,579 sq m.

Past Owner : Convent of Jesus and Mary.

Past Uses : This private cemetery was opened for the nuns of the Convent of Jesus and Mary. Some nuns of the Loreto order had also been buried here.

Present Owner : Convent of Jesus and Mary.

Present Use : Cemetery for the Nuns of Chelsea Estate.

Plants in Courtyard : *Berberis aristata, Cedrus deodara, Myrsine africana, Parthenocissus semicordata, Rhododendron arboreum, Rubus biflorus, Spiraea cantoniensis.*

OAK OVER CEMETERY

Features:

Year of Construction : 1828.

Location : Near Oak Over.

Extent : About 2,245 sq m.

Past Use : Used as Christian burial ground.

Present Use : It is now property of Himalaya New Testament Old Savior Church and not used for burial purposes.

Miscellaneous Aspects : This is the oldest cemetery in the town having dozen graves and monuments. First grave dates back to 1829 and last burial was of Captain Mathew Ford on March 17, 1841. All the graves were mostly of rough chiselled stones of the size of ordinary bricks with heavy slabs on the top.

Plants in Courtyard : *Agave americana*, *Cedrus deodara*, *Chamaerops humilis*, *Drepanostachyum falcatum*, *Myrsine africana*, *Pyrus pashia*, *Quercus oblongata*, *Solanum virginianum*.

CEMETERY SANJAULI

Features:

Year of Use : 1921.

Location : Near Dhalli tunnel.

Extent : About 55,015 sq m.

Past Use : As cemetery when Kanlog cemetery was found to be full.

Present Use : Still used as cemetery by Christian community. The ground holds over 600 graves.

Miscellaneous Aspects : The Rana of Koti granted the land on a 'perpetual lease' for this cemetery. It was dedicated by the Bishop of Lahore on July 29, 1921. The earliest grave in this cemetery was of Joseph Multani, an Indian Christian who was buried here on May 12, 1921.

Plants in Courtyard : *Aesculus indica*, *Agave americana*, *Berberis aristata*, *Berberis lycium*, *Cedrus deodara*, *Chamaerops humilis*, *Clematis buchananiana*, *Cornus macrophylla*, *Cupressus torulosa*, *Daphne papyracea*, *Debregeasia saeneb*, *Drepanostachyum falcatum*, *Elaeagnus umbellata*, *Eucalyptus globulus*, *Ficus palmata*, *Ficus sarmentosa*, *Girardinia diversifolia*, *Hedera nepalensis*, *Hypericum oblongifolium*, *Parthenocissus* *semicordata, Prinsepia utilis, Rhamnus persica, Rubus mecilentus, Rubus paniculatus, Smilax aspera, Spiraea canescens, Ulmus wallichiana, Viburnum mullaha.*

SOCIO-CULTURAL SPACES OF SHIMLA

ANNANDALE

Features:

Location : Annandale.

Extent : About 3,05,000 sq m.

Past Uses : Used by Britishers for organising fetes, polo, dog shows and many other public functions and as Gymkhana.

Present Owner : Indian Army.

Present Use : Used as a nine hole golf course, and training center and landing and parking area for the helicopter of Chief Minister.

Miscellaneous Aspects : The ground was extended by cutting into the hillside mainly through the efforts of Lord William Beresford, military secretary to Lord Dufferin (1884-1888) at the cost of ₹ 80,000/-. Durand football cup was once held here.

Plants in Courtyard : *Bougainvillea glabra, Cedrus deodara, Chamaerops humilis, Cupressus torulosa, Diospyros kaki, Eucalyptus globulus, Ficus benjamina, Hibiscus syriacus, Jasminum mesnyi, Juglans regia, Malvaviscus arboreus* Cav. var. *mexicanus, Pistacia khinjuk, Prunus armeniaca, Robinia pseudoacacia, Thuja orientalis.*

NALDEHRA GOLF COURSE

Features:

Location : Naldehra.

Extent : About 4,06504 sq m.

Past Use : Camping site and as Golf Course.

Present Owner : Managed by H.P. Tourism Department.

Present Use : Golf Course.

Miscellaneous Aspects : The 9 hole golf course at Naldehra was originally laid out by the British Viceroy, Lord Curzon who was so enchanted by this spot that he gave his daughter (Alexandra), Naldehra as her second name. At present, the golf course has been expanded to 18 holes.

Plants in Courtyard : *Cedrus deodara*, *Cupressus torulosa*, *Hibiscus syriacus*, *Hydrangea macrophylla*, *Myrsine africana*, *Solanum virginianum*, *Thuja occidentalis*, *Zanthoxylum armatum.*

LADIES PARK (RANI JHANSI PARK)

Features:

Year of Use : First quarter of the 19th Century.

Location : Below Ridge.

Extent : About 4,345 sq m.

Past Use : Hotel.

Present Owner : Nagar Nigam, Shimla.

Present Use : Cutting and Tailoring School for Women (National Council of Women).

Miscellaneous Aspects : It went on to hold Shimla's first hotel known as the 'Royal Hotel' which was renamed as 'Lowrie's Hotel' after the owner. Due to a fire, the hotel was reduced to ashes in 1930s. Afterwards, it was decided to use the space as the Municipal Park for Women and Children. The park was inaugurated by Raj Kumari Amrit Kaur on July 6, 1937.

Plants in Courtyard : *Cedrus deodara*, *Chamaerops humilis*, *Cupressus torulosa*, *Hibiscus syriacus*, *Juniperus pseudosabina*, *Lonicera japonica*, *Magnolia grandiflora*, *Parthenocissus semicordata*, *Prunus armeniaca*, *Pyrus pashia*, *Robinia pseudoacacia*, *Spartium junceum*, *Thuja orientalis*, *Yucca aloifolia.*

RIDGE

Features:

Location : Ridge.

Past Use : The place was used for various socio- cultural activities.

Present Owner : Nagar Nigam, Shimla.

Present Uses : It is a place of congregation and socio-cultural space for National and State level events.

Miscellaneous Aspects : The Ridge, located in the centre, commands a glorious panoramic view of the town and the snow covered peaks of the distant Great Himalayas with Scandel Point in West, Lakkar Bazaar, Library and

Chirst Church in the East, alongwith Town Hall and Goofa on the Southern side.

Plants in Courtyard : *Aesculus indica, Buddleja davidii, Buxus wallichiana, Cupressus torulosa, Diospyros lotus, Juniperus pseudosabina, Lagerstroemia indica, Platanus orientalis, Populus ciliata.*

ROTHNEY CASTLE GROUND

Features:

Location : Jakhu Hills.

Extent : About 3,750 sq m.

Past Owners : Colonel Rothney, Mr A.H. Mathews, Mr P. Mitchell, Mr A.O. Hume.

Past Use : Used for socio-cultural gatherings.

Present Owner : Lala Chunna Mull.

Present Use : No use.

Plants in Courtyard : *Cedrus deodara, Cryptomeria japonica, Euonymus tingens.*

SCANDAL POINT

Features:

Location : Mall Road.

Extent : About 220 sq m.

Past Owner : British Empire.

Past Use : Socio-Cultural Space.

Present Owner : Nagar Nigam, Shimla.

Present Use : Socio-cultural Space.

Miscellaneous Aspects : It is the point where the Mall road joins Ridge on the west side. The name arose from the commotion caused by the supposed elopement of the daughter of the British Viceroy with Maharaja of Patiala in 1905. This led British authorities to ban the entry of Maharaja in Shimla. He countered the move by establishing a new summer capital – now famous hill resort of Chail, 45 km from Shimla.

Plants in Courtyard : *Cedrus deodara, Cupressus torulosa.*

ICE SKATING RINK

Features:

Year of Construction : 1920.

Location : Near Lakkar Bazaar Bus Stand.

Extent : About 1,456 sq m.

Building Material : Bricks, Plaster, Stones, Cement.

Past Owner : Raja Jagat Bahadur of Nahan.

Past Use : Ice Skating Rink.

Present Owner : Ice Skating Club.

Present Use : Ice Skating Rink.

Miscellaneous Aspects : This is Asia's oldest natural ice skating rink. In the beginning, it accepted only Europeans for its membership, later it admitted Indians with certain reservations.

Plants in Courtyard : *Cedrus deodara*, *Robinia pseudoacacia.*

TOWN HALL SQUARE

Features:

Location : Mall Road.

Extent : About 1,480 sq m.

Past Use : Socio-Cultural Activities.

Present Owner : Nagar Nigam, Shimla.

Present Use : Socio-Cultural Activities.

Plants in Courtyard : *Hedera nepalensis*, *Juniperus pseudosabina*, *Ligustrum* X *vicaryi*, *Rosa chinensis*, *Solanum laxum*, *Thuja orientalis.*

•

Chapter 4

ETHNOBOTANICAL INFORMATION OF PLANTS AT HERITAGE SITES OF SHIMLA

Abies pindrow Royle

Syn.: *A. chiloensis* Carriere; *A. himalayensia* Lavallee; *Picea herbertiana* Madden; *P. pindrow* (Royle *ex* D. Don) Loudon; *Pinus pindrow* Royle *ex* D. Don.

Family: Pinaceae.

Vern.: Fir, Tosh.

English, Hindi and Sanskrit Names:

Eng. : Himalayan Silver Fir, West Himalayan Fir;
Hindi : Ragha, Talispatra;
Sans. : Dhaatipatra, Manduparni, Talispatra.

Distribution : India. Afghanistan - W. Nepal: 2,100-3,600m.

Description : Trees with dark grey bark divided into long, narrow scales by deep, vertical grooves. Leaves flat, linear, stiff, single, spirally arranged around the branches but spreading more or less in one plane with upper dark green, shining surface and lower with a silvery-white line along each side of the midrib. Stalk twisted. Tip of leaves notched. Male and female cones on the same tree. Male cones sessile usually clustered. Female cones terminal, solitary, ovoid.

Reproductive Cycle : April-May.

Materials Examined : HPU-EL-SKS-6148; Advance Study (2,128m), St. Edward's School (2,060m).

Propagation Notes : Seeds ripe in autumn and prefers a good moist heavy clay soil and usually take about 6 - 8 weeks for germination. Stratification said to produce a more even germination.

Part Used : Wood.

Folk Uses : Wood used for **house interiors**, **building works**, **cases**, **furniture**, **water troughs** and as **fuel**.

Uses in Literature : Reported to be used earlier as an antiflatulent, antiperiodic, antipyretic, antispasmodic, appetizer, astringent, carminative, contraceptive, diuretic, expectorant, febrifuge, fodder, paper pulp, purgative, slipper, stomachic, tonic, water containers; and for asthma, bronchitis, catarrh of bladder, chest infection of infants, cold, construction, cough, dyeing, floor boards, gonorrhoea, haemoptysis, headache, hoarseness, neuralgia, packing cases, phithisis, pulmonary, rheumatism, roofing of houses, ulcer, and aircraft work **(Agarwal, 2003; Ambasta, 1986; Anonymous, 1948-1976; Chatterjee & Pakrashi, 1997; Chauhan, 1999; Dey & Bahadur, 1973; Dwivedi, 2007; Farooq, 2005; Jain *et al.*, 1994; Kapur, 1996a; Kapur *et al.*, 1996; Kirn *et al.*, 2000; Mehra & Jain, 1976; Pande *et al.*, 2006; Samant *et al.*, 1996; Sharma, 1996; Singh, 2000b; Thakur *et al.*, 1989; Uphof, 1968; Warda, 1985; Watt, 1889-1896)**.

Abutilon X *hybridum* Voss

Family: Malvaceae.

Vern.: Chinese Lantern.

English Names:

Eng. : Chinese Bell Flower, Chinese Lantern, Flowering Maple, Indian Mallow, Parlour Maple.

Distribution : Indigenous in tropical and subtropical regions of C. & S. America.

Description : A branched evergreen shrub which is a hybrid between *A. darwinii* and *A. striatum*. Leaves alternate, 3-5 lobed, ovate-rounded. Flowers pendent and coloured variously in shades of yellow, orange, red, white. Fruit a capsule.

Reproductive Cycle : May-August.

Materials Examined : HPU-EL-SKS-6178; Advance Study (2,128m), Bemloe (2,010m).

Propagation Notes : Requires a sunny position or part day shade in a fertile well-drained soil. Seeds sown for at least the first winter in a greenhouse and planted out in late spring or early summer, after the last expected frosts. Germination takes place within a few weeks. As this species is a hybrid, so the seedlings will not be true to type. Additional plants can also be cloned by softwood cuttings taken at any time of the year.

Parts Used : Whole Plant. Flowers.

Folk Uses : Flowers eaten **raw** or **cooked** due its sweet flavour. Also, grown as a **hedge** for its **ornamental** value.

Uses in Literature : Described earlier in literature as edible (flower), and for ornamental purposes **(Bean, 1981; Davis, 1990; Grey-Wilson & Matthews, 1983; Hay *et al.*, 1983; Huxley, 1992; Paliwal, 2009)**.

Abutilon megapotamicum (Spreng.) St. Hil. & Naudin.

Syn.: *A. inflatum* Garcke & K. Schum; *A. vexillarium* E. Morren; *Periptera megapotamica* (Spreng.) G. Don.

Family: Malvaceae.

Vern.: Lantern Flower.

English Name:

Eng. : Trailing Abutilon.

Distribution : Plant native of Argentina, Brazil and Uruguay.

Description : A graceful evergreen shrub with bright green, ovate, 3-lobed, pointed leaves and lantern-like pendent flowers. Calyx inflated with emerging orange-yellow petals and a cluster of protruding red stamens. Fruit a capsule.

Reproductive Cycle : April-September.

Materials Examined : HPU-EL-SKS-6189; Advance Study (2,128m), Chhota Shimla (2,070m), Raj Bhawan (2,134m).

Propagation Notes : Requires a sunny position or part day shade in a fertile well-drained soil for germination. Besides propagating through seeds sown in spring, propagation through cutting of young shoots in June also undertaken.

Part Used : Whole Plant.

Folk Use : Grown for its **ornamental** flowers.

Uses in Literature : Used earlier as edible (flower) and for ornamental purposes **(Bean, 1981; Davis, 1990; Hay *et al.*, 1983; Huxley, 1992; Paliwal, 2009; Phillips & Rix, 1989)**.

Acacia dealbata Link

Syn.: *A. affinis* Sweet; *A. dewentii* Siebert & Voss; *A. puberula* Dehnh.; *Racosperma dealbatum* (Link) Pedley.

Family: Fabaceae.

Vern.: Khair.

English Names:

Eng. : Blue Wattle, Silver Wattle.

Distribution : Native to Australia and Tasmania; naturalized in Nilgiri and Palni hills.

Description : Evergreen trees or shrubs up to 30m tall. Leaves bipinnate, glaucous blue-green - silvery grey with each pinna divided into leaflets. Yellow flower heads in axillary and terminal panicles. Pods flat.

Reproductive Cycle : February-May.

Materials Examined : HPU-EL-SKS-619; Armsdell (2,102m), Chhota Shimla (2,078m), Nigam Bihar (2,110m), Raj Bhawan (2,132m).

Propagation Notes : Plant reproduces naturally through root-suckers, and prefers a sandy loam and sunny position sheltered from strong winds. For germination through seeds, these are first pre-soaked for 12 hours in warm water and then sown in March.

Parts Used : Whole Plant. Wood. Gum.

Folk Uses : Wood used as **fuel.** Exuded gum from trunk **chewed** as a substitute of **gum**. Planted alongside of road as an **avenue** tree.

Uses in Literature : So far described for its essential oil, gum, perfumery, tannins, timber and as fuel **(Agarwal, 2003; Ambasta, 1986; Anonymous, 1948-1976; Asolkar *et al.*, 1992; Bean, 1981; Bhattacharjee, 2001; Hay *et al.*, 1983; Huxley, 1992; Jaeger, 1961; Kumar *et al.*, 2000; Phillips & Rix, 1998; Sood *et al.*, 2010, 2013; Uphof, 1968; Usher, 1974; Watt, 1889-1896)**.

***Acer negundo* L.**

Syn.: *Acer fauriei* H. Lev. & Vaniot; *A. fraxinifolium* Raf.; *A. lobatum* Raf.; *A. trifoliatum* Raf.; *A. violaceum* (Booth *ex* G. Kirchn.) Simonk.; *Rulac negundo* (L.) Hitchc.

Family: Aceraceae.

Vern.: Maple.

English Names:

Eng. : Ash-leaved Maple, Box Elder.

Distribution : Plant native of America; introduced into Jammu & Kashmir, Dehra Dun and Ootacamund in Nilgiri hills.

Description : Large trees having pinnate leaves with 3-5 leaflets which are ovate or oblong - lanceolate and coarsely serrate. Terminal leaflet 3 - lobed, long and acuminate. Flowers appearing before leaves, yellowish-green; staminate in pendulous corymbs and pistillate in pendulous racemes. Fruits long, glabrous, wings diverging at acute angle.

Reproductive Cycle : April-June.

Material Examined : HPU-EL-SKS-6195; Advance Study (2,128m).

Propagation Notes : Plant prefers a rich moist, well-drained soil and sunny position for its growth. For propagation through seeds, these are pre-soaked for 24 hours and then stratified for 2 - 4 months at 1 - 8°C.

Part Used : Whole Plant.

Folk Use : Planted for its **ornamental** value.

Uses in Literature : Reported to be used for shelterbelt plantings, packaging apples (leaves); and as astringent, antitumorous, cheap furniture, drums, fuel, pulp, and vulnerary **(Ambasta, 1986; Asolkar *et al.*, 1992; Bailey, 1957; Chittendon, 1956; Jaeger, 1961; Pool, 1941; Riotte, 1978; Uphof, 1968; Usher, 1974)**.

Acer oblongum Wall. *ex* DC.

Syn.: *A. buzimpala* Buch.-Ham. *ex* D. Don; *A. hypoleucum* Hayata; *A. laikuanii* Ling; *A. lanceolatum* Molliard; *A. laurifolium* D. Don; *A. litseifolium* Hayata; *Euacer laevifolium* Opiz.

Family: Aceraceae.

Vern.: Padang.

English Name:

Eng. : Himalayan Maple.

Distribution : Kashmir-Arunachal Pradesh and Manipur, ascending to 2,500m.

Description : Evergreen trees having glabrous, oblong, entire, long-pointed leaves with upper surface dark green and lower white. Flowers greenish - white in hairy corymbs. Fruits hairless, winged.

Reproductive Cycle : February-April.

Materials Examined : HPU-EL-SKS-6198; Boileauganj Cemetery (1,198m), Chhota Shimla (2,069m).

Propagation Notes : Grows well in moist well drained, heavy clay soils. For germination through seeds, these are first pre-soaked for 24 hours followed by stratification for 2 - 4 months at 1 - 8°C. Layering which takes about 12 months also successful with most species in this genus. Cuttings of young shoots undertaken in June or July.

Part Used : Wood.

Folk Uses : Wood used for **construction** works and as **fuel**.

Uses in Literature : Used earlier for agricultural implements, turnery, minor construction works; and as contraceptive, fuel and plywood **(Agarwal, 2003; Ahmed & Borthakur, 2005; Ambasta, 1986; Chaudhuri, 1993; Dwivedi, 2007; Pande *et al*., 2006; Sharma, 1996; Sood *et al*., 2013)**.

Acer palmatum Thunb

Syn.: *A. amoenum* Carriere; *A. decompositum* Dippel; *A. dissectum* Thunb.; *A. formosum* Carriere; *A. incisum* Dippel; *A. ornatum* Carriere.

Family: Aceraceae.

Vern.: Japanese Maple.

English Name:

Eng. : Japanese Maple.

Distribution : E. Asia - Japan.

Description : Shrubs or small trees with 5-9 lobed leaves. Lobes acuminate, doubly serrate or incised, glabrous. Flowers small, purple and in erect corymbs. Fruits long with widely spreading wings.

Reproductive Cycle : May-July.

Material Examined : HPU-EL-SKS-6203; Advance Study (2,128m).

Propagation Notes : Prefers a good moist, well-drained acidic partial shaded, heavy clay soil. Seeds pre-soaked for 24 hours and then stratified for 2 - 4 months at 1 - 8°C. Layering takes about 12 months. Cutting of young shoots each comprising 2 - 3 pair of leaves having one pair of buds at the base performed in June or July.

Part Used : Whole Plant.

Folk Use : Planted as an **ornamental** for its beautiful leaves.

Uses in Literature : Recorded in literature as an edible (leaves) and ornamental tree; and for packaging apples (**Bailey, 1957; Bean, 1981; Chittendon, 1956; Hadfield, 1956; Huxley, 1992; Riotte, 1978; Thomas, 1992**).

Actinidia chinensis Planch.

Family: Actinidiaceae.

Vern.: Kiwi.

English Names:

Eng. : Chinese Goose Berry, Kiwi Fruit.

Distribution : Cultivated upto 2,100m.

Description : A woody vine with leaf apices at the end of branches sharp pointed. Fruits rusty brown, oblong and with hairy surface.

Reproductive Cycle : February-November.

Material Examined : HPU-EL-SKS-6238; Advance Study (2,126m).

Propagation Notes : Prefers a loamy acidic soil. Propagated through cuttings of softwood as well as through seeds. Fresh seed germinates in 2 - 3 months at 10°C whereas stored seeds take longer.

Part Used : Fruit.

Folk Uses : Fruits **eaten** raw or **cooked** as a good source of vit. C.

Uses in Literature : Known so far for preparing jams, marmalade, wine and decorating ice creams; and as edible **(Ambasta, 1986; Uphof, 1968)**.

Aesculus indica (Wall. *ex* Jacquem.) Hook.f.

Family: Hippocastanaceae.

Vern.: Kanhor, Khanor.

English and Hindi Names:

Eng. : Himalayan Chestnut, Indian Horse-chestnut;
Hindi : Bankhor, Gugu, Pangar, Pankar.

Distribution : W. Himalayas: 1,550-3,550m (Indus-Nepal).

Description : Trees with a massive trunk, a fairly large spreading crown and digitate, exstipular and deciduous leaves possessing a long petiole. Leaflets 5-9 in number, serrated, oblanceolate or oblong and acuminate with an acute base. Flowers creamy white, borne in showy pyramidal clusters. Capsule brown, smooth, ovoid, pretty rough from outside. Seeds dark brown, shining.

Reproductive Cycle : May-September.

Materials Examined : HPU-EL-SKS-6245; Advance Study (2,128m), Bantony (2,182m), Chhota Shimla (2,075m), Inverarm (2,150m), Ridge (2,250m), Summerhill (2,080m).

Propagation Notes : Natural germination of seeds occurs in spring following seed dispersal (about 4 months after fruits drop).

Parts Used : Leaves. Fruit. Seed. Wood.

Folk Uses : Wood used for **construction works** whereas leaves employed as **fodder**. Seed oil used for curing **arthritis**. Paste of fruit applied in nostrils of livestock to expel **leeches** during rainy season.

Active Constituents : Active properties due to aesculuside, aescigenin, escin, protoaescigenin, quercetin-3-xylosido-glucoside, kaempferol-3-xylosido-glucoside, quercetin-3-glucosido-rhamnoside, kaemp-ferol-3- xylosidoglucoside, kaemferol-3-glucosido-rhamnoside, querctin-3-glucoside, quercetin-3-diglucoside, quercetin-3,4-diglucoside, kaempferol-3-α-L-arabinofuranoside, kaempferol-3- α-L-rhamnofuranoside, avicularin and quercitin-3- α-L-rhamnofuranoside (**seeds**), castaprenol-11, castaprenol-12, castaprenol-13, quercitin, β-sitosterol, escin (**leaves**), rutin and astragalin (**stem**).

Uses in Literature : So far used in India as anthelmintic, colic (vet.), detergent for woollen clothes, diuretic, drinking cups of Tibetans, edible (flour kernels), edible (vet.), fodder, galactagogue (vet.), platters, psychoactive plant, shoe-heels, and wrapping paper; and for ague, arthritis, backache, building, constipation, cracks, dislocated joints, effusion of blood, frost bite, furniture, haemorrhoids, indigestion (vet.), intermittent fevers, leucorrhoea, loose motions (vet.), match-splints, mathematical instruments, migraine, obstinate constipation, packing cases, phlebitis, piles, pneumonia (vet.), preventing thrombosis, rheumatism, skin fissures, stomach diseases (vet.), ulcers, water troughs, wounds (vet.) and whooping cough **(Agarwal, 2003; Ambasta, 1986; Arora & Pandey, 1996; Bennet *et al.*, 1991; Chauhan, 1999; Chopra *et al.*, 1956; Das & Agarwal, 1991; Farooq, 2005; Jain, 1991; Jain *et al.*, 1994; Kapur & Singh, 1996; Kaul, 1997; Khare, 2004; Kirtikar & Basu, 1984; Lata *et al.*, 2008; Manandhar, 1980a; Negi & Gaur, 1991; Pande *et al.*, 2006; Parmar & Kaushal, 1982; Prajapati *et al.*, 2006; Prakash, 2001; Raju, 2000; Rana *et al.*, 1996; Roy *et al.*, 1998; Samant *et al.*, 1996, 2001; Sharma, 2003; Singh, 1996; Singh & Kumar, 2000a; Sood *et al.*, 2013; Uphof, 1968; Watt, 1889-1896)**.

Afrocarpus gracilior (Pilg.) C.N. Page

Syn.: *Decussocarpus gracilior* (Pilg.) de Laub.; *Nageia falcata* var. *gracilior* (Pilg.) Silba; *Podocarpus gracilior* Pilg.

Family: Podocarpaceae.

Vern.: Podocarpus.

English Names:

Eng. : African Fern Pine, East African Yellowwood.

Distribution : Indigenous to Kenya, Uganda and Ethiopia; introduced into India.

Description : A valuable timber tree having beautiful juvenile foliage, glossy dark green leaves and somewhat pendulous branches which carry deep bluish-green irregularly arranged leaves on older plants. Seed cones highly modified with a single, purple seed.

Reproductive Cycle : April-October.

Materials Examined : HPU-EL-SKS-6248; Advance Study (2,128m), Hainault (2,294m).

Propagation Notes : Propagated through seeds and cuttings of half-ripe terminal shoots of 5 - 10cm long in the month of July/August.

Part Used : Whole Plant.

Folk Uses : Grown as a **hedge** for its **ornamental** value.

Active Constituents : These are amentoflavone, podocarpusflavone A, isoginkgetin and kayaflavone (**leaves**).

Uses in Literature : Known to be useful as an attractive pot plant; and for doors, flooring, furniture and panelling **(Ambasta, 1986; Anonymous, 1948-1976; Bailey, 1957; Hay *et al.*, 1983; Rastogi & Mehrotra, 1995a)**.

Agave americana L.

Syn.: *A. cantula* Roxb.; *A. spectabilis* Salisb.; *Aloe americana* (L.) Crantz.

Family: Agavaceae.

Vern.: Keora.

English, Hindi and Sanskrit Names:

Eng. : Agave, American Aloe, Century Plant;
Hindi : Banskeora, Bara, Kantala, Kanwar;
Sans. : Kalakantala, Kantala,Wilyatikaitalu.

Distribution : Grown as an ornamental plant in tropical and subtropical area around the World.

Description : Shrubby rhizomatous plants possessing greyish-green, spiny, wax coated leaves with curved tips, and marginal hooked teeth crowded at base. Flowers dull greenish - yellow and in huge, terminal and spicate spike with a cyme. Seeds flat, disk- like and enclosed in capsule.

Reproductive Cycle : March-May.

Materials Examined : HPU-EL-SKS-6254; Advance Study (2,128m), Bantony (2,182m), Chhota Shimla (2,080m), Gorton Castle (2,094m), Holly Lodge (2,278m), Inverarm (2,150m), Kasumpti (2,120m).

Propagation Notes : Plant succeeds in light, well-drained soil. Propagation by seeds and suckers. Seeds germinate best at 24 - 26°C.

Parts Used : Whole Plant. Leaves. Peduncle. Rhizome.

Folk Uses : Grown as an **ornamental hedge** plant. Leaves used as **brush** for **painting**; also its fibres used for **cordages.** Peduncle relished as **vegetable**. Paste of its crushed rhizome, turmeric and rock salt tied with bamboo sticks to set **fractured bones.**

Active Constituents : Active constituents are hecogenin, chlorogenin, 10 steroidal saponins-agavoside A-I (**leaves**).

Uses in Literature : Used so far as an anthelmintic, antisyphilitic, blood purifier, brushes, diaphoretic, diuretic, emmenagogue, laxative for ascites, fire wood, fibre, fish poison, fodder, hedge, paper material, poultice, proof against ravages of white ants, resolvent, substitute for soap; and for anasarca, boils, bruises and contusions, burns, cancer, conjunctivitis, controlling exparasite (vet.), cordage, cut, diarrhoea, dropsy, dysentery, falling of hair, fever, fever (vet.), goitre, gonorrhoea, internal wounds, kidney problems, mats, paralysis, scrofula, scurvy and venereal sores, skin diseases (vet.), gastric and liver disorders, skin disorders, thatching, tuberculosis, warts (vet.), and wounds **(Agarwal, 2003; Ambasta, 1986; Baburaj *et al.*, 2000; Dey & Bahadur, 1973; Harborne & Baxter, 2001; Jain, 1991; Jain *et al.*, 1991; Kapur, 1996a, b; Khare, 2004; Kirtikar & Basu, 1984; Kurian, 1995; Lalramnghinglova, 1996; Paliwal, 2009; Pande *et al.*, 2006; Parrotta, 2001; Perry & Hay, 1982; Prajapati *et al.*, 2006; Rana *et al.*, 1996; Rao, 1996; Said, 1997; Saxena & Vyas, 1983; Singh, 1996, 2000a; Singh & Pandey, 1998; Singh & Srivastava, 2000; Sood & Thakur, 2004; Sood *et al.*, 2012b, 2013; Uphof, 1968; Watt, 1889-1896**).

Aloysia citriodora **Palau**

Syn.: *A. sleumeri* Moldenke; *A. triphylla* (L Her.) Britton; *Cordia microcephala* Willd. *ex* Roem. & Schult; *Lippia citriodora* (Lam.) Kunth; *L. triphylla* (L Her.) Kuntze; *Verbena citriodora* (Lam.) Cav.; *V. triphylla* L Her.; *Zappania citriodora* Lam.

Family: Verbenaceae.

Vern.: Verbena.

English Name:

Eng. : Lemon Verbena.

Distribution : Native to S. America.

Description : Deciduous shrubs up to 2m tall with strongly scented, yellowish – green, lance-shaped leaves and cluster of tubular, pale mauvish-white flowers in panicles. Fruits 2, dry, small, dark brown in tear drop-shaped nutlets.

Reproductive Cycle : August-September.

Materials Examined : HPU-EL-SKS-6256; Advance Study (2,128m), Wood Villa (2,108m).

Propagation Notes : Propagated by seeds and vegetative methods; plant thrives well on loamy soil mixed with mould.

Parts Used : Whole Plant. Leaves.

Folk Uses : Grown as a lemon - scented, **ornamental** shrub. Leaves eaten raw, also used for preparing **herbal tea** and **flavouring** dishes.

Uses in Literature : Used earlier in abdominal disorders, depression, digestion, lemon balm; and as febrifuge, neurotonic, sedative and conservatory boarder **(Ambasta, 1986; Anonymous, 1948-1976; Hay *et al.*, 1983; Prajapati *et al.*, 2006)**.

Amygdalus mira (Koehne) Ricker

Syn.: *Persica mira* (Koehne) Kovalev & Kostina; *Prunus mira* Koehne.

Family: Rosaceae.

Vern.: Vaimi.

Distribution : U.P. - Bhutan, S.W. China upto 2,800m.

Description : Trees having lanceolate, long pointed leaves and pink flowers. Fruits stony, pale-green with brown hair.

Reproductive Cycle : March-April.

Materials Examined : HPU-EL-SKS-6260; Chhota Shimla (2,092m), Kasumpti (2,098m).

Propagation Notes : Propagated by seeds; cold stratification required before germination taking sometimes 18 months to germinate. Cuttings of half-ripe wood with a heel performed in July/August in a frame. Layering carried out in spring.

Parts Used : Fruits. Seeds.

Folk Uses : Ripe fruits **edible.** Seed oil used for **cooking**.

Uses in Literature : Known for its edible fruits and seed's oil for cooking **(Sharma & Rana, 2005; Sood *et al.*, 2012b)**.

Araucaria bidwillii Hook.

Syn.: *Columbea bidwillii* (Hook.) Carriere; *Marywildea bidwillii* (Hook.) A.V. Bobrov & Melikyan.

Family: Araucariaceae.

Vern.: Bunya Araucaria.

English Name:

Eng. : Bunya-Bunya.

Distribution : Distributed from Australia - Queensland.

Description : Evergreen trees with lanceolate juvenile leaves which are thick, shining, sharp-pointed, contracted at base and spreading in two rows. Adult leaves spirally imbricated, ovate, long and woody. Staminate catkins upto 13 cm long. Fruiting cones globose - ovoid. Scales terminating in an acute edge.

Reproductive Cycle : June-September.

Material Examined : HPU-EL-SKS-6268; Advance Study (2,128m).

Propagation Notes : Plant succeeds in most soils. Seeds sown in February usually germinate in 1 - 2 months at 15°C.

Part Used : Whole Plant.

Folk Uses : Grown as an **ornamental** tree. Branches used for **fencing** purposes.

Uses in Literature : Reported earlier for cabinet making, carving, flooring, fencing, furniture, plywood; and as edible (seed), fuel, ornamental **(Ambasta, 1986; Bailey, 1957; Chittendon, 1956; Graf, 1980; Huxley, 1992; Manandhar, 2002; Rushforth, 1987; Uphof, 1968; Warda, 1985).**

Araucaria columnaris Hook.

Syn.: *A. cookie* R. Br. *ex* Lindl.; *A. excelsa* (Lamb.) R. Br.; *A. intermedia* R. Br. *ex* Vieill.; *Columbea excelsa* (Lamb.) Spreng.; *Eutacta excelsa* (Lamb.) Link; *E. humilis* Carriere; *E. minor* Carriere.

Family: Araucariaceae.

Vern.: Araucaria.

English Name:

Eng. : Monkey Puzzle Tree.

Distribution : Native of New Caledonia; introduced into Indian gardens.

Description : A slender columnar tree with older branches attached horizontally to the main trunk; leaves imbricated, lanceolate-ovate-triangular, obtuse possessing prominent midrib. Staminate catkins and fruiting cones ovoid. Seeds winged with a terminal thorny tip.

Reproductive Cycle : August-October.

Materials Examined : HPU-EL-SKS-6270; Advance Study (2,128m), Bantony (2,182m), Chhota Shimla (2,080m), Inverarm (2,150m), Kasumpti (2,120m), Ridge (2,250m), St. Edward's School (2,060m).

Propagation Notes : Propagated by sowing freshly collected seeds during June and July; seeds usually germinate in 1 - 2 months at 15°C. Planting done in a deep well-drained soil.

Parts Used : Whole Plant. Bark.

Folk Use : Plant grown for **ornamental** purposes.

Uses in Literature : Used in general carpentary work and as decorative garden plant in tubs, gardens, corners rockeries and edible (nut) purposes **(Ambasta, 1986; Bailey, 1957; Beri, 1992; Chittendon, 1956; Graf, 1980; Harborne & Baxter, 2001; Huxley, 1992; Paliwal, 2009; Perry & Hay 1982; Sood *et al.*, 2013; Warda, 1985; Warman, 1999)**.

Araucaria cunninghamii Aiton *ex* D. Don

Syn.: *Eutacta cunninghamii* (Aiton *ex* D. Don) Link; *Eutassa cunninghamii* (Aiton *ex* D. Don) G. Don; *E. cunninghamii* (Aiton *ex* D. Don) Spach.

Family: Araucariaceae.

Vern.: Hoop Pine.

English Names:

Eng. : Colonial Pine, Hoop Pine, Moreton Bay Pine.

Distribution : Native to Eastern Australia and New Guinea, S. Wales, Queensland.

Description : Slender trees with upper ascending branches, and lower horizontal ones. Bark rough, in horizontal bands, exfoliating in fine circular bands. Leaves quite stiff, sharp-pointed and nearly straight which on young trees are awl-shaped, scale-like, incurved and about 2 mm thick at the base and 1-2 cm long and 4 mm broad on mature trees. Cones ovoid, 8-10 cm long and 6-8 cm in diameter. Seeds winged.

Reproductive Cycle : June-October.

Materials Examined : HPU-EL-SKS-6271; Advance Study (2,127m).

Propagation Notes : Propagated through cuttings and seeds. Seeds sown in March in pots of sandy soil at 15-20Cº. Cuttings can be taken from trees that have been cut back; the cuttings are set in sand in March or April in a greenhouse or frame to form roots.

Parts Used : Whole Plant. Wood.

Folk Uses : Cultivated as a **decorative** garden tree. Wood employed for making low grade **furniture**.

Uses in Literature : Used in general carpentary and cabinet works, flooring, furniture, railway cars, agricultural implements; and as an ornamental **(Ambasta, 1986; Graf, 1980; Paliwal, 2009; Uphof, 1968; Warda, 1985; Warman, 1999)**.

Arundo donax L.

Syn.: *Aira bengalensis* (Retz.) J.F. Gmel.; *Amphidonax bengalensis* (Retz.) Steud.; *Arundo bambusifolia* Hook.f.; *A. bengalensis* Retz.; *A. coleotricha* (Hack.) Honda; *A. sativa* Lam.; *A. triflora* Roxb.; *A. versicolor* Mill.; *Cynodon donax* (L.) Raspail; *Donax arundinaceus* P. Beauv.; *D. bengalensis* (Retz.); *D. bifarius* (Ritz.) Spreng.; *D. sativa* (Lam.) J.Presl; *D. versicolor* (Mill.) P. Beauv.; *Scolochloa donax* (L.) Gaudin.

Family: Poaceae.

Vern.: Nal.

English Names:

Eng. : Giant Reed, Spanish-bamboo-reed.

Distribution : Grows on lower Himalayas from Kashmir -Assam and Nilgiris.

Description : Perennials with erect or creeping stem, drooping flat leaves in two opposite series which are tapering from a stem-clasping broad base to a fine point. Ligules very short, hairy. Spikelets 3-4 flowered, in a dense, erect panicle. Grains oblong, free within the persistent glumes.

Reproductive Cycle : September-October.

Materials Examined : HPU-EL-SKS-6273; Advance Study (2,128m), Armsdell (2,102m), Kanlog Cemetery (1,989m).

Propagation Notes : Propagation through seeds which usually takes 1 - 3 months time at 15°C. Plant prefers moist fertile soil in a sunny sheltered position, preferably by water.

Parts Used : Leaves. Culms. Panicles.

Folk Uses : Planted for checking **soil erosion** and as **wind-break screen**. Culm branches used as a **hedge**. **Brooms** made from the terminal panicles. Leaves woven into **mats**. Also, culms used for making **pens**, **screens**, **walking sticks**, **roofing**, walls of houses and in **basketry**.

Active Constituents : Active constituents are sterol, donoxarine and triterpenes (**leaves**), gramine and its methohydroxide, eleagnine, N,N-dimethyltryptamine methohydroxide and 3,3-bis(indolylmethyl) dimethyammonium hydroxide, 5-methoxy-N-methyltryptamine, bufotenine, N,N-dimethyltryptamine and gramine Nb-oxide (**flower**).

Uses in Literature : Useful as diuretic, emollient; and for asthma, basketry, diarrheoa, hats, herpes, mats, skin diseases, thatching, typhoid fever, fibres, stimulating menstrual discharge, worm infections, and diminishing secretion of milk **(Agarwal, 2003; Ahmed & Borthakur, 2005; Ambasta, 1986; Anonymous, 1948-1976; Borthakur *et al.*, 1999; Das *et al.*, 1996; Dwivedi, 2007; Farooq, 2005; Hajra & Baishya, 1997; Khare, 2004; Paliwal, 2009; Pande *et al.*, 2006; Rastogi & Mehrotra, 1991; Sharma *et al.*, 2003; Shrivastava & Singh, 2003; Singh & Pandey, 1998; Srivastava *et al.*, 2003a; Uphof, 1968)**.

Asparagus aethiopicus L.

Syn.: *Asparagopsis aethiopica* (L.) Kunth; *A. lancea* (Thunb.) Kunth; *Asparagus laetus* Salisb.; *A. lanceus* Thunb.; *A. maximus* Voss; *A. sprengeri* Regel; *Protasparagus aethiopicus* (L.) Oberm.

Family: Asparagaceae.

Vern.: Sprengeri Asparagus.

English Name:

Eng. : Foxtail Fern.

Distribution : Native of S. Africa.

Description : Commonly cultivated shrubs having drooping or scrambling glabrous stems with flat, linear, straight or only slightly curved cladodes. Scales on main stems spiniferous. Flowers pinkish, fragrant, in loose racemes. Berries red, somewhat 3-lobed.

Reproductive Cycle : April-June.

Materials Examined : HPU-EL-SKS-6275; Advance Study (2,128m), Cecil Hotel (2,090m), Gorton Castle (2,094m), Grand Hotel (2,105m).

Propagation Notes : Propagated through rootstock, mostly in July-August.

Part Used : Whole Plant.

Folk Uses : Cultivated as an **ornamental climber** due to its beautiful feathery leaves.

Uses in Literature : Known to be used in diarrheoa, gastrointestinal symptoms, skin diseases and vomiting; and as indoor as well as ornamental plant **(Asolkar *et al.*, 1992; Bailey, 1957; Graf, 1980, 1981**; **Paliwal, 2009)**.

Asparagus densiflorus (Kunth) Jessop

Syn.: *Asparagopsis densiflorus* Kunth; *Protasparagus densiflorus* (Kunth) Oberm.

Family: Asparagaceae.

Vern.: Asparagus.

English Name:

Eng. : Shatavari.

Distribution : Native of S. Africa.

Description : Tuberous-rooted plants with plume-like branch and needle-like, rich green foliage. Flowers small, pinkish, fragrant. Berries globose, black or red. Seeds black.

Reproductive Cycle : May-June.

Materials Examined : HPU-EL-SKS-6278; Advance Study (2,128m), Holly Lodge (2,278m), Raj Bhawan (2,132m).

Propagation Notes : Propagation by seeds as well as by division of clumps in spring. Loam based compost, good light and watering best for its growth.

Part Used : Whole Plant.

Folk Uses : Planted as an **ornamental climber** due to its graceful feathery foliage.

Uses in Literature : Used for adornment of pillars, posts, tree trunk and arches **(Bailey, 1957; Graf, 1980, 1981; Hay *et al.*, 1983; Paliwal, 2009)**.

Asparagus racemosus Willd.

Syn.: *A. acerosus* Roxb.; *A. dubius* Decne; *A. jacquemontii* Baker; *A. penduliflorus* Zipp. *ex* Span.; *A. zeylanicus* (Baker) Hook. f.; *Asparagopsis abyssinica* Kunth; *A. acerosa* Kunth; *A. brownei* Kunth; *A. decaisnei* Kunth; *A. floribunda* Kunth; *A. javanica* Kunth; *Protasparagus racemosus* (Willd.) Oberm.

Family: Asparagaceae.

Vern.: Sanspai, Satavar.

English, Hindi and Sanskrit Names:

Eng. : Indian Asparagus, Wild Asparagus;
Hindi : Chatwal, Satawar, Satawari, Satmuli, Shakakul;
Sans. : Abhiru, Atiraso, Atirasa, Bahusutaa, Indivari, Satamuli, Shatavirya, Shvetamuli, Sukshmapatra, Supatra, Supatrika, Svadurasa, Svetmuli, Tailavalli, Vaishnavi, Vari, Vasudevapriyankari, Vishvasya, Vrishya.

Distribution : Tropical and subtropical India; ascending upto 1,500m in Himalayas.

Description : Tall, climbing stems with more or less recurved prickles and short rootstock bearing numerous succulent tuberous roots. Cladodes curved, terete, spreading and in tufts of 2-6. Leaves reduced to spines. Flowers white, fragrant, small, crowded in simple and branched racemes. Berries globose, red when ripe.

Reproductive Cycle : April-August.

Materials Examined : HPU-EL-SKS-6280; Advance Study (2,128m), Grand Hotel (2,105m).

Propagation Notes : Propagation by seeds as well as through division of clumps. Seeds usually take 3 - 6 weeks at 25°C for germination.

Parts Used : Rhizome. Root.

Folk Uses : Rhizome regarded as a **soothing tonic** against threatened **miscarriage**, **menopausal problems** and **dysentery**. Roots employed in the treatment of **dysentery**.

Active Constituents : Main active constituents are steroidal glycosides, aglycones, sarasapogenin, arasosapogenin, sterol and diosgenin (**roots**).

Uses in Literature : Described earlier as an alternative, aphrodisiac, antidysenteric, antirabies, antispasmodic, astringent to bowl, demulcent, diuretic, galactogogue, nutritive, refrigerant, stomachic, tonic; and biliousness, dyspepsia, inflammations, insect bite, kidney problems, leucorrhoea, rheumatism, scabies, tumours, eye diseases, kidney and liver disorders, sexual weakness, and urinary problems **(Agarwal, 2003; Ahmed & Borthakur, 2005; Ambasta, 1986; Aminuddin & Girach, 1996; Anonymous, 1948-1976; Asolkar *et al.*, 1992; Badhwar & Fernandez, 2011; Bhatt & Mitaliya, 1999; Chaudhuri, 1993; Chopra *et al.*, 1956, 1969; Deokota & Chhetri, 2007, 2009; Farooq, 2005; Girach, 1992; Girach & Aminuddin, 1995, Gogoi & Das, 2003; Jamir *et al.*, 2011; Kirtikar & Basu, 1984; Kshirsagar & Singh, 2007; Kumar, 2003; Kumar *et al.*, 2000; Nautiyal *et al.*, 1997; Pande *et al.*, 2006; Parrotta, 2001; Prajapati *et al.*, 2006; Rajendran & Henry, 1994; Rana *et al.*, 1996; Sahoo & Mudgal, 1995; Sharma *et al.*, 2003; Singh & Pandey, 1998; Singh & Prakash, 1994; Singh *et al.*, 2003c; Sivakumar *et al.*, 2003; Sivarajan & Balachandran, 1994; Sood *et al.*, 2013; Thakur *et al.* 1989; Uphof, 1968; Yoganarasimhan, 1996).**

***Asparagus setaceus* (Kunth) Jessop**

Syn.: *Asparagopsis setacea* Kunth; *Asparagus lujae* De Wild; *A. plumosus* Baker; *A. zanzibaricus* Baker; *Protasparagus plumosus* (Baker) Oberm.; *Protasparagus setaceus* (Kunth) Oberm.

Family: Asparagaceae.

Vern.: Asparagus Fern.

English Name:

Eng. : Asparagus Fern.

Distribution : S. Africa, Southern Cape Province - Natal and north - Malawi.

Description : Commonly cultivated shrubs possessing tall, climbing, woody glabrous stem and filiform, bright green cladodes in fascicle. Flowers white on a short pedicels. Berries globose, purple black.

Reproductive Cycle : May-August.

Materials Examined : HPU-EL-SKS-6283; Advance Study (2,130m), Armsdell (2,102m), Bemloe (2,010m).

Propagation Notes : Easily grown in any good garden soil rich in sandy loam through seeds taking around 3 - 6 weeks at 25°C for germination.

Part Used : Whole Plant.

Folk Uses : Grown as an **ornamental climber** for its feathery leaves.

Active Constituents : Active constituents are two newfurostanol glycosides – 3 – O - [α – L - rhamnopyranosyl (1→2) – α – L - rhamnopyranosyl (1→3) - β – D - glucopyranosyl] – 26 – O – β – D –glucopyranosyl - (25S) 22 α – methoxyfurost – 5 – en - 3 β, 26 - diol (I) and 22 - hydroxy derivative (II) **(leaves)**.

Uses in Literature : Recorded in literature for adornment of pillars, posts, tree trunk and arches **(Chittendon, 1956; Graf, 1980, 1981; Hay *et al.*, 1983; Huxley, 1992; Paliwal, 2009; Phillips & Rix, 1998; Rastogi & Mehrotra, 1995a).**

Asparagus virgatus Baker

Syn.: *A. sylvaticus* Burch. *ex* Baker; *Protasparagus virgatus* (Baker) Oberm.

Family: Asparagaceae.

Vern.: Asparagus.

English Names:

Eng. : African Broom Fern, Tiki Fern, Tree Fern;

Distribution : Native of S.E. Africa.

Description : Shrubs having erect, bushy stems with many stiff branches above. Cladodes in 3's, not very abundant, stiff, long, angled. Flowers greenish-white, solitary at nodes on drooping pedicels. Berries orange-red.

Reproductive Cycle : April-June.

Materials Examined : HPU-EL-SKS-6287; Advance Study (2,127m), Bemloe (2,010m).

Propagation Notes : Propagation by seeds or by division of clumps in spring.

Part Used : Whole Plant.

Folk Use : Cultivated as an **ornamental** climber.

Uses in Literature : Grown as ornamental and for cut flowers **(Bailey, 1957; Chittendon, 1956; Graf, 1980, 1981; Huxley, 1992).**

Berberis aristata DC.

Syn.: *B. bussmul* K. Koch *ex* Miq.; *B. coccinea* K. Koch; *B. elegans* K. Koch; *B. undulata* K. Koch.

Family: Berberidaceae.

Vern.: Kashmal.

English, Hindi and Sanskrit Names:

Eng. : Indian Berberry, Tree turmeric;
Hindi : Chitra, Darhald, Daruhaldi, Kashmal;
Sans. : Daruharidra, Darvi, Kantakati, Kata, Kateri.

Distribution : Distributed in Temperate Himalayas: 1,829-3,048m, Nilgiri Hills and Sri Lanka.

Description : Shrubs with red young branches and sessile, broadly lanceolate leaves having a few distant teeth. Racemes long - stalked, compound, corymbose or umbellate, drooping, much longer than leaves. Berries oblong-ovoid, white-blue when ripe.

Reproductive Cycle : May-September.

Materials Examined : HPU-EL-SKS-6290; Advance Study (2,128m), Bantony (2,182m), Boileauganj Cemetery (1,992m), Chhota Shimla (2,083m), Inverarm (2,150m), Kanlog Cemetery (1,972m), Kasumpti (2,120m), Longwood (2,237m).

Propagation Notes : Propagation by seeds which germinate in late winter or early spring. Cold stratification occurs naturally in winter. Grows well in heavy clay soil.

Parts Used : Tender Shoot. Leaves. Root. Fruit. Wood.

Folk Uses : Ripe fruits, 'kashmal' **eaten** raw. Tender shoots **chewed** to quench thirst. Wood used as **fuel** and leaves as **fodder** for goats and sheep. 3-5 ml of root juice used as eye drops to cure **eye diseases**.

Uses in Literature : Used so for as an antiperiodic, antiscorbutic, blood purifier, cooling laxative to children, deobstruent, diaphoretic, fuel, psychoactive plant, stomachic, refrigerant, substitute for raisin, tonic; and for affections of the eye, ammenorrhoea, anorexia, apthous sores, bites of rat and snake, boils, cholera, diarrhoea, dyeing, dyspepsia, enlargement of spleen, edible fruits, sauce, eye complaints, infection in horses, intermittent and malarial fevers, intestinal catarrh, jaundice, liver (vet.), menorrhagia, painful micturition, piles, rheumatism, skin diseases, stomach disorders, sun-blindness, tannins, ulcers, uterine troubles, cicatrization, and vomiting during pregnancy **(Agarwal, 2003; Ambasta, 1986; Anonymous, 1948-1976; Arora & Pandey, 1996; Asolkar *et al.*, 1992; Badhwar & Fernandez, 2011; Chandra, 1997; Chauhan, 1999; Chhetri, 2005; Chopra *et al.*, 1956; Dey & Bahadur, 1973; Duke *et al.*, 2002; Farooq, 2005; Jain, 1968, 1991; Jain *et al.*, 1994; Kalakoti, 2001; Kapur & Nanda, 1996; Khare, 2004; Kirtikar & Basu, 1984; Nautiyal *et al.*, 2003; Pande *et al.*, 2006; Panthi & Chaudhary, 2003; Parmar & Kaushal, 1982; Prakash, 2001; Rao, 1990; Roy *et al.*, 1998; Said, 1997; Samant *et al.*, 1996, 2001; Sensarma, 1991; Sharma *et al.*, 2003; Shome *et al.*, 1996; Singh 1996; Singh & Kumar, 2000b; Sinha & Sinha, 2001; Sood *et al.*, 2012b; Uphof, 1968; Watt, 1889-1896)**.

Berberis lycium Royle

Syn.: *B. angustifolia* Roxb.; *B. insignis* Hook.f.
Family: Berberidaceae.

Vern.: Kashmal.

English, Hindi and Sanskrit Names:

Eng. : Raisin Berberry;
Hindi : Chitra, Kirmara, Kushmal;
Sans. : Chatori, Daruharidra.

Distribution : Pakistan- E. Nepal, India: 1,500-3,000m.

Description : Shrubs having pale grey bark, angular branches and sessile, tough, narrowly lanceolate acute leaves. Upper surface of leaves bright green and lower paler. Racemes shortly - stalked, simple, barely longer than leaves. Flowers pale yellow, stalk slender. Berries ovoid, violet, covered with bloom.

Reproductive Cycle : April-July

Materials Examined : HPU-EL-SKS-6291; Bantony (2,182m), Boileauganj Cemetery (1,998m), Chhota Shimla (2,082m), Inverarm (2,150m), Kanlog Cemetery (1,972m), Kasumpti (2,120m), Longwood (2,237m), Sanjauli (2,020m).

Propagation Notes : Seed germinate in late winter or early spring. Seeds from over-ripe fruits take longer to germinate and require cold stratification. Plant prefers a warm moist loamy soil and light shade.

Parts Used : Leaves. Stem. Tender Shoot. Fruit.

Folk Uses : Ripe fruits **edible**. Tender shoots **chewed**. Leaves provide good **forage** for sheep and goats. Stem used as **fuel wood**.

Uses in Literature : Used in India as an alterative, antidote for poisoning, antimalarial, antiseptic, antispasmodic, aperient, blood purifier, carminative, coagulant, febrifuge, forage, fuelwood, mild laxative, sedative, tonic; and for allaying irritability and thirst, bleeding piles, boils, chest and throat troubles, cholera, chronic diarrhoea, cough, disorders of urinary tract, dysentery, ear problems, edible purposes, enlargement of liver and spleen, epigastrium, eye diseases (chronic opthalmia), increasing appetite, jaundice, leprosy, malaria, menorrhagia, oral ulcers, oriental sores, periodic neuralgia, piles, promoting digestion, skin diseases, sores, spleen troubles, stomach disorders, throat pain, and in urogenital disorders **(Agarwal, 2003; Ambasta, 1986; Anonymous, 1948-1976; Arora & Pandey, 1996; Badhwar & Fernandez, 2011; Bennet *et al.*, 1991; Chandra, 1997; Chauhan, 1999; Chopra *et al.*, 1956; Dey**

& Bahadur, 1973; Farooq, 2005; Jain, 1991; Kapur *et al.*, 1996; Khare, 2004; Kirtikar & Basu, 1984; Kaul, 1997; Lal *et al.*, 1996; Nautiyal *et al.*, 2003; Negi & Gaur, 1991; Pande *et al.*, 2006; Rana *et al.*, 1996; Rao, 1990; Roy *et al.*, 1998; Said, 1997; Samant *et al.*, 1996; Singh, 1996, 1999; Singh & Kumar, 2000a; Sood & Thakur, 2004; Sood *et al.*, 2012b, Uphof, 1968; Verma & Chauhan, 2006; Watt, 1889-1896).

Bougainvillea glabra Choisy

Syn.: *B. brachycarpa* Heimerl; *B. rubicunda* Schott *ex* Rohrb.

Family: Nyctaginaceae.

Vern.: Boganvillea.

English and Hindi Names:

Eng. : Paper Flower;
Hindi : Bag Bahar.

Distribution : Originated in Brazil; widely planted in India.

Description : Glabrous shrubs with a woody trunk and stems with stout usually straight spines. Leaves oblong lanceolate or ovate-oblong - broadly ovate, alternate. Flowers mostly scattered on a long leafy wand – like branches with elliptic or elliptic-lanceolate bracts which are sub - equal to flowers. Achenes 5 - lobed, oblong, 1 cm long and 1- seeded.

Reproductive Cycle : April-October.

Material Examined : HPU-EL-SKS-6293; Annandale (1,860m).

Propagation Notes : Plant adapted to various well drained soils and wide climatic conditions. Propagation by cutting, ground or air layering and budding from July - September.

Part Used : Whole Plant.

Folk Uses : Planted as an **ornamental** shrub or **climber** as covers of ugly walls. Also, used as **hedge** plant.

Uses in Literature : Grown in Indian gardens for ornamental purposes and fences; and as boundary hedge and dye (flower) **(Bailey, 1957; Cowan, 1950; Hay *et al.*, 1983; Jaeger, 1961; Paliwal, 2009; Randhawa, 1957; Sood *et al.*, 2013; Vankar, 2007)**.

Buddleja asiatica Lour.

Syn.: *B. acuminatissima* Blume; *B. amentacea* Kraenzl.; *B. serrulata* Roth; *B. sundaica* Blume; *Vitex esquirolii* H. Lev.

Family: Buddlejaceae.

Vern.: Buddliya, Bana.

English and Hindi Names:

Eng. : Butterfly Bush, White Butterfly Bush;
Hindi : Jaintia, Neemda.

Distribution : Commonly distributed from Himalayas - S.E. Asia from 900-1,800m.

Description : Tall, erect shrubs covered with white or pale yellow tomentum. Leaves lanceolate, minutely toothed, long-pointed; upper ones usually entire. Flowers white and crowded in a long drooping spikes, often combined in terminal panicles. Capsule glabrous.

Reproductive Cycle : February-June.

Materials Examined : HPU-EL-SKS-6295; Advance Study (2,134m), Bantony (2,182m), Chhota Shimla (2,080m), Inverarm (2,150m).

Propagation Notes : Propagation done with branch cutting and through seeds sown in February/March. Germination usually takes 3 - 4 weeks at 21°C.

Parts Used : Whole Plant. Branches. Flower.

Folk Uses : Often planted in garden as an **ornamental hedge**. Branches used as **walking sticks**. Flowers raw edible and its juice applied against **skin diseases**.

Active Constituents : Active constituents are quercetin and linarin (acacetin-7-rutinoside) (**flower**).

Uses in Literature : Reported to be used as edible (flower), abortifacient, drink preparation (leaves, root), ornamental; and for headache, skin complaints, walking sticks, tool handles **(Ahmed & Borthakur, 2005; Ambasta, 1986; Badhwar & Fernandez, 2011; Bennet *et al.*, 1991; Chaudhuri, 1993; Dam & Hajra, 1997; Farooq, 2005; Lalramnghinglova, 1996; Paliwal, 2009; Panda, 1996; Pande *et al.*, 2006; Rajendran, 2003; Raju, 2000; Rastogi & Mehrotra, 1990; Singh & Kumar, 2000a; Siwakoti & Varma, 1996)**.

Buddleja crispa Benth.

Syn.: *B. tibetica* var. *grandiflora* C. Marquand.

Family: Buddlejaceae.

Vern.: Newarpati.

English and Hindi Names:

Eng. : Himalayan Butterfly Bush;
Hindi : Neemda.

Distribution : India. Pakistan - S.W. China: 1,200-4,000m.

Description : Shrubs or small trees with gnarled and crooked stem and extremely variable leaves, those at the end of flowering branches oblong-lanceolate and entire whereas those lower down hastate, deeply sinute-dentate, lobed. Stipules leafy. Petiole sometimes winged. Flowers nearly sessile, fragrant, pale mauve and in short dense spikes arranged in large terminal leafy panicles. Fruit a capsule.

Reproductive Cycle : March-May.

Materials Examined : HPU-EL-SKS-6297; Advance Study (2,128m), Bantony (2,182m), Chhota Shimla (2,070m), Inverarm (2,150m), Kanlog Cemetery (1,998m), Summerhill (2,080m).

Propagation Notes : Propagated by seeds.

Parts Used : Stem. Leaves. Roots.

Folk Uses : Stems used as **walking sticks**. Leaves used as **fodder**. Crushed roots fermented along with other aromatic herbs in preparation of local drink '**sur**'.

Uses in Literature : Known in India as abortifacient, edible (flowers), fermented liquor, sacred and for skin complaints **(Ambasta, 1986; Bennet *et al.*, 1991; Chhetri & Shrestha, 2004; Chopra *et al.*, 1956; Lalramnghinglova, 1996; Panda, 1996; Saklani & Jain, 1994; Singh & Kumar, 2000a)**.

Buddleja davidii Franch.

Syn. *B. shimidzuana* Nakai; *B. striata* Z.Y. Zhang; *B. variabilis* Hemsl.

Family: Buddlejaceae.

Vern.: Buddleya.

English Names:

Eng. : Butterfly Bush, Summer Lilac.

Distribution : Distributed from E. Asia-China.

Description : Deciduous shrubs having lance-shaped, pointed, mild green to grey green leaves and lilac-purple flowers in long panicles. Capsule elongated with two compartments having numerous seeds. Seeds winged.

Reproductive Cycle : April-September.

Materials Examined : HPU-EL-SKS-6299; Advance Study (2,128m), Bantony (2,182m), Chhota Shimla (2,090m), Gorton Castle (2,094m), Holly Lodge (2,278m), Ridge (2,250m).

Propagation Notes : Propagation done with branch cutting and seeds which for its germination usually take 3-4 weeks at 21°C. Plant prefers a rich loamy well-drained soil.

Part Used : Whole Plant.

Folk Use : Grown as an **ornamental** plant.

Uses in Literature : Known to be useful as ornamental plant and for yielding a black or green dye (flower) **(Ambasta, 1986; Asolkar *et al.*, 1992; Bean, 1981; Chittendon, 1956; Grae, 1974; Huxley, 1992; Phillips & Rix, 1998; Thomas, 1992)**.

Buxus wallichiana Baill.

Family: Buxaceae.

Vern.: Shamshad.

English and Hindi Names:

Eng.	:	Box-Wood Tree;
Hindi	:	Chikri, Papri.

Distribution : India. Afghanistan - C. Nepal: 1,800-2,700m.

Description : An evergreen shrub growing upto 2m. Leaves shining above and pale-green beneath. Flowers yellow - green, strongly scented. Capsule ovoid, wrinkled.

Reproductive Cycle : April-November.

Materials Examined : HPU-EL-SKS-6301; Chhota Shimla (2,085m), Dalziel House (2,082m), R.K.M.V. (2,256m), U.S. Club (2,290m).

Propagation Notes : Propagated through seeds for which stratification is not necessary. Germination usually takes 1 - 3 months at 15°C. Plant prefers to germinate in any soil that is well-drained.

Parts Used : Whole Plant. Wood. Leaves.

Folk Uses : Planted as a **hedge**. Wood used for making **agricultural implements**. Leaves **poisonous** to cattle; occasionally browsed by goats.

Active Constituents : Active properties of **wood** and **leaves** due to steroids alkaloids, buxine, cyclovirobuxine-D, cyclobuxine-D, cycloprotobuxine-C and buxtauine.

Uses in Literature : Used as diaphoretic, febrifuge, fodder in scarcity, hedge plant, poisonous (leaves), purgative; and for rheumatism, syphilis, burns (vet.), bone fracture, boxes, cabinet work, carving, combs, mathematical and other precision instruments, toys and turney **(Agarwal, 2003; Ambasta, 1986; Anonymous, 1948-1976; Asolkar *et al.*, 1992; Bean, 1981; Chauhan, 1999; Chittendon, 1956; Dwivedi, 2007; Jain, 1991; Jaiswal *et al.*, 2004; Mittre, 1997; Pande *et al.*, 2006; Sharma, 1996; Singh & Kumar, 2000a; Thomas, 1992; Uphof, 1968; Watt, 1889-1896)**.

Callistemon lanceolatus (Sm.) Sweet

Syn.: *C. flavescens* Regel; *C. lilacinus* Cheel; *C. marginatus* (Cav.) Sweet; *C. pendulus* Regel; *Metrosideros falcata* Dum. Cours.; *M. lanceolata* Sm.; *M. longifolia* Dum. Cours.; *M. marginata* Cav.; *M. myrtifolia* Hoffmanns.

Family: Myrtaceae.

Vern.: Bottle Brush.

English and Hindi Names:

Eng. : Bottle Brush;
Hindi : Boothal Brush Ped.

Distribution : Native of Australia; widely planted in the subtropics.

Description : A small, evergreen tree with graceful pendulous branches. Leaves linear - lanceolate, leathery, gland-dotted having a faint lemon odour. Flowers red and in terminal drooping spikes resembling a bottle brush. Pods filled with small fine seeds.

Reproductive Cycle : May-September.

Materials Examined : HPU-EL-SKS-6305; Armsdell (2,012m), Auckland House (2,257m), Bemloe (2,010m), Chhota Shimla (2,080m), Grand Hotel (2,105m), Raj Bhawan (2,172m).

Propagation Notes : Propagated from seeds, cutting or by air layering. Plant succeeds in any soil, tolerating both water-logging and drought.

Part Used : Whole Plant.

Folk Use : Planted as an **ornamental** due to its highly graceful and weeping habit.

Uses in Literature : Used as dye (flower), essential oil, fuelwood, tea (leaves), tool handle; and planted for providing beautiful site along the boundary, small gardens path, border of tanks, lakes and ponds and in perfumery **(Agarwal, 2003; Ambasta, 1986; Asolkar *et al.*, 1992; Bean, 1981; Beri, 1992; Bhattacharjee, 2001; Chittendon, 1956; Davis, 1990; Hay *et al.*, 1983; Huxley, 1992; Jaeger, 1961; Manandhar, 2002; Paliwal, 2009; Phillips & Rix, 1998; Raju, 2000; Randhawa, 1957, 1960; Rastogi & Mehrotra, 1993; Sood *et al.*, 2010, 2013; Thomas, 1992; Warman, 1999)**.

Camellia japonica L.

Syn.: *C. bonnardii* Berl. *ex* Lem.; *C. florida* Salisb.; *C.wabiske* (Makino) Kitam.

Family: Theaceae.

Vern.: Camellia.

English Names:

Eng. : Camellia, Japanese Rose, Rose of Winter.

Distribution : Distributed from E. Asia-Japan, Korea.

Description : Small trees or shrubs with purplish-brown young branches and greyish-brown older ones. Leaves alternate, leathery, dark green from upper side and pale from lower side, petiolated, cuneate; margins toothed, tip pointed. Flowers pink, single or in pairs. Capsule globose - shaped with two brown seeds.

Reproductive Cycle : November-April.

Materials Examined : HPU-EL-SKS-6307; Advance Study (2,128m), Chhota Shimla (2,080m), Grand Hotel (2,105m), Hainault (2,294m).

Propagation Notes : Plant prefers a woodland soil but can thrive in a open, well-drained loam supplemented with leaf mould. Propagated through seeds; usually take 1- 3 months times at 23°C.

Part Used : Whole Plant.

Folk Use : Grown as an **ornamental** plant due its beautiful flowers.

Uses in Literature : Reported to be used earlier as anticancer, antihaemorrhagic, astringent, dye, edible (flowers), haemostatic, oil (seeds), ornamental plant, salve, tea (leaves), tonic; and for cosmetic and beauty products **(Ambasta, 1986; Asolkar *et al.*, 1992; Bean, 1981; Chittendon, 1956; Davis, 1990; Grae, 1974; Harborne & Baxter, 2001; Hay *et al.*, 1983; Huxley,**

1992; Jaeger, 1961; Manandhar, 2002; Paliwal, 2009; Phillips & Rix, 1998; Randhawa, 1960; Rastogi & Mehrotra, 1995b; Thomas, 1992; Uphof, 1968; Usher, 1974).

Campsis radicans (L.) Seem.

Syn.: *Bignonia coccinea* Steud.; *B. florida* Salisb.; *B. radicans* L.; *Gelseminum radicans* (L.) Juss.; *Tecoma radicans* Juss.

Family: Bignoniaceae.

Vern.: Shehnai Phool.

English Names:

Eng. : Trumpet Climber, Trumpet-vine.

Distribution : Native of S.E. United States.

Description : Deciduous scandent plants with adventitious roots arising from almost every node and opposite compound leaves. Leaflets 9-11, toothed. Flowers funnel-shaped, orange coloured; 4-12 in a cluster. Pods with numerous brown paper-like seeds.

Reproductive Cycle : August-September.

Materials Examined : HPU-EL-SKS-6308; Advance Study (2,128m), Bemloe (2,010m), Chhota Shimla (2,080m), Convent of Jesus & Mary (2,104m), Wood Villa (2,108m).

Propagation Notes : April-August.

Part Used : Whole Plant.

Folk Uses : Used as an **ornamental climber** for adornment of pillars, posts, arches and walls.

Uses in Literature : Known earlier as home remedy for causing sweating and for healing wounds **(Ambasta, 1986; Bhattacharjee, 2001; Pool, 1941; Uphof, 1968).**

Cedrus deodara (Roxb. *ex* D. Don) G. Don

Syn.: *Abies deodara* (Roxb. *ex* Lamb.) Lindl.; *C. indica* Chambray; *Larix deodara* (Roxb. *ex* Lamb.) K. Koch; *Pinus deodara* Roxb.

Family: Pinaceae.

Vern.: Deaar, Devdar.

English, Hindi and Sanskrit Names:

Eng. : Cedor, Deodar, Himalayan Cedar, True Cedar;
Hindi : Debdar, Keyodar, Kilankaper;

Sans. : Bhadravata, Devadaru, Indravriksha, Svehavrikha.

Distribution : Himalayas, 1,800-3,000m.

Description : Robust evergreen trees having end of branches drooping in graceful curves and bluish green or dark green, 3-sided, acicular solitary leaves on elongated shoots which are in dense clusters. Male flowers in catkins about 3 - 4 cm long. Cones oblong. Seeds winged, broadly triangular and with rounded sides.

Reproductive Cycle : June-July.

Materials Examined : HPU-EL-SKS-6309; Advance Study (2132m), Annandale (1,860m), Bantony (2,182m), Bemloe (2,010m), Chhota Shimla (2,080m), Inverarm (2,150m), Jakhu (2,449m), Raj Bhawan (2,132m), Summerhill (2,080m), Wood Villa (2,108m).

Propagation Notes : Propagated in nature through seeds after winter stratification. Plant succeeds on most soils.

Part Used : Wood.

Folk Uses : Wood regarded good for **furniture** and **building works**.

Active Constituents : Active constituents are cedeodarin, cedrin, dihydromyricetin, cedrinoside, taxifolin meso-secoisolariciresinol and cedrusinin **(wood)**.

Uses in Literature : Reported to be used as an alterative with turmeric and guggul, anodyne, anthelmintic, astringent, carminative, cephalalgia, cardiotonic, digestive, diaphoretic, diuretic, laxative, stomachic, symbolic (abode of gods); and for ascites, bilious fevers, bridges, chronic skin diseases, construction, costiveness, diarrhoea, doors, dropsy, dysentery, indolent swellings, intermittent fevers, flatulence, furniture, gonorrhoea, gout, headache, heart palpitations, inflammation, leprosy, leucoderma, material requirements, packing cases, paralysis, piles, pulmonary complaints, railway sleepers, rheumatism, ring worm (vet.), shingles, skin diseases, syphilis, ulcers, urinary disorders, and window frames **(Agarwal, 2003; Ambasta, 1986; Anonymous, 1948-1976; Chatterjee & Pakrashi, 1997; Chauhan, 1999; Dey & Bahadur, 1973; Dhiman, 2003, 2005; Ghate, 1998; Jain, 1991; Kapur & Singh, 1996; Khare, 2004; Kirn *et al.*, 2000; Lata *et al.*, 2008; Nautiyal *et al.*, 2003; Pande *et al.*, 2006; Prajapati *et al.*, 2006; Sharma, 1996; Singh, 1991, 1999, 2000a; Singh & Kumar, 2000a; Sood *et al.*, 2010, 2013; Thakur *et al.*, 1989; Uphof, 1968; Warda, 1985; Warman, 1999; Watt, 1889-1896)**.

Celtis australis L.

Syn.: *C. alpinia* Royle; *C. eriocarpa* Decne; *C. serrata* Dippel.

Family: Ulmaceae.

Vern.: Khirak.

English Names:

Eng. : European Nettle Tree, The Honey Berry Tree.

Distribution : Temperate Himalayas; 1,350-2,650m.

Description : Large deciduous trees with pale ashy bark having numerous lenticels, pubescent young shoots and subulate stipules. Male flowers racemed on short leafless axillary branchlets. Sepals oblong, margins woolly. Female or bisexual flowers rather larger than the male. Drupes globose.

Reproductive Cycle : April-October.

Materials Examined : HPU-EL-SKS-6311; Advance Study (2,128m), Bantony (2,182m), Bemloe (2,010m), Chhota Shimla (2,088m), Inverarm (2,150m).

Propagation Notes : Cutting of half-ripe wood and seeds subjected to cold stratification are methods commonly employed for its propagation. Plant prefers a good fertile well-drained loamy soil.

Parts Used : Leaves. Wood.

Folk Uses : Wood employed for making **agricultural implements**. Leaves considered as good quality **fodder** for cattle.

Uses in Literature : Known to be useful for agricultural implements, ammenorrhoea, blow instruments, coach whips, cups, diarrhoea, dysentery, edible fruits, fatty oil, hay-forks, leprosy, oars, peptic ulcer, ram-rods, spoons, turnery, wagons, walking sticks, and whip handles; and as cattle fodder, colic **(Agarwal, 2003; Ambasta, 1986; Arora, 1997; Arora & Pandey, 1996; Badhwar & Fernandez, 2011; Bennet *et al.*, 1991; Bhattacharjee, 2001; Das & Agarwal, 1991; Dwivedi, 2007; Jain, 1991; Jain *et al.*, 1991; Kapur & Nanda, 1996; Kapur *et al.*, 1996; Lalramnghinglova, 2002; Maheswari *et al.*, 1981; Prajapati *et al.*, 2006; Roy *et al.*, 1998; Sharma *et al.*, 2000; Singh & Kumar, 2000a; Sood & Thakur, 2004; Uphof, 1968; Warman, 1999; Watt, 1889-1896).**

Cerasus cerasoides (D. Don) S.Ya. Sokolov

Syn.: *Cerasus puddum* Roxb. *ex* Ser.; *Prunus cerasoides* Buch.- Ham. *ex* D. Don; D. Don; *P. puddum* (Roxb. *ex* Ser.) Brandis.

Family: Rosaceae.

Vern.: Pajja.

English, Hindi and Sanskrit Names:

Eng. : Himalayan Wild Cherry;
Hindi : Paddam, Pama-kashtha, Paya;
Sans. : Padmaka, Padmaksh.

Distribution : Temperate Himalayas: 1,200-2,500m.

Description : A middle-sized tree having glossy, long-pointed, toothed leaves with glandular petiole. Flowers long-stalked, pink appearing before leaves in umbellate fascicles. Fruits ovoid, long, yellow or red.

Reproductive Cycle : November-May

Materials Examined : HPU-EL-SKS-6317; B.C.S. (2,023m), Bemloe (2,010m), Catholic Chruch (2,150m), Chhota Shimla (2,080m).

Propagation Notes : Naturally propagated by seeds, suckers and cuttings. Tree thrives in a well-drained moisture-retentive loamy soil.

Parts Used : Whole Plant. Leaves. Stem. Fruit. Branches. Wood. Bark.

Folk Uses : Plant used as a **rootstock** for propagation of cherry. Leaves lopped as **fodder** for livestock. Fruits **edible**. Wood used as **fuel**. Branches used for **decorating** marriage gates and also burnt as **incense** to appease local deities on 'Shivratri'. Stem used for making **walking sticks**, **dip pens** and **tooth brush**. Poultice of bark applied to heal **fracture**.

Active Constituents : Chief ingredients are chyrsophanol, emodin and their 8β-D-glucosides, orientalone, physcion, puddumin A, padmakastein and its derivatives, β-sitosterol behenate, tectochrysin, genistein, leucocyanidin and 4`-glucoside of genkwanin (**stem bark**), naringenin, apigenin, β-sitosterol, sakuranetin, prunetin and genkawanin (**stems**).

Uses in Literature : Described earlier as an analgesic, carminative, conceptive, edible (fruits), expectorant, febrifuge, fuel, furniture, hair oil, psychoactive plant, tonic; and for asthma, bone dislocation and fractures, building, burning sensation of the body, cold, cough, cuts, diarrhoea, hemicrania, hoof (vet), injury (vet.), internal injuries, kidney stones, liquor preparation, marriages, muscular pain, pregnancy, religious ceremonies, rheumatic pains, rosaries, seminal weakness, sprains, stomach problem, tanning, venereal diseases, walking sticks, and wounds **(Agarwal, 2003; Ambasta, 1986; Anonymous, 1948-1976; Ansari, 1991; Arora, 1997; Arora & Pandey, 1996; Badhwar & Fernandez, 2011; Bennet *et al.*, 1991; Bhattacharjee, 2001; Bhuyan, 2003; Chatterjee & Pakrashi, 1997; Chauhan, 1999; Chhetri, 2005; Chopra *et al.*, 1956; Das & Agarwal, 1991; Dash *et al.*, 2003; Dey, 1980; Dhiman, 2003; Dwivedi, 2007; Farooq, 2005; Jain, 1991; Jain *et al.*, 1991, 1994; Karnick, 1994; Khare, 2004; Kirtikar & Basu, 1984; Lal *et al.*, 1996; Lalramnhinglova, 1996; Lata *et al.*, 2008; Negi *et al.*, 1999; Pande *et al.*, 2006; Prajapati *et al.*, 2006; Raju, 2000; Roy *et al.*, 1998; Sharma, 1996, 2003; Singh & Kumar, 2000a; Sinha, 1996; Sood *et al.*, 2013; Uniyal, 1968; Uphof, 1968; Watt, 1889-1896)**.

Cestrum elegans (Brongn. *ex* Neumann) Schltdl.

Syn.: *C. nutans* Hort. *ex* Francey; *C. purpureum* (Lindl.) Standl.; *C. smithii* Coutts; *Habrothamnus elegans* Walp.; *H. huegelii* Regel; *Meyenia purpurea* (Lindl.) Heynh.

Family: Solanaceae.

Vern.: Raat Ki Chameli.

English and Hindi Names:

Eng. : Pink Cestrum, Purple Cestrum, Red Cestrum;
Hindi : Raat Ki Chameli.

Distribution : Native of Mexico; grown in Indian gardens throughout warm regions and on hills.

Description : A large shrub with pendulous, cylindrical branches covered with soft velvety pubescence. Leaves alternate, entire, ovate or ovate-lanceolate with pointed tips and pungent smelling. Leaf apex tapering to a point; leaf base acute or obscurely cordate. Flowers crimson to purplish red, tubular, gradually contracted below mouth. Berries deep red purple.

Reproductive Cycle : October-December.

Materials Examined : HPU-EL-SKS-6319; Advance Study (2,128m), St. Edward's School (2,060m), Wood Villa (2,108m).

Propagation Notes : Propagated by seeds and stem cutting in July-August.

Part Used : Whole Plant.

Folk Uses : Cultivated as an **ornamental** in gardens.

Uses in Literature : Reported earlier as an ornamental, reducing pollutant shrub and in perfumery **(Agarwal, 2003; Ambasta, 1986; Anonymous, 1948-1976)**.

Cestrum nocturnum L.

Syn.: *C. graciliflorum* Dunal; *C. leucocarpum* Dunal; *C. multiflorum* Roem. & Schult.; *C. suberosum* Jacq.; *Chiococca nocturna* Jacq.

Family: Solanaceae.

Vern.: Rat-ki Rani.

English and Hindi Names:

Eng. : Lady or Queen of Night, Night Jasmine;
Hindi : Rat Ki Rani.

Distribution : Indigenous to West Indies; largely cultivated both in plains and hills in India.

Description : Evergreen, subscandent, quick growing hardy, glabrous shrubs having slender, smooth often yellowish branches with numerous lenticels. Leaves alternate, exstipulate, entire, pale green, ovate-lanceolate. Flowers greenish yellow; scented at night, close during day. Berries blue or blackish. Seeds numerous, compressed.

Reproductive Cycle : August-October.

Material Examined : HPU-EL-SKS-6321; Bemloe (2,010m).

Propagation Notes : Plant succeeds in fertile, well-drained soils in sun or partial shade; propagated by cutting.

Parts Used : Whole Plant. Flowers.

Folk Uses : **Planted** for its pleasing fragrance. Flowers considered **sacred** for offering to deities.

Active Constituents : Active properties due to benzaldehyde, benzyl alcohol, benzyl acetate, methyl benzoate, phenyl acetaldehyde, phenyl ethyl alcohol, linalool and α- farnesene (**flowers**).

Uses in Literature : Recorded earlier as hallucinogens, ornamental perfumes; and for cosmetics, epilepsy **(Agarwal, 2003; Anonymous, 1948-1976; Asolkar *et al.*, 1992; Harborne & Baxter, 2001; Kaur & Sharma, 2004; Randhawa, 1957, 1960, 1965; Sharma, 2003; Sood *et al.*, 2013; Uphof, 1968)**.

Cestrum parqui L Her.

Syn.: *C. campestre* Griseb.; *C. foetidissimun* Jacq.; *C. poeppigii* Schltdl; *C. spurium* Dunal; *C. virgatum* Ruiz & Pav.

Family: Solanaceae.

Vern.: Rat-ki-Rani.

English Names:

Eng. : Chilean Cestrum, Green Cestrum, Green Poison berry, Willow-leaved Jasmine.

Distribution : Native of C. & N. America.

Description : Upright straggly woody perennial shrubs upto 3m tall with one or more brittle green stems. Leaves light green, alternate and produce foul rubbery smell when crushed. Flowers yellow-green, in clusters at the end of branches. Fruit a black berry.

Reproductive Cycle : November-January.

Material Examined : HPU-EL-SKS-6325; Sidhowal Lodge (2,254m).

Propagation Notes : Propagated by seeds and stem cutting in July-August.

Part Used : Whole Plant.

Folk Use : Grown as an **ornamental** plant in gardens.

Uses in Literature : Reported to be used as diaphoretic, hallucinogens, ornamental; and for skin troubles **(Ambasta, 1986; Anonymous, 1948-1976; Asolkar *et al.*, 1992; Harborne & Baxter, 2001; Pande *et al.*, 2006; Sood *et al.*, 2010)**.

Chamaerops humilis L.

Syn.: *Phoenix humilis* (L.) Cav.

Family: Arecaceae.

Vern.: Fan Palm, Khajuri.

English Names:

Eng. : Dwarf Fan Palm, European Fan Palm, Hair Palm.

Distribution : Indigenous to Europe and N. Africa.

Description : Evergreen shrubs or small trees possessing fan-shaped leaves which are divided almost to the base into long narrow segments. Flowers borne in dense, erect panicles (spadix) in axil of lower leaves. Drupes dark-yellow or reddish when mature.

Reproductive Cycle : February-March.

Materials Examined : HPU-EL-SKS-6327; Advance Study (2,128m), Bantony (2,182m), Bemloe (2,010m), Chhota Shimla (2,090m), St. Edward's School (2,060m).

Propagation Notes : Usually propagated from seeds or by division of suckers. Plant grows in well drained, moderately fertile soil in full sun.

Parts Used : Whole Plant. Leaves. Leaf Buds.

Folk Uses : Commonly planted as a **lawn plant** for **ornamental** purposes. **Fibre** obtained from leaves used for stuffing upholstery and weaving carpets. Young leaf buds cooked as **vegetable**.

Uses in Literature : Known in Indian literature as edible (leaves, fruit), astringent, cordage, and hedge plant **(Ambasta, 1986; Anonymous, 1948-1976; Bean, 1981; Chittendon, 1956; Hay *et al.*, 1983; Hill, 1951; Huxley, 1992; McMillian-Browse, 1993; Phillips & Rix, 1998; Sood *et al.*, 2013; Uphof, 1968; Usher, 1974)**.

Citrus limon (L.) Burm. f.

Syn.: *C. limonum* Risso.

Family: Rutaceae.

Vern.: Galgal.

English, Hindi and Sanskrit Names:

Eng. : Galgal, Hill lemons, Lemon;
Hindi : Baranibu, Janbira, Paharinimbu, Paharikaghzi;
Sans. : Maha Nimbu, Nimbuphala, Vanabijapura.

Distribution : Valley of Himalayas, on forest along streams, up to 1,200m.

Description : A small prickly tree having ovate, dark green shining leaves and light pinkish-white, fragrant young flower buds. Fruits oblong - round with a thick rind.

Reproductive Cycle : November-February.

Materials Examined : HPU-EL-SKS-6333; Bemloe (2,012m), Sanitarium (2,060m).

Propagation Notes : Propagated through seeds and by cuttings of half-ripe wood in July/August in a frame and layering in October. Germination of seeds occur within 2 - 3 weeks at 20-25°C. Plant succeeds in a moderately heavy loam with a generous amount of compost and sand.

Parts Used : Leaves. Fruit.

Folk Uses : Fruits roasted by burrowing in hot ash of fireplace eaten for **rheumatic pains**. Fruits also used for **pickling**; its extr. used for **souring** purposes and **offered** to appease Goddess 'Tulsi'. Also, leaves used for **scouring teeth** and checking **foul breath**.

Active Constituents : Fruit peel yields apigenin, luteolin, chrysoseriol, quercitin, isorhamnetin, new flavonols- limocitrin and limocitrol- sinapic acid, p-coumaric acid, scopoletin, umbelliferone, hesperidin, citrusins A, B, C and D, coniferin syringin and dehydro-diconeferyl alcohol-4β-glucoside.

Uses in Literature : Known to be useful as an anthelmintic (vet.), antidote (snake bite), arteriosclerosis, antiseptic, carminative, disinfectant, febrifuge, juice (vet.), laxative, lemon oil (obtained as bye-products), nervine tonic, stimulant; and for acidity, burns, citric acid, cold, corns, cough, culinary purposes and in preparation of beverages, dandruff, flavoring liqueurs, flu, headache, hypertrophy of spleen, indigestion, lowering body heat, pectin, pickles, preventing dandruff in hair, rheumatism, scabies, scurvy, stomachache, throat diseases, vitality to the body, vomiting and warts **(Agarwal, 2003; Ahmed & Borthakur, 2005; Ambasta, 1986; Anonymous, 1948-1976; Arora & Pandey, 1996; Bhattacharjee, 2001; Borthakur *et al.*, 1999; Chatterjee & Pakrashi, 1997; Chevallier, 1996; Chopra *et al.*, 1956; Dagar & Dagar, 1996; Harborne & Baxter, 2001; Jamir *et al.*, 2011; Khare, 2004; Kurian, 1995; Mishra *et al.*, 1996; Pande *et al.*, 2006; Prajapati *et al.*, 2006; Reddi *et al.*, 2005; Roy *et al.*, 1998; Sharma *et al.*, 2000; Shome *et al.*, 1996; Singh & Pandey, 1998; Sood *et al.*, 2010, 2012b, 2013; Tirkey *et al.*, 2003a; Uphof, 1968; Vishwanathan & Singh, 1996; Warman, 1999)**.

Citrus maxima (Burm.) Merr.

Syn.: *C. grandis* Osbeck; *C. decumana* L.; *Aurantium maximum* Burm.

Family: Rutaceae.

Vern.: Chakotra.

English, Hindi and Sanskrit Names:

Eng. : Forbidden Fruit, Paradies Apple, Pommel-Moose Shadock;
Hindi : Batavinebu, Botornebu;
Sans. : Madhukarkati.

Distribution : Cultivated.

Description : Trees with large, ovate-oblong leaflets, white flowers and large-sized, pale globose fruits with thick rind.

Reproductive Cycle : November-March.

Material Examined : HPU-EL-SKS-6330; Armsdell (2,102m).

Propagation Notes : Propagation by seeds, cutting and layering.

Part Used : Fruit.

Folk Uses : Fruits **edible**, **refreshing** and considered good against **vomiting**.

Active Constituents : These are 5 - geranyloxy – 7 - methoxycoumarin, isoimperatorin, bergapten, and 5 -geranyloxypsoralen (**fruits**), nomilin, limolin, deoxylimonin and obacunone (**juicy hair**).

Uses in Literature : So far useful as carminative, edible, nutritive, refrigerant; and for appetite, cholera, constipation, convulsive coughs, deserts, diabetes, diarrhoea (vet.), epilepsy, flatulent pains, heart disease, indigestion (vet.), inflamed nerves, jams and marmalades, insanity, morbid cancer, nervous disorders, offerings, pimple, painful swellings, rheumatism, scurvy, swellings, ulcer, vomiting, waist and back pains **(Agarwal, 2003; Ambasta, 1986; Borthakur, 1996a; Borthakur *et al.*, 1999; Chopra *et al.*, 1956; Chun-Lin & Jieru, 1995; Das & Agarwal, 1991; Gogoi & Borthakur, 1991; Kurian, 1995; Pande *et al.*, 2006; Prajapati *et al.*, 2006; Saikia *et al.*, 2010; Saklani & Jain, 1994; Sharma & Rana, 2005; Sharma *et al.*, 2000; Shenghi, 1991; Singh & Pandey, 1998; Sood *et al.*, 2010, 2012b, 2013; Uphof, 1968)**.

Clematis buchananiana Wall.

Syn.: *C. bucamara* Buch.-Ham. *ex* DC.; *C. loasifolia* D. Don.

Family: Ranunculaceae.

Vern.: Dhudki Bel.

English Name:

Eng. : Old Man Beard.

Distribution : Throughout Temperate Himalayas: 1,800-3,000m.

Description : Shortly hairy climbing shrubs with grooved stems and pinnate leaves. Leaflets 5-7, broadly ovate, cordate, often lobed, coarsely and irregularly crenately toothed. Flowers campanulate, pale yellow and in long leafy panicles. Sepals ribbed, narrowly oblong, erect, tip recurved pointed. Achenes densely hairy with long persistent feathery style.

Reproductive Cycle : July-October.

Materials Examined : HPU-EL-SKS-6335; Advance Study (2,128m), Bemloe (2,009m), Boileauganj Cemetery (2,001m), Chhota Shimla (2,090m), Kanlog Cemetery (1,998m).

Propagation Notes : Propagation by seeds, internodal cuttings of soft-semi-ripe wood in late spring and layering of old stems in late winter or early spring are commonly employed methods. Plant succeeds in a deep moist soil.

Parts Used : Whole Plant. Leaves.

Folk Uses : Plant considered **poisonous** to cattle. Fibres from plant good for **cordages**. Poultice of leaves applied for relief against pain in **sprain**.

Uses in Literature : Recorded earlier for cuts, eczema, pain in stomach due to tumour, scabs, skin diseases, sores, stomach disorders, toothache, wounds; and as head wash and poison (vet.), **(Chaudhuri, 1993; Farooq, 2005; Jain, 1991; Jain *et al.*, 1991; Kapur, 1996a; Kapur & Singh, 1996; Kumar, 2002; Manandhar, 1980a, 1996; Megoneitso & Rao, 1983; Pande *et al.*, 2006; Sood *et al.*, 2009)**.

Clematis montana Buch-Ham. *ex* DC.

Syn.: *Anemone curta* Wall.; *C. insularialpina* D. Don.

Family: Ranunculaceae.

Vern.: Garol, Kaunia-bali.

English Name:

Eng. : Anemone Clematis.

Distribution : Himalayas (Kashmir - Arunachal Pradesh): 1,600-4,000m and Khasi Hills: 1,200m.

Description : Glabrous or pubescent shrubs climbing by their leaf-stalks. Stems terete. Leaves of 3 - leaflets, stalked, clustered at nodes. Leaflets narrowly ovate, acute, margins more or less toothed, especially towards the tip. Flowers white, solitary on axillary stalks. Achenes almost sessile, glabrous and with long feathery tails (styles).

Reproductive Cycle : April-June.

Materials Examined : HPU-EL-SKS-6336; Bemloe (2,010m), Stirling Castle (2,250m).

Propagation Notes : Propagated by seeds, air layering or cutting.

Part Used : Whole Plant.

Folk Use : Commonly employed as an **adornment** for ugly sheds and walls.

Uses in Literature : Used earlier as anti-inflammatory, deleterious to skin, diuretic, ornamental climber, spasmolytic; and for diabetes and urinary troubles **(Asolkar *et al.*, 1992; Berry & Bradley, 2000; Chaudhuri, 1993; Farooq, 2005; Kapur, 1996a; Kapur & Srivastava, 1996; Miller Gault, 1976; Pal, 1972; Pande *et al.*, 2006; Rastogi & Mehrotra, 1995b)**.

Colquhounia coccinea Wall.

Family: Lamiaceae.

Vern.: Bhekuyi.

English and Hindi Names:

Eng. : Himalayan Mint Plant;
Hindi : Pansra.

Distribution : S.E. Asia: 1,200-3,000m.

Description : Tall rambling shrubs with branchlets, underside of leaves and leaf-stalks covered with thick woolly star-shaped hair and large tubular red flowers striped with orange within. Leaves opposite, long-pointed, toothed, rugose above, stalked and sometime white-woolly beneath. Flowers in dense axillary whorls or forming dense terminal spikes. Nutlets oblanceolate.

Reproductive Cycle : August-October.

Materials Examined : HPU-EL-SKS-6338; Advance Study (2,128m), Kanlog Cemetery (1,990m).

Propagation Notes : Propagated by seeds.

Parts Used : Whole Plant. Twigs.

Folk Uses : Planted as an **ornamental**. Twigs used as **fodder** and **fuel wood**.

Uses in Literature : Described earlier for its use as fodder, fuel, edible (leaves chewed with betel nut) and spasmolytic **(Ahmed & Borthakur, 2005; Asolkar *et al.*, 1992; Farooq, 2005; Nautiyal *et al.*, 2003; Samant *et al.*, 1996)**.

Coriaria nepalensis Wall.

Family: Coriariaceae.

Vern.: Kachog.

English and Hindi Names:

Eng. : Mussoorie Berry;
Hindi : Makola, Masuri.

Distribution : Himalayas; Pakistan-S.W. China: 1,000-2,700m.

Description : A glabrous shrub with 4-angled branches, red bark and opposite, nearly sessile, ovate, shortly pointed, entire leaves having prominent basal nerves. Flowers small, bisexual, regular, in lateral clustered racemes. Fruits small, 1-seeded with prominently ridged carpels and enclosed within much enlarged, purple succulent petals.

Reproductive Cycle : March-June.

Material Examined : HPU-EL-SKS-6343; Boileauganj Cemetery (2,001m), Kanlog Cemetery (1,990m).

Propagation Notes : Propagated by seeds.

Parts Used : Fruits. Wood.

Folk Uses : Fruits **edible**. Wood used for making **boxes** and other small **agricultural implements**.

Uses in Literature : Described earlier as edible (fruit), fuel, poisonous (leaves), purgative, tannin; and for boxes **(Agarwal, 2003; Ambasta, 1986; Arora, 1997; Farooq, 2005; Kirtikar & Basu, 1984; Pande *et al.*, 2006; Raju, 2000).**

Cornus capitata Wall.

Syn.: *Benthamidia capitata* (Wall). Nakai; *Cynoxylon capitatum* (Wall.) Nakai; *Dendrobenthamia capitata* (Wall.) Hutch.

Family: Cornaceae.

Vern.: Teelahan.

English and Hindi Names:

Eng. : Himalayan Strwberry Tree;
Hindi : Bamora, Tharmal, Tharbal.

Distribution : H.P. - S.W. China: 1,320-3,400m.

Description : Small trees having oblong or ovate, densely hairy leaves with short stalks. Flowers white, crowded in hemispheric terminal heads with conspicuous involucres of four pale yellow, petal-like ovate bracts. Mature drupes very small, tinged with red and coalescing in a succulent globose head.

Reproductive Cycle : May-June.

Materials Examined : HPU-EL-SKS-6345; Boileauganj Cemetery (2,001m), Chhota Shimla (2,080m).

Propagation Notes : Propagated by seeds and cuttings of half-ripe side shoots in July/August.

Parts Used : Fruits. Wood.

Folk Uses : Fruits **edible**. Wood used as **fuel**.

Uses in Literature : Fruits edible, cooked with sugar and pickled **(Agarwal, 2003; Ambasta, 1986; Arora, 1997; Arora & Pandey, 1996; Badhwar & Fernandez, 2011; Chandra, 1997; Chaudhuri, 1993; Farooq, 2005; Rao, 1990; Roy *et al.*, 1998; Samant *et al.*, 2001; Sharma, 1996; Singh, 1996; Sood *et al.*, 2012b)**.

Cornus macrophylla Wall.

Syn.: *Swida macrophylla* (Wall.) Sojak; *Thelycrania macrophylla* (Wall.) Pojark.

Family: Cornaceae.

Vern.: Nang.

English and Hindi Names:

Eng. : Large-Leaf Dogwood;
Hindi : Kachur, Kander, Kasia.

Distribution : Throughout Himalayas: 1,350-2,650m.

Description : Large-sized trees with broadly ovate leaves having upper surface glabrous and lower minutely hairy. Cymes terminal, compound, hairy across. Flowers white, in terminal compound cymes. Drupes pubescent, globose.

Reproductive Cycle : May-August.

Materials Examined : HPU-EL-SKS-6347 Advance Study (2,128m), Madan Kunj (2,078m).

Propagation Notes : Propagated by seeds and cuttings of half-ripe shoots in July/August.

Parts Used : Leaves. Wood.

Folk Uses : Leaves used as **fodder**. Wood used for **building works**, **furniture**, **agricultural implements** and making **walking sticks**.

Uses in Literature : Reported earlier as charcoal, edible (fruits, wood), fodder, gun powder, mallet heads, pulley, shuttles; and for bobbin and ulcerous sores **(Ambasta, 1986; Badhwar & Fernandez, 2011; Chaudhuri, 1993; Pande *et al.*, 2006; Samant *et al.*, 2001; Watt, 1889-1896)**.

Cotoneaster bacillaris Wall. *ex* Lindl.

Family: Rosaceae.

Vern.: Simbli.

Hindi Name:

Hindi : Khairz.

Distribution : W. Himalayas; Afghanistan-Bhutan: 1,800-3,000m.

Description : A large-sized shrub having long, slender branches and ovate or ovate-lanceolate, obtuse or acute leaves with lower often hairy along the nerves and flat margins. Flowers white, in large, hairy or pubescent, stalked, loosely branched cymes on short, lateral leafy branchlets. Fruits globose, red at first, nearly black when ripè.

Reproductive Cycle : May-October.

Material Examined : HPU-EL-SKS-6348; Sidhowal Lodge (2,254m).

Propagation Notes : Propagated by seeds and cuttings of half-ripe wood with a heel in July/August.

Part Used : Twigs.

Folk Uses : Twigs used as **fodder**; and for making **agricultural implements** and **walking sticks**.

Uses in Literature : Reported to be used as astringent; and for agricultural implements, bows, golf-clubs, rheumatism, scabies, umbrella-handles, tent pegs and walking sticks **(Agarwal, 2003; Ambasta, 1986; Asolkar *et al.*, 1992; Kirtikar & Basu, 1984; Pande *et al.*, 2006)**.

Cryptomeria japonica D. Don

Syn.: *C. araucarioides* Carriere; *C. compacta* Beissn.; *C. fortunei* Otto & A. Dietr.; *C. kawaii* Hayata; *C. lobbiana* Billain; *C. mucronata* Beissn.; *C. nigricans* Carriere; *C. pungens* Beissn.; *C. viridis* Beissn.; *Cupressus japonica* Thunb. *ex* L. f.; *Taxodium japonicum* (Thunb. *ex* L.f.) Brongn.

Family: Taxodiaceae.

Vern.: Cryptomeria.

English Name:

Eng. : Japanese Cedar.

Distribution : E. Asia - China, Japan; cultivated throughout the districts of Darjeeling, Shimla and occasionally in other hill station.

Description : Evergreen pyramidal trees with a straight trunk. Leaves bluntly keeled on the dorsal and sharply keeled on the ventral side, bluish-green. Cones globose reddish-brown.

Reproductive Cycle : March-October.

Materials Examined : HPU-EL-SKS-6350; Cedar Circuit House (2,156m), Holly Lodge (2,278m), Oak Over (2,154m), St. Edward's School (2,060m).

Propagation Notes : Plant succeeds in a deep rich moist alluvial soil and a sheltered position in full sun. Propagated through seeds and cuttings in autumn planted in a sandy soil in a cold frame.

Parts Used : Whole Plant. Leaves. Wood.

Folk Uses : Planted as an **ornamental** tree. Leaves aromatic and used as **incense sticks**. Wood strongly rot resistant and used for **buildings**, **furniture** and **utensils**.

Active Constituents : These are - five new hexacarbocyclic triterpenes-chamaecydin, 6 α- and 6β-hydroxychamaecydins, 10α- and 10β-hydroxycryptoquinones, four new abietane derivatives- compounds I, II, III and 6β-hydroxyferruginol and two new kaurane derivatives (IV&V) (**leaves**).

Uses in Literature : Used earlier as incense, lamp posts, ornamental; and for building works, furniture, gonorrhoea, shelterbelt, tanning and utensils **(Ambasta, 1986; Asolkar *et al.*, 1992; Bailey, 1957; Bean, 1981; Chaudhuri, 1993; Chittendon, 1956; Huxley, 1992; Manandhar, 2002; Randhawa, 1960; Rastogi & Mehrotra, 1995b; Rushforth, 1987; Sood *et al.*, 2010; Uphof, 1968; Usher, 1974; Warda, 1985)**.

Cunninghamia lanceolata Hook.

Syn.: *Abies batavorum* Siebold *ex* Carriere; *A. lanceolata* (Lamb.) Poir.; *Belis lanceolata* (Lamb.) Hoffmanns; *Cunninghamia sinensis* R.Br.; *C. unicanaliculata* D.Y. Wang & H.L.Liu; *Larix chinensis* Mill.; *Pinus lanceolata* Lamb.

Family: Taxodiaceae.

Vern.: Ghadhal, Thanala.

English Name:

Eng. : Chinese Fir.

Distribution : Native of China.

Description : Evergreen trees. Leaves linear-lanceolate, sharp-pointed having a broad decurrent base and finely serrated with 2 broad whitish bands beneath. Cones roundish-ovate with leathery, serrated and pointed scales.

Reproductive Cycle : March-October.

Material Examined : HPU-EL-SKS-6355; Chapslee (2,243m), Raj Bhawan (2,132m).

Propagation Notes : Propagated mostly through seeds, cuttings of half ripe wood, 5 - 8cm with a heel in July/August and division of suckers. Plant prefers a rich warm soil and a sheltered sunny position.

Parts Used : Whole Plant. Wood. Bark.

Folk Uses : Grown as an **ornamental** plant. 5-10 ml decoction of wood given twice daily to check **chronic ulcer**. Ash of bark applied externally to treat **burns, scalds** and **wounds**. Wood employed in **construction works** and for making **furniture**.

Active Constituents : Reported to contain robustaflavone, sequoiaflovone, hinokiflavone, isocryptomerin, amentoflavone, di-O-methylamentoflavone, kayaflavone and apigenin (**leaves**).

Uses in Literature : Described earlier as antidote, carminative, essential oil, and ornamental plant; and for bruises, construction, cough, furniture, hernia, pain, rheumatism, wounds **(Anonymous, 1948-1976; Bailey, 1957; Bean, 1981; Chittendon, 1956; Huxley, 1992; Rastogi & Mehrotra, 1995a; Sheat, 1948; Uphof, 1968; Usher, 1974; Warda, 1985)**.

Cupressus cashmeriana Royle *ex* Carriere

Syn.: *C. assamica* Silba; *C. darjeelingensis* (Silba) Silba; *C. himalaica* Silba; *C. pseudohimalaica* Silba.

Family: Cupressaceae.

Vern.: Kashmiri Cypress.

English Names:

Eng. : Bhutan Cypress, Kashmir Cypress.

Distribution : Native of E. Himalayas in Bhutan and adjacent areas of Arunachal Pradesh in N.E. India upto 2,800m.

Description : Medium - large coniferous trees upto 45m. Foliage comprises pendulous sprays of blue-green, very slender, flattened shoots. Leaves scale-like. Cones ovoid with 8-12 scales, dark green when mature.

Reproductive Cycle : May-August.

Material Examined : HPU-EL-SKS-6357; St. Edward's School (2,060m).

Propagation Notes : Grown from semi-hard cuttings and through seeds which are surface sown at 18-24°C; pre-chilling for 3 weeks may aid germination.

Parts Used : Whole Plant. Wood.

Folk Uses : Grown as an **ornamental** plant. Wood used in general **construction**.

Uses in Literature : Reported to be grown as avenue or ornamental plant; and for timber **(Farjon, 1994, 2005; Rastogi & Mehrotra, 1995a; Warda, 1985)**.

Cupressus torulosa D. Don

Syn.: *Athrotaxis joucadan* Carriere; *C. austrotibetica* Silba; *C. corneyana* Knight & Perry *ex* Carriere; *C. doniana* Hook.f.; *C. karnaliensis* Silba.

Family: Cupressaceae.

Vern.: Saru.

English Name:

Eng. : Himalayan Cypress.

Distribution : Outer range of Himalayas from Chamba - Nepal 2,000-3,000m.

Description : Conical trees with spreading branches and drooping ends. Young foliage blue-green, becoming much darker with age. Leaves opposite, scale-like, triangular, closely imbricate, margin white. Cones terminal on branchlets; male and female on the same tree. Fruit a globose head, consisting of woody peltate scales tightly closed along their edges at first, opening when mature. Seeds numerous, small, winged.

Reproductive Cycle : February-April.

Materials Examined : HPU-EL-SKS-6358; Advance Study (2,128m), Naldehra (2,127m), Oak Over (2,154m), Sidhowal Lodge (2,254m).

Propagation Notes : Thrives in a well-drained loamy or peaty soil. Propagated through seeds and cuttings of almost ripe wood in April/May.

Parts Used : Whole Plant. Leaves. Wood.

Folk Uses : Grown as an **ornamental hedge** plant. Leaves burnt as **incense**. Wood used in general **construction works**.

Uses in Literature : Used in India for construction, furniture, inlay work, perfumery, wine; and as diuretic, essential oil, fuel wood **(Agarwal, 2003; Ambasta, 1986; Anonymous, 1948-1976; Asolkar *et al.*, 1992; Bailey, 1957; Bean, 1981; Chittendon, 1956; Huxley, 1992; Manandhar, 2002; Nautiyal *et al.*, 2003; Rushforth, 1987; Samant *et al.*, 1996; Sharma, 1996; Sood *et al.*, 2013; Uphof, 1968; Usher, 1974; Warda, 1985)**.

Cycas revoluta Thunb.

Syn.: *C. miquelii* Warb.; *Epicycas miquelii* (Warb.) de Laub.
Family: Cycadaceae.

Vern.: Cycas.

English Names:

Eng. : China Fern Palm, China Sago Palm, Japan Fern Palm, Japan Sago Palm.

Distribution : E. Asia-China, Japan.

Description : Palm-like trees with simple or branched trunk and recurved leaves having numerous, narrow, stiff acute pinnae terminating in a spine-like tip and revolute margin. Staminate inflorescence cylindrical. Pistillate inflorescence a semi globose head. Carpophylls pectinate, broadly ovate and densely clothed with brownish, felt-like wool. Fruits ovate, compressed, red.

Reproductive Cycle : February-May.

Material Examined : HPU-EL-SKS-6360; Advance Study (2,128m).

Propagation Notes : Propagation chiefly done in September from seeds whereas it is done in February-March and September-October from suckers and bulbils. Succeeds in dry soils and a sunny position.

Parts Used : Whole Plant. Leaves.

Folk Uses : Commonly planted in gardens as an **ornamental.** Leaves used for **decorative** purposes.

Active Constituents : Active constituents are amentoflavone, 2,3-dihydroamentoflavone, hinokiflavone, 2,3-dihydrohinokiflavone, podocarpusflavone A and sotetsuflavone (**leaves**).

Uses in Literature : Reported to be used as anticancer, astringent, diuretic, edible (seeds), expectorant, tonic, ornamental in lawns, rockeries and in pots; and for rheumatism **(Agarwal, 2003; Ambasta, 1986; Anonymous, 1948-1976; Bailey, 1957; Beri, 1992; Brickell, 1990; Chittendon, 1956; Huxley, 1992; Paliwal, 2009; Phillips & Rix, 1998; Randhawa, 1957; Rastogi & Mehrotra, 1991; Sood *et al.*, 2013; Uphof, 1968; Warman, 1999).**

Daphne papyracea Wall. *ex* Steud.

Syn.: *D. cavaleriei* H. Lev.; *D. eriocephala* Royle; *D. laciniata* Lecomte; *D. mairei* H. Lev.

Family: Thymelaeaceae.

Vern.: Chamar-Chumb, Raabal.

English and Hindi Names:

Eng. : Nepal Paper Plant;
Hindi : Satpura, Setburusa, Setburwa.

Distribution : Temperate Himalayas and Khasia Hills: 1,500-2,100m.

Description : Glabrous shrubs having shortly stalked, lanceolate, obtuse leaves which are crowded near the end of branches. Flowers white or lilac, in heads surrounded by numerous lanceolate bracts. Drupes red or orange when ripe.

Reproductive Cycle : April-October.

Materials Examined : HPU-EL-SKS-6361; Advance Study (2,128m), Boileauganj Cemetery (1,999m), Cedar Circuit House (2,156), Chhota Shimla (2,070m), Hainault (2,294m).

Propagation Notes : Propagated by seeds under natural condition.

Parts Used : Aerial Parts. Fruits.

Folk Uses : Poultice of aerial parts applied to hasten **suppuration**. Fruits **edible**.

Active Constituents : Active constituents are taraxerone, genkwanin, daphenetin-8-α-glucoside, taraxerol and its α-sitosterol-D-glucoside (**aerial parts**).

Uses in Literature : Described so far as anticancerous, edible, febrifuge, purgative; and for cordages and fibres from bark, eczema, genital swellings, intestinal complaints, ornamental garlands, rheumatism, urinary disorders (**Agarwal, 2003; Ahmed & Borthakur, 2005; Ambasta, 1986; Anonymous, 1948-1976; Aswal, 1996; Chopra *et al.*, 1956; Dam & Hajra, 1997; Dhiman, 2003; Farooq, 2005; Jain, 1991; Lal *et al.*, 1996; Pande *et al.*, 2006; Singh & Kumar, 2000b; Sood & Thakur, 2004; Sood *et al.*, 2005, 2012b; Uphof, 1968; Watt, 1889-1896; Yonzone *et al.*, 1996**).

Datura stramonium L.

Syn.: *Datura bernhardii* Lundstr.; *D. cabanesii* P. Fourn.; *D. inermis* Juss. *ex* Jacq.; *D. laevis* L.f.; *D. microcarpa* Godr.; *D. praecox* Godr.; *D. tatula* L.; *D. walliichii* Dunal; *Strmonium spinosum* Lam.

Family: Solanaceae.

Vern.: Dhatura.

English, Hindi and Sanskrit Names:

Eng. : Apple of Peru, Devil's Apple, Jimson Weed, Madapple, Sting Weed, Stramonium;

Hindi : Dhatura;

Sans. : Devika, Dhattura, Dhurta, Dhultara, Ghanta-pusha, Kalama, Kantaphala, Madana, Shiv-apriya, Shivashekhara, Shyama, Turi, Unmatta, Unmattaka.

Distribution : Temperate Himalayas: Kashmir-Sikkim up to 2,700m.

Description : Shrubs having stalked, ovate, coarsely and irregularly lobed, toothed leaves and white, single, short flowers; usually axillary stalked. Capsule ovoid, covered with rigid, sharp prickles. Seeds wrinkled.

Reproductive Cycle : June-October.

Materials Examined : HPU-EL-SKS-6365; Bemloe (2,010m), Kanlog Cemetery (1,994m).

Propagation Notes : Propagated by seeds. Plant prefers a rich light sandy soil or a calcareous loam, and an open sunny position.

Parts Used : Leaves. Flowers. Fruits.

Folk Uses : Flowers and fruits considered **sacred** and offered to appease Lord Shiva. Poultice of leaves applied on **boils**, **sores** and **bites**. Inhalation of smoke emnating from burning leaves good against **asthma**. Fruit juice applied on scalp to cure **dandruff** and **hairfall**.

Active Constituents : Various constituents are vitanolide-vitastramonolide along with daturalactone and its keto derivative characterised as 5α,12α,27-trihydroxy-1-oxo-6α, 7α-epoxy-22R-vita 2,24-dienolide (**leaves**), hyoscine, 3α,6β-ditigloycoxytropane, 3α, 6β-ditigloyloxytropane-7β-OL, apoatropine, tigcoidine, tropine, meteliodine (Nigerian plant), 0.3 to 0.5% of alkaloids chiefly as hyoslyamine with smaller amount of hyoscine and atropine (**dried leaves, flowering tops**).

Uses in Literature : Reported earlier as anodyne, antirheumatic, antispasmodic, cigarettes (leaves), mydriatic, narcotic in treatment of bronchitis, psychoactive plant; and for asthma, boils, breast inflammation due to excessive formation of milk, controlling salivation and exparasite, dental caries, diarrhoea, dislocation of joints, dysmenorrhoea, enlarged testicles, epilepsy, haemorrhoids, headache, inflammatory pains, itch, jaundice, lactation (vet.), milching (vet.), nervous disorder (vet.), nymphomania, painful diseases of rectum, paralysis, post encephalitic parkinsonianism, purpurea, rheumatism (vet.), sciatica, skin diseases, stomach complaints, swelling of neck (vet.), tonsilitis, toothache, travel sickness, ulcers, and vicious indulgence **(Agarwal, 2003; Ahmed & Borthakur, 2005; Ambasta, 1986; Anonymous, 1948-1976; Beigh *et al.*, 2003; Borthakur *et al.*, 1999; Chauhan, 1999; Chhetri, 2005; Chopra *et al.*, 1956; Dey & Bahadur, 1973; Dhiman, 2003, 2005; Dwivedi, 2003, 2007; Dwivedi *et al.*, 2007; Farooq, 2005; Ganai & Nawachoo, 2003; Gogoi & Das, 2003; Harborne & Baxter, 2001; Jain, 1968, 1991; Jain *et al.*, 1994; Jha *et al.*, 1991; Kaul, 1997; Khare, 2004; Manandhar, 1996; Meena *et al.*, 2003; Pande *et al.*, 2006; Prajapati *et al.*, 2006; Prashantkumar & Vidyasagar, 2008; Rana *et al.*, 1996; Ranjan, 2000b; Said, 1997; Singh, 1991; Sood *et al.*, 2009, 2013; Srivastava *et al.*, 2003b; Subramani & Goraya, 2003; Upadhye *et al.*, 1994; Uphof, 1968; Verma & Chauhan, 2006; Viswanathan *et al.*, 2006; Watt, 1889-1896.**

Debregeasia saeneb (Forssk.) Hepper & J.R.I. Wood

Syn.: *Boehmaria hypoleuca* Hochst. *ex* A. Rich; *B. salicifolia* D. Don; *B. hypoleuca* Hochst. *ex* A. Rich.; *D. bicolor* (Roxb.) Wedd.; *D. hypoleuca* (Hochst. *ex* A. Rich.) Wedd.; *D. salicifolius* (D. Don) Rendle; *Morocarpus salicifolius* Blume; *Procris hypoleuca* Steud.; *Rhus saeneb* Forssk.; *Urtica bicolor* Roxb.

Family: Urticaceae.

Vern.: Siaru, Syaru.

Hindi Names:

Hindi : Seharu, Siaru.

Distribution : W. Temperate Himalayas: 1,000-2,400m.

Description : A softly pubescent shrub. Leaves alternate, short-stalked, long-pointed, base 3-nerved, teeth numerous, upper surface rough, lower white-tomentose. Stipules 2 - parted. Flowers unisexual interspersed with numerous bracteoles in rounded, sessile axillary clusters; the male and female on different plants. Drupes small, yellow, aggregated at a head.

Reproductive Cycle : April-August.

Materials Examined : HPU-EL-SKS-6368; Advance Study (2,128m), Boileauganj Cemetery (1,998m), Chhota Shimla (2,078m), Gorton Castle (2,094m), Sidhowal Lodge (2,254m).

Propagation Notes : Propagated through seeds and branch cutting. Plant succeeds in a fertile well-drained loam soil.

Parts Used : Leaves. Stem. Fruit. Wood.

Folk Uses : Fruits **edible** and also used for **flavouring** various dishes. Leaves considered as good **cattle feed**. Wood used as **fuel**. **Fibres** obtained from bark used in **cordages**.

Uses in Literature : Used so far for cordages (bark), edible purposes (fruits), stomach pain, ulcers, killing maggots in wounds of cattle and goats; and as flavouring agent, fishing lines and fuel **(Agarwal, 2003; Ambasta, 1986; Anonymous, 1948-1976; Arora & Pandey, 1996; Badhwar & Fernandez, 2011; Chandra, 1997; Kapur, 1996a; Kirtikar & Basu, 1984; Mehtani & Ingle, 1999; Pande *et al.*, 2006; Raju, 2000; Rama Rao & Henry, 1996; Roy *et al.*, 1998; Singh & Kumar, 2000a; Sood *et al.*, 2010, 2012b, 2013; Uphof, 1968; Watt, 1889-1896).**

Desmodium elegans DC.

Syn.: *D. cinerascens* Franch.; *D. esquirolii* H. Lev.; *D. forrestii* Schindl.; *D. glaucophyllum* Pamp.; *D.tiliaefolium* G. Don; *Hedysarum tiliaefolium* D. Don.

Family: Fabaceae.

Vern.: Chamla.

Distribution : Distributed in N. India, temperate and tropical region.

Description : Tall erect shrubs with tomentose or nearly glabrous stem and broadly ovate, entire or sinuate, obtuse or acute leaflets having upper surface thinly hairy or pubescent, and lower usually grey-tomentose or nearly glabrous and pale. Racemes numerous. Flowers pale pink. Pods sessile, pubescent or densely hairy, upper margin slightly and lower deeply indented.

Reproductive Cycle : August-November.

Material Examined : HPU-EL-SKS-6369; C.T.O. (2,228m), Kanlog Cemetery (1,990m).

Propagation Notes : Propagated naturally by seeds. Plant requires a sunny sheltered position.

Part Used : Leaves.

Folk Uses : Leaves used as **fodder**; its juice applied externally to cure **tonsils**.

Active Constituents : Active constituents are wax ester, hexacosanyl eicosanoate, four new coumaronochromones-desmoxyphyllins A and B and their 7-O-glucopyranosides, alcohols and n-aliphatic (**leaves**).

Uses in Literature : Known earlier as carminative, diuretic, tonic, fibres, fodder, fuel, ropes; and for biliousness complaints and epilepsy **(Agarwal, 2003; Ambasta, 1986; Bean, 1981; Chaudhuri, 1993; Kapur & Nanda, 1996; Kapur & Srivastava, 1996; Kirtikar & Basu, 1984; Manandhar, 2002; Nautiyal *et al.*, 2003; Pande *et al.*, 2006; Phillips & Rix, 1998; Rastogi & Mehrotra, 1995b; Samant *et al.*, 1996; Singh & Kumar, 2000a; Thomas, 1992)**.

Desmodium multiflorum DC.

Syn.: *D. angulatum* DC.; *D. dubium* Lindl.; *D. floribundum* (D. Don) Sweet; *D. nepalense* H. Ohashi; *D. sambuense* (D. Don) Walp.; *Hedysarum sambuense* D. Don; *Meibomia floribunda* (D. Don) Kuntze.

Family: Fabaceae.

Vern.: Martoi.

Distribution : Himalayas, from Kashmir-Bhutan and Assam: 1,800-2,600m.

Description : Large erect shrubs with hairy, branched stem and shortly angular twigs. Leaves trifoliate, hairy on both sides. Flowers pink or purple, in numerous axillary and terminal racemes. Pods sessile, hairy.

Reproductive Cycle : June-October.

Materials Examined : HPU-EL-SKS-6370; Bemloe (2,014m), Boileauganj Cemetery (1,998m), Chhota Shimla (2,070m).

Propagation Notes : Propagated by seeds.

Parts Used : Stem. Leaves. Wood.

Folk Uses : Leaves used as **fodder** for cattle. Wood used as **fuel** and stem for **scouring teeth**.

Use in Literature : Reported to be used against acidity **(Chaudhuri, 1993; Farooq, 2005; Manandhar, 2003)**.

Diospyros kaki Thunb.

Syn.: D. *amara* Perrier; *D. argyi* H.Lev.; *D. bertii* Andre; *D. costata* Carriere; *D. roxburghii* Carriere; *D. schi-tse* Bunge; *D. sinensis* Naudin; *D. sphenophylla* Hiern; *Embryopteris kaki* G. Don.

Family: Ebenaceae.

Vern.: Japani-phal.

English and Hindi Names:

Eng. : Japanese Persimmon, Kaki Persimmon;
Hindi : Halwa Tendu.

Distribution : Planted; Western & N. India.

Description : Small deciduous trees possessing alternate, simple, ovate glossy-green leaves and cream coloured (female) and pink-tinged (male) flowers. Fruit spherical to squarish, orange red.

Reproductive Cycle : April - July.

Material Examined : HPU-EL-SKS-6371; Annandale (1,860m).

Propagation Notes : Plant succeeds in most soils. Propagated by grafting (whip and tongue method) on seedlings of *Diospyros lotus* and *D. viriginiana*, budding, root cutting and through seeds.

Parts Used : Fruit. Wood.

Folk Uses : Fruits **edible**. Wood used for **ornamental purposes**.

Active Constituents : Active constituents are delphinidin, cyanidin, lycopene, catechin-3-gallate, gallocatechin and gallocatechin-3-gallate (**fruit**).

Uses in Literature : Known to be useful as an anthelmintic, antivinous, appetizer, astringent, edible (fruit), expectorant, febrifuge, laxative; and for boxes, dyeing, ornamental works and tannin (raw fruits) **(Agarwal, 2003; Ahmed & Borthakur, 2005; Ambasta, 1986; Anonymous, 1948-1976; Arora, 1997; Badhwar & Fernandez, 2011; Bean, 1981; Bennet *et al.*, 1991; Chittendon, 1956; Chopra *et al.*, 1986; Dwivedi, 2007; Farooq, 2005; Jain, 1991; Sheat, 1948; Uphof, 1968; Usher, 1974; Watt, 1889-1896).**

Diospyros lotus L.

Syn.: *Dactylus trapezuntinus* Forssk.; *Diospyros calycina* Dippel; *D. microcarpa* Siebold; *D. umlovok* Griff.

Family: Ebenaceae.

Vern.: Barain.

English and Hindi Names:

Eng. : Dateplum Persimmon;
Hindi : Amlok.

Distribution : N.W. Himalayas particularly in forests of Jammu & Kashmir.

Description : Deciduous trees with rough bark, alternate, acuminate leaves and subsessile male flowers, in cluster of 2-3. Calyx lobes ciliate. Female flowers nearly sessile, solitary. Staminodes 8, hairy. Fruits globose or ovoid, purple in colour.

Reproductive Cycle : April-August.

Material Examined : HPU-EL-SKS-6374; Ridge (2,250m).

Propagation Notes : Propagated through seeds and cuttings of half-ripe wood in July/August in a frame, and layering in spring. Plant requires a good deep loamy soil in sun or light shade.

Parts Used : Whole Plant. Fruit.

Folk Uses : Planted as an **ornamental** tree. Fruits **edible**; and its juice used for making '**sherbet**' (drink).

Active Constituents : Active constituents are lupeol, betulin, betulinic acid, 7 - methyljuglone, rotundiquinone and neodiospyrin (**fruit**).

Uses in Literature : Reported to be used as edible (fruit), febrifuge, sacred, sedative, syrups; and for preparing domestic items **(Agarwal, 2003; Ambasta, 1986; Anonymous, 1948-1976; Bennet *et al.*, 1991; Chaudhuri, 1993; Chittendon, 1956; Chopra *et al.*, 1986; Dwivedi, 2007; Huxley, 1992; Kirtikar & Basu, 1984; Rastogi & Mehrotra, 1995a; Sheat, 1948; Sood *et al.*, 2013; Uphof, 1968; Usher, 1974).**

Drepanostachyum falcatum **(Nees) Keng f.**

Syn.: *Arundinaria falcata* Nees; *A. gracilis* (Riviere & C. Riviere) Blanch.; *A. interrupta* Trin; A. *utilis* Cleghorn; *Bambusa gracilis* Riviere & C. Riviere; *Fargesia falcata* (Nees) Nakai; *Pleioblastus falcatus* (Nees) T.Q. Nguyen; *Sinarundinaria falcata* (Nees) C.S. Chao & Renvoize.

Family: Poaceae.

Vern.: Nagali.

English and Hindi Names:

Eng. : Himalayan Bamboo;
Hindi : Go-ningal, Gurwa, Spiag, Spikso.

Distribution : W. Himalayas: 1,200-2,100m.

Description : Shrubs possessing very slender culm and oblong-lanceolate, striate leaves having glabrous sheath. Flower racemes falcate, fascicled at the node, occurring on leafless stem. Glumes ribbed.

Reproductive Cycle : May-November; flowering sporadically.

Materials Examined : HPU-EL-SKS-6375; Advance Study (2,128m), Armsdell (2,102m), Bemloe (2,012m), Boileauganj Cemetery (1,998m), Kanlog Cemetery (1,990m).

Propagation Notes : Propagated by seeds as well as culms in July- August.

Parts Used : Whloe Plant. Branches. Stem. Leaves.

Folk Uses : Grown as a **hedge** plant. Peeled young stem **cooked** as a pot herb. Fresh branches used to make **welcome arches** in marriages. Culms used for making **baskets, mats, brooms, pen ('kalam'), cattle shed, agricultural implements, walking sticks** and **field's umbrella**; and as good **fodder** for animals.

Uses in Literature : Used earlier as base material for cattle shed, house, and fibre, fishing rods, fodder, hookahs; and for agricultural implements, baskets, brooms, containers for compost fertilizer, grains, and roofing **(Ambasta, 1986; Anonymous, 1948-1976; Dwivedi, 2007; Farooq, 2005; Negi *et al.*, 1993; Pande *et al.*, 2006; Raju, 2000; Singh & Kumar, 2000a; Uphof, 1968; Watt, 1889-1896).**

Duranta erecta L.

Syn.: *Darbyana integrifolia* A.M. Murr.; *Duranta angustifolia* Salisb.; *D. dentata* Pers; *D. ellisiae* Jacq.; *D. inermis* L.; *D. latifolia* Salisb.; *D. microphylla* Willd.; *D. plumieri* Jacq.; *D. repens* L.

Family: Verbenaceae.

Vern.: Neel Kanthi.

English, Hindi and Sanskrit Names:

Eng.	:	Creeping Skin Flower, Golden Dew Drop, Pigeon Berry, Sky Flower;
Hindi	:	Daranta;
Sans.	:	Vriti Kanta.

Distribution : Native of C. America (Mexico and Brazil) and W. Indies, W. Himalayas: 200-1,600m.

Description : Evergreen, more or less thorny perennial shrubs up to 6m in height. Stem erect, woody, branched, green, cylindrical and solid. Leaves simple, opposite decussate, broadly elliptic, entire. Spines stout. Flowers small, bright, lavender-blue with pale center, in 10-15 cm long axillary or terminal racemes on gracefully drooping branches. Fruits golden-yellow, globular.

Reproductive Cycle : April-August.

Material Examined : HPU-EL-SKS-6379; B.C.S. (2,025m).

Propagation Notes : Propagated readily from cuttings in July-August and by seeds. Plant thrives in moist but well drained moderately fertile soil in full sun.

Parts Used : Whole Plant. Wood.

Folk Uses : Cultivated as an impenetrable **ornamental hedge** known for its **soothing** effects in summer. Stem used as **fuel-wood**.

Active Constituents : Active constituent are β-sitosterol, α-amyrin, scuttellaein, 4-methoxy-scutellarein, scutellarein, scutellaerin-7-O-rhamnoside, pectolinarigenin and its 7-O-rutinoside (**leaves**).

Uses in Literature : Described earlier as insecticidal, poisonous, and for wounds, sore throat, turnery and hedge **(Agarwal, 2003; Ahmed & Borthakur, 2005; Ambasta, 1986; Anonymous, 1948-1976; Devi & Singh, 2008; Diwakar & Ansari, 2003; Kaur & Sharma, 2004; Kshirsagar & Singh, 2007; Paliwal, 2009; Pande *et al.*, 2006; Rajendran, 2003; Rajendran & Sikarwar, 2003; Singh & Pandey, 1998; Sood *et al.*, 2013; Srivastava; 1989)**.

Elaeagnus umbellata Thunb.

Syn.: *E. crispa* Thunb.; *E. padifolia* K. Koch; *E. praematura* (Koidz.) Araki.

Family: Elaeagnaceae.

Vern.: Ghain.

Distribution : Afghanistan, S.W. China, from 900-3,100m.

Description : Thorny shrubs with shoots and young branches clothed with silvery scale. Leaves simple, entire, alternate, shortly - stalked, oblong-lanceolate, obtuse, glabrous or stellately hairy on the upper surface and silvery-scaly on the lower. Flowers fragrant, white, stalked, in axillary clusters appearing with the leaves. Nuts ovoid, bony, ribbed, covered inside with white hair.

Reproductive Cycle : May-October.

Materials Examined : HPU-EL-SKS-6381; Armsdell (2,102m), Boileauganj Cemetery (1,998m), Chhota Shimla (2,080m), Kanlog Cemetery (1,986m), Raj Bhawan (2,132m).

Propagation Notes : Propagated by seeds, cutting, grafting and layering.

Parts Used : Whole Plant. Fruit. Wood.

Folk Uses : Grown as protective **hedge** around fields, building, houses and gardens. Ripe fruits **eaten** as well as **pickled**. Wood used as a **fuel**.

Uses in Literature : Recorded as an edible (fruits), astringent, diuretic, fuel, haemostatic, hedge, stimulant; and for bronchitis, cardiac disorders, cough, curries, dysentery, pickling, and pulmonary affections **(Agarwal, 2003; Ambasta, 1986; Anonymous, 1948-1976; Arora, 1997; Arora & Pandey, 1996; Badhwar & Fernandez, 2011; Bennet *et al.*, 1991; Das & Agarwal, 1991; Farooq, 2005; Guha Bakshi *et al.*, 1999; Jain, 1991; Kapur & Nanda, 1996; Kapur & Singh, 1996; Kirtikar & Basu, 1984; Manandhar, 1995; Nautiyal *et al.*, 2003; Negi & Gaur, 1991; Negi *et al.*, 1993; Pande *et al.*, 2006; Parmar & Kaushal, 1982; Roy *et al.*, 1998; Sharma & Rana, 2005; Sharma *et al.*, 2000; Singh, 1996, 2000a; Singh & Kumar, 2000a; Singh *et al.*, 1996a; Sinha, 1996; Sood & Thakur, 2004; Sood *et al.*, 2009b, 2010; Uphof, 1968; Watt, 1889-1896).**

Eriobotrya japonica (Thunb.) Lindl

Syn.: *Crataegus bibas* Lour.; *Mespilus japonica* Thunb.

Family: Rosaceae.

Vern.: Lokat, Lugat, Lukat.

English and Hindi Names:

Eng. : Chinese Medlar, Japan Medlar, Japan Quince, Loquat;

Hindi : Logat, Lokat.

Distribution : Native of China and Japan; cultivated in N. India.

Description : Small evergreen trees with densely woolly inflorescence. Leaves oblong-elliptic, toothed, green above and rusty tomentose beneath. Flowers white, in terminal panicles and sweet scented. Fruit pyriform, yellow to deep orange.

Reproductive Cycle : July-September.

Materials Examined : HPU-EL-SKS-6385; Armsdell (2,102m), Oak Over (2,154m), Spring Field (2,075m).

Propagation Notes : Propagated by seeds, cuttings of half-ripe wood in July/August and air layering.

Parts Used : Fruit. Leaves.

Folk Uses : Fruits eaten **raw** or **cooked** or **preserved**. Due to acidic nature these are also cooked in **pies**, **sauces**, **jellies** etc. 10-20ml juice of leaves given twice daily till cure for **diarrhoea**.

Active Constituents : These are leucocysnin - loquatoside along with fructose, glucose sucrose and β-carotene, cryptoxanthin (**fruit pulp**), a new sesquiterpene glycoside – loquatifolin A - characterized as α-L-rhamnopyranosyl (1→4) – α – L - rhamnopyranosyl (1→2) - [α-L-rhamnopyranosyl (1→6)]-β-D-glucopyranosyl-6,7-trans-neloridol, sorbitol-6-phosphate dehydrogenase and quercetin-3-O-galactoside (**leaves**).

Uses in Literature : Known earlier as analgesic, astringent, antibacterial, antiemetic, antispasmodic, antitussive, diuretic, expectorant, sedative; and for edible purposes (fruits) and in drinks, jellies, salads **(Agarwal, 2003; Ahmed & Borthakur, 2005; Ambasta, 1986; Anonymous, 1948-1976; Asolkar *et al.*, 1992; Bean, 1981; Chopra *et al.*, 1986; Davis, 1990; Farooq, 2005; Harborne & Baxter, 2001; Huxley, 1992; Kirtikar & Basu, 1984; Pande *et al.*, 2006; Parrotta, 2001; Perry & Hay, 1982; Prajapati *et al.*, 2006; Sheat, 1948; Sood *et al.*, 2013; Srivastava, 1989; Uphof, 1968; Warman, 1999; Watt, 1882-1896).**

Erythrina arborescens Roxb.

Syn.: *Corallodendron arborescens* (Roxb.) Kuntze; *E. tienensis* Wang & T. Tang.

Family: Fabaceae.

Vern.: Pariyara.

English Name:

Eng. : Coral Tree.

Distribution : Outer Himalayas from Kumaon - Sikkim and Khasia Hills: 1,500-3,000m.

Description : Trees with a few prickles on stem and leaf stacks and long erect spike-like cluster of drooping vivid scarlet flowers with large trifoliar leaves. Leaflets broad and long, brown-velvety when young, becoming hairless. Pods lanceolate, curved, brown-hairy.

Reproductive Cycle : August-October.

Materials Examined : HPU-EL-SKS-6387; Kanlog Cemetery (1,994m), Yates Place (2,208m).

Propagation Notes : Propagated by seeds and stem-cutting in July-August.

Part Used : Whole Plant.

Folk Uses : Grown as an **ornamental hedge**. Wood used in **minor construction works** and as **fuel**.

Active Constituents : Active constituents are erythrabine, erysotramidine, erysodine, erysotrine, erythratidine, N-nororientaline, erybidine (**leaves**), sitosterol, stigmasterol and campesterol (**bark**).

Uses in Literature : Reported to be used as anthelmintic, hedge, toothbrush; and for dislocation of bone, eye diseases (vet.), fever, intestinal worms, rheumatism, skin diseases, toothache, chewing (bark) **(Agarwal, 2003; Ahmed & Borthakur, 2005; Ambasta, 1986; Anonymous, 1948-1976; Bhattacharjee, 2001; Chaudhuri, 1993; Cowan, 1950; Manandhar, 2003; Mishra & Chandra, 1998; Pande *et al.*, 2006; Rastogi & Mehrotra, 1991)**.

Eucalyptus globulus Labill.

Syn.: *E. maidenii* subsp. *globulus* (Labill.) J.B. Kirkp.

Family: Myrtaceae.

Vern.: Eucalyptus, Safeda.

English, Hindi and Sanskrit Names:

Eng. : Blue Gum Eucalypt;
Hindi : Yukeliptas;
Sans. : Nilaniryasa, Tailaparnah.

Distribution : Planted extensively in Nilgiris, Annamalai and Palni hills in S. India, Shimla hills and Shillong.

Description : Large-sized trees having a clean straight bole, smooth trunk, easily detachable bark and oblong-ovate, slightly falcate leaves which are unequal at the base. Flowers large, white, 1-3 together in axils; operculum thick and warty with numerous exserted stamens. Capsule hard and dehisces longitudinally at the mouth. Seeds very small.

Reproductive Cycle : February-October.

Materials Examined : HPU-EL-SKS-6389; Boileauganj Cemetery (1,998m), St. Edward's School (2,060m).

Propagation Notes : Plant succeeds in a moderately fertile well-drained moisture retentive circum-neutral soil. Propagated by seeds.

Parts Used : Whole Plant. Leaves. Wood.

Folk Uses : Grown as an **avenue** plant. Infusion of fresh leaves given internally for **aphthous ulceration** in mouth and throat of children; also inhalation of fumes emnating from hot infusion good againt **respiratory affections**. Wood used in **construction** and as **fuel**.

Active Constituents : Found to contain aromadendrene, λ-cadinene, 1,8-cineole, α-gurjunene, globulol, linalool oxide, eremorphilene, β-pinene, piperitone, α-, β- and λ-terpinene, terpinen-4-ol, alloaromadendrene (**leaves, shoot**).

Uses in Literature : Known so far as anodyne, anthelmintic, antibacterial, antipyretic, antiseptic, astringent, cardiotonic, carminative, deodorant, digestive, diuretic, expectorant, essential oil, fuel, insect repellent, rubefacient, stimulant, sudorific, avenue plant; and for asthma, bronchitis, burns, cephalalgia, chronic cough, chronic cystitis, diarrhoea, dyspepsia, fever, flatulence, gum, halitosis, headache, hoarseness, pharyngodynia, pneumonosis, pyorrhoea, rheumatism, skin diseases, strangary, throat infections, tuberculosis, vomiting, ulcers, whooping cough **(Agarwal, 2003; Ambasta, 1986; Anonymous, 1948-1976; Baburaj *et al.*, 2000; Bean, 1981; Bhattacharjee, 2001; Chatterjee & Pakrashi, 1997; Dey & Bahadur, 1973; Dwivedi, 2007; Grae, 1974; Harborne & Baxter, 2001; Hay *et al.*, 1983; Huxley, 1992; Jaeger, 1961; Kayal *et al.*, 2003; Khare, 2004; Kumar *et al.*, 2000; Nautiyal *et al.*, 1997; Pande *et al.*, 2006; Pool, 1941; Prajapati *et al.*, 2006; Randhawa, 1960; Sood *et al.*, 2010; Uphof, 1968; Usher, 1974; Warman, 1999)**.

Euonymus fimbriatus Wall.

Syn.: *E. lacerus* Buch.-Ham. *ex* D. Don; *E. micranthus* D. Don.

Family: Celastraceae.

Vern.: Euonymus.

Distribution : E. Asia-Himalayas: 1,700-2,100m.

Description : Deciduous trees with sturdy branches and twig. Leaf blade leathery, elliptic-ovate, base semi-rotund, margin sharply double-serrulate, apex acuminate. Peduncle from new branches only, slender, dichotomously branched with several to many greenish-yellow flowers. Capsule pink -red when fresh and brown to greenish when dry; nearly compressed globose with wings.

Reproductive Cycle : April-June.

Materials Examined : HPU-EL-SKS-6391; Advance Study (2,128m), Sanitarium (2,060m).

Propagation Notes : Propagated through seeds and cuttings of 5 - 8 cm long half-ripe wood taken at a node or with a heel in July/August. Plant thrives in almost any soil, including chalk.

Parts Used : Seeds. Wood.

Folk Uses : Seeds used as **beads**. Wood employed for minor **construction** works and **carving**.

Uses in Literature : Recorded earlier for minor construction and carving **(Bean, 1981; Chittendon, 1956; Huxley, 1992; Sheat, 1948)**.

Euonymus japonicus Thunb.

Syn.: *Cassine fortune* (Turcz.) Kuntze; *Elaeodendron fortune* Turcz.; *Euonymus kewensis* Hesse; *E. repens* Carriere.

Family: Celastraceae.

Vern.: Japani Euonymus.

English Name:

Eng. : Japanese Spindle Tree.

Distribution : E. Asia - Japan. Locally naturalized in S. Europe.

Description : Densely leafy, upright evergreen shrubs with broadly oval leaves which are wedge-shaped at base and obtusely serrate, shining above. Flowers greenish-white, in stout - peduncled cymes. Capsules globose with orange aril.

Reproductive Cycle : May-August.

Materials Examined : HPU-EL-SKS-6393; Advance Study (2,128m), Raj Bhawan (2,132m), Gorton Castle (2,094m), Grand Hotel (2,105m), Walsingham (2,266m).

Propagation Notes : Propagation by seeds, air-layering during February-March and July-August and cuttings of 5 - 7cm long half-ripe wood taken at a node or with a heel in July/August are the useful methods. Plant particularly suited to dry shaded areas.

Part Used : Whole Plant.

Folk Use : Grown as a **hedge** plant.

Uses in Literature : Used earlier as antirheumatic, diuretic, hedge plant, non elastic rubber, tonic; and for mensturation problems **(Bailey, 1957; Bean, 1981; Beri, 1992; Chittendon, 1956; Graf, 1981; Huxley, 1992; Paliwal, 2009; Phillips & Rix, 1998; Randhawa, 1960; Thomas, 1992; Uphof, 1968; Usher, 1974)**.

Euonymus lucidus D. Don

Syn.: *E. pendulus* Wall. *ex* M.A. Lawson.
Family: Celastraceae.

Vern.: Bharmela.

Hindi and Sanskrit Names:

Hindi : Bhillotak, Kumkum;
Sans. : Bhillotak.

Distribution : Himalayas from Kashmir - Bhutan and N. E. India: 1,800-2,600m.

Description : A middle-sized tree having shining reddish-brown young shoots and sharply serrated leaves. Cymes frequently arise from the axil of bracts below the leaves. Peduncles long, branches short. Fruits 3- 4 lobed, angles often winged.

Reproductive Cycle : April-August.

Material Examined : HPU-EL-SKS-6395; Advance Study (2,128m).

Propagation Notes : Plant suited to dry shaded areas. Propagated through seeds and by cuttings of half-ripe wood taken at a node in July/August.

Part Used : Whole Plant.

Folk Use : Grown as an **ornamental** plant.

Active Constituents : Main constituents are isohexacosane, cerotic acid, dulcitol, friedelin, epifriedelinol and β-sitosterol (**twigs**).

Uses in Literature : Described in literature for constipation, dyspepsia, eye diseases, tannin, carving and turney work; and as dye, sacred **(Agarwal, 2003; Ambasta, 1986; Asolkar *et al.*, 1992; Dhiman, 2003; Pande *et al.*, 2006; Rastogi & Mehrotra, 1991; Singh & Kumar, 2000a; Uphof, 1968)**.

Euonymus tingens Wall.

Family: Celastraceae.

Vern.: Marmakoul.

English, Hindi and Sanskrit Names:

Eng. : Spindle Wood;
Hindi : Bar-phali, Bhillotak, Kumkum, Kungku;
Sans. : Bhillotak.

Distribution : Temperate Himalayas (H.P.-Nepal): 2,100-3,350m.

Description : Evergreen shrubs or small trees with dark ash-grey corky bark, dull green, leathery, wrinkled, closely toothed, elliptic - ovate-lanceolate leaves and flowers in long-stalked axillary clusters. Flowers yellowish-white and strongly veined with purple tinge. Fruits top-shaped, 3-5 angled, not winged. Seeds half encircled in orange aril.

Reproductive Cycle : April-August.

Materials Examined : HPU-EL-SKS-6396; Cedar Circuit House (2,156m), Eagle Mount (2,152m), Gorton Castle (2,094m), Holly Lodge (2,278m), Sanitarium (2,060m).

Propagation Notes : Propagated by seeds for which cold stratification is required before germination. Plant thrives in almost any soil, including chalk.

Parts Used : Wood. Bark

Folk Uses : Juice of bark added as drops against **eye affections**. Also, its 15-20 ml decoction taken before bedtime to check **constipation**. Wood used in minor **construction** and as **fuel**.

Active Constituents : Active constituents are tingenone and 20-hydroxytingenone (**bark**).

Uses in Literature : Recorded so far as dye, purgative, fuelwood; and for antitumours, constipation, cosmetics (fruit), dyspepsia, dysentery, eye diseases, headache, and carving **(Agarwal, 2003; Ambasta, 1986; Anonymous, 1948-1976; Asolkar *et al.*, 1992; Bean, 1981; Chopra *et al.*, 1986; Farooq, 2005; Huxley, 1992; Kirtikar & Basu, 1984; Manandhar, 2002; Pande *et al.*, 2006; Rastogi & Mehrotra, 1991; Sharma, 1996; Uphof, 1968; Usher, 1974; Watt, 1889-1896)**.

Fagus sylvatica L.

Syn.: *Castanea fagus* Scop.; *F. aenea* Dum. Cours.; *F. cochleata* (Dippel) Domin; *F. crispa* Dippel; *F. cristata* Dum. Cours.; *F. cuprea* Hurter *ex* A.DC.; *F. salicifolia* A.DC.; *F. variegata* A.DC.

Family: Fagaceae.

Vern.: Rang Badalane Bala Plant.

English Names:

Eng. : Beech, European Beech.

Distribution : Native of Europe, Asia minor and C. Asia.

Description : Tall deciduous tree with a smooth, grey, straight trunk and male and female flowers (globular) separate on the same tree. Nuts enclosed in a prickly case.

Reproductive Cycle : April-October.

Material Examined : HPU-EL-SKS-6398; Kanlog Cemetery (1,992m).

Propagation Notes : Plant ill-adapted for a heavy wet soil. Propagated by seeds.

Parts Used : Whole Plant. Wood.

Folk Uses : Grown as an **ornamental** plant. Wood employed for **construction**, **furniture** and **agricultural implements**.

Active Constituents : Active constituents are squalene, cyclortenyl acetate, β-amyrin acetate, acetylmethyl betulinate, sitosterol, stigmasterol and dihydrisitosterol (**wood**).

Uses in Literature : Recorded in India for agricultural implements, furniture, ship building, soap-making, tools, turned articles, wood pulp, wooden shoes; and as poisonous to human being and cattle (seeds) and substitute of tobacco (leaves) **(Ambasta, 1986; Anonymous, 1948-1976; Asolkar *et al.*, 1992; Bhattacharjee, 2001; Hadfield, 1956; Jaeger, 1961; Pool, 1941; Seth, 2004; Uphof, 1968)**.

Ficus auriculata Lour.

Syn.: *Covellia macrophylla* Miq.; *F. roxburghii* Steud.; *F. oligodon* Miq.; *F. pomifera* Wall. *ex* King; *F. regia* Miq.; *F. sclerocarpa* Griff.

Family: Moraceae.

Vern.: Timbal, Traimbal.

Hindi Names:

Hindi : Timla, Tirmal.

Distribution : Outer Himalayas, (Kashmir-Bhutan), ascending to 1,800m.

Description : Medium - low height trees possessing spreading crown, large, stalked and broadly ovate leaves and stalked figs clustered on short, thick leafless branches. Figs red or dull purple when ripe.

Reproductive Cycle : April-August.

Material Examined : HPU-EL-SKS-6399; Cecil Hotel (2,090m), Kanlog Cemetery (2,002m).

Propagation Notes : Propagated by seeds.

Parts Used : Leaves. Fruits. Bark.

Folk Uses : Plant used as **fodder**. Fruits **edible**; unripe ones cooked as **vegetable**. Bark paste applied for suppurating **boils**.

Uses in Literature : Recorded earlier as edible (fruits also used in curries, jams) fodder, galactagogue (vet.), vegetable; and for diabetes, dysentery, eye diseases, high blood pressure, hyperacidity, indigestion, breast ulcers, delivery (vet.), leucoderma, meal plates for feasts in villages, ring worm, stomachache and ropes (bark) **(Agarwal, 2003; Ambasta, 1986; Anonymous, 1948-1976; Arora & Pandey, 1996; Badhwar & Fernandez, 2011; Bennet *et al.*, 1991; Borthakur *et al.*, 1999; Deokata & Chhetri, 2007; Dwivedi, 2007; Jain, 1991; Pande *et al.*, 2006; Parmar & Kaushal, 1982; Rama Rao & Henry, 1996; Rana *et al.*, 1996; Roy *et al.*, 1998; Sharma, 1996; Singh & Kumar, 2000a; Sood *et al.*, 2012b, 2013; Watt, 1889-1896).**

***Ficus benjamina* L.**

Syn.: *F. comosa* Roxb.; *F. haematocarpa* Blume *ex* Decne.; *F. lucida* Aiton; *F. nitida* Thunb.; *F. nuda* (Miq.) Miq.; *F. pyrifolia* Salisb; *F. umbrina* Elmer; *F. xavieri* Merr.; *Urostigma nudum* Miq.

Family: Moraceae.

Vern.: Kabra.

English Names:

Eng. : Java Fig, Java Willow, Weeping Fig.

Distribution : Sub-Himalayan tract and Peninsular India; planted along roadsides and in gardens.

Description : Glabrous trees with drooping branches. Leaves small, shining, moderately thick, oval-ovate-elliptic, abruptly narrowed into a short point, many-nerved and with entire margins. Figs in sessile axillary pairs, globular or ovoid, becoming dark red.

Reproductive Cycle : February-June.

Material Examined : HPU-EL-SKS-6403; Annandale (1,860m).

Propagation Notes : From woody stem cutting and air layering.

Part Used : Whole Plant.

Folk Use : Grown as an **ornamental** plant.

Uses in Literature : Reported to be used as edible, street tree or hedge, poisonous, diuretic; and for eye problems, leprosy, liver complaints, lumbago, pain, rheumatism, scabies, ulcers **(Abraham, 1997; Agarwal, 2003; Ambasta, 1986; Ahmed & Borthakur, 2005; Bailey, 1957; Benthall, 1946; Caius, 1986; Chaudhuri, 1993; Chhetri, 2005; Hay *et al.*, 1983; Kirtikar & Basu, 1984; Mccann, 1950; Paliwal, 2009; Parrotta, 2001; Tigga *et al.*, 2003; Warman, 1999; Watt, 1889-1896)**.

Ficus elastica Roxb.

Syn.: *F. clusiifolia Summerh.*; *Macrophthalma elastica* (Roxb. *ex* Hornem.) Gasp.; *Visiania elastica* (Roxb. *ex* Hornem.) Gasp.

Family: Moraceae.

Vern.: Rubber Plant.

English Names:

Eng. : Indian Rubber Bush, Rubber Bush, Rubber Fig, Rubber Plant.

Distribution : Indigenous in the moist evergreen forests of Assam and Myanamar; also cultivated.

Description : A large glabrous tree having very thick, deep glossy green leaves with many fine parallel side-nerves. Petioles stout, long. Fruits in pair in old axils, sessile, oblong and yellowish.

Reproductive Cycle : April-July.

Materials Examined : HPU-EL-SKS-6405; Advance Study (2,128m), Convent of Jesus & Mary (2,104m), St. Edward's School (2,060m).

Propagation Notes : Propagated from stem cuttings during rains or by air-layering.

Part Used : Whole Plant.

Folk Use : Commonly planted as **ornamental** tree.

Uses in Literature : Grown for indoor decoration; and as fibre, fodder, ornamental rubber tree **(Agarwal, 2003; Ahmed & Borthakur, 2005; Ambasta, 1986; Anonymous, 1948-1976; Bailey, 1957; Benthall, 1946;**

Caius, 1986; Chaudhuri, 1993; Cowan, 1950; Dey & Bahadur, 1973; Harborne & Baxter, 2001; Paliwal, 2009; Pool, 1941; Sood *et al.*, 2013; Uphof, 1968; Warman, 1999).

Ficus neriifolia Sm

Syn.: *F. densa* Miq.; *F. diadena* Steud.; *F. fieldingii* Miq.; *F. neriifolia* var. *nemoralis (*Wall. *ex* Miq.) Corner; *F. trilepis* Miq.; *F. wardii* C.E.C. Fisch.

Family: Moraceae.

Vern.: Dudhia.

Distribution : Outer Himalayas from Punjab – E. India upto 2,400m.

Description : Trees possessing rounded, brown stem and branches. Leaves narrow shining with prominent veins, pointed tip, small leaf stalk, and trinerved base. Figs globose, smooth, reddish when ripe, solitary or in pair.

Reproductive Cycle : April-November.

Materials Examined : HPU-EL-SKS-6407; Armsdell (2,102m), B.C.S. (2,025m).

Propagation Notes : Propagated by seeds and cutting.

Folk Uses : Planted as **hedge** plant. Fresh leaves given as forage to domestic animals to enhance **lactation.**

Uses in Literature : Used for boils; and as edible (fruit), fodder and hedge **(Ahmed & Borthakur, 2005; Ambasta, 1986; Arya *et al.*, 1999; Bennet *et al.*, 1991; Chaudhuri, 1993; Deokota & Chhetri, 2009; Dwivedi, 2007; Paliwal, 2009).**

Ficus palmata Forssk.

Syn.: *F. forskalaei* Vahl; *F. malabarica* Miq.; *F. petitiana* A. Rich.; *F. pseudosycamorus* Decne.

Family: Moraceae.

Vern.: Fegura, Phegura.

Hindi Names:

Hindi : Anjiri, Bedu, Beru, Gular, Khabara.

Distribution : N.W. India, ascending to 2,300m in the Himalayas.

Description : A low-branching, densely foliated tree with broadly ovate, serrated leaves which are toothed towards the top. Figs axillary, stalked, solitary, pear-shaped and purple coloured. Male flowers numerous in the upper part of receptacles.

Reproductive Cycle : April-September.

Material Examined : HPU-EL-SKS-6409; Boileauganj Cemetery (1,998m), Cecil Hotel (2,090m), Cedar Circuit House (2,154m), Chhota Shimla (2,078m), Pari Mahal (2,105m).

Propagation Notes : Requires a well-drained medium to light loam with some lime rubble in the soil. Usually, propagated by seeds and cuttings of 10 - 12cm long mature wood with a heel in winter.

Parts Used : Leaves. Fruit. Latex.

Folk Uses : Ripe fruits **edible**, unripe ones and tenders **cooked** and used for checking **cough** and **constipation**. Latex from leaves used to **exude spine** from skin. Leaves used as **fodder**.

Active Constituents : β-sitosterol and bergapten isolated from **leaves**.

Uses in Literature : Recorded in India as edible (fruits, leaves), cattle fodder, demulcent, jam, laxative, squash; and for boils, bladder, building purposes, diabetes, diseases of lungs, dysentery, high blood pressure, indigestion, stomachache, pimples, and to draw out thorns **(Agarwal, 2003; Ambasta, 1986; Anonymous, 1948-1976; Ansari, 1991; Arora & Pandey, 1996; Aswal, 1996; Badhwar & Fernandez, 2011; Bennet *et al.*, 1991; Chopra *et al.*, 1956; Das & Agarwal, 1991; Farooq, 2005; Jain, 1991; Kapur & Nanda, 1996; Kirtikar & Basu, 1984; Lal *et al.*, 1996; Murthy *et al.*, 2003; Negi *et al.*, 1993; Pande *et al.*, 2006; Parmar & Kaushal, 1982; Rana *et al.*, 1996; Roy *et al.*, 1998; Saini, 1996a; Samant *et al.*, 2001; Singh, 1996; Singh & Kumar, 2000a; Singh & Pandey, 1998; Singh *et al.*, 2003c; Sinha, 1996; Sood *et al.*, 2012b, 2013; Warman, 1999; Watt, 1889-1896)**.

Ficus religiosa L.

Syn.: *F. caudata* Stokes; *F. peepul* Griff.; *F. rhynchophylla* Steud.; *Urostigma religiosum* Gasp.

Family: Moraceae.

Vern.: Peepal, Pipal.

English, Hindi and Sanskrit Names:

Eng. : The Peepal Tree;
Hindi : Pipal;
Sans. : Asvattha, Aswaththamu, Chaityavriksha, Pavitraka, Vriksharaja, Yajnika.

Distribution : Wild in the Sub-Himalayan forests, Bengal and C. India; planted elsewhere.

Description : Large-sized glabrous trees with long-petioled leaves which are alternate, coriaceous, ovate, pendulous, shining above and dry beneath. Receptacles dark purple. Figs sessile, in axillary pair, depressed-globose, black or purple at maturity. Male flowers few, sessile, only near the mouth of some recepatacles. Female ones sessile or pedicellate.

Reproductive Cycle : March- October.

Material Examined : HPU-EL-SKS-6411; Peterhoff (2,098m).

Propagation Notes : Propagated by seeds.

Parts Used : Whole Plant. Branches. Latex. Bark. Fruit.

Folk Uses : Planted as **ornamental** tree. Plant considered **sacred** and **worshipped** as incarnation of Lord 'Brahma'. Dried branches used in '**havans**'. Ash of burnt branches applied on tongue to heal **sores**. One tsp paste of its ash and that of *Ficus palmata* taken with honey every morning and at bedtime to relieve **cough**. Latex used against **toothache**. Paste of bark applied on **scorpion bite**. Fruits **edible**.

Uses in Literature : Known earlier as an edible (fruits and tender leaf buds eaten in times of scarcity), abortifacient, absorbent in inflammatory swelling, alterative, antidote to snake bite, astringent, colouration material and fixative, contraceptive, fodder, laxative, psychoactive, sacred (worshipped- cutting tree more sinful than killing a saint), tonic, tree of fertility, vermifuge (vet.); and for anaemia, after-pain (postnatal), aphthae, asthma, bird -lime, blisters, boils, bowels, bronchitis (vet.), burns, carbuncle, cholera, chronic liver disorders, constipation (vet.), cough (vet.), dyeing, dysentery, eczema, enhancing memory, fever, fibre, fistula, foot and mouth disease (vet.), gonorrhoea, gum, heamaturia, match-boxes, obstinate hiccups, over intoxication by opium, packing cases, paralysis, rheumatism, scabies, skin affections, small pox, sores in mouth, spermatorrhoea, spoons, tanning, throat infection, toothache, ulcer, urinary complaint, whooping cough, and wounds (vet.) **(Ambasta, 1986; Anonymous, 1948-1976; Arora & Pandey, 1996; Bhatt *et al.*, 2001; Bhattacharya *et al.*, 2005; Beg *et al.*, 2005; Bennet *et al.*, 1991; Borthakur, 1996a; Borthakur *et al.*, 1999; Chatterjee & Pakrashi, 1997; Chopra *et al.*, 1956; Cowan, 1950; Das & Agarwal, 1991; Dhiman, 2003; D'Rozario *et al.*, 1999; Dwivedi, 2003; Farooq, 2005; Ghate, 1998; Gogoi & Borthakur, 1991; Gogoi *et al.*, 2003; Hosagoudar & Henry, 1996b; Jain, 1991; Jain *et al.*, 1991, 1994; Jeyapardha *et al.*, 2011; Jha & Basak, 1994; Khanna, 2002; Khanna *et al.*, 1996a,c; Khare, 2004; Kirtikar & Basu, 1984; Kumar, 2003;**

Lata *et al.*, 2008; Masih, 2003; Mccann, 1950; Mishra & Chandra, 1998; Mishra *et al.*, 1996; Mitaliya & Bhatt, 2003; Molla & Roy, 1985; Nayak *et al.*, 2003; Pande *et al.*, 2006; Panthi & Chaudhary, 2003; Prajapati *et al.*, 2006; Rama Rao & Henry, 1996; Reddi *et al.*, 2005; Roy *et al.*, 1998; Sahni, 1998; Saini, 1996a,b; Samant *et al.*, 2001; Sathapathy & Brahmam, 1996; Sharma, 1996; Shenghi, 1991; Sikarwar, 1996; Sikarwar & Kaushik, 1992; Singh & Pandey, 1998; Singh & Prakash, 1996; Singh & Zaheer, 1996; Sinha, 1996; Sinha & Sinha, 2001; Sivakumar *et al.*, 2003; Sivaperumal *et al.*, 2009; Siwakoti & Varma, 1996; Sood *et al.*, 2009b, 2012b, 2013; Srivastava *et al.*, 2003a; Thomas & Britto, 2003; Tigga *et al.*, 2003; Uphof, 1968; Virjee *et al.*, 1984; Warman, 1999; Watt, 1889-1896).

Ficus sarmentosa Buch.-Ham. *ex* Sm.

Syn.: *F. cabur* Buch.-Ham. *ex* Sm.; *F. foveolata* (Wall. *ex* Miq.) Miq.; *F. nepalensis* Spreng.; *F. oblongifolia* D. Don.

Family: Moraceae.

Vern.: Phaegdi.

Distribution : Outer Himalayas: 700-2,400m.

Description : A shrub creeping over rocks and forming mats or climbing over trees. Leaves pubescent or glabrous. Figs axillary, solitary. Achenes oblong-reniform.

Reproductive Cycle : March-July.

Materials Examined : HPU-EL-SKS-6415; Kanlog Cemetery (1,990m), Sanjauli Cemetery (2,068m).

Propagation Notes : Propagated by seeds.

Parts Used : Leaves. Fruits.

Folk Uses : Leaves used as **fodder**. Fruits **edible**.

Uses in Literature : Reported earlier for edible purposes, cold and as fodder **(Ambasta, 1986; Jain, 1991; Roy *et al.*, 1998; Sood & Thakur, 2004; Sood *et al.*, 2012b; Watt, 1889-1896)**.

Ficus subincisa Buch.-Ham. *ex* Sm.

Syn.: *F. chincha* Roxb.; *F. clavata* Wall. *ex* Miq.

Family: Moraceae.

Vern.: Ban Feguda.

Distribution : Lower hills of Himalayas: ascending upto 1,500m.

Description : Erect shrubs having leaves with a narrow long pointed, axillary, shortly-stalked figs which are pale yellow on ripening.

Reproductive Cycle : June-September.

Materials Examined : HPU-EL-SKS-6417; Bemloe (2,010m), Boileauganj Cemetery (1,998m).

Propagation Notes : Propagated naturally by seeds.

Parts Used : Leaves. Fruits.

Folk Uses : Fruits **edible** and leaves used as occasional cattle **feed**.

Active Constituents : Active constituents are two-new triterpenoids-rhoiptelenol and 3α-hydroxy-isohop-22 (29)-en-24-oic acid, lupeol, glutinol, ursolic and betulinic acids (**leaves**).

Uses in Literature : Known to be used as fodder and vegetable **(Ambasta 1986; Jain, 1991; Rastogi & Mehrotra, 1995b; Sood *et al.*, 2012b).**

Forsythia X *intermedia* Zabel

Syn.: *F.* X *densiflora* (Koehne) Koehne; *F.* X *intermedia* var. *densiflora* Koehne; *F.* X *intermedia* var. *primulina* Rehder.

Family: Oleaceae.

Vern.: Forsythia.

English Name:

Eng. : Boarder Forsythia.

Distribution : Cultivated in cooler part of India.

Description : A deciduous hybrid shrub of garden origin between *F. suspensa* x *F. viridissima*. Branches with lamellate pith. Leaves narrower oblong -ovate lanceolate, tapering from base. Flowers bright yellow, crowded at the base of branch.

Reproductive Cycle : February-April.

Material Examined : HPU-EL-SKS-6419; B.C.S. (2,025m), Raj Bhawan (2,132m), R.K.M.V. (2,256m).

Propagation Notes : Propagated through cuttings of half-ripe wood 10 – 15 cm taken at a node in July/August in a frame. Plant prefers a well drained ordinary garden soil.

Part Used : Whole Plant.

Folk Uses : Planted as an **ornamental hedge** shrub for its bright yellow flower.

Uses in Literature : Reported to be grown as ornamental and hedge plant **(Bean, 1981; Brickell, 1990; Chittendon, 1956; Dirr, 2009; Donald, 1950; Huxley, 1992; Jaeger, 1961; Sheat, 1948; Shepherd, 1974)**.

Forsythia ovata Nakai

Syn.: *Rangium ovatum* (Nakai) Ohwi.

Family: Oleaceae.

Vern.: Forsythia.

English Names:

Eng. : Early Forsythia, Korean Forsythia, Yellowbells.

Distribution : Native to Korea.

Description : Deciduous ornamental shrubs having ovate - broad-ovate, serrated or nearly entire leaves which are abruptly acuminate, truncate or broadly cuneate at base. Flowers solitary, amber-yellow. Calyx half as long as corolla tube. Capsules 2 - celled with dehiscent caps containing several winged seeds.

Reproductive Cycle : March-April.

Material Examined : HPU-EL-SKS-6423; Advance Study (2,128m).

Propagation Notes : Seeds sown in spring in a cold frame. Propagated through seeds, cuttings of half-ripe of wood 10 - 15cm taken at a node in July/ August in a frame and layering in spring or summer.

Part Used : Whole Plant.

Folk Use : Cultivated as an **ornamental** tree.

Use in Literature : Reported to be grown as ornamental tree **(Bailey, 1957; Bean, 1981; Chittendon, 1956; Donald, 1950)**.

Fuchsia X *hybrida* Voss

Family: Onagraceae.

Vern.: Doll Flower.

English Name:

Eng. : Lady's Eardrops.

Distribution : Native to S. America.

Description : Shrubby, glabrous or somewhat pubescent plant which is a hybrid derived from *F. fulgens* and *F. magellanica*. Leaves ovate, opposite, deccusate, dark green. Flowers showy, solitary, pendulous. Calyx usually crimson. Hypanthium twice or more as long as the ovary. Petals purple, shorter than calyx lobe. Fruits small, dark reddish -green, deep red or purple berry.

Reproductive Cycle : March-November.

Material Examined : HPU-EL-SKS-6429; Advance Study (2,128m), Grand Hotel (2,105m).

Propagation Notes : Propagated by seeds, stem-tip cutting and soft-wood cuttings which usually succeed at any time during the growing season. Plant succeeds in any fertile well-drained circum-neutral soil rich in humus.

Part Used : Whole Plant.

Folk Use : Cultivated as an **ornamental** shrub for its beautiful various coloured doll - shaped flowers.

Uses in Literature : Grown as potted flowering plant or in hanging baskets **(Bailey, 1957; Beri, 1992; Graf, 1980, 1981; Paliwal, 2009; Pool, 1941)**.

Fuchsia magellanica Lam.

Syn.: *F. araucana* F. Phil.; *F. chonotica* Phil.; *F. discolor* Lindl.; *F. globosa* Lindl.; *F. gracilis* Lindl.; *F. pumila* (Voss) Voss; *Nahusia coccinea* Schneevoogt.

Family: Onagraceae.

Vern.: Doll Flower, Fuschia.

English Names:

Eng. : Doll Flower, Fuchsia.

Distribution : Native of Peru, Chile; naturalized in S.W. England.

Description : Glabrous shrubs upto 2m in height. Leaves opposite or ternate, glabrous - puberulent beneath, usually serrulate. Flower in upper axils, pedicellate. Hypanthium narrow, cylindrical, and red. Calyx lobe red scarcely connate at base. Petals purplish. Stamen exserted, reddish.

Reproductive Cycle : April -November.

Materials Examined : HPU-EL-SKS-6433; Advance Study (2,128m), St. Edward's School (2,060m).

Propagation Notes : Propagated through seeds and inter-nodal cuttings of greenwood of 5 - 8cm long in May/June in a frame. Plant succeeds in any fertile well-drained circum-neutral soil rich in humus.

Part Used : Whole Plant.

Folk Use : Cultivated as an **ornamental** bush.

Uses in Literature : Reported to be used as diuretic, dye, edible (fruit), febrifuge and as an ornamental hedge plant **(Bailey, 1957; Bean, 1981; Beri, 1992; Chittendon, 1956; Graf, 1980, 1981; Hay *et al.*, 1983; Huxley, 1992; Paliwal, 2009; Phillips, 1954; Sheat, 1948; Shepherd, 1974; Thomas, 1990)**.

Ginkgo biloba L.

Syn.: *G. macophylla* K. Koch; *Pterophyllus salisburiensis* J. Nelson; *Salisburia biloba* (L.) Hoffmanns.; *S. ginkgo* Rich.; *S. macrophylla* Reyn.

Family: Ginkgoaceae.

Vern.: Balkumari.

English Name:

Eng. : Maidenhair Tree.

Distribution : E. Asia - N. China.

Description : Sparsely branched glabrous trees. Leaves rhomboidally - ovate, fan-shaped, striated with numerous parallel veins and on long slender petiole. Flowers in spike; pistillate and staminate ones borne on a separate trees. Fruits orange, obovoid or ellipsoid, yellowish with ill smelled pulp surrounding thin-shelled, 2-angled creamy-white nut.

Reproductive Cycle : February-March.

Material Examined : HPU-EL-SKS-6435; Advance Study (2,128m).

Propagation Notes : Propagated through seeds, softwood cuttings in a frame in spring and cuttings of half-ripe wood in July/August. Root cutting facilitated by application of hormones. Plant usually succeeds in most soil types.

Parts Used : Whole Plant. Leaves.

Folk Uses : Planted for **ornamental** purposes. 1-3g powdered leaves with honey given twice daily in **cardiovascular disorders**.

Active Constituents : These are shikimic acid and flavonoids, DL-threo-guaiacyglycerol, DL-threo – p-hydroxyphenylglycerol, sesquiterpene, bilobalide, β-sitosterol, ginkgolide A, ginkgolide B and ginkgolide C (**leaves**).

Uses in Literature : So far known as antiasthmatic, antibacterial, anticancerous, antifungal, astringent, brain tonic, edible (seeds), expectorant, ophthalmic, sedative, vermifuge, ornamental and sacred tree; and for allergy, blood circulation, bronchitis, cardiovascular disorders, cough, hypertension, leucorrhoea, sputum **(Ambasta, 1986; Anonymous, 1948-1976; Asolkar *et al.*, 1992; Bailey, 1957; Bean, 1981; Chauhan, 1999; Chittendon, 1956; Harborne & Baxter, 2001; Jaeger, 1961; Pande *et al.*, 2006; Prajapati *et al.*, 2006;**

Randhawa, 1957, 1960; Rushforth, 1987; Sheat, 1948; Sood *et al.*, 2005; Uphof, 1968; Warda, 1985; Warman, 1999).

Girardinia diversifolia (Link) Friis.

Syn.: *G. adoensis* Wedd.; *G. erosa* Decne.; *G. formosana* Hayata *ex* Yamam; *G. heterophylla* Decne; *G. javanica* Wedd.

Family: Urticaceae.

Vern.: Bicchu Buti.

English Names:

Eng. : Himalayan Nettle, Nilgiri Nettle, Stinging Nettle.

Distribution : Pakistan-Bhutan, India, China, S. E. Asia: 1,200-3,000m.

Description : Erect, coarse robust shrubs covered with rigid, sharp, stinging bristles. Leaves alternate, stalked, usually cordate, 3-nerved from the base, broadly ovate, margins sharply toothed. Flowers small, green, mono-sexual, sessile, densely crowded. Male flowers in slender, paniculately branched spike. Female ones in short, oblong, thick spikes. Achenes flat, ovate, black.

Reproductive Cycle : July-August.

Materials Examined : HPU-EL-SKS-6437; Advance Study (2,128m), Boileauganj Cemetery (1,998m), Chapslee (2,243m), Chhota Shimla (2,087m), Corner House (2,155m).

Propagation Notes : Propagated by seeds.

Parts Used : Leaves. Inflorescence. Branches.

Folk Uses : Young leaves and inflorescence cooked as a green **vegetable**. Stem yields **fibre** for **cordages**. Branches used as **fodder**. Fresh juice of leaves applied externally for checking **headaches** and **swollen joints**. Ash of plant applied externally against **ringworm** and **eczema**.

Uses in Literature : Reported earlier for constipation, fever, gonorrhoea, headache, malignant boils, rheumatism, skin diseases, spermatorrhoea, sprain, urinary disorders; and as fibre, fodder, vegetable (leaves, young shoots) (**Ahmed & Borthakur, 2005; Ambasta, 1986; Bennet *et al.*, 1991; Chaudhuri, 1993; Chopra *et al.*, 1986; Deokota & Chhetri, 2009; Dwivedi, 2007; Farooq, 2005; Manandhar, 2002; Pande *et al.*, 2006; Rana *et al.*, 1996; Samant *et al.*, 1996; Sharma & Rana, 2000; Singh, 2000b; Uphof, 1968; Usher, 1974**).

Grevillea robusta A. Cunn *ex* R. Br.

Syn.: *G. umbratica* A. Cunn. *ex* Meisn.; *G. venusta* A. Cunn. *ex* Meisn.

Family: Proteaceae.

Vern.: Silver Fir, Sliver Oak.

English Names:

Eng. : Australian Silk Oak, Silver Oak.

Distribution : Indigenous to Queensland and New South Wales; often planted along roadsides in Himalayas.

Description : Tall, partially deciduous trees with a straight columnar growth. Young shoots rusty tomentose. Leaves decompound; leaflets entire, lanceolate, green above white silky beneath, each leaf appearing fern-like. Flowers golden orange, in dwarf 1 - sided fascicled racemes. Follicles oblique, 1- or 2- seeded.

Reproductive Cycle : March-July.

Materials Examined : HPU-EL-SKS-6439; B.C.S. (2,025m), Bemloe (2,012m), Gorton Castle (2,094m), Grand Hotel (2,105m).

Propagation Notes : Mostly through seeds and cuttings of half-ripe wood of 5 - 8cm size with a heel in June/July. Tree grows in a well-drained slightly acidic soil and a sunny position.

Parts Used : Whole Plant. Leaves. Wood.

Folk Uses : Planted for its **aesthetic** value. Also, considered an excellent source of **nectar** for bee keeping. Wood extensively employed for **ornamental panelling, furniture**, **toys** and **plywood**. Leaves used as **green manure**.

Active Constituents : Active constituents are grevillol, sitosterol, epi-lyeniresinol, 2-O-glucoside, rhamnocitrin-3-O-rutinoside and kaempferol-3-O-rutinoside **(wood)** and a mixture of macrocyclic phenols robustol, rutin, 2,5-dihydroxycinnamic acid, 6-hydroxycoumarin, methyl 4-hydroxycinnamate, quercetin, hydroquinone and arbutin derivatives **(leaves)**.

Uses in Literature : Useful as dye, edible, fuel, green manure, gum, oil, poison; and for floors, furniture, ornamental panelling, parquet, plywood, tannin, toys, and veneering **(Ambasta, 1986; Anonymous, 1948-1976; Beri, 1992; Chittendon, 1956; Chopra *et al.*, 1986; Deokata & Chhetri, 2007; Dwivedi, 2007; Hay *et al.*, 1983; Huxley, 1992; Manandhar, 2002; Mishra & Chandra, 1998; Paliwal, 2009; Phillips & Rix, 1998; Raju, 2000; Randhawa,**

1957, 1960; Sheat, 1948; Singh & Pandey, 1998; Sinha, 1996; Sood *et al.*, 2013; Warman, 1999).

Hedera canariensis Willd.

Syn.: *H. grandifolia* Hibberd; *H. sevillana* Sprenger; *H. viridis* (Hibberd) G. Nicholson.

Family: Araliaceae.

Vern.: Banbat Kari, Banda, Kuri.

English Names:

Eng. : Algerian Ivy, Canary Island Ivy, North African Ivy.

Distribution : Native of Algeria, Tunisia, Azores.

Description : Evergreen glabrous climbers having a few aerial rootlets and glossy, pale green, mostly triangular, usually shallowed 3 - lobed juvenile leaves. Fruits black.

Reproductive Cycle : August-November.

Materials Examined : HPU-EL-SKS-6440; Advance Study (2,128m), St. Edward's School (2,060m).

Propagation Notes : Propagated from aerial roots, softwood cutting or rooted layers. Plant requires good light and sunshine for its healthy growth.

Part Used : Whole Plant.

Folk Uses : Planted for its attractive **climbing** foliage as a **ground cover** for sheltered wall.

Use in Literature : Grown as an evergreen climber **(Bailey, 1957; Graf, 1980, 1981; Hay *et al.*, 1983)**.

Hedera colchica (K. Koch) K. Koch

Syn.: *H. acuta* G. Nicholson; *H. amurensis* Hibberd *ex* Lawr. & Schulze; *H. caucasica* Carriere; *H. cordifolia* Hibberd; *H. regneriana* Hibberd; *H. rhombifolia* Rupr. *ex* Regel; *H. roegneriana* Boiss.

Family: Araliaceae.

Vern.: Kiraya.

English Name:

Eng. : Cholchis Ivy.

Distribution : Native of Caucasus, N. Iran.

Description : Climbers having pale green stout twigs. Aerial rootlets occur at most nodes. Juvenile leaves leathery, broadly ovate, 3-lobed, base cordate, emitting celery-like odour when crushed. Fruits blue-black.

Reproductive Cycle : September-November.

Materials Examined : HPU-EL-SKS-6445; Advance Study (2,128m), Holly Lodge (2,278m).

Propagation Notes : Ivy succeeds in all soil types. Propagated through seeds, cuttings of half-ripe wood in July/August in a shady position in a frame and layering.

Part Used : Whole Plant.

Folk Use : Grown as an adornment **climber** for walls.

Uses in Literature : Grown as ornamental hedge plant **(Asolkar *et al.*, 1992; Bailey, 1957; Graf, 1980, 1981)**.

***Hedera helix* L.**

Family: Araliaceae.

Vern.: Dakari.

English Names:

Eng. : Barren Ivy, Beene Tree, Bent Wood, Black Ivy, Creeping Ivy, Nepal Ivy, Small Ivy, Wood Blind.

Distribution : N.W. Himalayas: 1,800-3,000m.

Description : Climbers, climbing by aerial adhesive rootlets. Leaves simple, thick, shining, ovate, angular or 3-5 lobed and very variable in size. Umbels stellately hairy, solitary or in clusters of 2 or 3, sometimes panicled. Flowers yellow-green, polygamous. Fruits black or yellow, sometimes red, smooth and globose.

Reproductive Cycle : September-November.

Materials Examined : HPU-EL-SKS-6449; Advance Study (2,128m), Walsingham (2,266m), Wood Villa (2,108m).

Propagation Notes : Propagation by seeds, aerial roots, softwood cutting or rooted layers.

Parts Used : Whole Plant. Leaves.

Folk Uses : Cultivated as a **showy climbing** foliage plant. Leaves lopped for **fodder**. Leaves and bark constituent of **local wine 'sur'**. Herb used externally for **skin eruptions**, **swollen tissue**, **painful joints**, **burns** and **suppurating cuts**.

Active Constituents : These are rutin, caffeic acid, 3,5-dicaffeoylquinic acid and kizuta saponin K5 (**leaves**).

Uses in Literature : Known earlier as antibacterial, antiseptic, antispasmodic, astringent, cataract, cathartic, diaphoretic, emetic, emmenagogue, fodder, house plant, purgative, stimulant, vasodilator, vermifuge; and for inflammation, mumps, psoriasis, skin infections, sores, swelling **(Agarwal, 2003; Ambasta, 1986; Anonymous, 1948-1976; Baburaj *et al.*, 2000; Bailey, 1957; Bean, 1981; Chaudhuri, 1993; Chhetri, 1994; Chittendon, 1956; Hadfield, 1956; Harborne & Baxter, 2001; Hay *et al.*, 1983; Paliwal, 2009; Pool, 1941; Rajendran, 2003; Uphof, 1968; Usher, 1974; Vickery, 1994; Watt, 1889-1896)**.

Hedera nepalensis K. Koch

Syn.: *Cissus wallichiana* Turcz.; *H. cinerea* (Hibberd) Bean; *H. himalaica* (Hibberd) Carriere.

Family: Araliaceae.

Vern.: Berbel.

English Name:

Eng. : Nepal Ivy.

Distribution : Afghanistan - S.W. China; Myanmar upto 3,000m.

Description : A woody climber upto 30m, climbing by aerial roots with simple lanceolate - ovate, entire or variously lobed, leathery, dark green glossy leaves. Flowers tiny, many, yellowish-green in stalked globular umbels arranged in dome cluster. Fruits globular, shining yellow, then black.

Reproductive Cycle : September-November.

Materials Examined : HPU-EL-SKS-6450; Boileauganj Cemetery (1,998m), Rothney Castle (2,350m), Wood Villa (2,108m), Yates Place (2,208m).

Propagation Notes : Propagated by seeds, aerial roots, softwood cutting or rooted layers.

Part Used : Whole Plant.

Folk Uses : Decoction of plant used for treating **skin diseases** and killing **hair lice**.

Active Constituents : Active constituents are saponins and lactones (**leaves**).

Uses in Literature : Known earlier as anticancerous, carthartic, diaphoretic, fodder, purgative, stimulant; and for dyspepsia, mumps, rheumatism, ulcer, skin diseases, stomachache **(Ambasta, 1986; Asolkar *et al.*, 1992; Bean,**

1981; Chhetri, 1994; Chittendon, 1956; Huxley, 1992; Manandhar, 2002; Pande *et al.*, 2006; Singh & Kumar, 2000a).

***Hibiscus rosa-sinensis* L.**

Syn.: *H. boryanus* DC.; *H. festalis* Salisb.; *H. storckii* Seem.

Family: Malvaceae.

Vern.: China Rose.

English, Hindi and Sanskrit Names:

Eng.	:	Chinese Hibiscus, Bengal rose, China rose, Rose of China, Scarlet Rose-Mallow, Shoe Flower;
Hindi	:	Gulhar, Jasum, Jasut, Odhul;
Sans.	:	Arkapriya, Aruna, Harivallabha, Japa-pushpam, Java, Joba, Odhrapushpa, Pratiko, Raktapushpi, Rudra Pushpam.

Distribution : Native of China.

Description : Shrubs having woody branched stem. Leaves ovate, coarsely toothed at apex and entire at base. Corolla tubular below, red. Capsule many seeded.

Reproductive Cycle : February-October.

Material Examined : HPU-EL-SKS-6451; Grand Hotel (2,105m).

Propagation Notes : Propagated from cuttings, layering, budding or grafting. Plant succeeds in a well-drained humus rich fertile soil in a sheltered position in full sun.

Parts Used : Whole Plant. Leaf. Flower.

Folk Uses : **Cultivated** in gardens for its beautiful flowers. Flowers **sacred** and offered to deities in religious **worshipping**. Flowers **edible**; its paste with curd applied to check **baldness**. Leaf paste applied for hastening **suppuration** of **boils**.

Active Constituents : Active properties due to quercetin - 3 -diglucoside 3,7 - diglucoside, cyaniding - 3, 5 -diglucoside, delphinidin, esculetin, flavones and cyanidin-3-sophoroside - 5 – glucoside (**flowers**).

Uses in Literature : Described earlier as abortifacient, anodyne, demulcent, emollient, hair dye, laxative, refrigerant drink in fevers, symbolic (offering in chronic diarrhoea), vegetable (flowers eaten raw or pickled); and for antifertility, appendicitis, asthma, bleeding piles, boils, bronchitis, burn spots, cancerous swellings, child birth, colouring syrups, consumption, cough, developing foetus (vet.), diarrhoea, flu, gonorrhoea, headache, impotency, inflammation, irritable conditions of genito-urinary tract, kidney diseases, marriage ceremonies (flowers), menstrual complaints, mumps, offerings and adornments, postnatal haemorrhage, bronchitis, whooping cough, tuberculosis, scabies (vet.), skin diseases, soothing internal and external wounds, sores, and syphilis **(Agarwal, 2003; Ambasta, 1986; Anonymous, 1948-1976; Banerjee, 1996; Banerjee & Pal, 1996; Beri, 1992; Caius, 1986; Chatterjee & Pakrashi, 1997; Chowdhury *et al.*, 1996; Cowan, 1950; Datt, 1996; Dey & Bahadur, 1973; Dhiman, 2003; Gill *et al.*, 1993; Girach, 1992; Girach *et al.*, 1996; Huidrom, 1996; Huxley, 1992; Islam & Jha, 2003; Jadhav, 2009; Jain, 1991; Jeyapardha *et al.*, 2011; Jha & Basak, 1994; Jha *et al.*, 1996; Khare, 2004; Kirtikar & Basu, 1984; Kumar *et al.*, 2003; Kurian, 1995; Maliya, 2004; Manandhar, 2002; Nath & Maiti, 2003; Paliwal, 2009; Pande *et al.*, 2006; Pandey, 2003; Pawar & Patil, 2004; Prajapati *et al.,* 2006; Prasad *et al.*, 1996; Ranjan, 2003; Rawat & Chowdhury, 1998; Saikia & Nath, 2003; Saikia *et al.*, 2010; Sen & Behera, 2003; Sharma, 2003; Singh & Pandey, 1998; Sood *et al.*, 2013; Srivastava, 1989; Thomas & Britto, 2003; Uphof, 1968; Vihari, 1995; Watt, 1889-1**

Hibiscus syriacus **L.**

Syn.: *Althaea frutex* Mill.; *H. acerifolius* Salisb.; *H. rhombifolius* Cav.; *Ketmia arborea* Moench; *K. syrorum* Medik.

Family: Malvaceae.

Vern.: Brahm Phool.

English Names:

Eng. : Athaea, Rose of Sharon, Shrub Althaea.

Distribution : Native of Syria.

Description : Erect, nearly glabrous shrubs or small trees upto 3m tall. Leaves triangular - rhombic ovate, narrowly 3-lobed and coarsely toothed. Flowers on short pedicels, whitish purple with crimson base and in upper leaf axils. Involucral bracts 6-8, mostly lanceolate. Capsule 5-celled, each cell with 3- or more seeded.

Reproductive Cycle : July-September.

Materials Examined : HPU-EL-SKS-6452; Advance Study (2,128m), Auckland House (2,257m), Bemloe (2,014m), St. Edward's School (2,060m), Raj Bhawan (2,132m).

Propagation Notes : Propagated mainly by stem cuttings and layering in mid summer to early autumn. For its growth, plant prefers a well-drained humus rich fertile soil in a sheltered position in full sun.

Part Used : Whole Plant.

Folk Uses : Cultivated around buildings as a **hedge** plant. Flowers **eaten** raw.

Active Constituents : Reported to possess cartotenoid pigments -cryptoxanthin, chrysanthemaxanthin and antheraxanthin and taxifolin-3-O-β-D-glucopyranoside, herbacetin-7-β-D-glucopyranoside and kaempferol-3 α-L-arabinoside-7α-L-rhamnopyranoside (**flowers**).

Uses in Literature : Reported to be used for abdominal pain, anthelmintic, anticomplementary, antifungal, antipyretic, antiulcer, border hedge, diuretic, edible (flower, tender leaves), fibre, hypoglycaemic, and ornamental; and for dysentery, itches, leucorrhoea, omalgia, skin troubles **(Ahmed & Borthakur, 2005; Ambasta, 1986; Anonymous, 1948-1976; Bean, 1981; Chittendon, 1956; Chopra *et al.*, 1986; Cowan, 1950; Huxley, 1992; Paliwal, 2009; Prajapati *et al.*, 2006; Randhawa, 1957; Rastogi & Mehrotra, 1991; Sharma & Sood, 1997; Sood *et al.*, 2013)**.

Himalrandia tetrasperma (Wall. *ex* Roxb.) T. Yamaz

Syn.: *Aidia tetrasperma* (Wall.) T. Yamaz.; *Gardenia densa* Wall.; *G. rigida* Wall.; *G. tetrasperma* Wall. *ex* Roxb.; *Randia tetrasperma* (Wall. *ex* Roxb.) Benth. & Hook.f. *ex* Brandis.

Family: Rubiaceae.

Vern.: Daadey.

Hindi Names:

Hindi : Bharani, Bhotiya - ghingaru, Katul, Panar, Paniah, Pendar, Pindalu.

Distribution : Distributed in all tropical and subtropical regions; most abundant in Asia, Africa, Sub-Himalayan tracts from Pakistan - S.E Tibet: 1,200-2,400m.

Description : Small, glabrous rigid shrubs having short branchlets usually ending in a spine. Leaves opposite, crowded towards the end of branchlets, ovate or oblanceolate. Stipules broadly lanceolate, long-pointed. Flowers solitary, axillary, sessile, white tinged with green. Berries purple, globose.

Reproductive Cycle : July-October.

Material Examined : HPU-EL-SKS-6454; Kanlog Cemetery (1,992m).

Propagation Notes : Propagated naturally by seeds.

Parts Used : Leaves. Branches. Fruits.

Folk Uses : Leaves used as **fodder** for goats. Straight branches used as **walking sticks**. Powdered fruits used as detergent to **wash** clothes and hair.

Active Constituents : Main ingredients are scopoletin, its β-D-galactoside, β-sitosterol, its glucoside, oleanolic acid and mannitol (**fruit**).

Uses in Literature : Reported to be used as febrifuge, fodder; and for making walking sticks **(Ambasta, 1986; Kaur & Sharma, 2004; Rastogi & Mehrotra, 1995a; Samant *et al.* 2001; Watt, 1889-1896)**.

***Hydrangea macrophylla* (Thunb.) Ser.**

Syn.: *Hortensia opuloides* Lam.; *Hydrangea hortensis* Sm.; *H. maritime* Haw.-Booth.

Family: Hydrangeaceae.

Vern.: Hydrangea.

English and Hindi Names:

Eng. : Common Hydrangea;
Hindi : Hydrangea.

Distribution : Native of Japan and China.

Description : Bushy shrubs with brownish branches and oblong-ovate to elliptic-lanceolate leaves. Inflorescence ball-shaped; flowers range from white through pink - violet. Heads globular (mainly of sterile flowers) or 'lace-caps' with outer sterile flowers and the inner fertile ones.

Reproductive Cycle : May-July.

Materials Examined : HPU-EL-SKS-6457; Advance Study (2,128m), Cedar Circuit House (2,154m), Chapslee (2,243m), Raj Bhawan (2,132m), Ridge (2,250m).

Propagation Notes : Easily propagated by cuttings. Plant succeeds in a well-drained loamy soil.

Part Used : Whole Plant.

Folk Uses : Mainly **planted** around buildings for their green foliage, large showy cluster of flowers and as a **hedge** plant.

Active Constituents : These are cis- and trans-febrifugines, phyllodulcin, hydrangenol, its 8-O-galactoside, p-aminophenyl-α-D-glucose, and 4-O-glucoside (**leaves**) and hydrangenol glucoside (**flower**).

Uses in Literature : Known to be useful as diuretic, edible (leaves), ornamental, hedge; and for tea **(Anonymous, 1948-1976; Bean, 1981; Chittendon, 1956; Chopra *et al.*, 1986; Hay *et al.*, 1983; Huxley, 1992; Lunardi,**

1987; Paliwal, 2009; Rastogi & Mehrotra, 1990, 1991, 1995b; Sood *et al.*, 2013; Thomas, 1992; Trivedi, 1990).

Hypericum hookerianum **Wight & Arn.**

Syn.: *Norysca hookeriana* (Wight & Arn.) Wight.

Family: Clusiaceae.

Vern.: Phiunli.

English Name:

Eng. : Golden lotus, Hooker's St. Johnwort.

Distribution : Native of Bangladesh, Bhutan, China (Xizang and Yunnan), India, Nepal, Myanmar and northern Thailand.

Description : Shrubs with large, 3-6 cm across, cup - shaped yellow flowers. Sepals rounded with a blunt apex. Leaves broadly lanceolate - elliptic, blunt or acute with conspicuous venation beneath. Branches red-brown, rounded. Capsule ovoid, about 1 - 1.7cm long, containing reddish brown seeds.

Reproductive Cycle : July-October.

Material Examined : HPU-EL-SKS-6459; Advance Study (2,128m).

Propagation Notes : Multiplied by seeds and cuttings.

Part Used : Whole Plant.

Folk Uses : Planted for **hedge** purposes and as an **ornamental.**

Uses in Literature : Known so far as ornamental and hedge; and also used for burns and wounds (vet.) **(Chaudhuri, 1993; Huxley, 1992; Jaiswal *et al.*, 2004; Rehder, 1949; Robson, 1970).**

Hypericum oblongifolium **Choisy**

Syn.: *H. cernuum* Roxb. *ex* D. Don; *H. speciosum* Wall.; *H. govanianum* Wall.

Family: Clusiaceae.

Vern.: Lyoon.

English Name:

Eng. : St. John's Wort.

Distribution : W. Temperate Himalayas (Kumaon - Sirmour): 1,700-2,350m.

Description : Shrubs having slender, terete, green branches which are tinged with reddish-purple. Leaves ovate or oblong-lanceolate; each pair of leaves placed at right angles to the next pair above or below it. Flowers golden-yellow in terminal cymes. Capsules conical.

Reproductive Cycle : February-October.

Materials Examined : HPU-EL-SKS-6460; Advance Study (2,128m), Boileauganj Cemetery (1,998m), Yates Place (2,208m).

Propagation Notes : Propagated by seeds as well as through cuttings of half-ripe wood of 10 - 12 cm long with a heel in July/August. Plant loves to grow in any reasonably good well-drained but moisture retentive soil.

Parts Used : Whole Plant. Leaves. Flowers. Twig. Seed.

Folk Uses : Grown as an **ornamental** plant. Twigs used as **fodder**. Juice of crushed flowers and leaves good against **wounds** and **boils**. Seeds used as condiments for **flavouring** curries.

Uses in Literature : Known earlier as an aromatic, stimulant; and for boils, fodder, old sores, eczema, toothache and wounds **(Asolkar *et al.*, 1992; Dean, 1964; Kapur, 1996a; Kapur & Srivastava, 1996; Pande *et al.*, 2006)**.

Hypericum patulum Thunb.

Syn.: *Komana patula* (Thunb.) Y. Kimura *ex* Honda; *Norysca patula* (Thunb.) Voigt.

Family: Clusiaceae.

Vern.: Phulanu.

English and Hindi Names:

Eng. : St. John's Wort;
Hindi : Tumbhul.

Distribution : Distributed in N. temperate regions from Kashmir - Bhutan, China, Japan, Myanmar, Thailand, Sumatra: 900-3,600m.

Description : Shrubs possessing erect, glabrous cylindrical arching stems, 2-ridged, reddish spreading branches and ovate or oblong-lanceolate leaves with each pair placed at right angles to the next pair above or below it. Lower surface of leaves pale, black dotted. Flowers yellow, in terminal cymes. Capsules obtuse or conical.

Reproductive Cycle : August-September.

Materials Examined : HPU-EL-SKS-6461; Aira Holme (2,015m), Kanlog Cemetery (1,990m).

Propagation Notes : Multiplied through seeds and cuttings of half-ripe wood of 10 - 12 cm long with a heel in July/August. Plant preferably grows in any reasonably good well-drained but moisture retentive soil.

Parts Used : Leaves. Roots

Folk Uses : Juice of leaves applied to heal **cuts** and **wounds**. Poultice of crushed roots useful against **boils**.

Uses in Literature : Useful earlier as an abortifacient, anthelmintic, aromatic (seeds), astringent, diuretic, food poisoning, stimulant; and against boil and blisters **(Ambasta, 1986; Anonymous, 1948-1976; Asolkar *et al.*, 1992; Chaudhuri, 1993; Chittendon, 1956; Chopra *et al.*, 1986; Farooq, 2005; Huxley, 1992; Kirtikar & Basu, 1984; Pande *et al.*, 2006; Prajapati *et al.*, 2006; Singh & Kumar, 2000a; Sood *et al.*, 2009; Watt, 1889-1896)**.

Ilex dipyrena Wall.

Syn.: *I. dentonii* Hort. *ex* Loudon; *I. monopyrena* G. Watt *ex* Loes.

Family: Aquifoliaceae.

Vern.: Charkha, Holly Tree.

English Name:

Eng. : Himalayan Holly.

Distribution : N.W. Himalayas: 1,500-3,000m.

Description : A middle-sized nearly glabrous tree having shining, elliptic-lanceolate leaves generally with large strong spine. Flowers bisexual, white, in axillary subglobose clusters; pedicels very short, stout. Bracts small. Drupes with 2 stones.

Reproductive Cycle : April-October.

Materials Examined : HPU-EL-SKS-6463; Eagle Mount (2,152m), Holly Lodge (2,278m).

Propagation Notes : Prefers a good deep fertile loam for its growth. Propagated by seeds.

Parts Used : Leaves. Wood.

Folk Uses : Leaves used as **fodder**; and its paste applied as poultice against **nail disorders**. Wood used for low grade **furniture**.

Uses in Literature : Reported earlier for ascariasis; and as diuretic, occasional fodder, purgative, inferior quality timber, toothbrush **(Ambasta, 1986; Chaudhuri, 1993; Kapur, 1996a; Pande *et al.*, 2006; Singh & Kumar, 2000a; Watt, 1889-1896)**.

Indigofera dosua Buch.-Ham. *ex* D. Don

Family: Fabaceae.

Vern.: Kathee.

Distribution : H.P. - Arunachal Pradesh: 600-3,000m.

Description : Densely hairy shrubs with deep pink or red flowers in long slender erect spikes and silky haired calyx. Flowers 1-1.3 cm long, standard hairy on back. Bracts narrow-lanceolate, long pointed, silky-haired, much longer than the flower buds. Leaflets opposite, elliptic, densely adpressed-hairy. Pods straight, hairy.

Reproductive Cycle : April-July.

Materials Examined : HPU-EL-SKS-6465; Boileauganj Cemetery (1,988m), Kanlog Cemetery (1,992m).

Propagation Notes : Propagated by seeds.

Parts Used : Twigs. Flower.

Folk Uses : Twigs used as **fodder**. Flowers used as **vegetable**. 15-20ml decoction prepared from flowers given twice daily against **fever**.

Uses in Literature : Reported to be used as vegetable (flowers), dye, fodder **(Agarwal, 2003; Ambasta, 1986; Borthakur, 1997; Chaudhuri, 1993; Kapur & Srivastava, 1996).**

Inula cappa (Buch.-Ham *ex* D. Don) DC.

Syn.: *Blumea arnottiana* Steud.; *Conyza cappa* Buch.-Ham. *ex* D. Don; *C. oblonga* Wall. *ex* DC.; *Vernonia congesta* Benth.

Family: Asteraceae.

Vern.: Basanti.

English Name:

Eng. : Sheep's Hair.

Distribution : E. Asia-Himalayas from H.P.-S.W. China: 1,000-2,400m.

Description : Aromatic shrubs with robust stem and tomentose branches and inflorescence. Leaves thick, shortly stalked, oblong-lanceolate; upper surface hairy, lower softly and densely silky. Heads very many, in numerous crowded rounded corymbs and with a few and very short ligules. Achenes small, cylindric, silky.

Reproductive Cycle : August-February.

Materials Examined : HPU-EL-SKS-6466; Advance Study (2,128m), Boileauganj Cemetery (1,998m), Chhota Shimla (2,068m), Kanlog Cemetery (1,999m), Green Gate (2,096m).

Propagation Notes : Propagated by seeds.

Parts Used : Root. Flowers.

Folk Uses : 5-10ml juice of roots given on empty stomach against **peptic ulcers** and **gastric disorders**. Flowers used for **decorative** purposes.

Uses in Literature : Known in India for cough, dysentery, fever, indigestion, stomachache; and as anodyne, anthelmentic, carminative, diuretic, dye, cancer, expectorant, tonic **(Ahmed & Borthakur, 2005; Chaudhuri, 1993; Gamble, 1972; Manandhar, 1980b, 1990, 2002; Pande *et al.*, 2006; Sood *et al.*, 2005**).

Ipomoea purpurea (L.) Roth.

Syn.: *Convovulus purpureus* L.; *I. diversifolia* Lindl.; *I. hirsutula* J. Jacq.; *Pharbitis hispida* (Zuccagni) Choisy; *P. purpurea* (L.) Voigt.

Family: Convolvulaceae.

Vern.: Gonani.

English Name:

Eng. : Morning Glory.

Distribution : India; ascending to 1,000-2,300m in Khasia; W. Deccan Peninsula.

Description : Annual hairy climbers. Leaves ovate, cordate, acute. Flowers large, pink or purple, in small, stalked clusters which sometimes are solitary. Fruits a capsule. Seeds glabrous.

Reproductive Cycle : June-August.

Materials Examined : HPU-EL-SKS-6468; Advance Study (2,128m), Boileauganj Cemetery (1,998m), Green Gate (2,098m).

Propagation Notes : Multiplied through seeds.

Part Used : Whole Plant.

Folk Use : Grown as a beautiful **climber** on walls or trees.

Uses in Literature : Reported to be used as anthelmintic, antibacterial, antisyphilitic, diuretic, hallucinogenic, laxative, purgative and ornamental **(Ambasta, 1986; Anonymous, 1948-1976; Asolkar *et al.*, 1992; Borthakur *et al.*, 1999; Cowan, 1950; Dean, 1964; Farooq, 2005; Harborne & Baxter,**

2001; Hay *et al.*, 1983; Huxley, 1992; Jaeger, 1961; Paliwal, 2009; Pande *et al.*, 2006; Phillips, 1954; Phillips & Rix, 1998; Pool, 1941; Rastogi & Mehrotra, 1995b).

Jacaranda acutifolia Bonpl.

Family: Bignoniaceae.

Vern.: Jackranda, Nili Gulmohra.

English and Hindi Names:

Eng. : Jacaranda;
Hindi : Jacaranda, Nili-gulmohar.

Distribution : Outer foothills of W. Himalayas.

Description : A medium - sized deciduous tree attaining 8-10m in height. Leaves opposite, bipinnate. Flowers bluish-purple, in loose pyramidal panicles. Capsule flat, woody. Seeds winged.

Reproductive Cycle : March-June.

Materials Examined : HPU-EL-SKS-6469; B.C.S. (2,025m), St. Edward's School (2,060m).

Propagation Notes : Propagated by sowing seeds in April-June.

Parts Used : Whole Plant. Leaves. Wood.

Folk Uses : Planted as an **avenue** plant and in **gardens** for its decorative floral panicles. Wood used for low-grade **furniture**, **construction wood** and **tool handles**. Finely powdered leaves dusted on **wounds** for quick healing.

Active Constituents : These are scutellarein, hydroxyquinone, jacaranose, jacaranone, verbascoside and phenylacetic-b-glucoside (**leaves**), and 3-monoglucoside and 3,5, diglucoside of delphinidin (**flower**).

Uses in Literature : Used earlier for blennorrhagia, syphilis, venereal diseases; and as furniture, garden tree, pectoral, tool handles and vulnerary **(Ambasta, 1986; Anonymous, 1948-1976; Asolkar *et al.*, 1992; Bhattacharjee, 2001; Cowan, 1950; Jain & Sharma, 1996; Paliwal, 2009; Pool, 1941; Raju, 2000; Randhawa, 1957, 1960, 1965; Rastogi & Mehrotra, 1990; Sood *et al.*, 2013; Uphof, 1968; Warman, 1999).**

Jasminum grandiflorum L.

Syn.: *J. aureum* Don; *J. officinale* var. *grandiflorum* (L.) Stokes.

Family: Oleaceae.

Vern.: Chameli, Malti.

English, Hindi and Sanskrit Names:

Eng.	:	Catalonian Jasmine, Common White Garden Jasmine, Spanish Jasmine, White Jasmine;
Hindi	:	Chameli, Jati;
Sans.	:	Bahugandha, Gajavhaya, Jati, Punyagandha, Shankha-yuthika.

Distribution : N.W. Himalayas (600-2,500m); wild as well as cultivated.

Description : A climbing shrub. Leaves opposite, leaflets ovate, often partially united with the uppermost pair. Flowers white, often tinged with purple outside, in lax axillary and terminal cyme. Fruit a berry.

Reproductive Cycle : April-September.

Materials Examined : HPU-EL-SKS-6471; Advance Study (2,129m) Raj Bhawan (2,132m), U.S. Club (2,290m).

Propagation Notes : Multiplied through seeds, cuttings, layering or suckers. Plant succeeds in fertile, well drained soil, in full sun or partial shade.

Parts Used : Whole Plant. Leaves. Flower.

Folk Uses : Planted as an **ornamental**; flowers offered to **deities**. 5-10ml decoction of flowers taken on empty stomach against **piles** and **skin eruptions**. Leaves chewed for relief from **toothache** and **mouth ulcers**.

Active Constituents : Active properties of **flowers** due to cis-3-hexenol, 2-vinylpyridine, indole, myrcene, linalool, geranyl linalool, α-terpineol, geraniol, linalyl acetate, nerolidol, phytol isophytol, farnesol, engenol, benzyl alcohol, p-cresol, methyl benzoate benzyl cyanide, benzyl acetate, methyl dihydrojasmonate methyl anthranitate, lis-jasmone, methyl N-methyl anthranitate, vanillin, cis-3-hexenyl benzoate, benzyl benzoate, methyl palmalte and methyl linolate.

Uses in Literature : Recorded earlier as an anthelmintic, aphrodisiac, audient, cooling, deobstruant, diuretic, emmenagogue, hair oils, perfumed waters, psychoactive plant, social taboo (girls place flames in their hair when they became pregnant); and for apthae, attars, corns between the toes, dermatitis, headache, offerings, otorrhoea, stopping breast milk production, mouth cleansing, sexual impotency, skin eruptions, toothache, treatment of ear discharge and ulceration of mouth **(Ambasta, 1986; Bhattacharjee, 2001; Chatterjee & Pakrashi, 1997; Dey & Bahadur, 1973; Dhiman, 2003; Ghate, 1998; Jain *et al.*, 1994; Kapur & Singh, 1996; Khare, 2004; Kirtikar & Basu, 1984; Nautiyal *et al.*, 1997; Paliwal, 2009; Pande *et al.*, 2006; Prajapati *et al.*, 2006; Raju, 2000; Rana *et al.*, 1996; Randhawa, 1957; Shome *et al.*, 1996; Singh & Pandey, 1998; Sood *et al.*, 2013; Srivastava *et al.*, 2003c; Thakur, 2005; Thomas & Britto, 2003; Uphof, 1968; Vihari, 1995; Watt, 1889-1896)**.

Jasminum humile L.

Family: Oleaceae.

Vern.: Peeli Chameli, Sunajui.

English, Hindi and Sanskrit Names:

Eng. : Golden Jasmine, Italian Jasmine, Nepal Jasmine, Yellow Jasmine;
Hindi : Malto, Peeli Chameli, Pitmalti;
Sans. : Basantee, Gandadhya, Haima, Hemapushpi, Hemapushpika, Pitika, Svarnajuthika.

Distribution : Subtropical Himalayas: 1,500-3,000m.

Description : Erect shrubs with numerous green angular branches. Leaves pinnate; leaflets elliptic-acute at both ends. Flowers solitary or in dense cymes at the end of twigs, yellow. Fruits a juicy black berry.

Reproductive Cycle : May-September.

Materials Examined : HPU-EL-SKS-6473; Bemloe (2,014m), Benmore (2,114m), Grand Hotel (2,105m).

Propagation Notes : Propagated easily by seeds or cuttings of half ripe wood in June-August. Plant prefers a good well-drained loam in a sunny position.

Parts Used : Whole Plant. Leaves. Root. Flowers

Folk Uses : Plant grown as an **ornamental**. Flowers considered **sacred** and offered to **deities**. Leaves chewed for relief to check **toothache** and its poultice applied to heal **skin diseases**, **wounds** and **ulcers**. Decoction of roots (5-10ml) prescribed twice daily for 5-10 days to cure **skin ailments**.

Active Constituents : Active properties due to α-amyrin, betulin, friedelin, lupeol, betulinic, ursolic acids, β-sitosterol, 10-cinnamoyo-ieoside-7-methyester and secoiridoid glucoside (**leaves**).

Uses in Literature : Described in Indian literature as antidote in poison cases, aromatic essential oil, astringent, tonic to heart; and for chronic sinuses, dental complaints, dyeing, skin diseases, wounds, ulcers, stomatitis, toothache, earache, eye diseases, fistulas, headache, ringworm affections, skin diseases, stomatitis, throat infection, ulceration and wound **(Ambasta, 1986; Anonymous, 1948-1976; Asolkar *et al.*, 1992; Bhattacharjee, 2001; Chatterjee & Pakrashi, 1997; Chauhan, 1999; Chopra *et al.*, 1965; Deokota & Chhetri, 2009; Farooq, 2005; Jain, 1991; Paliwal, 2009; Pande *et al.*, 2006; Raju, 2000; Ranjan, 2000b; Sharma, 2003; Sood *et al.*, 2009, 2010, 2013; Watt, 1889-1896)**.

Jasminum mesnyi Hance

Syn.: *J. primulinum* Hemsl. *ex* Baker.

Family: Oleaceae.

Vern.: Chameli.

English Names:

Eng. : Jasmine, Primrose Jasmine.

Distribution : Native of Yunnan (China); cultivated throughout the tropical and subtropical parts of the world.

Description : Evergreen twiggy shrubs having 4-angled, stiff branches. Leaves opposite, 3-foliolate; leaflets almost sessile, glabrous, thick in texture. Flowers solitary, axillary, primrose in colour. Fruits a berry.

Reproductive Cycle : March-May.

Materials Examined : HPU-EL-SKS-6474; B.C.S. (2,025m), Bemloe (2,014m), Benmore (2,114m), Chapslee (2,243m), Grand Hotel (2,105m), Raj Bhawan (2,132m).

Propagation Notes : Propagation by cutting, root suckers and layering. It can be grown in any fertile, well drained soil in full sun or partial shade.

Part Used : Whole Plant.

Folk Use : Commonly cultivated in gardens as an **ornamental** against some support.

Active Constituents : These are ceryl alcohol, α-amyrin, β-sitosterol, mannitol, quercetin, rutin, flavonoids and jasminin (**leaves**).

Uses in Literature : Known to be grown as an ornamental in gardens **(Ambasta, 1986; Anonymous, 1948-1976; Asolkar *et al.*, 1992; Graf, 1980; Hay *et al.*, 1983; Rastogi & Mehrotra, 1993; Satyal *et al.*, 2012b; Sood *et al.*, 2013)**.

Jasminum multiflorum Andrews

Syn.: *J. gracillimum* Hook.f.; *J. pubescens* (Retz.) Willd; *Mogorium multiflorum* (Burm.f.) Lam.; *M. pubescens* (Retz.) Lam.; *Nyctanthes multiflora* Burm. f.; *N. pubescens* Retz.

Family: Oleaceae.

Vern.: Chameli.

English, Hindi and Sanskrit Names:

Eng.	:	Common Jasmine, Downy Jasmine, Musk Jasmine;
Hindi	:	Chameli, Kunda, Kundo, Kundphul;
Sans.	:	Atimukha, Attahasa, Attapuspaka, Damana, Kunda, Madhya, Mahamoda, Makaranda, Manodna, Manohara, Manorama, Mughya, Muktapushpa, Tarapushpa, Vanahasa

Distribution : Throughout India.

Description : Densely villous scandent shrubs with young branches clothed with velvety pubescence. Leaves simple, opposite, ovate, more or less pubescent beneath and densely villous petiole. Calyx teeth with yellow hair. Flowers white in dense cymes. Fruits a berry.

Reproductive Cycle : November-April.

Material Examined : HPU-EL-SKS-6475; Grand Hotel (2,105m).

Propagation Notes : Propagated by stem cutting or by layering.

Parts Used : Whole Plant. Flower. Leaves.

Folk Uses : Grown as an **ornamental** plant. Flowers offered to various **deities**. Leaf-paste applied for **suppuration of boils**.

Active Constituents : These are friedelin, lupeol, betulin, α-amyrin and betulinic, ursolic and oleanolic acids (**leaves**), jasmolactones A, B, C and D, multifloroside, multiroside and 10-hydroxyoleoside-11-methyl ester, jasmultiside and multiflorin (**aerial parts**).

Uses in Literature : Known to be useful earlier as an antidote to snake-bite, emetic; and for burns, chest pain, cough and cold, eczema, hounding dogs (vet.), offerings and adornments, pimples, pin-worms, religious and ceremonial functions, scabies, stomachache, ulcer, and urine blockage **(Ambasta, 1986; Anonymous, 1948-1976; Dagar, 2003; Datt, 1996; Gogoi & Borthakur, 1991; Jain, 1991; Kapur & Nanda, 1996; Paliwal, 2009; Pande *et al.*, 2006; Prajapati *et al.*, 2006; Raju, 2000; Randhawa, 1957; Sharma, 2003; Sood *et al.*, 2010; Watt, 1889-1896)**.

Juglans regia L.

Syn.: *J. duclouxiana* Dode; *J. fallax* Dode; *J. kamaonia* (C.DC.) Dode; *J. orientis* Dode; *J. sinensis* (C.DC.) Dode.

Family: Juglandaceae.

Vern.: Akhrot, Khor.

English, Hindi and Sanskrit Names:

Eng. : Common Walnut Tree, European Walnut, Himalayan Walnut, Walnut Tree;
Hindi : Akhor, Akhrot, Khor;
Sans. : Akhoda, Akschota.

Distribution : Kashmir-S.E. Tibet: 1,500-3,000m.

Description : Large deciduous trees having velvety shoot, imparipinnate leaves which are about 25 - 30 cm long. Leaflets opposite or sub-opposite, elliptic, usually entire. Flowers in catkin, male and females ones appearing separately. Nuts ovoid ribbed with a hard shell.

Reproductive Cycle : April-October.

Materials Examined : HPU-EL-SKS-6477; Chhota Shimla (2,080m), Eagle Mount (2,152m), Sanitarium (2,060m).

Propagation Notes : Multiplication through seeds. Plant requires a deep, well-drained loam for its growth.

Parts Used : Bark. Wood. Fruit.

Folk Uses : Bark used for **scouring** teeth and **colouring** lips; also decoction of its green bark boiled with 'harad' (*Terminalia chebula*) and mustard oil considered an excellent **hair dye.** Kernels **edible**. Wood used for preparing high grade **furniture**.

Active Constituents : Active properties due to 5-hydroxy-1, 4-naptho-quinone, α-tetralone-(-) regiolone, berberine (**stem bark**); oxalic acid, resin and fixed oil of drying nature (**fruit**).

Uses in Literature : Recorded in Indian literature as edible (kernels), alexipharmic, alterative in scrofula, anthelmintic, aphrodisiac, astringent, cattle feed, cholagogue, detergent, ice-creams, insect repellent for storage of grains, laxative, nerve tonic, oil for culinary purposes and illuminant, oil paints, psycho-active plant, tooth brush, tooth powder; and for bed wetting, carving, chewing to color lips, confectionary (pastries), decolourising charcoal and activated carbon, diarrhoea, dispelling flies, drawing instruments, dyeing, eczema, foot and mouth disease (vet.), frames, frost bite, furniture, gun-stocks, handlooms, herpes, hoof (vet), lacquer work, leucoderma, menorrhagia, musical instruments, ornamental works, paints, ploughs, printing inks, rheumatism, rickets, scrofula, shafts, sore throat, sores of toes, stupefying fish, syphilis, and tanning **(Agarwal, 2003; Ambasta, 1986; Anonymous, 1948-1976; Aswal, 1996; Badhwar & Fernandez, 2011; Bennet *et al.*, 1991; Chakraborty, 2003; Chauhan, 1999; Chopra *et al.*, 1956; Das & Agarwal, 1991; Dey & Bahadur, 1973; Harborne & Baxter, 2001; Jain, 1991; Jain *et al.*, 1994; Kapur & Singh, 1996; Khare, 2004; Kirtikar & Basu, 1984; Lalramnghinglova, 1996; Negi & Pant, 1990; Negi *et al.*, 1993; Pande *et al.*, 2006; Pandey & Joshi, 2001; Prajapati *et al.*, 2006; Rana *et al.*, 1996; Saini, 1996a,b; Samant *et al.* 2001; Sharma, 1996; Sharma *et al.*, 2000, 2003; Singh, 1991, 1999; Singh & Kumar, 2000a; Sood *et al.*, 2010, 2012b, 2013; Uphof, 1968; Verma & Chauhan, 2006; Vishwanathan & Singh, 1996; Warman, 1999; Watt, 1889-1896).**

Juniperus chinensis L.

Syn.: *J. cabiancae* Vis.; *J. cernua* Roxb.; *J. dimorpha* Roxb.; *J. flagelliformis* Loudon; *J. fortunei* Carriere; *J. gaussenii* W.C. Cheng; *J. jacobii* Beissn.; *J. reevesiana* Endl.; *J. sphaerica* Lindl.; *J. virginica* Thunb.; *Sabina chinensis* (L.) Antoine; *S. dimorpha* (Roxb.) Antoine; *S. sphaerica* (Lindl.) Antoine; *S. struthiacea* (Knight) Antoine.

Family: Cupressaceae.

Vern.: Chinese Juniperus.

English Name:

Eng. : Chinese Juniper.

Distribution : Native of China, Magnolia, Manchuria, Korea and Japan; introduced in India.

Description : Shrubby trees with two kind of leaves. Juvenile ones needle - shaped and decurrent whereas adults scale - like, opposite, obtuse and appressed. Fruits globose, brownish-violet, 2-3 seeded.

Reproductive Cycle : April-October.

Material Examined : HPU-EL-SKS-6478; Bemloe (2,018m), Grand Hotel (2,105m).

Propagation Notes : Propagated through seeds, cuttings of mature wood of 5 - 10cm long with a heel in September/October in a cold frame and layering in September/October. Plant prefers a neutral or slightly alkaline soil for its growth.

Part Used : Whole Plant.

Folk Uses : Planted as an **ornamental hedge**.

Uses in Literature : Recorded to be used as antirheumatic, antitumorous, haemostatic, incense, ornamental; and for burns, scalds, hepatitis, skin diseases and cosmetic preparations **(Ambasta, 1986; Anonymous, 1948-1976; Asolkar *et al.*, 1992; Bailey, 1957; Bean, 1981; Chittendon, 1956; Huxley, 1992; Randhawa, 1960; Rushforth, 1987; Sheat, 1948; Sood *et al.*, 2010, 2013; Warda, 1985)**.

Juniperus horizontalis Moench

Syn.: *J. hudsonica* Forbes; *J. prostrata* Pers.; *J. racemosa* Risso; *J. repens* Nutt.; *Sabina prostrata* (Pers.) Antoine; *S. racemosa* (Risso) Antoine.

Family: Cupressaceae.

Vern.: Juniper.

English Name:

Eng. : Creeping Juniper.

Distribution : Northern America - Newfoundland to British Columbia.

Description : Procumbent shrubs with long trailing branches and leaves of two kind. Juvenile leaves needle- like whereas adult ones scale-like, acute- apiculate, mostly opposite, green or sometime bluish. Fruits blue-black, somewhat glaucous and having recurved pedicels. Seeds 1-6.

Reproductive Cycle : March-September.

Materials Examined : HPU-EL-SKS-6482; Bemloe (2,014m), Grand Hotel (2,105m), Raj Bhawan (2,132m).

Propagation Notes : Propagated through seeds, cuttings and layering. Seed requires a period of cold stratification. Plant succeeds in a sunny position in most well drained soils.

Parts Used : Whole Plant. Twigs.

Folk Uses : Plant used as a good **ground cover**. Young twigs burnt as **incense**.

Uses in Literature : Used earlier as beverages (fruit), diaphoretic, febrifuge, incense, ornamental; and for body care, cold, cough **(Bailey, 1957; Bean, 1981; Chittendon, 1956; Huxley, 1992; Rusforth, 1987; Sheat, 1948; Warda, 1985)**.

Juniperus procera Hochst. *ex* Endl.

Syn.: *J. abyssinica* K. Koch; *J. hochstetteri* Antoine; *Sabina procera* (Hochst. *ex* Endl.) Antoine.

Family: Cupressaceae.

Vern.: African Juniper.

English Names:

Eng. : African Juniper, East African Juniper.

Distribution : Native of E. Africa from E. Sudan to Zimbabwe; introduced in India in Nilgiris.

Description : Medium-sized trees reaching upto 25m. Crown broadly conical-rounded or irregular. Leaves of two forms, juvenile needle-like and adult scale-like and arranged in decussate pair or whorls of three. Cones berry-like, blue-black with a whitish waxy bloom and 2-5 seeded.

Reproductive Cycle : March-September.

Materials Examined : HPU-EL-SKS-6485; Bemloe (2,014m), Gorton Castle (2,094m), Raj Bhawan (2,132m).

Propagation Notes : Propagated by seeds requiring cold stratification and through cuttings and layering.

Part Used : Whole Plant.

Folk Use : Grown as an **ornamental** plant.

Active Constituents : These are five sesquiterpenes-epicedranediol, 4-ketocedrol, 3β-hydroxycedrol, isocedrolic acid, β-chamigrenic acid (**heartwood**), monoterpene hydrocarbons, sesquiterpene hydrocarbons, oxygenated sesquiterpene and oxygenated monoterpene (**essential oil**).

Uses in Literature : Reported to be grown as an ornamental; and for building houses, construction, furniture, making beehives (bark), pencils and essential oil **(Adams, 2004; Ambasta, 1986; Anonymous, 1948-1976; Asolkar *et al.*, 1992; Bailey, 1957; Farjon, 2005; Uphof, 1968; Warda, 1985)**.

Juniperus pseudosabina Fisch. & C.A.Mey.

Syn.: *J. centrasiatica* Kom.; *J. dumosa* Lindl. & Gordon; *J. lycia* Siev.; *J. turkestanica* Kom; *Sabina centrasiatica* (Kom.) Kom.

Family: Cupressaceae.

Vern.: Juniper.

English Name:

Eng. : Turkestan Juniper.

Distribution : Native of mountains of C. Asia, N.E Afghanistan, Tajikistan, W. China, W. Mongolia and S. C. Russia upto 4,100m.

Description : Shrubs or small-sized trees reaching 1-12m. Leaves of two forms, juvenile needle-like and adult ones scale-like on shoots. Cones berry-like, blue-black with a whitish waxy bloom and 1-seeded.

Reproductive Cycle : March-August.

Materials Examined : HPU-EL-SKS-6487; Gorton Castle (2,094m), Peterhoff (2,098m), Town Hall (2,250m).

Propagation Notes : Propagated by seeds, cuttings of mature wood of 5 - 10cm long with a heel in September/October in a cold frame and layering in September/October.

Part Used : Whole Plant.

Folk Use : Grown for **ornamental** purposes.

Active Constituents : These are amentoflavone, cupressuflavone, hinokiflavone and isocryptomerin (**leaves**), and seven new labdanediterpenes-(+)3-oxomanool, (+) 3α-hydroxymannol and its isomer (I), 3α-hydroxy-9β-10β-8(17), 12,(E)-labdadiene (II), 3α-hydroxy-14, 15-dihydroxy-8β, 13β-manoyloxide (III) and 3-oxo-12,13 (Z)-biformene (**aerial parts**).

Uses in Literature : Useful earlier as an incense, ornamental; and for kidney affections **(Adam, 2004; Ambasta, 1986; Chaudhury, 1993; Farjon, 2005; Mehra & Jain, 1976; Rastogi & Mehrotra, 1995a; Uphof, 1968; Wangchuk *et al.*, 2008; Warda, 1985)**.

Juniperus squamata **Buch.-Ham.**

Syn.: *J. densa* (Carriere) Gordon; *J. franchetiana* H. Lev. *ex* Kom; *Sabina squamata* (Buch.-Ham. *ex* D. Don) Antoine.

Family: Cupressaceae.

Vern.: Juniper, Thanara.

English Name:

Eng. : Flaky Juniper.

Distribution : Himalayas and W. China.

Description : Decumbent, wide spreading shrubs having needle-like leaves in whorls of 3 which are decurrent, spreading or loosely appressed, and slightly incurved. Fruits reddish-brown, becoming purplish black, 1-seeded.

Reproductive Cycle : April-October.

Materials Examined : HPU-EL-SKS-6488; B.C.S. (2,025m), Bemloe (2,010m), Cecil Hotel (2,090m), Convent of Jesus & Mary (2,104m), Raj Bhawan (2,132m).

Propagation Notes : Preferably grown in a well drained, neutral or slightly alkaline soil. Propagated mainly through seeds requiring cold stratification, cuttings and layerings in September-October.

Parts Used : Whole Plant. Wood.

Folk Uses : Grown as an **ornamental** plant. Infusion of plant in water applied for **suppurating skin diseases**. Wood used as **fuel** and **incense**.

Uses in Literature : So far described as fuel, incense; and for kidney inflammation, skin diseases, excess accumulation of watery fluid in body joints and ornamental purposes **(Ambasta, 1986; Anonymous, 1948-1976; Asolkar *et al.*, 1992; Bailey, 1957; Bean, 1981; Brickell, 1990; Chittendon, 1956; Huxley, 1992; Manandhar, 2002; Mehra & Jain, 1976; Rushforth, 1987; Sharma & Rana, 2000; Sheat, 1948; Wangchuk *et al.*, 2008; Warda, 1985)**.

Kerria japonica (L.) DC.

Syn.: *Rubus japonicus* L.

Family: Rosaceae.

Vern.: Japani Rose.

English Names:

Eng. : Bachelor's Button, Japanese Rose.

Distribution : Native of Japan; a rare garden escape in Britain.

Description : Deciduous shrubs upto 3m possessing weak arching stems often scrambling over other vegetation or rocks and alternate, simple leaves with a sharply serrated margin. Flowers bright yellow. Achenes dry, 1-seeded.

Reproductive Cycle : March-May.

Material Examined : HPU-EL-SKS-6490; Advance Study (2,128m), Bemloe (2,018m).

Propagation Notes : Plant prefers good loamy soil. Multiplication through seeds, division of suckers removed from established plants during dormant season, cuttings of young shoot, layering and root division also successful.

Part Used : Whole Plant.

Folk Use : Grown as an **ornamental** plant.

Uses in Literature : Reported for cough, edible purposes as a rich source of vit.C (leaves), women's complaints; and as an ornamental **(Ambasta, 1986; Anonymous, 1948-1976; Asolkar *et al.*, 1992; Bailey, 1957; Chittendon, 1956; Davis, 1990; Huxley, 1992; Phillips & Rix, 1989; Randhawa, 1960; Rastogi & Mehrotra, 1993; Thomas, 1992)**.

Lagerstroemia indica L.

Syn.: *L. chinensis* Lam.; *L. elegans* Wall. *ex* Paxton; *L. minor* Retz.; *Velaga globosa* Gaertn.

Family: Lythraceae.

Vern.: Har-Shingar.

English and Hindi Names:

Eng.	:	China Privet, Crape, Crape Myrtle, Indian Lilac, Myrtle;
Hindi	:	Phurush, Saoni, Telinga-china.

Distribution : Throughout India; originated from China.

Description : Deciduous shrubs or trees with elliptical leaves, pedicelled flower which are solitary or in penduncled cymes at the end of branchlets. Petals crumpled and curled. Capsules globose.

Reproductive Cycle : June-October.

Materials Examined : HPU-EL-SKS-6491; Hainault (2,294m), Spring Field (2,075m), St. Edward's School (2,060m).

Propagation Notes : Succeeds in most well-drained soils in a sunny sheltered position. Propagated through seeds which are sown as soon as ripe, cuttings of half-ripe wood in July/August, and root cuttings of 4cm long in December.

Parts Used : Whole Plant. Flower.

Folk Uses : Grown as an **ornamental**. Paste of flowers applied externally to heal **cuts** and **wounds**.

Active Constituents : Active properties due to delphinidin-3-arabinoride, petnnidin-3-arabinoside, malridin-3-arasinoside, gallic acid, mrthyl gallate and ellagic acid (**flower**).

Uses in Literature : So far reported as astringent, depurative, diuretic, febrifuge, hydrogogue, narcotic, purgative, stimulant, styptic; and for apathae of mouth, ornamental purposes and timber **(Agarwal, 2003; Ambasta, 1986; Anonymous, 1948-1976; Bean, 1981; Bhattacharjee, 2001; Chatterjee & Pakrashi, 1997; Chittendon, 1956; Chopra *et al.*, 1986; Cowan, 1950; Davis,**

1990; Farooq, 2005; Hay *et al.*, 1983; Huxley, 1992; Mishra & Chandra, 1998; Paliwal, 2009; Phillips & Rix, 1989, 1998; Randhawa, 1957; Sheat, 1948; Sood *et al.*, 2013; Watt, 1889-1896).

Leea asiatica (L.) Ridsdale

Syn.: *L. aspara* Edgew.; *L. crispa* L.; *L. edgeworthii* Sant.; *L. herbacea* Buch.-Ham.; *L. pinnata* Andrews; *L. pumila* Kurz; *Phytolacca asiatica* L.

Family: Leeaceae.

Vern.: Gan-Gola.

Hindi and Sanskrit Names:

Hindi : Kurkurjihava;
Sans. : Chatri, Karkatajihava, Kukkurajihva.

Distribution : Tropical Himalayas: ascending to 1,700m.

Description : Stout, spreading pubescent shrubs having grooved stems and branches. Leaves pinnate, rarely bipinnate. Leaflets ovate-oblong, acuminate, upper surface often scabrous. Flowers yellow-green, in spreading leaf opposed cymes. Berries black when ripe.

Reproductive Cycle : July-August.

Materials Examined : HPU-EL-SKS-6493; Bemloe (2,010m), Kanlog Cemetery (1,984m),

Propagation Notes : Propagated naturally by seeds and vegetative methods.

Parts Used : Leaves. Stem. Fruit.

Folk Uses : Fruit and juvenile stems cooked as **vegetable**. Leaves used as **fodder**; its poultice applied to heal **cuts** and **wound**.

Uses in Literature : Recorded to be used as anthelmintic, astringent, edible (fruits, leaves, roots), febrifuge, sudorific; and for cardiac diseases, cuts, diarrhoea, dysentery, erysipelas, leprosy, rheumatism, skin diseases, snake bites, urinary complaints, vegetable, wounds and against guinea worms **(Agarwal, 2003; Ambasta, 1986; Bennet *et al.*, 1991; Chandra, 1997; Chatterjee & Pakrashi, 1997; Chaudhuri, 1993; Dwivedi, 2007; Jain, 1991; Kirtikar & Basu, 1984; Kumar, 2003; Murthy *et al.*, 2003; Pande *et al.*, 2006; Parrotta, 2001; Prajapati *et al.*, 2006; Roy *et al.*, 1998; Singh *et al.*, 1996b; Sinha, 1996; Sood & Thakur, 2004; Sood *et al.*, 2012b; Watt, 1889-1896).**

Ligustrum lucidum W.T. Aiton

Syn.: *Esquirolia sinensis* H. Lev.; *L. esquirolii* H. Lev.; *L. magnoliifolium* Dippel; *L. roxburghii* Blume; *L. wallichii* Vis.; *Olea clavata* G. Don; *Phillyrea paniculata* Roxb.; *Visiania paniculata* (Roxb.) DC.

Family: Oleaceae.

Vern.: Pataal Ped.

English Names:

Eng. : Chinese Privet, Glossy Privet.

Distribution : E. Asia-China; Khasia hills upto 1,500m.

Description : Evergreen shrubs possessing grey bark, coriaceous glossy leaves and creamy white fragrant flowers in panicles. Fruits oblong, blue black when ripe.

Reproductive Cycle : July-November.

Material Examined : HPU-EL-SKS-6495; St. Edward's School (2,060m).

Propagation Notes : Propagated through seeds and cuttings of half-ripe wood with a heel in July/August. Plant succeeds in any soil which is not water-logged.

Part Used : Whole Plant.

Folk Uses : Grown for **ornamental purposes** and as a **hedge** plant.

Active Constituents : These are mannitol, rutin, luteolin-7-β-D-glucoside and quercetin (**flower**), cyanidin-3-glucoside, cyaniding-3-rutinoside, oleanolic acid and malvidin-3-rutinoside-5-glucoside (**fruit**).

Uses in Literature : Used so far as anodyne, antiseptic, antitumorous, cardiotonic, diaphoretic, diuretic, edible (leaves), febrifuge, ophthalmic, tonic, hedge plant; and for hypertension, insomnia, menopausal problems, respiratory problems rheumatic pain, and making handles, insect box and tools **(Ahmed & Borthakur, 2005; Ambasta, 1986; Anonymous, 1948-1976; Bean, 1981; Chittendon, 1956; Davis, 1990; Huxley, 1992; Rastogi & Mehrotra, 1991, 1995a,b; Sheat, 1948; Sood *et al.*, 2013; Thomas, 1992; Uphof, 1968; Usher, 1974)**.

Ligustrum X vicaryi Rehder

Family: Oleaceae.

Vern.: Privet.

English Names:

Eng. : Golden Privet, Vicary Previt.

Distribution : Cultivated and of garden origin.

Description : A shrubs of hybrid origin between California privet *Ligustrum ovalifolium* Hassk. and European or common privet *Ligustrum vulgare* L. Leaves golden yellow, lance – shaped. Flowers white, in panicles. Fruits round, blue-black.

Reproductive Cycle : May-July.

Materials Examined : HPU-EL-SKS-6497; Advance Study (2,129m), Oak Over (2,154m), St. Edward's School (2,060m).

Propagation Notes : Propagated by seeds and cuttings of half-ripe wood with a heel in July/August.

Part Used : Whole Plant.

Folk Use : Cultivated as a golden **hedge**.

Uses in Literature : Used for hedging and as shrub borders **(Agate, 1998; Brickell, 2003; Davis, 1990; Graf, 1981)**.

Liriodendron tulipifera L.

Syn.: *L. procerum* Salisb.; *L. truncatifolium* Stokes; *Tulipifera liriodendrum* Mill.

Family: Magnoliaceae.

Vern.: Tulip Tree.

English Name:

Eng. : Tulip Tree.

Distribution : Eastern N. America - Nova Scotia to Florida.

Description : Huge trees with spreading branches and glabrous unbranched trunk. Leaves broad with one apical and 1-4 basal lobes on each side. Flowers greenish- yellow, orange within. Stamens shorter than petals. Fruits cone - like, brown.

Reproductive Cycle : June-August.

Material Examined : HPU-EL-SKS-6498; Advance Study (2,129m).

Propagation Notes : Propagated through seeds requiring cold stratification and layering in spring. Plant succeeds in a deep rich soil and a sheltered position.

Part Used : Whole Plant.

Folk Use : Cultivated as an **ornamental** plant for its handsome foliage and large flowers.

Uses in Literature : Recorded to be used as anthelmintic, antipyretic, aphrodisiac, diuretic, diaphoretic, stimulant, tonic, ornamental, avenue and hedge; and for cardiac problems, boils, burns, cough, dysentery, fever, flavouring beer (roots), hysteria, indigestion, inflammation, nervine, rheumatism, wounds, cabinet box, interior finish, shingles **(Ambasta, 1986; Anonymous, 1948-1976; Bailey, 1957; Bean, 1981; Bhattacharjee, 2001; Brickell, 1990; Chittendon, 1956; Cowan, 1950; Hill, 1951; Huxley, 1992; Jaeger, 1961; Pool, 1941; Rastogi & Mehrotra, 1991; Sheat, 1948; Uphof, 1968)**.

Lonicera japonica Thunb.

Syn.: *Caprifolium chinense* Wats. *ex* Loudon; *C. japonicum* (Thunb.) Dum. Cours.; Don; *C. roseum* Lam.

Family: Caprifoliaceae.

Vern.: Lushai.

English Names:

Eng. : Honeysuckle, Japanese Honeysuckle.

Distribution : C. Nepal-Bhutan, China and S.E Asia: 1,500-2,400m.

Description : Semi-evergreen shrubby climbers having opposite coriaceous leaves; opposite pairs often connate at the base, shortly - stalked, dark green on upper side and glaucous beneath, entire, acute or obtuse at tip. Peduncles in subterminal panicles. Berries black.

Reproductive Cycle : April-August.

Materials Examined : HPU-EL-SKS-6499; Advance Study (2,129m), Bemloe (2,012m), Grand Hotel (2,105m), Kasumpti (2,102m).

Propagation Notes : Propagated by seeds, layering and semi-ripe cuttings. Plant can be grown in any well-drained soil in full sun or partial shade.

Parts Used : Whole Plant. Leaves. Flower.

Folk Uses : Commonly planted as a favourite **climbing** plant in gardens because of its exceedingly beautiful and sweet - scented flowers.

Parboiled leaves used as a **vegetable**. Flowers **sucked** for their sweet nectar or its decoction made into a **syrup** and **pudding**.

Active Constituents : Reported to possess lonicerin (**leaves**), (-) cis 2,6,6-trimethyl - 2 - vixyl - 5 -hydroxytetrahydropyran, linalool, pinene, hexane-1, cis-3-hexa-nol-1, cis and trans - 2 – methyl – 2 – vinyl – 5 - (a -hydroxyisopropyl) tet-rahydrofuran, geraniol, α-terpineol, benzyl alcohol, β-phenyl ethyl alcohol carvacrol, eugenol (**flower essential oil**), loniceroside A, B, epivogeloside, secologanin dimethyl, acetal, coganin and voeloside (**aerial parts**), chlorogenic and isochlorogenic acids (**flower**).

Uses in Literature : Known to be used as antiallergic, antibacterial, anti-inflammatory, antipyretic, antiviral, diuretic, edible (flowers and leaves), febrifuge, fodder, stomachic, quick groundcover; and for boils, cold, cordage, diarrhoea, dysentery, erythema, impetigo, malaria, measles, pain, rheumatism, syphilis, urticaria and fencerows **(Ambasta, 1986; Anonymous, 1948-1976; Brickell, 1990; Davis, 1990; Huxley, 1992; Jain, 1991; Lalramnghinglova, 1996; Paliwal, 2009; Phillips & Rix, 1989; Prajapati *et al.*, 2006; Randhawa, 1957, 1960; Rastogi & Mehrotra, 1990, 1993, 1995a; Sood *et al.*, 2013; Thomas, 1992; Yang & Blunden, 1993)**.

Lonicera quinquelocularis Hardw.

Syn.: *Caprifolium quinqueloculare* Kuntze.

Family: Caprifoliaceae.

Vern.: Bakru, Chaunsh.

English Name:

Eng. : The Himalayan Honey Suckle.

Distribution : Himalayas: 1,800-3,200m.

Description : Erect deciduous shrubs possessing shortly -stalked, ovate or broadly lanceolate, more or less pubescent leaves and yellow sessile flowers in leaf-axil. Berries distinct, ovoid, white, translucent.

Reproductive Cycle : April-September.

Materials Examined : HPU-EL-SKS-6501; Bemloe (2,012m), Boileauganj Cemetery (1,998m), Chhota Shimla (2,080m).

Propagation Notes : Propagated by seeds and through cuttings of half ripe wood. Plant prefers a good moist soil in a sunny position.

Parts Used : Leaves. Branches. Wood.

Folk Uses : Leaves serve as an excellent **fodder**. Branches good for making **walking sticks** and **tool handles**. Wood used as **fuel**.

Active Constituents : These are sitosterol, hexacosanol, octacosanol, nonacosane and L-inositol (**aerial parts**).

Uses in Literature : Used so far as laxative, fibres, fodder; and for carving, food utensils and turney **(Ambasta, 1986; Anonymous, 1948-1976; Kapur, 1996a; Kapur & Singh, 1996; Rastogi & Mehrotra, 1991; Sood & Thakur, 2004; Watt, 1889-1896)**.

Lyonia ovalifolia **(Wall.) Drude**

Syn.: *Pieris ovalifolia* D. Don; *Xolisma ovalifolia* (Wall.) Rehder.

Family: Ericaceae.

Vern.: Bhelar.

Distribution : S.E. Asia: 1,500-3,000m.

Description : A small glabrous tree. Leaves shortly - stalked, ovate or oblong, entire, acute. Flowers white, in racemes. Capsule globose.

Reproductive Cycle : April-May.

Materials Examined : HPU-EL-SKS-6504; Eagle Mount (2,152m), Kanlog Cemetery (1,997m).

Propagation Notes : Propagated by seeds. Plant grows in dappled shade in moist humus-rich, lime-free soils.

Part Used : Leaves.

Folk Uses : Paste of leaves applied as poultice to heal **skin eruptions** and to check **itching**. Leaves **poisonous** to cattle.

Active Constituents : These are putatine, β-sitosterol-b-d-glucoside, ursolic acid, lyofolic acid, hexacasane, hexacosanol, hyperoside, β-sitosterol, taxaxerol and quercetin (**leaves**).

Uses in Literature : Known to be useful as edible (flower and fruits eaten), fuel, insecticidal, poison (vet.); and for boils, cutaneous troubles, pimples, scabies (vet.), skin eruptions, worms, wound and turnery work **(Ahmed & Borthakur, 2005; Ambasta, 1986; Anonymous, 1948-1976; Ansari, 1991; Arya *et al.*, 1999; Bhattacharjee, 2001; Chaudhuri, 1993; Dash *et al.* 2003; Farooq, 2005; Jain, 1991; Kapur & Nanda, 1996; Lalramnghinglova, 1996; Manandhar, 1990; Mao, 1993; Panda, 1996; Pande *et al.*, 2006; Samant *et al.*, 1996; Sharma, 1996, 2003; Singh & Kumar, 2000a; Sood *et al.*, 2009)**.

Magnolia grandiflora L.

Syn.: *M. angustifolia* Millais: *M. elliptica* Link; *M. exoniensis* Millais; *M. foetida* (L.) Sarg.; *M. lanceolata* Link; *M. longifolia* Sweet; *M. rotundifolia* Millais.

Family: Magnoliaceae.

Vern.: Magnolia.

English and Hindi Names:

Eng. : Bull Bay, Southern Magnolia;
Hindi : Andachampa, Him Champa.

Distribution : Indigenous to S. E. United States (N. Carolina, Florida and Texas).

Description : Evergreen trees possessing stiff leaves bearing large wax-like, fragrant white flowers with leaves issuing from great silky-hairy conical buds. Leaves oval, oblong - obovate, tapering both ways, shining green above and rusty tomentose beneath. Flowers large with petaloid sepals; petals 6-12, creamy-white, obovate; filament purple and extremely fragrant. Fruits brown, cylindrical and heavy.

Reproductive Cycle : April-August.

Materials Examined : HPU-EL-SKS-6506; Armsdell (2,102m), Grand Hotel (2,105m), Raj Bhawan (2,132m), Sanitarium (2,060m).

Propagation Notes : Propagated by cuttings, air layering in early spring and seeds. Plant succeeds in a warm position in a moderately rich free soil of an open texture.

Part Used : Whole Plant.

Folk Use : Planted for **ornamental** purposes.

Uses in Literature : Known earlier as condiments, diaphoretic, edible (flower), fuel, hypotensive, ornamental stimulant, tonic; and for baskets, crates, furniture, malaria, rheumatism, woodenware **(Ambasta, 1986; Anonymous, 1948-1976; Bailey, 1957; Brickell, 1990; Chittendon, 1956; Chopra *et al.*, 1986; Davis, 1990; Paliwal, 2009; Pool, 1941; Randhawa, 1957, 1965; Sood *et al.*, 2013; Thomas, 1992; Uphof, 1968; Usher, 1974; Warman, 1999)**.

Magnolia obovata Thunb.

Syn.: *Liriodendron liliiflorum* Steud.; *M. honogi* P. Parm.; *M. hypoleuca* Siebold & Zucc.

Family: Magnoliaceae.

Vern.: Japani Magnolia.

English Names:

Eng. : Japanese Big Leaf Magnolia, Japanese White Bark Magnolia.

Distribution : Native of Japan.

Description : Trees with purplish branchlets. Leaves obovate - oblong, nearly or quite obtuse, glaucous and pubescent beneath at least while young. Flowers white appearing with foliage, cup-shaped, fragrant. Fruits oblong-cylindric, scarlet.

Reproductive Cycle : January-April.

Material Examined : HPU-EL-SKS-6507; Advance Study (2,128m).

Propagation Notes : Best grown in a warm position in a moderately rich free soil of an open texture. Multiplied by seeds and layering in early spring.

Part Used : Whole Plant.

Folk Use : Grown as an **ornamental** plant.

Uses in Literature : Used so far as a substitute of Boxwood; and for cabinet-work, engraving, furniture and utensils **(Ambasta, 1986; Anonymous, 1948-1976; Bailey, 1957; Randhawa, 1960)**.

Magnolia X *soulangeana* Hort.

Syn.: *M.*X *brozzonii* Millais; *M.*X *lenneana* (Lem.) Koehne; *Yulania* X *lenneana* Lem.; *Y.* X *soulangeana* (Soul.-Bod.) D.L.Fu.

Family: Magnoliaceae.

Vern.: Pink Magnolia.

English Names:

Eng. : Chinese Magnolia, Saucer Magnolia, Tulip Magnolia.

Distribution : Cultivated in cooler parts of India.

Description : Large shrubs or small trees of hybrid origin between *M. denudata* and *M. liliflora*. Leaves broad ovate - obovate abruptly short pointed. Flowers open well - shaped, purplish or rose coloured outside and white within. Petals few, oblong, ovate - obovate.

Reproductive Cycle : January-March.

Material Examined : HPU-EL-SKS-6509; Hainault (2,294m).

Propagation Notes : Propagated by seeds as well as through cuttings in July-August.

Part Used : Whole Plant.

Folk Use : Planted for **ornamental** purposes.

Use in Literature : Grown as an ornamental tree in garden **(Bailey, 1957; Pool, 1941)**.

Malus pumila Mill.

Syn.: *Malus communis* Poir; *M. domestica* Borkh.; *Pyrus malus* L; *P. pumila* (Mill) Steud.

Family: Rosaceae.

Vern.: Seb, Seu.

English, Hindi and Sanskrit Names:

Eng.	:	Apple;
Hindi	:	Seb, Seo, Sev;
Sans.	:	Seba, Seva.

Distribution : Cultivated in cooler parts of India.

Description : Trees upto 10m tall with ovate, obtusely serrated leaves and white flowers tinged with red. Calyx lobes persistent. Petals with a tomentose claw. Fruits globose, intruded at both ends.

Reproductive Cycle : May-October.

Materials Examined : HPU-EL-SKS-6510; Chhota Shimla (2,080m), Hainault (2,294m).

Propagation Notes : Propagated mainly by shield budding, bench grafting and tongue grafting on seedlings raised from seeds.

Part Used : Fruit.

Folk Uses : Fruits raw **edible**, **pickled** and cooked as **vegetable**.

Active Constituents : Fruits yield vomifoliol-4-O-β-D-xylopyranoside-6-O-β-D-glucoside, quercetin-3-O-α-D-glactoside, 3-O-xyloside, 3-O-α-L-arabinofuranoside, 3-O-rhamnoside and 3-O- β-D-glucoside, rectin and phloretamide.

Uses in Literature : Known earlier as anthelmintic, edible (fruit), heart tonic, hypnotic, refrigerant; and for apple sauce, cider, cog-wheels, fever, jam, jelly, knobs, mallet heads, marmalade, tool handles and turnery **(Ambasta,**

1986; Badhwar & Fernandez, 2011; Bennet *et al.*, 1991; Jaeger, 1961; Kirtikar & Basu, 1984; Manral & Pande, 2004; Pande *et al.*, 2006; Singh *et al.*, 1996a; Singh & Kumar, 2000a; Sood *et al.*, 2013; Uphof, 1968; Watt, 1889-1896).

Malvaviscus arboreus Cav. var. *mexicanus* Schltdl.

Syn.: *Achania pilosa* Sw.; *M. conzattii* Greenm.; *M. grandiflorus* Kunth; *M. pulvinatus* A. Rich.; *M. sagraeanus* A. Rich.; *M. sepium* Schltdl.; *Pavonia spiralis* Cav.

Family: Malvaceae.

Vern.: Gudhal.

English and Hindi Names:

Eng. : Gaint Turk's Cap;
Hindi : Mirchi.

Distribution : Native of America; cultivated in tropical gardens.

Description : Shrubs upto 2.5m tall. Leaves alternate, simple, broadly ovate, coarsely-toothed with upper surface rough. Flowers usually solitary, axillary, scarlet. Fruits red, tomato-like.

Reproductive Cycle : Major part of the year.

Material Examined : HPU-EL-SKS-6511; Annandale (1,859m).

Propagation Notes : Propagated by cutting, transplants or seeds. Plant prefers moist soil, but once established it is fairly drought tolerant.

Part Used : Flowers.

Folk Uses : Planted as an **ornamental**; flowers considered **sacred** and offered to various **deities**. Flower base juice considered **refrigerant**.

Active Constituents : These are β-sitosterol and its glucoside, allantoin, betaine, kaempferol and their 3β-glucoside and 3β-sophoroside (**flower**).

Uses in Literature : Grown as an ornamental plant **(Paliwal, 2009; Rastogi & Mehrotra, 1993; Sood *et al.*, 2013)**.

Morus alba L.

Syn.: *M. acidosa* Griff.; *M. intermedia* Perr.; *M. multicaulis* Perr.; *M. tatarica* L.

Family: Moraceae.

Vern.: Toot.

English, Hindi and Sanskrit Names:

Eng. : Mulberry, White Mulberry;

Hindi : Chinni, Chun, Tul, Tulklu, Tut;
Sans. : Brahmadaru, Brahmakashtha, Brahmaneshta, Brahmanya, Kramuka, Madasdra, Madhupippali, Mridupippali, Mridusara, Nilarangaka, Nilavrintaka, Palashika, Puna, Pusha, Shalmali, Supushpa, Surupa, Tuda, Tula, Tuta, Yusha.

Distribution : Cultivated in Punjab; N.W. Himalayas and W. Tibet, ascending to 3,300m.

Description : A middle-sized deciduous tree. Leaves acute, often cordate, teeth small, nearly uniform. Male spikes puberulous; flowers minutely pedicellate. Female spikes ovoid, pedunculate. Fruits white or red, sweet.

Reproductive Cycle : March-May.

Materials Examined : HPU-EL-SKS-6512; Advance Study (2,128m), B.C.S. (2,025m), Boileauganj Cemetery (1,998m), Kanlog Cemetery (1,990m).

Propagation Notes : Propagated by seeds. Also, multiplied through cuttings of half-ripe wood of 7 - 10cm with a heel in July/August. Plants for its growth prefer a well-drained loamy soil in a sunny position

Parts Used : Leaves. Fruit. Wood.

Folk Uses : Fruits **edible**. Tender leaves relished as **vegetable** and also used for **culinary preparations**. Leaves used as a good quality **fodder**. Wood used for **furniture** and **fuel**.

Active Constituents : Main ingredients are β - sitosterol, stigmasterol, campesterol, cholesterol, heptanoic, caprylic, pelargonic, capric, myristic, palmitic, palmitoleic, methyl heptadecanoic, stearic, oleic, linoleic, nonodecanoic, linilenic acids (**fruit**), astragalin, moracin C and N, β–amyrin, β–sitosterol, cis-5-hydroxy-L-pipecolic acid, bergapten, scopoletin, umbelliferone, isoquercetin, astragalin (**leaves**).

Uses in Literature : Reported in literature as anthelmintic, edible (fruits), diaphoretic, expectorant, febrifuge, fodder, purgative, refrigerant, vegetable, vermifuge; and for agricultural implements, boats, buildings, dyspepsia, feeding silk worms, furniture, headache, inflamed eyes, melancholia, paper, silk worm and sore throat **(Agarwal, 2003; Ambasta, 1986; Anonymous, 1948-1976; Arora & Pandey, 1996; Bennet *et al.*, 1991; Chevallier, 1996; Das & Agarwal, 1991; Dey & Bahadur, 1973; Dwivedi, 2007; Farooq, 2005; Hadfield, 1956; Harborne & Baxter, 2001; Jain, 1991; Kaushik & Dhiman, 1999; Kirtikar & Basu, 1984; Kshirsagar & Singh, 2000; Negi & Gaur, 1991; Parrotta, 2001; Pool, 1941; Prajapati *et al.*, 2006; Reddi *et al.*, 2005;**

Roy *et al.*, 1998; Saini, 1996a; Samant *et al.*, 2001; Sharma & Rana, 2005; Singh & Kumar, 2000a; Sood & Thakur, 2004; Sood *et al.*, 2012b, 2013; Uphof, 1968; Warman, 1999; Watt, 1889-1896).

Morus nigra L.

Family: Moraceae.

Vern.: Toot.

English and Hindi Names:

Eng. : Black Mulberry Tree, Common Mulberry;
Hindi : Shahtut.

Distribution : Throughout India.

Description : Trees having firm, thick, fine – nerved and broad cordate leaves. Flowers greyish-white, in catkins. Sepals and styles hairy. Sorosis purple.

Reproductive Cycle : March-June.

Material Examined : HPU-EL-SKS-6514; Advance Study (2,128m).

Propagation Notes : Plant for its growth prefers a warm moist but well-drained loamy soil in a sheltered sunny position. Propagated through seeds and cuttings of half-ripe wood with a heel in July/August and layering in autumn.

Parts Used : Leaves. Fruit. Wood.

Folk Uses : Leaves used as **fodder** for cattle. Fruits **relished**. Wood employed for **furniture** and as **fuel**.

Uses in Literature : Reported earlier as an edible (fruits-jams, jellies), coolant, laxative, purgative, refrigerant and vermifuge; and good for anaemia, conjunctivitis, cosmetics, expelling worms **(Agarwal, 2003; Ambasta, 1986; Anonymous, 1948-1976; Hadfield, 1956; Harborne & Baxter, 2001; Kirtikar & Basu, 1984; Kurian, 1995; Pool, 1941; Samant *et al.*, 2001; Sood *et al.*, 2012b; Uphof, 1968; Watt, 1889-1896).**

Morus serrata Roxb.

Syn.: *M. alba* var. *serrata* (Roxb.) Bureau; *M. gyirongensis* S.S. Chang; *M. pabularia* Decne.

Family: Moraceae.

Vern.: Kimu.

English and Hindi Names:

Eng. : Himalayan Mulberry;
Hindi : Kimu, Himu.

Distribution : Punjab - Kumaon, inner range of Himalayas: 1,200-2,700m.

Description : Trees having broadly ovate, long - pointed, usually cordate leaves with large unequal teeth. Male and female flowers usually on different trees. Sorosis purple.

Reproductive Cycle : April-May.

Materials Examined : HPU-EL-SKS-6516; Advance Study (2,128m), Chhota Shimla (2,084m), Gorton Castle (2,094m).

Propagation Notes : Propagated by seeds as well as cuttings of half-ripe wood with a heel in July/August.

Parts Used : Twig. Fruits. Wood.

Folk Uses : Ripe fruits **edible**. Twigs used as **fodder**. Wood used for **furniture work** and as **fuel**.

Uses in Literature : Described earlier for agricultural implements, carving, cheap type of guns and rifles, edible purposes, furniture, ornamental panelling, sport goods, toys, rearing silk-worms; and as fodder **(Agarwal, 2003; Ahmed & Borthakur, 2005; Ambasta, 1986; Anonymous, 1948-1976; Arora & Pandey, 1996; Badhwar & Fernandez, 2011; Bennet *et al.*, 1991; Deokota & Chhetri, 2009; Kapur, 1996a; Mishra & Chandra, 1998; Nautiyal *et al.*, 2003; Pande *et al.*, 2006; Rana *et al.*, 1996; Roy *et al.*, 1998; Sahni, 1998; Samant *et al.*, 2001; Sharma, 1996; Sood *et al.*, 2012b, 2013).**

Musa paradisiaca L.

Syn.: *Musa sapientum* L.

Family: Musaceae.

Vern.: Kela.

English, Hindi and Sanskrit Names:

Eng. : Adam's fig, Banana, Fig of India, Plantain;
Hindi : Kach-kula, Kela, Maoz-kulo;
Sans. : Alabu, Bhanupriya, Kadali, Rambha, Varavriksha.

Distribution : Indigenous to tropical Asia; cultivated throughout India.

Description : A monoecious perennial pseudostem tree with rhizome and false aerial stem upto 10m tall. Leaves oblong. Inflorescence drooping and appears at top of aerial stem having lower bracts with female flowers, above these are male flowers. Fruits oblong, 3-gonous. Seeds numerous, black or brownish black.

Reproductive Cycle : Throughout the year.

Material Examined : HPU-EL-SKS-6517; Boileauganj Cemetery (1,998m).

Propagation Notes : Suckers from rhizome employed as planting material for vegetative propagation.

Parts Used : Leaves. Fruit.

Folk Uses : Leaves used for decoration in various **religious** ceremonies. Unripe fruits cooked as **vegetable**; ripe ones **eaten** as well as used in preparation of various **dishes**, **jam** and **jellies**.

Active Constituents : Sterylacylglucosides – sitoindosides 1 and 11, 3 – methyl butanoate, 3 – methyl pentanoate, acetate, butanoate, 3 – methyl butanoate and pentanoate esters isolated from **fruits**.

Uses in Literature : Known so far as anthelmintic, antidote to opium, astringent, coolant, edible (flowers, fruits, vegetable, jams, jellies, pulp flour), laxative, purgative, tonic; and good for ascites, asthma, bronchitis, burns, cholera, colour extraction, diabetes, diarrhoea, dropsy, dysentery, dyspepsia, goitre, ear diseases, gonorrhoea, heart burn, hypertension, intestinal disorders, kidney stones, leprosy, leucorrhoea, nephritis, ophthalmia, otalgia, piles, scabies, snake bite, swellings, throat infection, ulcer, cuts and wounds, vaginal and urinary discharges, and other vascular diseases **(Agarwal, 2003; Ambasta, 1986; Anonymous, 1948-1976; Bennet *et al.*, 1991; Bose, 1984; Chakrabarty & Balakrishnan, 2003; Chopra *et al.*, 1956; Chowdhury *et al.*, 1996; Cowan, 1950; Dagar & Dagar, 1996; Dhiman, 2003; Dwivedi, 2007; Ghate, 1998; Gill *et al.*, 1993, 1997; Hembrom, 1996; Jha & Basak, 1994; Khanna, 2002; Kirtikar & Basu, 1984; Kurian, 1995; Lalramnghinglova, 1996; Mohanty & Padhy, 1996; Nandanakunjidam & Abirai, 2005; Pande *et al.*, 2006; Pandey & Joshi, 2001; Prajapati *et al.*, 2006; Rastogi & Mehrotra, 1993; Rawat & Chowdhury, 1998; Sahoo & Satapathy, 2009; Saikia & Nath, 2003; Saini, 1996a; Saklani & Jain, 1994; Sen *et al.*, 2002; Sharma *et al.*, 2000, 2003; Singh & Pandey, 1998; Siwakoti & Siwakoti, 2000; Sood *et al.*, 2013; Thomas & Britto, 2003; Uphof, 1968; Verma *et al.*, 1995; Viswanathan *et al.*, 2006; Watt, 1889-1896)**.

Myrica esculenta Buch.-Ham. *ex* D. Don

Syn.: *M. sapida* Wall.

Family: Myricaceae.

Vern.: Kaphal.

English, Hindi and Sanskrit Names:

Eng.	:	Box Myrtle;
Hindi	:	Kaiphal;
Sans.	:	Aranya, Bhadra, Kaitarya, Mahakumbhi, Purusha.

Distribution : Kashmir-Bhutan, Myanmar, China, S. E. Asia: 1,000-2,300m.

Description : Small glabrous trees possessing oblong -lanceolate, leathery leaves which are crowded towards the end of branches. Lower surface of leaves pale or rust coloured, minutely gland-dotted and aromatic. Pedicel short, pubescent. Male flowers in catkins, solitary in leaf-axils or sessile on a common drooping, axillary stalk. Female ones in axillary erect spikes. Drupes sessile, scaly, ovoid, flesh red, stone wrinkled and pitted.

Reproductive Cycle : November-May.

Material Examined : HPU-EL-SKS-6519; Boileauganj Cemetery (1,998m).

Propagation Notes : Propagated by seeds, suckers and layering.

Parts Used : Leaves. Fruit.

Folk Uses : Leaves used as **fodder** and its decoction beneficial in **purifying blood**. Fruits considered **carminative** and as rich source of **vit. C**; also added in preparation of local wine, '**sur**'.

Active Constituents : These are embelfin (3%) and quercitol (**fruit**).

Uses in Literature : Described in literature as antipyretic, antiseptic, astringent, bronchodilator, carminative, diuretic, edible (fruit), fish-poison, pectoral, refreshing drink, sedative, stimulant, stomachic, tooth brush, tonic; and for affections of the chest and throat, anaemia, asthma, catarrh, cholera, chronic cough and dysentery, cold, dental problems, diarrhoea, earache, eye diseases, gonorrhoea, gum troubles, headache, heating, inflammation, lung affections, malaria, menorrhagia, nasal congestion, paralysis, phlegm, piles, poisons suffering from overdoses of mercury, putrid sores, rheumatism, tanning, tumours, ulcers, and urinary discharges **(Agarwal, 2003; Ahmed & Borthakur, 2005; Ambasta, 1986; Anonymous, 1948-1976; Ansari, 1991; Arora, 1997; Arora & Pandey, 1996; Badhwar & Fernandez, 2011; Bora *et al.*, 2003; Borthakur,**

1996b; Chandra, 1997; Chatterjee & Pakrashi, 1997; Chauhan, 1999; Chopra *et al.*, 1956; Deokata & Chhetri, 2007; Dey, 1980; Dey & Bahadur, 1973; Dwivedi, 2007; Farooq, 2005; Jain, 1991; Jamir *et al.*, 2011; Karnick, 1994; Kirtikar & Basu, 1984; Lal *et al.*, 1996; Lalramnghinglova, 1996; Lata *et al.*, 2008; Mahato & Chaudhary, 2005; Negi & Gaur, 1991; Negi *et al.*, 1993; Pande *et al.*, 2006; Parmar & Kaushal, 1982; Raju, 2000; Rao, 1990; Rawat & Chowdhury, 1998; Ringmichon *et al.*, 2011; Roy *et al.*, 1998; Said, 1997; Samant *et al.*, 2001; Sarin, 1996; Sharma, 1996, 2003; Singh, 1999; Singh & Kumar, 2000a; Sinha, 1996; Sood *et al.*, 2011, 2012b; Watt, 1889-1896).

Myrsine africana L.

Syn.: *M. bifaria* Wall.; *M. microphylla Hayata; M. potama* D. Don; *Rhamnus myrtillus* H. Lev.

Family: Myrsinaceae.

Vern.: Ghungri.

English and Hindi Names:

Eng. : African Boxwood, Cape Myrtle;
Hindi : Bebrang, Chapra, Chupra, Pahari Cha.

Distribution : Temperate Himalayas: 900-1,800m.

Description : A small evergreen bushy shrub with dark green, nearly sessile, lanceolate, finely toothed leaves. Flowers nearly sessile. Berries dotted with red glands.

Reproductive Cycle : April-October.

Materials Examined : HPU-EL-SKS-6521; Advance Study (2,128m), Boileauganj Cemetery (1,998m), Kanlog Cemetery (1,990m), Nav Bahar Cemetery (2,031m).

Propagation Notes : Succeeds in any well-drained fertile circum-neutral soil in full sun or semi-shade. Propagated by seeds and cutting.

Parts Used : Whole plant. Leaves. Fruits.

Folk Uses : Grown as a **hedge** plant. Shrub used as **fodder**, **fuel** and for making **brooms**. Decoction of leaves used as **blood purifier**. Fruits **edible**, considered good for cattle **digestion**, **laxative** and **anthelmintic**.

Active Constituents : Main constituents are embelin and quercitol (**fruit**).

Uses in Literature : Recorded in India as anthelmintic, antiageing, blood purifier, colic, edible (fruits), hedge, laxative; and for diarrhoea, dysentery, dysmenorrhoea, menstrual disorders, skin affections **(Ambasta, 1986;**

Anonymous, 1948-1976; Badhwar & Fernandez, 2011; Chauhan, 1999; Chopra *et al.*, 1986; Dey & Bahadur, 1973; Dhiman, 2005; Farooq, 2005; Hay *et al.*, 1983; Kapur & Singh, 1996; Pande *et al.*, 2006; Raju, 2000).

Myrtus communis L.

Family: Myrtaceae.

Vern.: Myrtle.

English and Hindi Names:

Eng. : Common Myrtle;
Hindi : Murad, Vilayiti Mehndi.

Distribution : Indigenous from Mediterranean - N.W. Himalayas; cultivated in gardens throughout India.

Description : Evergreen shrubs upto 3m tall with sessile or short - petioled, gland dotted, strongly scented, small, ovate or lanceolate, entire, smooth, and shining coriaceous leaves. Penducles solitary. Flowers in leaf axils, white and reddish brown pedicels with two linear bractlets. Stamens numerous, larger than petals. Berries ellipsoid, blue black. Seeds white, kidney - shaped.

Reproductive Cycle : June-October.

Material Examined : HPU-EL-SKS-6522; Spring Field (2,075m), Wood Villa (2,108m).

Propagation Notes : Propagated through seeds, cuttings and layering. Plant succeeds in any well-drained good soil.

Part Used : Whole Plant.

Folk Uses : Cultivated as an **ornamental** and **hedge** plant.

Active Constituents : These are esculin esculetin, caffeic acid, kaempferol, myricetin-3-galactoside, myricetin-rhamnoside, 3-rhamnoglucoside, 3,3-digalactoside, quercetin-3-galactoside and 3-glucoside (**leaves**), myricetin, kaempferol, ellagic acid and gallic acid (**flower**), delphinidin, petunidin, malvidin, peonidin, cyaniding-3-mono-and 3,3-digalactoside (**berries**).

Uses in Literature : Described earlier as antidiarrhoeal, antidysenteric, aromatic, astringent, edible (flower), carminative, disinfectant, laxative, ornamental hedge, perfume, tonic; and for bronchitis, blisters, bouquets, cerebral affections, cystitis, dyspepsia, eczema, epilepsy, garlands, haemorrhage, incipient phthisis, leucorrhoea, liver diseases, nose bleeding, pyelitis, rheumatism, scorpion bites, ulcer, skin diseases, wounds **(Ambasta, 1986; Anonymous, 1948-1976; Chatterjee & Pakrashi, 1997; Chevallier, 1996; Harborne & Baxter, 2001; Kapur *et al.*, 1996; Kirtikar & Basu, 1984; Mehtani & Ingle,**

1999; Phillips & Rix, 1989; Pool, 1941; Prajapati *et al.*, 2006; Rastogi & Mehrotra, 1995a; Sood *et al.*, 2010; Thomas, 1992; Uphof, 1968).

Nerium oleander L.

Syn.: *N. carneum Dum.* Cours; *N. flavescens* Spin; *N. grandiflorum* Desf.; *N. indicum* Mill.; *N. latifolium* Mill.; *N. odorum* Sol.; *N. odoratum* Lam.; *N. verecundum* Salisb.; *Oleander indica* (Mill.) Medik.; *O. vulgaris* Medik.

Family: Apocyanaceae.

Vern.: Kaner.

English, Hindi and Sanskrit Names:

Eng.	:	Indian Oleander, Sweet-scented Oleander;
Hindi	:	Kanel, Kaner, Karber, Kuruvira;
Sans.	:	Asvamaraca, Chandata, Hayamaraca, Karavira, Partihasa, Sataprasa, Virahuha, Vishavrykshanka.

Distribution : W. Himalayas (Nepal westwards, ascending to 2,000m); extensively cultivated throughout the greater part of India as well as in China and Japan.

Description : Erect glabrous shrubs with milky juice and leathery, shortly - stalked, narrowly lanceolate acute leaves. Flowers red or pink, fragrant, in terminal cymes. Follicles narrow, straight. Seeds numerous, crowned with a tuft of short hair.

Reproductive Cycle : April-November.

Materials Examined : HPU-EL-SKS-6524; Gorton Castle (2,094m), Peterhoff (2,098m), Spring Field (2,075m).

Propagation Notes : Prefers a heavy soil. Propagated by seeds; also successfully propagated by cuttings and layering.

Parts Used : Whole Plant. Leaves.

Folk Uses : Cultivated as an **ornamental** plant for its flower; Also, planted as **screens** and informal **hedge**. Poultice of leaves applied externally against **scabies** and to check **inflammation**. A pinch of powdered dried leaves taken as snuff for relief against **headache**.

Active Constituents : Reported to possess adynerin, dambonitol, quercetin, folinerin, isoquercetin, neriodin, neriumol, nerifol, neriin, oleandrin, betulin, betulinic, oleanolic, ursolic acid, urechitoxin, kaneroside, kanerodione, neriumoside, a compound similar to rutin, β-D-diginoside, β-D-digitaloside, 16-dehydro-adynerigenin - β - D - diginoside, oleandrigenin - β - D - glucopyranosyl – β – D - diginopyranoside, digitoxigenin (**leaves**).

Uses in Literature : So far described as an abortifacient, anthelmintic, attenuant, carminative, diaphoretic, febrifuge, fodder, gunpowder, insect repellent, insecticide, ophthalmic, ornamental garlands, powerful heart poison, rat poison, resolvent, symbolic (sacred), vermic (vet.); and for asthma, blisters, boils, bronchitis, copious lachrymation, dysentery, gangrene, epilepsy, gum troubles, hooka tubes, leprosy, menorrhagia, pthisis, ringworm, skin eruptions like herpes, snake bite, swellings, ulcers, worms, and wounds **(Ambasta, 1986; Anonymous, 1948-1976; Caius, 1986; Chatterjee & Pakrashi, 1997; Chauhan, 1999; Dastur, 1970; Girach *et al.*, 1996; Jain, 1991; Kapur & Singh, 1996; Kaur & Sharma, 2004; Khanna *et al.*, 1996a; Kirtikar & Basu, 1984; Kurian, 1995; Paliwal, 2009; Perry & Hay 1982; Prajapati *et al.*, 2006; Randhawa, 1957; Ranjan, 2003; Saini, 1996a; Sharma, 2003; Shukla *et al.*, 2003; Singh & Pandey, 1998; Sood *et al.*, 2013; Thakur *et al.*, 1989; Thomas & Britto, 2003; Watt, 1889-1896)**.

Opuntia dillenii Haw.

Syn.: *O. melanosperma* Svenson; *O. stricta* var. *dillenii* (Ker Garl.) L.D. Benson.

Family: Cactaceae.

Vern.: Vilayati Suri.

English, Hindi and Sanskrit Names:

Eng.	:	Prickly Pear, Slipper Thorn;
Hindi	:	Haththathoria, Nagphani;
Sans.	:	Bahudagdhika, Kandashaka, Nagaphana.

Distribution : Native of American tropics and subtropics; introduced in India.

Description : Shrubs possessing thick, flat branches dotted with hairy tuft of short spinescent bristles. Flowers orange-red, large. Fruits purple, berry-like, fleshy, depressed at apex, deep reddish when ripe. Seeds flat, comma - shaped.

Reproductive Cycle : April-June.

Materials Examined : HPU-EL-SKS-6528; Bemloe (2,014m), Boileauganj Cemetery (1,998m), Cecil Hotel (2,090m).

Propagation Notes : Propagated by seeds as well as through cuttings of leaf pads at any time in growing season. Falling fruits and joints also readily take root.

Parts Used : Whole Plant. Flower. Fruit.

Folk Uses : Cultivated as a **hedge** and **ornamental** plant. 3-5 ml of flower juice given on empty stomach against **asthma**. Ripe fruits boiled **eaten** after removing spines. Ash of plant massaged on affected parts to cure **pain** due to **injury**.

Active Constituents : These are isoquercetin, β-sitosterol, opuntiol, opuntiol acetate and glucoside narcissin (**flower**).

Uses in Literature : Useful in India as antidandruff, antipyretic, carminative, demulcent, edible (fruits), expectorant, hedge, laxative, purgative, refrigerant, stomachic, symbolic in religious ceremonies of 'Santhals' (spines); and for allaying heat, asthma, baldness, biliousness, boils, bronchitis, chest pains, controlling spasmodic cough, dyeing, enlargement of spleen, expelling worms, fibre as paper material, food, gonorrhoea, inflammation, leucoderma, loss of consciousness, ophthalmia, pains, piles, pimples, promoting flow of urine, scorbutic ulcers, skin diseases, sores, swellings, syphilis, tumours, ulcers, urinary complaints, and whooping cough **(Ambasta, 1986; Anonymous, 1948-1976; Arora & Pandey, 1996; Billore *et al.*, 1998; Chatterjee & Pakrashi, 1997; Chopra *et al.*, 1956; Das & Agarwal, 1991; Dey & Bahadur, 1973; Jain, 1991; Jamir *et al.*, 2011; Karnick, 1994; Kirtikar & Basu, 1984; Kurian, 1995; Parmar & Kaushal, 1982; Parrotta, 2001; Raju, 2000; Rastogi & Mehrotra, 1990; Singh, 1996; Sood & Thakur, 2004; Sood *et al.*, 2012b; Sudhakar & Vedavathy, 2000; Watt, 1889-1896; Yoganarasimhan, 1996)**.

Padus cornuta Carriere

Syn.: *Cerasus cornuta* Wall. *ex* Royle; *Prunus cornuta* (Wall. *ex* Royle) Stued.

Family: Rosaceae.

Vern.: Jangli jamun, Krun.

English and Hindi Names:

Eng. : Himalayan Bird Cherry;
Hindi : Jamana.

Distribution : Afghanistan - S.W. China, Myanmar: 2,100-3,500m.

Description : Deciduous trees with brown coloured bark and saw-toothed, oblong - lanceolate leaves. Flowers white, small, drooping. Ripe fruits nearly black, globular.

Reproductive Cycle : April-May.

Materials Examined : HPU-EL-SKS-6529; Bemloe (2,014m), Kanlog Cemetery (1,990m).

Propagation Notes : Propagated by seeds.

Parts Used : Foliage. Fruit. Wood.

Folk Uses : Ripe fruits **edible**. Wood used as **fuel**. Foliage cut for **fodder**.

Uses in Literature : Known earlier for edible fruits, burns, body swelling, cough, cuts, diarrheoa, fever, injury, liquor preparation, muscular pain and wounds; and as substitute for almond oil **(Agarwal, 2003; Ambasta, 1986; Anonymous, 1948-1976; Arora & Pandey, 1996; Arya *et al.*, 1999; Badhwar & Fernandez, 2011; Bennet *et al.*, 1991; Chopra *et al.*, 1956; Das & Agarwal, 1991; Farooq, 2005; Kapur & Nanda, 1996; Kirtikar & Basu, 1984; Pande *et al.*, 2006; Roy *et al.*, 1998; Samant *et al.*, 1996; Singh, 1996; Singh & Kumar, 2000a; Sood *et al.*, 2001, 2012b; Uphof, 1968)**.

Parthenocissus semicordata Planch

Syn.: *Amplelopsis himalayana* Royle; *Parthenocissus himalayana* Brandis; *Vitis himalayana* Brandis; *V. rubrifolia* H. Lev & Vaniot.

Family: Vitaceae.

Vern.: Mahamay, Philuna.

Distribution : W. Himalayas upto 3,400m.

Description : Climbers with digitately compound leaves. Leaflets three, shortly-stalked, ovate, long-pointed; terminal leaflet large and lateral ones unequal-sided. Flowers yellow-green, in spreading cymes, numerous. Fruit black, pea-sized.

Reproductive Cycle : July-November.

Materials Examined : HPU-EL-SKS-6532; Advance Study (2,128m), Boileauganj Cemetery (1,998m), Chhota Shimla (2,080m), Retreat (2,092m).

Propagation Notes : Propagated by seeds.

Parts Used : Leaves. Fruits.

Folk Uses : Leaves used as **fodder**. Fruits **edible**.

Uses in Literature : Reported earlier as edible (fruit), fodder and wild relative of *Vitis vinifera* Linn. **(Ambasta, 1986; Badhwar & Fernandez, 2011; Samant *et al.*, 1996; Sharma & Rana, 2005; Sood *et al.*, 2012b)**.

Paulownia tomentosa (Thunb.) Steud

Syn.: *Bignonia tomentosa* Thunb.; *P. imperialis* Siebold & Zucc.

Family: Scrophulariaceae.

Vern.: Videshi Phool.

English Names:

Eng. : Empress Tree, Foxglove Tree, Princess Tree.

Distribution : Native of E. Asia and distributed from E. Asia -China upto 2,000m.

Description : Elegant rapid growing trees up to 15m with large, cordate leaves and beautiful panicles of fragrant, violet or white flowers appearing before leaves. Capsule dry, egg-shaped, 3-4cm long with numerous seeds.

Reproductive Cycle : May-September.

Materials Examined : HPU-EL-SKS-6533; B.C.S. (2,025m), St. Edward's School (2,060m).

Propagation Notes : Multiplied through seeds, cuttings of half-ripe wood and root. Plant requires a deep moderately fertile moisture retentive but well-drained soil in a sunny sheltered position.

Part Used : Whole Plant.

Folk Use : Cultivated as an **ornamental** plant.

Active Constituents : These are hydroquinone dimethyl ether, 1-octen-3-ol, cis-3-hexen-1-ol, i-linalool, benzyl alcohol, phenethyl alcohol, benzaldehyde, anisaldehyde, ether palmitate, phenol,o-,m- and p-cresols, 3,4-dimethoxyphenol and unidentified volatile acids (**essential oil of flower**).

Uses in Literature : Reported earlier as astringent, edible (flower, leaves), ornamental, vermifuge; and for fever, promoting hair growth, skin diseases, making boxes, clogs, furniture, musical instruments **(Ambasta, 1986; Anonymous, 1948-1976; Bean, 1981; Chittendon, 1956; Duke & Ayensu, 1985; Foster & Duke, 1990; Huxley, 1992; Rastogi & Mehrotra, 1990, 1993, 1995b; Sheat, 1948; Uphof, 1968; Usher, 1974)**.

Picea smithiana Boiss.

Syn.: *Abies khutrow* (Royle *ex* Turra) Loudon; *A. morinda* (Link) Wender.; *A. smithiana* (Wall.) Lindl.; *Picea khutrow* (Royle *ex* Turra) Carriere; *P. morinda* Link; *Pinus smithiana* Lamb.

Family: Pinaceae.

Vern.: Rai.

English and Hindi Names:

Eng. : West Himalayan Spruce;
Hindi : Rai, Rau, Re, Riar, Kachal, Kachhlu, Sall, Tosh.

Distribution : N.W. Himalayas: 2,100-3,600m.

Description : Tall conical trees with horizontal, hanging, tassel-like branchlets. Foliage blue or grey-green. Leaves needle-like, green, 4-sided, scattered around the branches. Male and female cones on the same tree. Mature cones cylindric, obtuse, pendent from the end of branches, dark brown when ripe.

Reproductive Cycle : April-May.

Materials Examined : HPU-EL-SKS-6537; Advance Study (2,128m), Wood Villa (2,108m).

Propagation Notes : Propagated through seeds.

Parts Used : Wood. Bark. Cones.

Folk Uses : Cones used for making **decorative art pieces**. Bark used for **roofing** and making **water troughs**; also valued for its use in **pulp** industry to make **paper**. Wood used for **construction**, **crates**, and **household** purposes etc. Dried wood used as **fuel**.

Active Constituents : These are α-pinene and β-pinene, piceatannol, isorhapontin, (E) 4, 3, 5 – trihydroxy – 3 -methoxystilbene (I) and (E)3-β-D-glucosyloxy-3,4,5-trihydroxystilbene (II), astringin, resveratrol, isorhapontigenin, isorhapontin and piceatannol **(bark)**.

Uses in Literature : Known earlier as deodorant, room sprays, rubefacient, timber and fuel; and for house construction, furniture, roofing, bodyache, cracks, cuts, sores and wounds **(Agarwal, 2003; Ambasta, 1986; Anonymous, 1948-1976; Bean, 1981; Gamble, 1972; Kapur, 1996b; Kirn *et al.*, 2000; Manandhar, 2002; Pande *et al.*, 2006; Rushforth, 1987; Samant *et al.*, 1996; Sharma, 1996; Warda, 1985)**.

Pinus roxburghii Sarg.

Syn.: *P. longifolia* Roxb. *ex* Lamb.

Family: Pinaceae.

Vern.: Chil, Chir.

English, Hindi and Sanskrit Names:

Eng.	:	Chir Pine, Long-leaved Pine, Three-leaved Pine;
Hindi	:	Chil, Chir, Chirkagond, Dhupasarala, Gandha biroza, Salla, Saral;
Sans.	:	Bhadradaru, Dhupavrikshaka, Manojna, Marichapatraka, Pitadaru, Putikashtha, Sarala, Sarala-drava, Snighadarusanjnaka, Srivasa, Surabhidaruka.

Distribution : Outer Himalayan ranges from Indus - Bhutan: 500-2,500m.

Description : Trees with a rough corky bark and branches whorled up on the trunk. Leaves acicular in cluster of three, 22-30 cm long. Male cones cylindrical, small about 2 cm long. Female ones ovoid or conical on short stiff stalks, solitary, 10-15 cm long. Seeds with a thin membranous wing and 10-14 cotyledons.

Reproductive Cycle : March-May.

Materials Examined : HPU-EL-SKS-6538; B.C.S. (2,025m), Boileau-ganj Cemetery (1,980m), Cecil Hotel (2,090m).

Propagation Notes : Propagated by seeds naturally.

Parts Used : Needles. Seed. Cone. Wood.

Folk Uses : Pine needles used for making **packaging** boxes and **brooms**. Cones employed for making decorative **art pieces**. Roasted seeds **edible**. Wood used for **furniture**, **fuel**, and **house building purposes**.

Uses in Literature : Recorded in literature as an antiseptic, carminative in flatulent colic, charcoal, colouring matter, diaphoretic, denaturants, diuretic, edible (seeds), expectorant, fumigants, fuel wood, insecticide, psycho-active plant, rubefacient in various thematic affections such as lumbago, arthritis and neuralgia, stimulant; and for abscess, bangles, baskets, boils, bone fracture, buboes, burning of body, chronic bronchitis, constipation, cough, cracks in sole of feet, fever, furniture, gangrene of lungs, general care of hovers, gonorrhoea, house building, inflammation of abdomen, itching, leprosy, mange (vet.), match industry, minor haemorrhages on tooth-sockets, musical instruments, nose, painting, perfumery, phthisis, pine oil in various pharmaceuticals, preparation of plasters-ointments rafters, resin, rheumatism, worm infestations, shingles, skin diseases, snake bite, sport-goods, sprain, sprain (vet.), swelling, synthetic pine oils, tanning, tea chests, torches, turpentine, tympanitis, typhoid, and ulcers **(Agarwal, 2003; Ambasta, 1986; Anonymous, 1948-1976; Chatterjee & Pakrashi, 1997; Chauhan, 1999; Chopra *et al.*, 1956; Dey & Bahadur, 1973; Dhiman, 2005; Dwivedi, 2007; Farooq, 2005; Jain, 1968, 1991; Jain *et al.*, 1994; Kaur & Sharma, 2004; Kaushik & Dhiman, 1999; Khare, 2004; Kirn *et al.*, 2000; Kirtikar & Basu, 1984; Paliwal, 2009; Pande *et al.*, 2006; Prajapati *et al.*, 2006; Roy *et al.*, 1998; Saklani & Jain, 1994; Sensarma, 1991; Sharma, 1996, 2003; Singh, 2001; Singh & Kumar, 2000a; Sood & Thakur, 2004; Sood *et al.*, 2009, 2011; Srivastava *et al.*, 2003a; Uphof, 1968; Warman, 1999; Watt, 1889-1896)**.

Pinus wallichiana A.B. Jacks.

Syn.: *Leucopitys excelsa* Nieuwl.; *P. chylla* Lodd.; *P. excelsa* Wall. *ex* D. Don.

Family: Pinaceae.

Vern.: Kail.

English and Hindi Names:

Eng. : Bhutan Pine, Five-leaved Pine, Himalayan Blue Pine, Indian Blue Pine;

Hindi : Kail.

Distribution : Himalayas: 1,800-4,300m.

Description : Huge evergreen trees possessing rough, corky bark and drooping bluish green leaves in clusters of five. Male cones ovoid, small. Female ones in group of 2 or 3; cylindrical, erect in early stages of growth, later pendulous about 20 cm long, beaked. Seeds brackish, ovoid, acute at both ends.

Reproductive Cycle : April-June.

Materials Examined : HPU-EL-SKS-6540; Bemloe (2,009m), Chhota Shimla (2,080m), Kanlog Cemetery (1,990m), Raj Bhawan (2,132m).

Propagation Notes : Plant regenerates from seeds.

Part Used : Wood.

Folk Uses : Wood extensively used for **furniture** and **building works**.

Active Constituents : These are α-pinene, undecane, dodecane, tridecane, abietic, isopimaric and lambertianic (**wood**).

Uses in Literature : Used earlier for abscess, bone fracture (vet.), cough, dislocation of joints, gonorrhoea, headache, internal injury, pimples, ulcer, unconsciousness, resin, rheumatism, skin diseases, wounds and house construction and plastering walls; and as stimulant **(Agarwal, 2003; Ambasta, 1986; Anonymous, 1948-1976; Chauhan, 1984, 1999; Dam & Hajra, 1997; Dwivedi, 2007; Jain, 1991; Khare, 2004; Kirn *et al.*, 2000; Nautiyal *et al.*, 2003; Pande *et al.*, 2006; Rawat & Chowdhury, 1998; Samant *et al.*, 1996; Sharma, 1996; Singh & Kumar, 2000a; Warman, 1999)**.

Pistacia khinjuk Stocks

Syn.: *P. chinensis* Bunge; *P. integrimma* Stewart; *Rhus integrimma* Wall; *R. kakra-singee* Royle.

Family: Anacardiaceae.

Vern.: Kakad, Kakadsingi.

English, Hindi and Sanskrit Names:

Eng. : Crab's Claw;
Hindi : Kakra;
Sans. : Chakrangi, Chandraspada, Ghosha, Kakratakhya, Karkati, Kasavinashini, Kulingi, Vakra.

Distribution : N.W. Himalayas: 600-2,400m.

Description : A medium-sized glabrous tree possessing odd or even pinnate leaves. Leaflets 7-11, nearly opposite, shortly - stalked, lanceolate, long pointed, entire, base unequal; young foliage red. Flowers dioecious, pedicelled, red coloured, in axillary racemes. Drupes glabrous, wrinkled, globose, grey when ripe.

Reproductive Cycle : March-May.

Materials Examined : HPU-EL-SKS-6541; Boileauganj Cemetery (1,998m), R.K.M.V. (2,256m), U.S. Club (2,290m).

Propagation Notes : Propagated naturally by seeds. Tree grows on hot dry slopes with shallow soil or on open rocky ground and limestone soils.

Parts Used : Twig. Galls. Nuts. Wood.

Folk Uses : Twigs used as **fodder**. 2-3g powdered fruits or galls ('kakarsingi') prescribed with honey for **respiratory disorders**, **cough**, **asthma** and **lung diseases**; its paste in mustard oil applied externally to relieve **pain** in muscle. Nuts **eaten**. Wood used in **construction** and **furniture** works.

Active Constituents : Reported to yield α-pinene (21.8), β-pinene (16.2), α-phellandrene (15.5), car-3-ene (11.1%) β-phellandrene, terpinee, limonene, α-and β-ocimenes, α- and-β-terpineols along with three more tetracyclic triterpenes-pistacigemones D, E and F (**galls**), pistacin and pistacinin (**drupes**).

Uses in Literature : Known so far as an aphrodisiac, astringent, edible (kernels), expectorant, tonic, antidote to scorpion-sting, anti-inflammatory, fodder, flavouring plant; and for asthma, cough, dysentery, dyspepsia, fever, leprosy, phthisis, ulcer, skin diseases, snake-venom, wood for construction and furniture works **(Agarwal, 2003; Ambasta, 1986; Anonymous, 1948-1976; Chandra, 1997; Chauhan, 1999; Chopra *et al.*, 1956; Dey & Bahadur, 1973; Dhiman, 2005; Dwivedi, 2007; Farooq, 2005; Gupta, 1981; Kapur & Nanda, 1996; Kirtikar & Basu, 1984; Negi *et al.*, 1985; Prajapati *et al.*, 2006; Raju, 2000; Sarin, 1996; Sharma, 1996; Sharma & Rana, 2005; Singh, 1999, 2001; Sinha & Sinha, 2001; Sood & Thakur, 2004; Sood *et al.* 2009, 2013; Uphof, 1968)**.

Platanus orientalis L.

Syn.: *P. digitata* Gord.; *P. digitifolia* Palib.; *P. orientalior* Dode.

Family: Platanaceae.

Vern.: Chinar.

English and Hindi Names:

Eng. : Oriental Plane Tree, Plane Tree;
Hindi : Chanaar, Chinar.

Distribution : N.W. Himalayas: 1,500-2,800m, especially Kashmir.

Description : A large magnificent deciduous tree with a wide spreading crown, dull-greyish bark, palmatifid, 5-7 lobed, coarsely dentate, rarely entire leaves. Flowers monoecious, densely crowded in unisexual globose heads on a slender peduncle. Achenes 1-seeded.

Reproductive Cycle : April-July.

Materials Examined : HPU-EL-SKS-6543; Advance Study (2,128m), B.C.S. (2,025m), Retreat (2,092m), Ridge (2,250m).

Propagation Notes : Requires a sunny position in a deep fertile well-drained soil. Propagated through seeds and cuttings of mature wood of the current year's growth and layering of stools in spring or autumn.

Part Used : Whole Plant.

Folk Use : Planted as an **ornamental** tree.

Active Constituents : These are n-hentriacontane, 16-hentriacontanone, n-hentriacontanol, 12-tricosanol, β-sitosterol, stearate, ceryl stearate and amurensin (noricarintin-7-glucoside) (**fruit**).

Uses in Literature : Reported to be used as antidote to poisonous bites of animals, antirheumatic, astringent, ophthalmic, purgative, ornamental; and for diarrhoea, dysentery, hernias, inflammed surfaces, leucoderma, kidney and lung problems, sores, toothache, wounds; and in construction work, curved furniture, wood pulp **(Ambasta, 1986; Anonymous, 1948-1976; Bean, 1981; Brickell, 1990; Chaudhuri, 1993; Chopra *et al.*, 1986; Huxley, 1992; Kapur & Singh, 1996; Kirtikar & Basu, 1984; Pool, 1941; Randhawa, 1957, 1960; Sood *et al.*, 2013; Uphof, 1968; Usher, 1974; Warman, 1999)**.

Populus ciliata Wall. *ex* Royle

Family: Salicaceae.

Vern.: Popular, Safeda.

English and Hindi Names:

Eng. : Himalayan Poplar;
Hindi : Paharipipal, Sharphara, Tilaunja.

Distribution : N.W. Himalayas: 1,500-3,000m.

Description : Trees having sticky leaf buds which are ovate -lanceolate, cordate, acute, leaves minutely pubescent along the edges. Flowers in drooping raceme - like catkins appearing before or with the young leaves. Male flowers possess oblanceolate bracts fringed with long hair and many stamens whereas female ones have 3-4, very large stigmas. Capsule ovoid, 3-4 valved. Seeds numerous.

Reproductive Cycle : April-July.

Materials Examined : HPU-EL-SKS-6545; B.C.S. (2,023m), Bemloe (2,008m), Chhota Shimla (2,090m), Kanlog Cemetery (1,990m), Pari Mahal (2,105m), Ridge (2,250m).

Propagation Notes : Propagated through seeds which are sown as soon as ripe in spring due to extremely short period of viability. Other means are by cuttings of mature wood of the current season's growth in November/December in a sheltered outdoor bed and suckers in early spring.

Parts Used : Whole Plant. Leaves. Bark. Wood.

Folk Uses : Plant grown as **avenue**. Decoction of bark along with black pepper useful as **blood purifier**. Leaves used as **fodder** and wood for cheap **furniture** and **building construction**.

Uses in Literature : Reported earlier to be used as avenue plant, blood purifier, fodder, fuel, stimulant, tonic; and for agricultural tools, bone fracture, constructions, cricket bats, fruit crater, furniture, limbs, match boxes, packing, splints, sports good, turnery, veneers, water troughs **(Agarwal, 2003; Ambasta, 1986; Anonymous, 1948-1976; Bailey, 1957; Bhattacharjee, 2001; Chaudhuri, 1993; Chora *et al.*, 1986; Farooq, 2005; Gamble, 1972; Jain, 1991; Kapur & Nanda, 1996; Kapur & Singh, 1996; Kirtikar & Basu, 1984; Manandhar, 2002; Nautiyal *et al.*, 2003; Pande *et al.*, 2006; Purohit & Samant, 1995; Samant *et al.*, 1996; Sharma, 1996; Singh, 2000a; Singh *et al.*, 1996a; Sood *et al.*, 2013; Uphof, 1968; Watt, 1889-1896)**.

Populus nigra L.

Syn.: *P. sosnowskyo* Grossh.

Family: Salicaceae.

Vern.: Poplar, Popular.

English Names:

Eng. : Black Poplar, Lombardy, Poplar.

Distribution : N.W. Himalayas.

Description : Trees of medium - large size with grey bark, more or less viscid buds and somewhat triangular- rhomboid leaves. Blades mostly somewhat longer than broad, gradually or abruptly acuminate, margins with small coarse teeth. Petiole glandless, both surface glabrous and green. Staminate catkins 5-7.5 cm long, pistillate shorter than it.

Reproductive Cycle : April-July.

Materials Examined : HPU-EL-SKS-6547; U.S. Club (2,290m), Yates Place (2,208m).

Propagation Notes : Through seeds and cuttings of mature wood of the current season's growth, 20 - 40cm long in November/December and suckers in early spring.

Parts Used : Whole Plant. Leaves. Wood.

Folk Uses : Plant grown as a **shade tree**. Leaves lopped as **fodder**. Wood used in **match-stick** and **plywood** industries, **building works** and as **fuel**.

Active Constituents : These are salicin, salicortin, nigracin, caffeic acid-3-glucoside (**leaves**), 3-O-methylkaempferol, 3-O-methygalangin, chrysin, galangin, pinocembrin, izalpinin and pinostrobin (**bud**).

Uses in Literature : Useful so far as anti-inflammatory, astringent, avenue plant, coacher, depurative, diaphoretic, diuretic, expectorant, febrifuge, fodder, stimulant, tonic; and for bone fracture, carving, colds, cross-beams, furniture, haemorrhoids, hair growth, matches, poles, round boxes for grapes, small boxes, timber, truck and barrio trays **(Ambasta, 1986; Anonymous, 1948-1976; Bailey, 1957; Bean, 1981; Bhattacharjee, 2001; Chittendon, 1956; Harborne & Baxter, 2001; Jain, 1991; Kirtikar & Basu, 1984; Pande *et al.*, 2006; Pool, 1941; Rastogi & Mehrotra, 1995a; Singh *et al.*, 1996a; Sood *et al.*, 2013; Uphof, 1968; Usher, 1974; Watt, 1889-1896)**.

Prinsepia utilis Royle

Family: Rosaceae.

Vern.: Bekhal.

Hindi Names:

Hindi : Bekhra, Bhekal, Cherara, Dhatila, Jhatela, Karanga.

Distribution : Temperate Himalayas: 1,200-2,700m.

Description : Spinescent deciduous shrubs with stout, axillary, leaf-bearing thorns and coriaceous leaves. Leaves lanceolate, minutely toothed, long pointed. Flowers white, in short axillary racemes. Drupes purple, obliquely oblong.

Reproductive Cycle : June-October.

Materials Examined : HPU-EL-SKS-6549; Bemloe (2,010m), Boileau-ganj Cemetery (1,998m), Grand Hotel (2,105m), Rothney Castle (2,350m).

Propagation Notes : Propagated by seeds.

Parts Used : Whole Plant. Branch. Fruit. Root. Seed.

Folk Uses : Commonly grown as **hedge** plant. Branches hung on main door to ward off **evil spirits**. Ripe fruits **eaten**. Root paste along with *Piper nigrum* and cow's urine applied to hasten **suppuration** of **boils**. Seed oil applied locally to relieve **rheumatic pain**.

Active Constituents : These are new lignan-prinsepiol-along with (–) epicatelhin and β-sitosterol-β-glucoside (stems), 1.8% myristic, 15.2% palmitic, 4.5% stearic, 0.9% lignoceric, 32.6% oleic and 45.6% linoleic acids (**seed oil**), new lignin-prinsepiol, (-) epicatechin and β-sitosterol-β-glucoside (**stem**).

Uses in Literature : Recorded earlier as cooking seed's oil, edible, fuelwood, illuminant, insect repellent, rubefacient, tonic; and for bodyache, boils, burns, cuts, hydrogenation, pain due to fatigue, rheumatism, stomach disorders, soap making, walking sticks, and wounds **(Ambasta, 1986; Anonymous, 1948-1976; Badhwar & Fernandez, 2011; Bennet *et al.*, 1991; Chauhan, 1999; Chopra *et al.*, 1956; Dwivedi, 2007; Farooq, 2005; Jain, 1991; Kapur & Srivastava, 1996; Kirtikar & Basu, 1984; Negi *et al.*, 1999; Panda, 1996; Pande *et al.*, 2006; Raju, 2000; Rana *et al.*, 1996; Roy *et al.*, 1998; Samant *et al.*, 2001; Sharma, 2003; Sharma & Rana, 2000; Singh & Kumar, 2000a; Sood & Thakur, 2004; Verma & Chauhan, 2006; Watt, 1889-1896).**

Prunus amygdalus Stokes

Family: Rosaceae.

Vern.: Badam.

English, Hindi and Sanskrit Names:

Eng. : Almond Tree;
Hindi : Badam;
Sans. : Badama, Badmitte.

Distribution : N.W. Himalayas, cultivated in cooler parts of India.

Description : Trees with light coloured glabrous branchlets and firm, oblong - lanceolate, acuminate, very closely crenate-serrated glabrous leaves which are more or less shining above and lighter beneath. Petioles usually gland-bearing above. Flowers pink, solitary, sessile, appearing before or with early foliage. Fruits oblong-ovoid, compressed, pubescent and splitting at maturity to disclose shallow pitted stone.

Reproductive Cycle : February-September.

Material Examined : HPU-EL-SKS-6550; Chhota Shimla (2,070m).

Propagation Notes : Propagation by seeds, budding and grafting.

Part Used : Seed.

Folk Uses : Seeds **edible**. Also, 5-10 seeds (soaked in water overnight) prescribed with 'misri' (crystalline sugar) and butter in morning on empty stomach as **brain tonic**, **body tone** and for **vitality**. Decoction of seeds prepared in milk along with grates said to be good **aphrodisiac** preparation; and the paste used as lotion to cure **burns**. Seed oil given to check **constipation**, and massaged against **joint pains** and **skin ailments**.

Active Constituents : Reported to possess prunasin, daucosterin, sitosterol (**bitter seeds**), daucosterin, sitosterol (**sweet almonds**), a new biflavone-4-methoxy-2, 3 - dehydroflavenol - 4, 4 - oxymethyl -5, 7dimethoxy flavones, coumaric acid, 2-methylnonacosan-3-one, n-octacosanol, n-triacentane, β-sitosterol, procyamidin dimer (II), (+) catechin and (-) epicatechin (**seed coat**).

Uses in Literature : Useful earlier as an alternative, aphrodisiac, astringent, attenuant, demulcent, detergent, discutient, diuretic, edible (kernels), laxative, lithnotropic, nervine tonic, poultice in irritable sores, stimulant, symbolic-offering, tooth powder; and for amenorrhoea, confectionary, cosmetic preparation, cough, diabetes, gums, intestinal pain, marquetry, obstructions of liver and spleen, painful menstruation, peptic ulcers, perfumery, pharmaceutical, skin eruptions, sores and turnery **(Agarwal, 2003; Ambasta, 1986; Anonymous, 1948-1976; Athar & Nasir, 2005; Bhattacharjee, 2001; Chaudhuri, 1993; Cowan, 1950; Dey & Bahadur, 1973; Jain, 1991; Kapoor, 2005; Khare, 2004; Kirtikar & Basu, 1984; Prajapati *et al.*, 2006; Said, 1997; Shome *et al.*, 1996; Singh, 1969; Singh & Kumar, 2000a; Sood *et al.*, 2013; Uphof, 1968; Watt, 1889-1896)**.

***Prunus armeniaca* L.**

Syn.: *Armeniaca vulgaris* Lam.

Family: Rosaceae.

Vern.: Chuli, Khumani.

English and Hindi Names:

Eng. : Apricot, Common Apricot, Mishmus, Moon of the Faithful;

Hindi : Chilu, Chuari, Chulu, Khubani, Sardalu.

Distribution : Naturalized in N.W Himalayas, particularly in the valleys of Kashmir, Chenab, Kullu and Shimla hills upto 3,000m.

Description : Trees with reddish bark, glabrous twigs and ovate - round, abruptly short - pointed, obtusely serrated leaves which are glabrous above and pubescent on veins beneath. Petioles with conspicuous glands. Flowers pinkish or nearly white. Fruits roundish but usually somewhat flattened, yellow overlaid with red; the flesh firm and sweet, mostly free from the flat ridged stone.

Reproductive Cycle : February-August.

Material Examined : HPU-EL-SKS-6551; B.C.S. (2,025m), Bemloe (2,013m), Chhota Shimla (2,080m), R.K.M.V. (2,256m), Sidhowal Lodge (2,254m), U.S. Club (2,290m).

Propagation Notes : Plant succeeds in a well-drained moisture retentive fertile soil in a warm sunny position. Propagated by shield budding on wild apricot stock.

Parts Used : Fruit. Seed.

Folk Uses : Ripe, dried fruits and seeds **edible**; seed oil used as **tonic** and for **cooking**.

Active Constituents : Found to possess p - cymen - 8 - ol, butanol, isobutanol, isovaleraldehyde, butyl acetate, hexyl acetate, phenylacetaldehyde, 2, 2, 6-trimethyl-2-vinyltetrahydropyran, ionene, toluene, butyric, palmitic and linoleic acid, linalool, hexanal, γ- decalactone, 3, 4, dihydroxybenzoic, chlorogenic and vanillic acids, quercetin, quercetrin, rutin, kaempferol (**fruits**), β – ionene, 24-dehydrox-cholesterol, cholesterol, estrone, α-estradiol (estradiol-β-17-oL) (**seeds**), myrsitic, palmitic, stearic, oleic and linoleic acids (**seed kernels**).

Uses in Literature : Known so far as anthelmintic, antidiarrhoeal, antipyretic, cattle feed, demulcent, edible (fruit, oil), emetic, firewood, fodder cakes, illuminant, laxative, manure, nervine tonic, psychoactive plant, refrigerant, stimulant; and for allaying thirst, blood clotting, burning, cosmetic industry, earache, gum of commercial importance, jam, liver troubles, oil for massage, ophthalmia, pharmaceutical, piles and wounds **(Agarwal, 2003; Ambasta, 1986; Anonymous, 1948-1976; Arora & Pandey, 1996; Bennet *et al.*, 1991; Chopra *et al.*, 1956; Das & Agarwal, 1991; Duke *et al.*, 2002; Dwivedi, 2007; Farooq, 2005; Harborne & Baxter, 2001; Jain, 1991; Jain *et al.*, 1994; Kapur & Singh, 1996; Kirtikar & Basu, 1984; Pande *et al.*, 2006; Parmar & Kaushal, 1982; Prajapati *et al.*, 2006; Rajendran *et al.*, 1996; Raju, 2000; Rana *et al.*, 1996; Roy *et al.*, 1998; Sharma & Rana, 2000, 2005; Sharma *et al.*, 2000; Singh & Kumar, 2000a; Singh *et al.*, 1996b; Warman, 1999)**.

Prunus domestica L.

Syn.: *P. communis* Huds.; *P. oeconomica* Borkh.

Family: Rosaceae.

Vern.: Aloo Bukhara, Plum.

English, Hindi and Sanskrit Names:

Eng.	:	Plum Tree, Prum Tree, Common Plum;
Hindi	:	Alubukhara, Alucha;
Sans.	:	Aruka.

Distribution : Throughout Sub-Himalayan region.

Description : Trees with thick twigs which are pubescent when young. Leaf blade ovate - obovate, thick and rugose, dull green, short pointed - obtuse, coarsely and irregularly serrate, thin hairy or scabrous above and pubescent beneath. Flowers mostly clustered on spurs but sometimes single or 2-3 together and with pubescent pedicels. Fruits blue purple, ovoid - oblong.

Reproductive Cycle : April-August.

Material Examined : HPU-EL-SKS-6554; Bemloe (2,007m), Chhota Shimla (2,070m), Crow-Brough (2,048m), Sanitarium (2,060m).

Propagation Notes : Plums usually propagated by shield-budding on wild apricot or common peach stock. Cutting of half-ripe wood in July-August, air layering, and by seeds another ways of propagation.

Parts Used : Stem. Fruit.

Folk Uses : Ripe fruits **edible**; also these are **pickled** and used for making **jams** and **candies**. Stem employed as **fuel-wood**.

Active Constituents : Chief constituents are galacturmic acid, arabinose, galactose, xylose and rhamnose (**fruits**).

Uses in Literature : Known earlier as an aperient, astringent, demulcent, digestive, edible (fruit), jams, laxative, nutritious, purgative, refreshing, refrigerant; and for bilious states of the system, cabinet work, cathartic decoctions, cold, confectionary, desserts, fever, illuminating purposes, inlay and turnery, skeleton of paper and match boxes **(Agarwal, 2003; Ahmed & Borthakur, 2005; Ambasta, 1986; Anonymous, 1948-1976; Bhattacharjee, 2001; Bora *et al.*, 2003; Chopra *et al.*, 1956; Dey & Bahadur, 1973; Farooq, 2005; Kirtikar & Basu, 1984; Prajapati *et al.*, 2006; Singh & Kumar 2000a; Uphof, 1968; Warman, 1999; Watt, 1889-1896).**

Prunus laurocerasus L.

Family: Rosaceae.

Vern.: Cherry-laurel.

English, Hindi and Sanskrit Names:

Eng. : Common Cherry-Laurel.

Distribution : Native of Asia Minor and S.E. Europe and cultivated widely as an ornamental.

Description : Evergreen trees with handsome evergreen foliage. Leaves oblong, short-stalked, small toothed, and abruptly short-pointed. Racemes slender but not exceeding the leaves. Fruits conic and pointed, black purple.

Reproductive Cycle : April-September.

Material Examined : HPU-EL-SKS-6556; Advance Study (2,128m).

Propagation Notes : Requires a well-drained moisture retentive soil. Propagated by seeds, cutting and layering.

Part Used : Whole Plant.

Folk Uses : Cultivated as an **ornamental** plant. Leaves **poisonous** to cattle.

Uses in Literature : So far described as anodyne, antispasmodic, custard, dye, edible, flavouring boiled milk, hedge, insecticide, ornamental, poisonous (vet.), sedative; and for asthma, coughs, dyspepsia, indigestion, turnery work, whooping cough **(Ambasta, 1986; Anonymous, 1948-1976; Bailey, 1957; Bhattacharjee, 2001; Hadfield, 1956; Uphof, 1968; Vickery, 1994)**.

Prunus persica Stokes

Syn.: *Amygdalus colinus* Wall.; *A. persica* L.; *Persica vulgaris* Mill.; *P. saligna* Royle.

Family: Rosaceae.

Vern.: Adu, Aru.

English and Hindi Names:

Eng. : Peach Tree, Nectarine;
Hindi : Aru, Shaftalu.

Distribution : Cooler parts of India, upto 3,350m in N.W. Himalayas.

Description : Small trees having fimbriate stipules. Leaves oblong-lanceolate with petiolar and marginal glands. Flowers pink or red usually appearing before leaves. Drupes downy, succulent; stones deeply furrowed.

Reproductive Cycle : February-May.

Materials Examined : HPU-EL-SKS-6564; Chhota Shimla (2,080m), Eddleston (2,060m).

Propagation Notes : Propagated by seeds requiring 2 - 3 months cold stratification, ring budding, shield budding and stem cutting.

Parts Used : Fruit. Kernel.

Folk Uses : Plant used as a stock for **grafting** scion of almond (*Prunus amygdalus*). Fruits and kernels **edible** as well as **pickled**. Seed oil used as **body massage** for **rheumatic pains**.

Active Constituents : These are galacturonic acid, arabinose, galactose, xylose and rhamnose (**fruits**).

Uses in Literature : Reported earlier as an abortifacient, anthelmintic, antiemetic, antiscorbutic, cattle feed, demulcent, dessert, diuretic, dye, eczema, edible (fruits, kernels), emmenagogue, insecticidal, illuminant (oil), peach nectar, psychoactive plant, purgative, table fruit, tonic, vegetable; and for beverages, biliousness, blood diseases, building purposes, canning, carbonizing, cough, deftness, earache, foot, eye and mouth disease, headache, hoof rot of cattle, itching and scabies, lessening thirst, leucoderma, peach brandy, pickling, piles, removing bad smell of mouth and sputum, rheumatism, scabies, stomach troubles, and whooping cough **(Agarwal, 2003; Ahmed & Borthakur, 2005; Ambasta, 1986; Anonymous, 1948-1976; Arora & Pandey, 1996; Bennet *et al.*, 1991; Chopra *et al.*, 1956; Dwivedi, 2007; Farooq, 2005; Harborne & Baxter, 2001; Jain, 1991; Jain *et al.*, 1994; Kapur & Singh, 1996; Kirtikar & Basu, 1984; Kumar, 2003; Lal *et al.*, 1996; Parmar & Kaushal, 1982; Rawat & Chowdhury, 1998; Roy *et al.*, 1998; Said, 1997; Sharma, 1996; Sharma & Rana, 2005; Singh & Kumar, 2000a; Sood & Thakur, 2004; Sood *et al.*, 2012b, 2013; Warman, 1999; Watt, 1889-1896)**.

Prunus salicina Lindl.

Family: Rosaceae.

Vern.: Shakarpara.

English Name:

Eng. : Japanese Plum.

Distribution : N. E. & N.W. Himalayas upto 2,600m.

Description : Small trees with reddish or brown, shining twig. Leaves oblong - obovate, shining above, abruptly but prominently pointed, closely blunt serrate, dull and usually glabrous beneath. Flowers white appearing clustered on spurs; pedicels slender glabrous. Drupes depressed ovoid, yellow or light red.

Reproductive Cycle : January-June.

Materials Examined : HPU-EL-SKS-6566; Bemloe (2,014m), Chhota Shimla (2,080m).

Propagation Notes : Propagated by seeds.

Part Used : Fruit.

Folk Use : Fruits **edible**.

Uses in Literature : Reported earlier as edible (fruits), maggoticide; and for jaundice **(Ambasta, 1986; Anonymous, 1948-1976; Arora & Pandey, 1996; Bennet *et al.*, 1991; Bora *et al.*, 2003; Chopra *et al.*, 1956; Dwivedi, 2007; Pande *et al.*, 1996; Roy *et al.*, 1998; Sharma & Rana, 2005; Sood et al., 2012b; Uphof, 1968; Watt, 1889-1896).**

Punica granatum L.

Syn.: *P. florida* Salisb.; *P. nana* L.; *P. spinosa* Lam.

Family: Lythraceae.

Vern.: Anar, Daadu.

English, Hindi and Sanskrit Names:

Eng. : Pomegranate;
Hindi : Anar, Dhalim, Dharimb;
Sans. : Bijapura, Dadima, Dantabijaka, Suphala.

Distribution : Cultivated throughout India; wild in N.W. regions.

Description : Small shrubby deciduous trees having glossy green opposite leaves clustered at the end of branches. Veins red. Flowers terminal, deep red, solitary or in cluster of 2 - 4. Fruits globose, containing red juice around seeds.

Reproductive Cycle : May-September.

Materials Examined : HPU-EL-SKS-6568; Boileauganj Cemetery (1,998m), Retreat (2,092m).

Propagation Notes : Plant shows a preference to calcareous soils and semi-arid climate. Propagated by seeds and hard wood cuttings.

Parts Used : Whole Plant. Seeds. Rind.

Folk Uses : Plant considered **sacred** and worshipped during **religious ceremonies**. Seeds **edible** as a rich source of vit. C. Dried seeds sold as 'anardana' for '**chutney**' preparation and extr. of its crushed seeds considered a **refreshing drink**. Powdered rind taken one tsp every morning with honey for seven days against **piles** and **cough**.

Active Constituents : Constituents isolated are cyanidin-3-glucoside delphinidin-3-glucoside and 3,5-diglucoside (**seed**), glucose, fructose, maltose, starch, carotene, thiamine, riboflavin, nicotinic acid, vitamin, pectin and amino acid (**edible portion of fruit**).

Uses in Literature : Recorded in Indian literature as an anthelmintic, anti-vomiting agent, astringent, cardiac, colouration material in 'Mithila' paintings, condiment, edible (seeds 'anardana'), expectorant, good source of vit. C, laxative, refreshing drink, souring agent, stomachic, styptic and cicatrizing, symbolic (offering), tonic; and for bleeding after child birth and miscarriage, brain affections, bronchitis, chest troubles, cholera, congestion of liver, conjunctivitis, controlling cholesterol level, cough, diarrhoea, dysentery, dyeing, dyspepsia, enlarged spleen, epistaxis, eye complaint, fever, gastric, general debility, gums, tone to the stomach, heart disorders, dysentery, increasing sexual vigour, inflammation of urinary tract, haemorrhage, iron, jaundice, kidney stones, leucorrhoea, malaria, menstrual disorders, nose bleeding, pimples, sore throat, tannin, uterine diseases, and wine **(Ahmed & Borthakur, 2005; Ambasta, 1986; Anonymous, 1990; Arora & Pandey, 1996; Badhwar & Fernandez, 2011; Borthakur *et al.*, 1999; Chatterjee & Pakrashi, 1997; Chauhan, 1999; Chevallier, 1996; Chopra *et al.*, 1956; Dagar & Dagar, 1996; Das & Agarwal, 1991; Dey, 1980; Dey & Bahadur, 1973; Dhiman, 2003; Ghate, 1998; Ghosh *et al.*, 1996; Hosagoudar & Henry, 1996a, c; Jain, 1991; Jha & Basak, 1994; Kapur & Singh, 1996; Katewa *et al.*, 2001; Khanna *et al.*, 1996b; Kirtikar & Basu, 1984; Kulkarni & Kumbhojkar, 1993; Kumar & Rao, 2001; Kurian, 1995; Lalramnghinglova, 1996; Mahato & Mahato, 1996; Mohanty & Padhy, 1996; Pandey & Joshi, 2001; Parmar & Kaushal, 1982; Parrotta, 2001; Perry & Hay 1982; Prajapati *et al.*, 2006; Pruthi, 1976; Raju, 2000; Rana *et al.*, 1996; Reddi *et al.*, 2005; Roy *et al.*, 1998; Said, 1997; Saklani & Jain, 1994; Sanyal, 1994; Seetharam *et al.*, 1999; Sharma & Rana, 2005; Shome *et al.*, 1996; Singh, 1996, 1999; Singh & Kumar, 2000a; Singh & Pandey, 1996; Sood & Thakur, 2004; Sood *et al.*, 2012b, 2013; Thakur *et al.* 1989; Tirkey *et al.*, 2003b; Uphof, 1968; Verma & Chauhan, 2006; Warman, 1999; Watt, 1889-1896)**.

Pyracantha crenulata **(D. Don) M. Roem.**

Family: Rosaceae.

Vern.: Gingaru, Ringu.

English Name:

Eng. : Napalese White Thorn.

Distribution : Satluj valley - Bhutan: 1,000-2,600m.

Description : Large spinescent shrubs with linear, shining leathery, narrow-oblong or obovate crowded leaves and white flowers in corymbs. Fruits globose, orange red, crowned with erect calyx lobes.

Reproductive Cycle : March-June.

Material Examined : HPU-EL-SKS-6569; Boileauganj Cemetery (1,998m), Chapslee (2,243m).

Propagation Notes : Propagated by seeds requiring cold stratification before germination. Cuttings of almost mature wood, 5 - 10cm with a heel in mid-August another way of propagation.

Parts Used : Whole Plant. Fruits. Wood.

Folk Uses : Grown as a **hedge** plant. Fruits **edible**. Powdered dried fruit mixed with yoghurt good against **dysentery**. Wood used for making **walking stick**.

Active Constituents : Flavanols – kaempferol, quercetin and saponins extracted from **fruits**.

Uses in Literature : Described earlier as cardiac vasodilator, cardiotonic, edible (fruits), tonic; and for arteriosclerosis and burgor's diseases, constipation, hypertension myocardial weakness **(Ambasta, 1986; Arora & Pandey, 1996; Bailey, 1957; Chandra, 1997; Chauhan, 1999; Pande *et al.*, 2006; Samant *et al.*, 2001; Sharma, 2003)**.

Pyrus communis L.

Family: Rosaceae.

Vern.: Nashpati.

English, Hindi and Sanskrit Names:

Eng. : The Common Pear, Pear;
Hindi : Nashpati;
Sans. : Amritaphala.

Distribution : N.W. Himalayas.

Description : Small trees possessing oblong or ovate-acute, entire or somewhat crenulated leaves of variable size. Leaf stalk slender, long. Flowers white and borne profusely in corymbs, mostly on upper portion of the branches. Fruits top-shaped or sub-globose intruded at the top, green in colour.

Reproductive Cycle : June-August.

Materials Examined : HPU-EL-SKS-6571; Advance Study (2,128m), Chhota Shimla (2,078m), Inverarm (2,150m), Pari Mahal (2,105m).

Propagation Notes : Pear usually propagated by shield-budding in June-July. Cuttings and air layering also successful. Prefers a good well-drained loam in full sun.

Part Used : Fruits.

Folk Uses : Fruits **eaten** raw and considered **astringent** and **laxative** in nature; also employed for making **jams** and **pickles**.

Active Constituents : These are chlorogenic acid (anti-oxidant), procyanidins, flavonol, glycosides (**fruits**).

Uses in Literature : Used so far as astringent, edible, febrifuge, psychoactive plant, sedative; and for acute fever, beverages, carving, engraving purposes, malaria, mathematical and drawing instruments, turning, and wines **(Agarwal, 2003; Ahmed & Borthakur, 2005; Ambasta, 1986; Anonymous, 1948-1976; Chaudhuri, 1993; Dwivedi, 2007; Farooq, 2005; Gill *et al.*, 1993; Hadfield, 1956; Harborne & Baxter, 2001; Jain *et al.*, 1994; Kirtikar & Basu, 1984; Sood *et al.*, 2013; Srivastava, 1989; Uphof, 1968; Warman, 1999; Watt, 1889-1896).**

Pyrus pashia Buch.-Ham. *ex* D. Don

Family: Rosaceae.

Vern.: Kainth, Segal.

Hindi Names:

Hindi : Mehal, Mol.

Distribution : Temperate Himalayas (Kashmir-Bhutan): 800-2,700m; Khasia mountains 1,800m.

Description : Small trees having barren branchlets usually ending in a spine and simple, ovate or ovate-lanceolate, crenate, long-pointed leaves. Flowers white tinged with pink, in simple corymbs. Fruits yellow-brown, globular, warted, rough with small white spots.

Reproductive Cycle : May-August.

Materials Examined : HPU-EL-SKS-6573 B.C.S. (2,025m), Boileauganj Cemetery (1,998m), Cecil Hotel (2,090m), Chhota Shimla (2,080m), Gorton Castle (2,094m).

Propagation Notes : Propagated naturally by seeds.

Part Used : Leaves. Fruit. Wood.

Folk Uses : Leaves considered good **fodder** for goats. Fruits **edible**. Mashed fruits appilied on head as **coolant** and to check **nose bleeding**. Extr. of overnight soaked fruits used as hair shampoo to check **dandruff**. Also, unripe ones chewed for healing **mouth ulcer**. Wood used for **small implements**, **walking sticks** and as **fuel**.

Active Constituents : These are friedelin, β-sitosterol (**bark**), n-triacontane, myricyl alcohol as chloride and sitosterol (**leaves**).

Uses in Literature : Used earlier for acidity, asthma, bobbins, catarrh (vet.), combs, edible purposes, eye complaint, indigestion, injury (vet.), mouth infection, textile mills, tobacco pipes, walking sticks; and as fodder, fuel **(Ahmed & Borthakur, 2005; Ambasta, 1986; Anonymous, 1948-1976; Arora, 1997; Arora & Pandey, 1996; Badhwar & Fernandez, 2011; Bennet *et al.*, 1991; Dwivedi, 2007; Farooq, 2005; Jain, 1991; Kapur, 1996b; Lal *et al.*, 1996; Lata *et al.*, 2008; Pande *et al.*, 2006; Raju, 2000; Rana *et al.*, 1996; Rao, 1990; Roy *et al.*, 1998; Sahni, 1998; Samant *et al.*, 2001; Sharma & Rana, 2000; Singh, 1996; Singh & Kumar, 2000b; Sood *et al.*, 2012b, 2013; Watt, 1889-1896)**.

Pyrus pyrifolia Nakai

Syn.: *P. serotina* Rehder.

Family: Rosaceae.

Vern.: Zarainth.

English Names:

Eng. : Asian Pear, Korean Pear, Japanese Pear, Sand Pear.

Distribution : Throughout mid - hill region.

Description : Trees with rough, light brown coloured bark and cordate, stipulated, serrated acuminate leaves. Flowers white. Fruits globular, brown. Seeds dark brown, pyriform.

Reproductive Cycle : March-June.

Materials Examined : HPU-EL-SKS-6575; Advance Study (2,128m), Grand Hotel (2,105m), Wood Villa (2,108m).

Propagation Notes : Propagated by seeds.

Part Used : Fruit.

Folk Uses : Ripe fruits **edible** and unripe ones used for making **pickle** and **jams**.

Uses in Literature : Known so far as edible (fruits); and for bronchitis, phthisis **(Ahmed & Borthakur, 2005; Ambasta, 1986; Parmar & Kaushal, 1982; Prajapati *et al.*, 2006; Sharma & Rana, 2005; Sood *et al.*, 2012b)**.

Quercus floribunda Wall.

Syn.: *Q. dilatata* Lindl. *ex* A. DC. *Q. dilatata* Royle; *Q. himalayana* Bahadur.

Family: Fagaceae.

Vern.: Moru.

English and Hindi Names:

Eng.	:	Green Oak, Moru Oak;
Hindi	:	Moru.

Distribution : N.W. Himalayas: 2,000-2,700m.

Description : Trees with bark peeling off in longitudinal scales and glabrous leaves which are entire or with sharp spinescent teeth. Male flowers distant from each other; female ones in cluster in drooping catkins. Acorns usually solitary; cup covering ovoid, pointed, smooth, brown nut.

Reproductive Cycle : April-October.

Materials Examined : HPU-EL-SKS-6576; Advance Study (2,128m), Raj Bhawan (2,132m).

Propagation Notes : Propagated by seeds.

Parts Used : Shoot. Leaves. Wood.

Folk Uses : Leaves and young shoot lopped for **fodder**. Wood used as **fuel** and **building material**; and for **sleepers**, **axe-handles**, **walking sticks**, and **agricultural implements**.

Uses in Literature : Reportedly used as astringent, building materials, charcoal, diuretic, edible (galls), fodder, tannin; and for asthma, diarrhoea, indigestion, gonorrhoea treatment and agricultural implements, umbrella-handles and walking sticks **(Ambasta, 1986; Anonymous, 1948-1976; Dwivedi, 2007; Kapur, 1996b; Raju, 2000; Samant *et al.*, 1996; Sharma, 1996; Singh, 2000b)**.

Quercus incana W. Bartram

Syn.: *Dryopsila cinerea* (Michx.) Raf.; *D. oligodes* Raf.; *D. verrucosa* Raf.; *Q.cinerea* Michx.; *Q. ilexoides* Raf.; *Q. oligodes* Raf.; *Q. verrucosa* Raf.

Family: Fagaceae.

Vern.: Banj.

English and Hindi Names:

Eng.	:	Grey Oak, Himalayan White Oak, Kumaon Oak;
Hindi	:	Ban, Banj.

Distribution : Temperate Himalayas: 1,200-2,400m.

Description : Evergreen trees possessing dark grey bark and hoary or woolly young shoots and petiole. Leaves cuspidate-serrate, coriaceous, glabrous above and densely white-tomentose beneath. Male spike softly hairy. Female flowers solitary or clustered in leaf-axils on short stout peduncles. Acorns half-enclosed by the campanulate cup when ripe.

Reproductive Cycle : April-October.

Materials Examined : HPU-EL-SKS-6578; Holly Lodge (2,278m), Kanlog Cemetery (1,990m), Gorton Castle (2.094m).

Propagation Notes : Propagated by seeds.

Parts Used : Leaves. Wood.

Folk Uses : Leaves lopped for **fodder**. Wood used as **building material**, **fuel**; and for **agricultural implements** and **tool handles**.

Active Constituents : Active constituents are flavonoids, quercetin, quercetin-3-disaccaharide, galactose and arabinose (**leaves**).

Uses in Literature : Useful earlier as astringent, pen, diuretic, dye, fodder, fuel; and for charcoal formation, diarrheoa, gonorrheoa, household goods, pencil industry, tannin and tea (gum) **(Agarwal, 2003; Ambasta, 1986; Anonymous, 1948-1976; Bennet *et al.*, 1991; Chaudhuri, 1993; Dwivedi, 2007; Kirtikar & Basu, 1984; Manral & Pande, 2004; Singh, 2000b; Uphof, 1968)**.

Quercus oblongata D. Don

Syn.: *Q. leucotrichophora* A. Camus.

Family: Fagaceae.

Vern.: Ban.

English and Hindi Names:

Eng. : Ban Oak, Grey Oak, Himalayan White Oak, Kumaon Oak;

Hindi : Ban, Banji.

Distribution : Pakistan - C. Nepal. Sri Lanka. Myanmar: 1,200-2,400m.

Description : A large-sized tree possessing stalked, ovate-lanceolate leaves which are spinous towards the tip. Male and female flowers on separate spikes. Male spikes slender, drooping, 4-10 cm long, often much interrupted. Female ones axillary, sessile, solitary or clustered. Acorns usually solitary; cup at first enclosing the nut, glabrate, rough, woody.

Reproductive Cycle : April-October.

Materials Examined : HPU-EL-SKS-6579; Advance Study (2,128m), Armsdell (2,102m), Auckland House (2,257m), Chhota Shimla (2,090m), Gorton Castle (2,094m).

Propagation Notes : Propagated by seeds.

Parts Used : Leaves. Wood.

Folk Uses : Leaves considered an excellent **fodder** for enhancing **lactation** in milching cattle. Wood good for **house construction**, **furniture** and **agricultural implements**. Also, used as **fuelwood**.

Active Constituents : These are quercetin and its 3-O-disaccharide (**leaves**).

Uses in Literature : Known to be useful for asthma, confectionary, diarrhoea, dyspepsia, fever, gonorrohea, indigestion, scabies, skin diseases, snake bite, stomachache, tanning, urinary complaints; and as astringent, ploughs, building materials, charcoal, diuretic, edible (acorns), fodder, fuel, hardboards, handles **(Ambasta, 1986; Bennet *et al.*, 1991; Dwivedi, 2007; Farooq, 2005; Jain, 1991; Jain *et al.*, 1991; Kapur & Nanda, 1996; Negi & Gaur, 1991; Pande *et al.*, 2006; Samant *et al.* 2001; Singh & Kumar, 2000a; Uniyal & Chauhan, 1973; Watt, 1889-1896)**.

Quercus robur L.

Syn.: *Q. pedunculata* Ehrh.

Family: Fagaceae.

Vern.: English Oak.

English Names:

Eng. : Common Oak, English or European Oak, Pedunculate Oak.

Distribution : Grows throughout the northern hemisphere.

Description : Deciduous trees with dark grey bark, dark green leaves possessing 3-7 deep lobes per side. Flowers yellowish-green in separate male and female catkins. Fruit brown.

Reproductive Cycle : April-October.

Materials Examined : HPU-EL-SKS-6583; Advance Study (2,128m), Wood Villa (2,108m).

Propagation Notes : Propagated by seeds and vegetative method. Seeds quickly loose viability if allowed to dry out. Prefers a good deep fertile loam.

Part Used : Whole Plant.

Folk Use : Grown as an **ornamental** plant.

Uses in Literature : Reported to be used as edible (seed), ointment for annual fissures and haemorrhoids, antiseptic, astringent, charcoal, tonic; and for carpentry, construction, diarrhoea, dysentery, eczema, frames, furniture, nasal polyps, rectal bleeding, ship building, skin problems, small burns, sore throats, tanning, tonsillitis and cultivation **(Anonymous, 1948-1976; Chej, 1994; Harborne & Baxter, 2001; Huxley, 1992; Khare, 2004; Krisa, 1982; Kunkel, 1984; Pool, 1941; Prajapati *et al.*, 2006; Uphof, 1968; Usher, 1974)**.

Rhamnus persica Boiss.

Syn.: *R. pentapomica* Parker.

Family: Rhamnaceae.

Vern.: Thalta.

English and Hindi Names:

Eng. : Buckthorn;
Hindi : Chirla.

Distribution : Temperate Himalayas: 650-2,400m.

Description : Small trees with leaves which are acute at the apex and minutely tomentose beneath. Flowers few. Fruit obovoid - globose. Seeds with a wide open groove, the edge thickened.

Reproductive Cycle : April-October.

Material Examined : HPU-EL-SKS-6585; Sanjauli Cemetery (2,068m).

Propagation Notes : Propagated naturally by seeds.

Parts Used : Leaves. Fruits. Wood.

Folk Uses : Leaves used as **fodder**. Fruits **eaten** during scarcity. Wood used for **agricultural implements** and as **fuel**.

Uses in Literature : Reported in India for edible purposes, dyeing, turnery; and as fodder **(Ambasta, 1986; Arora & Pandey, 1996; Badhwar & Fernandez, 2011; Bennet *et al.*, 1991; Roy *et al.*, 1998)**.

Rhamnus virgatus Roxb.

Family: Rhamnaceae.

Vern.: Reushi.

English and Hindi Names:

Eng. : Himalayan Buck Thorn;

Hindi : Chato.

Distribution : Hilly districts throughout the India, ascending to 2,770m.

Description : Spinous nearly glabrous shrubs or small tree, having membranous, almost opposite leaves which are frequently fascicled on arrested branchlets. Flowers greenish on slender pedicels, in axillary clusters. Drupes obovoid.

Reproductive Cycle : April-June.

Materials Examined : HPU-EL-SKS-6587; Boileauganj Cemetery (1,992m), Green Gate (2,098m), Kanlog Cemetery (1,990m).

Propagation Notes : Propagated by seeds.

Parts Used : Whole Plant. Leaves. Wood.

Folk Uses : Plant considered **sacred** to appease Lord Shiva. Leaves lopped as **fodder** for sheep and goats. Wood used for **agricultural implements**.

Active Constituents : These are physcion - 8 - O - gentiobioside (**wood**), 5 - methoxy - 7 - hydroxyphthalide, 6 -O - methylalaterin, maesopsin (**stem**).

Uses in Literature : Known earlier as emetic, fodder, purgative; and for eczema, infection of spleen, ringworm infection **(Ambasta, 1986; Dwivedi, 2007; Farooq, 2005; Pande *et al.*, 2006; Rana *et al.*, 1996; Rastogi & Mehrotra, 1995a)**.

Rhapis excelsa (Thunb.) A. Henry

Syn.: *Chamaerops excelsa* Thunb.; *R. divaricata* Gagnep.; *R. major* Blume; *Trachycarpus excelsus* (Thunb.) H. Wendl.

Family: Arecaceae.

Vern.: Lady Palm.

English Names:

Eng. : Broadleaf Lady Palm, Dwarf Ground-Rattan, Lady Palm.

Distribution : Originated from S. China & Japan.

Description : Bamboo - like shrubs with stem forming clump from underground suckers and densely covered with coarse fibres. Leaves leathery glossy green, and deeply cut into several (3-10) segments which are uniformly curving or drooping at apex. Spinules small, present along margins. Mid-veins cross-wrinkled. Leaf-stalk about 45 cm long, slender, flattened, obscurely toothed in the margins and with small fibres at the base. Inflorescence small, at top of the plant with spirally-arranged fleshy flowers containing three petals fused at the base. Ripe fruits fleshy and white.

Reproductive Cycle : May-July.

Materials Examined : HPU-EL-SKS-6588; Advance Study (2,128m), Grand Hotel (2,105m).

Propagation Notes : Propagated by suckers during July-August.

Part Used : Whole Plant.

Folk Use : Grown as an **ornamental**.

Uses in Literature : Suitable for planting in groups for mass effect as well as in pots and tubs and also for umbrella handles, walking sticks and thatching **(Ambasta, 1986; Beri, 1992; Paliwal, 2009; Patil, 2008)**.

Rhododendron arboreum Sm.

Family: Ericaceae.

Vern.: Braah, Brass.

English and Hindi Names:

Eng. : Rhododendron Tree, Rose Tree;
Hindi : Burans.

Distribution : Temperate Himalayas: 1,500-3,600m.

Description : Evergreen, much branched trees. Leaves oblong-lanceolate with upper surface glabrous, lower silvery-scaly and crowded towards the end of branches. Flowers blood red, crowded in large head-like corymbs. Capsule scaly, curved.

Reproductive Cycle : March-May.

Materials Examined : HPU-EL-SKS-6588; Advance Study (2,128m), Boileauganj Cemetery (1,992m), Eagle Mount (2,152m), Kanlog Cemetery (1,992m), Nav Bahar Cemetery (2,031m).

Propagation Notes : Propagated naturally by seeds and cuttings. Plant succeeds in most humus rich lime free soils.

Parts Used : Branches. Flowers. Wood.

Folk Uses : Flowers **eaten** raw or made into **culinary preparations** like '**pakoras**', '**chutneys**' and **squash**. Its sugary decoction considered **astringent** to **bowels**, **refreshing cardiac drink**; and good against **nose bleeding** and **burning sensation** of **soles**. Also, it is an important constituent of **local**

wine, '**sur**'. Wood used for making **agricultural implements**, **tool handles**, and as **fuel**. Branches used for **decorating** arches of gates and temples etc.

Active Constituents : These are ursolic acid and quercetin (**flowers**).

Uses in Literature : Useful as fuel, refreshing drink, tool handles, vegetable; and for amoebic dysentery, boxes, chutneys, jam, jelly, diarrhoea, dysentery, fever, gum-stocks, headache, kukri, nose bleeding, park-saddles, plywood, posts, rheumatism, smoking pipe, wounds, decoration of temples, clearing fish bone stuck in throat, poisoning rats, and wood for building and plates **(Agarwal, 2003; Ambasta, 1986; Arora & Pandey, 1996; Bennet *et al.*, 1991; Bhattacharjee, 2001; Chauhan, 1984, 1999; Chhetri, 2005; Chopra *et al.*, 1956; Dash *et al.*, 2003; Dhiman, 2003; Dwivedi, 2007; Farooq, 2005; Jain, 1991; Jamir *et al.*, 2011; Kirtikar & Basu, 1984; Lal *et al.*, 1996; Lata *et al.*, 2008; Mahato & Chaudhary, 2005; Manandhar, 1990; Manral & Pande, 2004; Panda, 1996; Raju, 2000; Randhawa, 1957, 1960; Ranjan, 2000b; Roy *et al.*, 1998; Sahni, 1998; Samant *et al.*, 2001; Sharma, 1996; Singh, 1996; Singh & Kumar, 2000a; Sinha, 1996; Sood *et al.*, 2012b, 2013; Watt, 1889-1896)**.

Rhododendron simsii var. *vittatum* Wils.

Family: Ericaceae.

Vern.: Ajalea.

English Name:

Eng. : Azalea.

Distribution : Cultivated in cooler part of India upto 2,550m.

Description : Evergreen or half-evergreen ornamental shrubs with ovate - lanceolate, acute, hairy leaves. Flowers varying from white through all shades of pink and red - crimson; and in umbel - like racemes. Fruits a capsule.

Reproductive Cycle : January-March.

Material Examined : HPU-EL-SKS-6589; Advance Study (2,128m).

Propagation Notes : By seeds and layering in late July. Plant succeeds in a moist humus-rich lime-free soils.

Part Used : Whole Plant.

Folk Use : Planted as an **ornamental** shrub for its beautiful flowers.

Uses in Literature : Grown as an ornamental (**Bailey, 1957; Graf, 1981; Hay *et al.*, 1983; Jaeger, 1961; Paliwal, 2009**).

Ricinus communis L.

Syn.: *Cataputia major* Ludw.; *C. minor* Ludw.; *Croton spinosus* L.; *R. africanus* Mill.; *R. armatus* Andr.; *R. badius* Rchb.; *R. cambodgensis* Benary; *R. compactus* Huber ; *R. inermis* Mill.; *R. laevis* DC.; *R. macrocarous* Popova; *R. ruber* Miq.; *R. urens* Mill.; *R. zanaibarinus* Popova.

Family: Euphorbiaceae.

Vern.: Arind, Arindi.

English, Hindi and Sanskrit Names:

Eng.	:	Caster-oil Plant, Castorseed, Palma Christi;
Hindi	:	Arand, Arandi, Arend, Erand, Erandi, Erend, Ind, Rand;
Sans.	:	Amanda, Amangala, Bhanda, Chankuka, Chitrabija, Chitraka, Dirghadantaka, Eranda, Gandharvahastaka, Ishta, Kanta, Panchangula, Panjangula, Ruvuka, Shukla, Shulashatru, Shweteranda, Svehaprada, Vatari, Vranaha, Vuka, Vyadatvaka, Vyaghradala,. Vyaghrapuchha

Distribution : Native of Tropical Africa; naturalized near habitation upto 2,000m throughout India.

Description : Evergreen bushes or small trees possessing palmately lobed leaves and monoecious, rather large flowers in terminal subpaniculate racemes; male flowers crowded in upper portion of inflorescence, the female below. Capsule prickly, and each with a three, 2-valved cocci. Seeds oblong, smooth, mottled.

Reproductive Cycle : March-May (intermittent flowering).

Materials Examined : HPU-EL-SKS-6591; Bemloe (2,016m), Boileauganj Cemetery (1,992m).

Propagation Notes : Propagated by seeds.

Parts Used : Leaves. Stem. Flowers. Seed.

Folk Uses : Extr. of leaves appilied in anus for expelling **intestinal worms**. Oil smeared warm leaves applied over furuncles to hasten **suppuration**. Stem used for **scouring teeth**. Flowers hung on the main entrance of house for warding off **evil spirit**; also offered to **deities**. Seed paste prepared in butter good for **cutaneous** diseases. Mustard oil boiled with its seeds considered as **hair tonic**. Seed oil used against **constipation** and also massaged for relief from **rheumatic pains**.

Active Constituents : Various constituents isolated are hyperoside, rutin, ricinine (**flowers**), apigenin, chlorogenic (**endosperm**), β - amyrin, stigmasterol, β -sitosterol, ricinine, quercetin, rutin, hyperoside, n - dimethylricinine, 3 - 0 - β - d - xylopyranosides, 3 - 0 - β - d -glucopyranosides and 3 - 0 - β - D - rutinosides of kaempferol and quercetin (**leaves**).

Uses in Literature : Recorded earlier as an anthelmintic (vet.), antidote to food poison, antipyretic (vet.), anti-rheumatic, cathartic, contraceptive jellies, edible, emmenagogue, fish poison, fly-paper, fodder, fuel, heat-sealing adhesives, insulating materials, lactagogue, laxative, lubricant, lumbago, luminant, manure, psychoactive plant, silkworm host, soothing agent used in ointments, symbolic (wards off evil), typewriter ink, vermicide; and for abortion, amenorrhoea, anasarca, asthma, backache, body swelling, boils, bone fracture (vet.), burns, candles, cholera, chronic ulcer, constipation (vet.), cough, culinary purposes, cutaneous affections, dental caries, dermatitis, dropsy, dyeing, dysentery, dysentery (vet.), eczema, facilitating child birth, fever, flatulence, gum trouble, hard and dry wax in ears, headache, heat stroke, hydrocele, imitation leather, inflammation of breast during lactation, inflammation of eyes, injury, jaundice, loosening of limbs, muscular pain, nervous disorders, night blindness, plastic moulding composition, pneumonia, poultice to boils and sores, preservation of fruits and vegetables, preserving leather and hair oil, producing sterility for one month, rheumatism (vet.), scorpion and snake-bites, soap making, skin and hair dressing, sores, sprain and muscle cramps, swelling, synthetic nylon, transparent soaps, ulcers, warts, wound (vet.), wounds, and in bone sprain of horses **(Agarwal, 2003; Ambasta, 1986; Bhandary *et al.*, 1996; Bhatt *et al.*, 2003; Borthakur, 1993; Borthakur *et al.*, 1999; Chauhan, 1999; Dey & Bahadur, 1973; Dixit & Kumar, 2003; Dwivedi, 2007; Hembrom, 1996; Islam & Jha, 2003; Jain, 1968, 1991; Jain & Saklani, 1992; Jain *et al.*, 1994; Jamir *et al.*, 2011; Jha & Kumar, 2003; Jha *et al.*, 1991; Joshi & Joshi, 2005; Katewa & Arora, 1997; Khanna *et al.*, 1996a; Kharkonger & Joseph, 1981; Kshirsagar & Singh, 2007; Kumar *et al.*, 2004; Lalramnghinglova, 1996; Mahato & Chaudhary, 2005; Mahato & Mahato, 1996; Manandhar, 1996; Mao, 1993; Masih, 2003; Mishra *et al.*, 1996, 1997; Mokat & Deokule, 2004; Nayak *et al.*, 2003; Padhy & Dash, 2003; Panda, 1996; Pande *et al.*, 2006; Pandey, 2003; Parrotta, 2001; Pawar & Patil, 2004; Rajendran *et al.*, 2003; Raju, 2000; Rama Rao & Henry, 1996; Ranjan, 2003; Rawat & Chowdhury, 1998; Saini, 1996a; Satapathy & Brahmam, 1996; Sharma *et al.*, 1979; Shome *et al.*, 1996; Singh & Pandey, 1998; Singh *et al.*, 1996b, 1997; Siwakoti & Varma, 1996; Srivastava *et al.*, 2003b; Tirkey *et al.*, 2003b; Tiwari**

& Tiwari, 1996; Uphof, 1968; Vihari, 1995; Viswanathan & Singh, 1996; Watt, 1889-1896).

Robinia pseudoacacia L.

Syn.: *R. pringlei* Rose; *R. pseudoacacia* var. *rectissima* Raber.
Family: Fabaceae.

Vern.: Rabinia.

English Names:

Eng. : Black Locust, False Acacia, Robonia.

Distribution : Himalayas upto 2,400m.

Description : Trees with odd pinnate leaves and entire leaflets. Flowers white, fragrant in pendulous racemes. Pods brown, flat, hairless with 4-10 seeds.

Reproductive Cycle : April-May.

Materials Examined : HPU-EL-SKS-6594; Advance Study (2,128m), B.C.S. (2,025m), Chhota Shimla (2,090m), Grand Hotel (2,105m), Retreat (2,092m), St. Edward's School (2,060m).

Propagation Notes : Propagation by seeds. Plant succeeds in any well-drained soil.

Parts Used : Whole Plant. Twigs. Wood.

Folk Uses : Planted on roadsides for shade **fencing**. Twigs used as **fodder** and **toothbrush** for **scouring teeth**. Wood used for making **agricultural implements**, **furniture** and as **fuel**.

Active Constituents : Plant yields coumarin, scopoletin (**leaves**), 2,4-dihydroxybenzoic acid, its methyl, i-(4-hydroxycinnamoyl) esters and 1-(3,4,5-trihydroxycinnamoyl) esters, liquiritigenin, butein, butin, robinetin and dihydrorobinetin (**heartwood**).

Uses in Literature : Recorded earlier as antispasmodic, axle-trees, diuretic, edible (seeds), emetic, fuel, insulator for telegraph and telephone wires, laxative, shoe-lasts; and for agricultural implements, boast ribs, brackets, cart-wheels, digestive disorders, fence posts, flowers, furniture, gastric problems, gates, hard and heavy wood for construction, leucorrhoea, paper making, renovating exhausted soil and instructing poor land, ship-building, sleepers, sports goods, tool-handles, tree nails, turnery and wagon-hubs **(Agarwal, 2003; Ambasta, 1986; Anonymous, 1948-1976; Baburaj *et al.*, 2000; Dwivedi, 2007; Farooq, 2005; Singh, 2000a; Singh & Kumar, 2000a; Singh *et al.*, 1996a; Sood *et al.*, 2010, 2013; Uphof, 1968; Watt, 1889-1896)**.

Rosa alba L.

Syn.: *R. glandulifera* Roxb.
Family: Rosaceae.

Vern.: Gulab.

English, Hindi and Sanskrit Names:

Eng. : Common White Dog Rose, White Cottage Rose;
Hindi : Gulab;
Sans. : Bahupatrika, Bhrigavallabha, Gandhadhya, Shatapatri, Shivaballabha.

Distribution : Throughout India.

Description : Strong upright bushes with scattered uniform more or less hooked prickles and sometimes with bristles. Leaflets usually 5 sometimes 7, oblong - broad elliptic, short-pointed, simply serrate, greyish somewhat wrinkled, glabrous above and pubescent beneath. Flowers white or soft pink, single or often double, fragrant and with glandular-hispid pedicels. Receptacle smooth. Fruits oblong or ovoid, bright red.

Reproductive Cycle : June-August.

Materials Examined : HPU-EL-SKS-6597; Advance Study (2,128m), Hainault (2,294m).

Propagation Notes : Propagated by cuttings in July-August.

Part Used : Whole Plant.

Folk Uses : Cultivated as an **ornamental**; flowers **offered** to local deities.

Uses in Literature : Reported to be used as coolant in fever, edible (petals), essential oil, laxative, ophthalmic, ornamental; and for skin ailments **(Ambasta, 1986; Bailey, 1957; Kirtikar & Basu, 1984; Ranjan, 2000a; Saikia *et al.*, 2010; Sood *et al.*, 2009b, 2013)**.

Rosa banksiae R. Br.

Syn.: *R. banksiae* var. *alboplena* Rehder; *R. inermis* Roxb.
Family: Rosaceae.

Vern.: Banksia Rose, Gulabri.

English Name:

Eng. : Banksian Rose.

Distribution : Indigenous to China and Japan; cultivated in India.

Description : High climbing evergreen, slender glabrous twigs with a few or no prickles, and 3 or 5 oblong - lanceolate, sharp - serrulated, acute or obtuse, shining leaflets which are glabrous above and hairy on midrib beneath. Stipules free from petiole, subulate and caducous. Flowers small, few - many on very slender pedicels in umbel-like clusters, white or yellow, single or usually variously doubles. Fruits pea-size, sepals caducous.

Reproductive Cycle : April-June.

Materials Examined : HPU-EL-SKS-6599; Advance Study (2,128m), B.C.S. (2,025m), Chapslee (2,243m), Chhota Shimla (2,070m), Hainault (2,294m).

Propagation Notes : Propagated by seeds and stem-cutting in June-August.

Part Used : Whole Plant.

Folk Use : Planted as an **ornamental hedge** around boundary walls.

Uses in Literature : Reported to be used as anthelmintic, astringent, tanning (bark, root), vulnerary and ornamental climber **(Ambasta, 1986; Bailey, 1957; Kirtikar & Basu, 1984; Sood *et al.*, 2013; Uphof, 1968)**.

Rosa chinensis Jacq.

Syn.: *R. fragrans* Redoute; *R. indica* L.; *R. sinica* L.

Family: Rosaceae.

Vern.: Gulab.

English and Hindi Names:

Eng. : China Rose, Monthly Rose, Rose;
Hindi : Gulab.

Distribution : Indigenous to China.

Description : Evergreen glabrous shrubs having stipules which are adnate nearly to the apex. Flowers large on long slender pedicels; sepals reflexed. Fruits a red hip, 1–2 cm in diameter.

Reproductive Cycle : June-September.

Materials Examined : HPU-EL-SKS-6605; Advance Study (2,128m), Armsdell (2,102m), Bantony (2,182m), St. Edward's School (2,060m), Wood Villa (2,108m).

Propagation Notes : Propagated by seeds and stem cutting in June-August.

Parts Used : Whole Plant. Petals.

Folk Uses : Planted as an **ornamental** hedge. Flowers **offered** to deities; and its petals used for preparing **'gulkand'** which is considered good against digestive disorders.

Active Constituents : These are linalool acetate, limonene, methoxymethylbenzene, citronellol, β-caryophyllene, hexanol acetate, linalool and hexanol (**flowers**).

Uses in Literature : Known earlier for treating injuries, foul ulcers, sprains, wounds; and as ornamental plant **(Agarwal, 2003; Ambasta, 1986; Bailey, 1957; Bhattacharjee, 2001; Kirtikar & Basu, 1984; Paliwal, 2009)**.

Rosa damascena Mill.

Family: Rosaceae.

Vern.: Gulabadi.

English, Hindi and Sanskrit Names:

Eng.	:	Bussora Rose, Damascus Rose, Damask Rose, Summer Damask Rose;
Hindi	:	Bussorah, Fasli Gulab, Gulab, Gulab Ka Phul, Sudburg;
Sans.	:	Atimanjula, Pakshapushpa, Mahakumari, Shatadala, Shatapatri, Soumyagandra, Sumana, Sushita, Suoritta.

Distribution : Throughout India.

Description : Prickly shrubs having serrated leaflets and ovate-oblong stipules. Prickles dilated, uniform or all alike. Flowers blush-white - red, 6-12 in a corymb, very fragrant. Fruits a hip, obovate.

Reproductive Cycle : April-July.

Materials Examined : HPU-EL-SKS-6607; Advance Study (2,128m), Bemloe (2,010m), Grand Hotel (2,105m), R.K.M.V. (2,257m).

Propagation Notes : Propagated by seeds and stem cutting in July-August.

Parts Used : Whole Plant. Flower.

Folk Uses : Planted for **ornamental** purposes. Flowers **offered** to appease various deities. Petals consumed raw as **tonic** or as '**gulkand**' (a conserve made from petals and sugar); its decoction also considered good for **leucorrhoea** and **constipation.**

Active Constituents : Active properties due to 2 - hydroxyursolic acid, β - amyrin, β - phenethyl - β - D – glucopyranoside, methyl ursate, geraniol, β-phenylethanol and its glucoside, eugenol, chlorogenic acid, cyanin, cyaniding, carotene, 3-5-diglucoside, quercetin and sugar (**flowers**).

Uses in Literature : So far considered as antispasmodic, astringent, cardiotonic, expectorant, laxative, liquor, perfume, rose oil, rose water for confectionary, tonic, vasoconstrictor; and for cardiac, chest, kidney, liver, lung and stomach disorders, chronic fever, drinks, and headache **(Ambasta, 1986; Bailey, 1957; Bhatt *et al.*, 2001; Caius, 1986; Chatterjee & Pakrashi, 1997; Chauhan, 1999; Dey & Bahadur, 1973; Ghate, 1998; Harborne & Baxter, 2001; Kapur *et al.*, 1996; Kirtikar & Basu, 1984; Methani & Ingle, 1999; Nautiyal *et al.*, 1997; Said, 1997; Sanyal, 1994; Sasi *et al.*, 2010; Sharma, 2003; Sood *et al.*, 2011; Uphof, 1968).**

Rosa jacquemontii Crep. *ex* Hook. F.

Family: Rosaceae.

Vern.: Gulab.

Distribution : Kashmir-Kumaon: 1,500-4,100m.

Description : Prickly shrubs with white, sweet-scented flowers either solitary or in dense clusters. Leaflets coarsely toothed. Sepals lanceolate. Petals obcordate. Fruits flask-shaped.

Reproductive Cycle : June-September.

Material Examined : HPU-EL-SKS-6609; Advance Study (2,128m).

Propagation Notes : Propagated by seeds and stem cutting in July-August.

Part Used : Whole Plant.

Folk Use : Planted for **ornamental** purposes.

Uses in Literature : Reported to be grown for its ornamental flowers **(Bailey, 1957; Sood *et al.*, 2013)**.

Rosa moschata Mill.

Syn.: *R. arborea* Pers.; *R. brunonii* Lindl.; *R. lindleyi* Harb.; *R. pubescens* Roxb.

Family: Rosaceae.

Vern.: Kuja, Kuji.

English, Hindi and Sanskrit Names:

Eng. : Himalayan Musk Rose;
Hindi : Karer, Kuji, Kunja, Kwiala;
Sans. : Sewati.

Distribution : Temperate C. & W. Himalayas: 1,200-2,500m.

Description : Climbing prickly glabrous shrubs with small, slightly hooked pickles. Leaflets elliptic-ovate - oblong-ovate, acute or acuminate, finely serrate, glabrous. Petioles and rachis usually prickly. Flowers white, about 7 in a compound corymb. Fruits ovate, small, dark brown.

Reproductive Cycle : April-October.

Materials Examined : HPU-EL-SKS-6611; Bemloe (2,016m), Kanlog Cemetery (1,990m), Stirling Castle (2,250m).

Propagation Notes : Propagated naturally by seeds.

Parts Used : Whole Plant. Leaves. Flowers. Stem.

Folk Uses : Grown as a **hedge**. Leaves used as **fodder**. Flowers **eaten** raw, and its tender stems **chewed**. Flower paste considered a good **astringent** for cuts.

Uses in Literature : Described so far as edible (fruits), fodder, tonic; and for asthma, bilious affections, constipation, diarrheoa, eye troubles, skin diseases, headache, phthisis, perfumery, and walking sticks **(Ambasta, 1986; Bailey, 1957; Bhattacharjee, 2001; Farooq, 2005; Kapur & Srivastava, 1996; Lata *et al.*, 2008; Pande *et al.*, 2006; Raju, 2000; Singh & Kumar, 2000a; Sharma, 2003; Sood & Thakur, 2004; Sood *et al.*, 2010)**.

Rosa multiflora Thunb.

Family: Rosaceae.

Vern.: Gulabari.

English and Hindi Names:

Eng. : Baby Rose, Japanese Rose;
Hindi : Bunchy Gulab.

Distribution : Indigenous to Japan and China; cultivated in India.

Description : Deciduous shrubs with long, recurving or climbing branches, prominent stipules which are pectinate and with terminal conspicuous point. Leaflets slightly obovate and abruptly short pointed. Flowers in many-flowered pyramidal corymbs, rosy. Petioles and inflorescence with long soft hair. Fruits globular or obovoid, brown-red.

Reproductive Cycle : June- July.

Materials Examined : HPU-EL-SKS-6617; Advance Study (2,128m), Jakhu (2,338m).

Propagation Notes : By seeds and vegetative methods.

Parts Used : Whole Plant. Leaves. Fruits.

Folk Uses : Grown as a **hedge** plant. Fruit paste applied externally on **ulcerating wounds.** Tender leaves **chewed** as a rich source of vit. C.

Active Constituents : These are eugenol, phenylethanol and heneicosane (**floral oil**), β-sitosterol, 5α-stigmastin-3,6-dione, scoparone, salicylic acid, new-glycosides-multiflorin A and B characterized as kaempferol-3β – D - glucopyranosyl (1→4) – α – L -rhamnopyranoside monoacetate and kaempferol-3-β-D-glucopyranosyl (1→4) – α – L – rhamnopyranoside respectively and kaempferol-3α-L-rhamnopyranoside, methyl gallate (**fruit**).

Uses in Literature : Used earlier as anticularpain, antiperglycaemic, cathartic, edible (young leaves), ornamental, purgative; and for eye disorders, injuries, sprains, ulcers and wounds **(Ambasta, 1986; Bailey, 1957; Jaeger, 1961; Kirtikar & Basu, 1984; Paliwal, 2009; Prajapati *et al.*, 2006; Raju, 2000; Rastogi & Mehrotra, 1991; Sood *et al.*, 2013).**

Rubia cordifolia L.

Syn.: *Galium cordifolium* (L.) Kuntze.

Family: Rubiaceae.

Vern.: Manjith.

English, Hindi and Sanskrit Names:

Eng. : Indian Madder;
Hindi : Majit, Manjit;
Sans. : Aruna, Bhandi, Bhandiralatika, Bhanditaki, Chhatra, Chitrangi, Gandhira, Harini, Hemapushpi, Janani, Javarihantri, Kalabhandika, Kandhira, Latayastni, Lohitalata, Manijishta, Tamramuli, Vashrabhusana, Vitasa, Yojnaparnika.

Distribution : Throughout Hilly regions of India, ascending to 2,700m.

Description : A climber having rigid, 4-angled, minutely prickly, elongate stem and branches and long -stalked, cordate-ovate and long - pointed leaves in whorls of 4. Flowers small, dark red, in numerous small cymes forming large, bracteate panicles. Fruits succulent and globose.

Reproductive Cycle : July-October.

Material Examined : HPU-EL-SKS-6621; Bantony (2,182m).

Propagation Notes : Prefers a loose moist leafy soil in some shade. Propagated by seeds.

Parts Used : Roots. Stem. Leaves. Bark.

Folk Uses : 1-3g powdered roots given with cow's milk twice daily for 5-7 days as **tonic** and for **colouring** medicinal oils and regulating **menstruation**. Poultice of leaves, bark and roots useful against **swelling**, **ringworm** and **skin ailments**.

Active Constituents : Active constituents are purpurin, munjisthin, xanthopurein, psuedopurpurin, nordamnacanthal, physcion, alizarin, ruberythric acid, quinones and 1,3-dihydroxy-2-ethoxymethyl-9,10-anthaquinone (**roots**).

Uses in Literature : Known to be useful as an alexipharmic, antidote to scorpion sting, antiseptic, astringent, cosmetics, deobstruent, diuretic, dye, nasal drops, tonic, vermifuge; and for amenorrhoea, bone fracture to reduce inflammation and swelling, chest complaints, colouring medicinal oils, dyeing, expelling placenta (vet.), giddiness, headache, hepatic obstructions, jaundice, leucoderma, menstrual disorders, obstruction in urinary passages, paralysis, poison stings of insect, cobras and scorpions, postnatal problem (vet.), removal of freckles, rheumatism, rhinosinal infections, skin discolouration, stomachache, ulcers and urinary complaints **(Agarwal, 2003; Ahmed & Borthakur, 2005; Ambasta, 1986; Aminuddin & Girach, 1991; Chatterjee & Pakrashi, 1997; Chauhan, 1999; Dam & Hajra, 1997; Dey & Bahadur, 1973; Dhiman, 2005; Farooq, 2005; Haridasan *et al.*, 2001; Henry *et al.*, 1996; Hosagoudar & Henry, 1996a, b; Jain, 1991; Khare, 2004; Kirtikar & Basu, 1984; Kshirsagar & Singh, 2007; Kumar, 2002; Lalramnghinglova, 2003; Masih, 2003; Mukherjee *et al.*, 1986; Nadeem *et al.*, 2001; Nautiyal *et al.*, 1997; Nisha & Sivadasan, 2007; Prajapati *et al.*, 2006; Rama Rao & Henry, 1996; Rana *et al.* 1996; Rastogi & Mehrotra, 1993; Rawat & Chowdhury, 1998; Sharma & Sood, 1997; Singh, 1999; Singh & Kumar, 2000a; Singh *et al.*, 2003c; Sood *et al.*, 2009b; Thakur *et al.*, 1989; Tiwari & Tiwari, 1996; Uniyal, 1968; Uphof, 1968; Verma & Chauhan, 2006; Watt, 1889-1896)**.

Rubus biflorus **Buch.-Ham. *ex* Sm.**

Family: Rosaceae.

Vern.: Dher.

Distribution : Temperate Himalayas from Sirmour - Sikkim and Bhutan.

Description : A large spreading, nearly erect shrub with white bloom covering stems and branches and numerous prickles. Leaves compound, leaflets 3 or 5, ovate-lanceolate, upper surface green, pubescent, lower white-tomentose. Flowers white, long-stalked, solitary or in small clusters. Drupelets yellow.

Reproductive Cycle : June-August.

Materials Examined : HPU-EL-SKS-6625; Kanlog Cemetery (1,980m), Nav Bahar Cemetery (2,031m).

Propagation Notes : Propagated naturally by seeds.

Part Used : Fruits.

Folk Use : Ripe fruits **edible**.

Uses in Literature : Known in India for its edible fruits, diarrhoea; and as local drinks **(Agarwal, 2003; Ambasta, 1986; Arora, 1997; Arora & Pandey, 1996; Badhwar & Fernandez, 2011; Bisht & Jain, 2006; Pande *et al.*, 2006; Roy *et al.*, 1998; Samant *et al.*, 1996, 2001; Sood *et al.*, 2013)**.

Rubus ellipticus Sm.

Syn.: *R. flavus* D. Don.

Family: Rosaceae.

Vern.: Akhe.

English and Hindi Names:

Eng. : Himalayan Yellow-Raspberry;
Hindi : Anehhu, Hinsalu.

Distribution : S.E. Asia: 600-2,300m.

Description : A large grey tomentose shrub with trailing stems and branches which are shaggy and with long tawny bristles; prickles numerous and curved. Leaves compound. Leaflets-3, orbicular-ovate, upper surface green, pubescent, lower grey-tomentose. Flowers white, crowded in axillary and terminal panicles. Fruits globose, yellow, succulent.

Reproductive Cycle : January-May.

Materials Examined : HPU-EL-SKS-6629; Advance Study (2,127m), Boileauganj Cemetery (1,980m), Kanlog Cemetery (1,978m).

Propagation Notes : Propagated naturally by seeds.

Parts Used : Leaves. Fruits.

Folk Uses : Leaves used as **fodder** for sheep and goats. Fruits **edible** and **laxative** in nature.

Active Constituents : These are saponin (I), characterized as 3β-hydroxyurs - 12, 18 – diene - 28-oic acid – 3 – O –β – D - glucopyranosyl (1→4) – α – L -arabinopyranoside, tormentic acid, β-sitosterol and its glucoside, 24-deoxysericoside (**leaves**).

Uses in Literature : Known so far for cracked heals, cuts and wounds, dyeing, dysentery, indigestion, malaria, stomachache, worms; and as astringent, edible (ripe fruits) and fodder **(Ahmed & Borthakur, 2005; Ambasta, 1986; Arora & Pandey, 1996; Badhwar & Fernandez, 2011; Bennet *et al.*, 1991; Bisht & Jain, 2006; Brandis, 1921; Dash *et al.*, 2003; Farooq, 2005; Jain, 1991; Kumar *et al.*, 2000; Lal *et al.*, 1996; Lata *et al.*, 2008; Manral & Pande, 2004; Mao, 1993; Megoneitso & Rao, 1983; Murthy *et al.*, 2003; Panda, 1996; Raju, 2000; Ranjan, 2000b; Rao, 1990; Rastogi & Mehrotra, 1993; Ringmichon *et al.*, 2011; Roy *et al.*, 1998; Saklani & Jain, 1994; Samant *et al.*, 2001; Sharma & Rana, 2005; Sood & Thakur, 2004; Sood *et al.*, 2012b; Uphof, 1968; Watt, 1889-1896)**.

Rubus macilentus Camb.

Family: Rosaceae.

Vern.: Heere.

Distribution : Temperate Himalayas.

Description : Trailing prickly shrubs possessing glabrous, shining, red-brown stems and branches. Prickles numerous. Leaves compound; leaflets 3, glabrous. Flowers white, solitary or in clusters of 2 or 3. Druplets yellow or orange.

Reproductive Cycle : April-September.

Materials Examined : HPU-EL-SKS-6631; Kanlog Cemetery (1,992m), Sanjauli Cemetery (2,068m).

Propagation Notes : Propagated by seeds.

Parts Used : Stem. Fruit.

Folk Uses : Tender stems **chewed**. Ripe fruits **edible**.

Use in Literature : Known so far for edible fruits **(Ambasta, 1986; Arora & Pandey, 1996; Badhwar & Fernandez, 2011; Bisht & Jain, 2006; Brandis, 1921; Roy *et al.*, 1998; Sood *et al.*, 2012b)**.

Rubus niveus Thunb.

Syn.: *R. albescens* Roxb.; *R. bonatii* H. Lev.; *R. foliolosus* D. Don; *R. micranthus* D. Don; *R. micranthus* D. Don.

Family: Rosaceae.

Vern.: Kala Akha, Klenchu.

English Name:

Eng. : Mysore Raspberry.

Distribution : Temperate Himalayas: 1,000-2,350m.

Description : A large rambling shrub possessing purple glabrous or pubescent stem and branches. Prickles small. Leaves compound. Leaflets 3 rarely 5. Flowers pink, solitary or in small clusters. Drupelets succulent, black.

Reproductive Cycle : March-May.

Materials Examined : HPU-EL-SKS-6637; Chapslee (2,243m), Nav Bahar Cemetery (2,031m), Kanlog Cemetery (1,990m).

Propagation Notes : Propagated by seeds.

Parts Used : Whole Plant. Leaves. Twig. Fruit.

Folk Uses : Plant grown as **hedge**. Leaves given as **fodder** for sheep and goats. Ripe fruits **edible** and considered **refrigerant**. Leafy twigs swayed over the body of person bitten by snake as an **antidote**.

Active Constituents : These are new labdane-type diterpene glycosides-goshonosides F1, F2, F3, F4, F5 and 18-O-α-L-arabinofuranosyl (1→6)-β-D-glucopyranoside of 13 (E)ent-labda-8(17), 13-diene-3β, 15, 18-triol and 3β, 15-di-O-β-D-glucopyranoside of 13(E)ent-labda-8(17), 13-diene-3β, 15-diol (**fruit**).

Uses in Literature : Recorded useful as an antidote to snake bite, diuretic, edible (berries), fodder, laxative; and for cuts, stomachache, infected wounds, nocturnal micturition and swellings **(Agarwal, 2003; Ahmed & Borthakur, 2005; Ambasta, 1986; Arora & Pandey, 1996; Badhwar & Fernandez, 2011; Bisht & Jain, 2006; Hosagoudar & Henry, 1996c; Jain, 1991; Kapur, 1996b; Kapur & Srivastava, 1996; Kumar *et al.*, 2000; Lal *et al.*, 1996; Mao, 1993; Negi & Gaur, 1991; Negi *et al.*, 1999; Rana *et al.*, 1996; Rastogi & Mehrotra, 1995b; Roy *et al.*, 1998; Saklani & Jain, 1994; Samant *et al.*, 1996, 2001; Sharma & Rana, 2005; Singh & Kumar, 2000a; Uphof, 1968; Watt, 1889-1896).**

Rubus paniculatus Sm.

Family: Rosaceae.

Vern.: Anchu.

Hindi Names:

Hindi : Anchu, Kala-hisalu; Karailo.

Distribution : Temperate Himalayas.

Description : A rambling climber having simple, broadly ovate, usually cordate leaves which are long pointed, more or less lobed and upper surface nearly glabrous, lower white tomentose. Prickles hooked. Flowers white, in terminal panicles. Drupes large, black; stony rugose.

Reproductive Cycle : April-August.

Material Examined : HPU-EL-SKS-6639; Chhota Shimla (2,080m), Kanlog Cemetery (1,990m).

Propagation Notes : Propagated naturally by seeds.

Part Used : Fruit.

Folk Use : Ripe fruits **edible**.

Uses in Literature : Useful in India as an edible (ripe fruits); and for pudding, stomachache, whooping cough and skin diseases **(Ambasta, 1986; Arora & Pandey, 1996; Badhwar & Fernandez, 2011; Bisht & Jain, 2006; Rawat & Chowdhury, 1998; Roy *et al.*, 1998; Samant *et al.*, 1996, 2001)**.

Salix alba L.

Syn.: *S. vitellina* L.

Family: Salicaceae.

Vern.: Benusa, Biyunsh.

English Names:

Eng. : Huntingdon Willow, White Willow.

Distribution : N.W. Himalayas; W. Tibet.

Description : Trees with yellowish-brown branches which are somewhat pendent at ends. Leaves lanceolate, acuminate, serrate, ashy-grey and silky throughout the season. Catkins on leafy stalk, appearing with the leaves. Female catkins a little longer than the male. Capsule nearly sessile, pubescent.

Reproductive Cycle : March-May.

Materials Examined : HPU-EL-SKS-6641; Raj Bhawan (2,132m), Spring Field (2,075m).

Propagation Notes : Propagated by seeds and cutting.

Parts Used : Whole Plant. Stem. Bark.

Folk Uses : Grown as an **ornamental** tree. Poultice of bark applied against **skin eruptions**. Stems flexible and used in **basket** making.

Active Constituents : These are amentoflavone, cupressuflavone and apigenin (**leaves**).

Uses in Literature : Reported to be used as analgesic, anodyne, anti-inflammatory, antipyretic, antirheumatic, antiseptic, astringent, diuretic, febrifuge, fodder, ornamental, sedative, tonic; and for building construction, combs, cricket-bats, diarrheoa, dysentery, furniture gout, and rheumatism **(Ambasta, 1986; Bailey, 1957; Bean, 1981; Bhattacharjee, 2001; Chej, 1984; Chittendon, 1956; Hadfield, 1956; Harborne & Baxter, 2001; Khare, 2004; Kirtikar & Basu, 1984; Pool, 1941; Prajapati *et al.*, 2006; Singh *et al.*, 1996a; Sood *et al.*, 2010, 2013; Uphof, 1968; Usher, 1974)**.

Salix babylonica L.

Syn.: *S. matsudana* Koidz.

Family: Salicaceae.

Vern.: Majnu.

English and Hindi Names:

Eng. : The Weeping Willow;
Hindi : Majnun.

Distribution : Plains of India; Himalayas, ascending to 3,000m.

Description : Small trees with long fragile hanging greenish - brown branches. Leaves narrow-lanceolate, margin serrulate, glabrous above slightly downy under beneath. Flowers in separate male and female catkins appearing with leaves. Catkins very slender, terminating in short, lateral leafy branchlets. Capsules small and glabrous. Female catkins as long the male ones.

Reproductive Cycle : February-May.

Materials Examined : HPU-EL-SKS-6643; Cecil Hotel (2,090m), Raj Bhawan (2,132m), Retreat (2,092m), Western Command Complex (2,174m).

Propagation Notes : Prefers wet, ill-drained or intermittently flooded soils. Propagated by seeds and stem cuttings during rainy season.

Parts Used : Whole Plant. Twigs. Root. Wood.

Folk Uses : Plant grown as **ornamental**; its wood considered suitable for **basketry** and **building works**. Twigs used as **fodder**. Paste of roots applied as an **emollient** for **wrinkles** and **cracked heels**.

Active Constituents : These are benzyl ester of gentisic acid-2-O-acetyl-β-D-glucoside, apigenin-7-O-galactoside, kaempferol-7-O-glucoside, luteolin-4-O-glucoside, trichocarpin, salicin and an ester of terephthalic acid (**leaves**).

Uses in Literature : Recorded earlier to be used as an anodyne, anthelmintic, astringent, antipyretic, cattle feed, charcoal, febrifuge, firewood, gate fixtures, ornamental, tonic; and for basketry, bone fracture, brake blocks, building works, cooling, dams, fences, furniture, intermitted and remitted fevers, rheumatism, and sports goods **(Ambasta, 1986; Bailey, 1957; Bean, 1981; Brickell, 1990; Chopra *et al.* 1986; Dhiman, 2003; Gamble, 1972; Jain, 1991; Kaushik & Dhiman, 2000; Kirtikar & Basu, 1984; Farooq, 2005; Manandhar, 2002; Mishra & Chandra, 1998; Pande *et al.*, 2006; Pool, 1941; Randhawa, 1957, 1960, 1965; Rastogi & Mehrotra, 1995a; Singh *et al.*, 1996a; Sood *et al.*, 2009b, 2013; Watt, 1889-1896)**.

Salix pycnostachya Andersson var. *oxycarpa* (Andersson) Y.L.Chou & C.F. Fang

Syn.: *S. oxycarpa* Andersson.

Family: Salicaceae.

Vern.: Changma Beeuins.

Distribution : W. Himalayas: 2,000-3,500m.

Description : Shrubs or small trees with pubescent young shoots and lanceolate, pointed leaves having small teeth which are gland-tipped, upper surface nearly glabrous and lower covered with a pale or white bloom. Flowers appearing with or a little before leaves and in separate male and female catkins. Male catkins nearly erect, densely hairy with black bract whereas female ones are drooping and with brown bracts. Capsule glabrous and shortly - stalked.

Reproductive Cycle : May-September.

Material Examined : HPU-EL-SKS-6647; Chhota Shimla (2,118m).

Propagation Notes : Propagated naturally by seed.

Parts Used : Leaves. Root-Bark.

Folk Uses : Leaf poultice used to check **headaches**. Fresh root-bark chewed for instant relief from **toothaches**.

Uses in Literature : Used for dental caries, pyorrhoea and toothaches **(Sood & Thalur, 2004)**.

Salvia leucantha Cav.

Syn.: *S. bicolor* Sesse & Moc.

Family: Lamiaceae.

Vern.: Velvet Salvia.

English Names:

Eng. : Mexican Bush Sage, Mexican Sage, Velvet Sage.

Distribution : Native of C. & E. Mexico.

Description : Evergreen subshrubs with numerous erect stems, often arching at tips and with a long velvety purple coloured spike. Leaves linear-lanceolate, soft mild - green and with whitish, hairy undersides, heavily ridged with prominent venation. Fruits four, dark brown, glabrous ovoid, 3-angled smooth nutlets.

Reproductive Cycle : August-October.

Materials Examined : HPU-EL-SKS-6649; Advance Study (2,128m), Grand Hotel (2,105m), Holly Lodge (2,278m), Summerhill (2,080m).

Propagation Notes : Propagated by seeds or division of plant. Cutting in July-August also successful.

Part Used : Whole Plant.

Folk Use : Grown as an **ornamental** plant.

Uses in Literature : Known earlier as an ornamental and for yielding essential oil **(Ambasta, 1986; Rastogi & Mehrotra, 1995b)**.

Salvia splendens Sellow *ex* Schult.

Syn.: *Fenixanthes splendens* (Sellow *ex* Roem. Schult.) Raf.; *Jungia splendens* (Sellow *ex* Roem. & Schult.) Sojak; *S. brasiliensis* Spreng.

Family: Lamiaceae.

Vern.: Salvia.

English Name:

Eng. : Scarlet Sage.

Distribution : Native of Brazil.

Description : Subshrubs with quadrangular branched stem and opposite, decussate, petiolate leaves having tubular scarlet flowers borne in a long terminal spikes with a flat lower lip and helmet - shaped upper lip. Calyx bell-shaped with 3 broad-ovate teeth. Corolla tube exserted, lower lip short. Fruits with four, ovoid, dark brown, 3-angled nutlets.

Reproductive Cycle : Summer onwards.

Materials Examined : HPU-EL-SKS-6655; Advance Study (2,128m), B.C.S. (2,023m), Convent of Jesus & Mary (2,104m), Holly Lodge (2,278m).

Propagation Notes : Multiplied by seeds and cutting.

Part Used : Whole Plant.

Folk Use : Grown as an **ornamental** plant.

Active Constituents : These are salvianin and its demalonyl derivatives along with monardaein and its demalonyl derivatives, pelargonidin-3-caffeoylglucosido – 5 - dimalonylglucoside, pelargonidin – 3 – p - coumaroylglucosido – 5 - dimalonylglucoside, salviadelphin, its demalonyl derivatives and 4,6-dimalonylawobanin (**flower**).

Uses in Literature : Known so far as an ornamental garden plant **(Dean, 1964; Paliwal, 2009; Phillips, 1954; Pool, 1941; Rastogi & Mehrotra, 1995a, 1995b; Sood *et al.*, 2013)**.

Sambucus nigra L.

Syn.: *S. graveolens* Willd.; *S. peruviana* Kunth.

Family: Caprifoliaceae.

Vern.: Salja, Sambucus.

English Names:

Eng. : Bour Tree, Box Tree, Common Elder, Elder Berry, European Elder.

Distribution : Native of Europe and Asia Minor.

Description : Large shrubs or small trees with serrated leaves having an acute apex. Flowers tiny white and gathered in bouquet-like, flat - topped clusters at the apex of branches. Berries blue-black.

Reproductive Cycle : March-August.

Materials Examined : HPU-EL-SKS-6657; Chapslee (2,243m), Grand Hotel (2,105m), Holly Lodge (2,278m), R.K.M.V. (2,256m).

Propagation Notes : Propagated by seeds but requiring cold stratification before germination, cuttings of half-ripe wood of 7 - 10cm long with a heel in July/August and division of suckers in dormant season are commonly employed methods for its multiplication. Plant grows well on chalk, but prefers a moist loamy soil.

Parts Used : Whole Plant. Flower. Berries.

Folk Uses : Grown as an **ornamental** and **hedge** plant. Flowers used as home remedy for **catarrh**, **mouth wash** and **gargling**. Berries used for soothing **burns** and **scalds**.

Active Constituents : These are flavonoids - rutin, phenolic acids, triterpenes, sterols, volatile oil, isorhamnetin-3-O-glucoside, cysteine, 2,6,6-trimethyl-2-vinyl-5-hydroxytetrahydropyran, 2-phenylethyl alcohol, isorhamnetin-3-O-rutinoside, mucilage and tannin (**flowers**).

Uses in Literature : Considered in India as anti-inflammatory, depurative, diaphoretic, diuretic, dye, edible (flower, fruit), emetic, emollient, laxative, ornamental; and for asthma, bouquet preparation, burns, catarrh, cold, cough, gargles, rheumatism, scalds, increasing sweating and wine preparation (fruit) **(Ambasta, 1986; Bean, 1981; Chej, 1984; Chevallier, 1996; Conway, 1973; Hadfield, 1956; Harborne & Baxter, 2001; Holtom & Hylton, 1979; Huxley, 1992; Krisa, 1982; Prajapati *et al.*, 2006; Rastogi & Mehrotra, 1993, 1995a; Shepherd, 1974; Sood *et al.*, 2010; Thomas, 1992; Uphof, 1968; Usher, 1974; Vickery, 1994; Warman, 1999)**.

Sarcococca pruniformis Lindl.

Syn.: *Buxus saligna* D. Don.; *S. salicifolia* Baill.; *S. saligna* Mull. Arg.; *S. sumatrana* Blume.; *S.trinervia* Wight.

Family: Buxaceae.

Vern.: Raabal.

English Names:

Eng. : Christmas Box, Sarcocca, Sweet Box.

Distribution : Afghanistan-W. Nepal: 1,200-2,400m.

Description : Glabrous shrubs having alternate, shortly- stalked, narrowly lanceolate, long-pointed entire leaves. Flowers pale - yellow, in short, erect axillary racemes. Drupes dark purple, ovoid, containing 2-3 stones.

Reproductive Cycle : May-October.

Materials Examined : HPU-EL-SKS-6659; Advance Study (2,128m), Armsdell (2,102m), Chhota Shimla (2,070m), Corner House (2,155m).

Propagation Notes : Propagated by seeds.

Parts Used : Whole Plant. Leaves. Root.

Folk Uses : Slightly warmed leaves applied on joints for relief from **pains**. Branches used for making **welcome arches**. 2-3g dose of powdered roots prescribed with honey (twice daily for 10-15 days) for **gonorrohea** and its poultice applied for **suppuration of boils**.

Active Constituents : These are salignine, saracodine, betulin, saracodinine (kurchessine), sarcodine and saracocine (**leaves**).

Uses in Literature : Recorded earlier for fever, itching, joint pains, rheumatism, skin eruptions and making brushes **(Ambasta, 1986; Farooq, 2005; Kapur & Nanda, 1996; Pande *et al.*, 2006; Rastogi & Mehrotra, 1995a; Sood &Thakur, 2004)**.

Schefflera arboricola (Haryata) Merr.

Syn.: *Heptapleurum arboricola* Hayata.

Family: Araliaceae.

Vern.: Umbrella Plant.

English Names:

Eng. : Dwarf Umbrella Tree, Star Leaf.

Distribution : Native of China.

Description : An evergreen shrub upto 4m tall. Leaves palmate, glossy green. Leaflets 7-9, obovate. Flowers in a panicle of small umbels. Fruit a reddish-yellow berry.

Reproductive Cycle : August-September.

Materials Examined : HPU-EL-SKS-6670; Inverarm (2,150m), Oak Over (2,154m), Holly Lodge (2,278m), Kasumpti (2,102m).

Propagation Notes : Propagated by seeds, stem cuttings and air layering during July-August.

Part Used : Whole Plant.

Folk Use : Grown as an **ornamental** plant.

Uses in Literature : Known earlier as an excellent indoor plant **(Paliwal, 2009; Sood *et al.*, 2013)**.

Skimmia laureola Siebold & Zucc. *ex* Walp.

Syn.: *Limonia laureola* DC.

Family: Rutaceae.

Vern.: Patti Dhup.

Distribution : Temperate Himalayas: 2,000-3,500m.

Description : Gregarious, evergreen glabrous aromatic shrubs with alternate, simple, entire leaves having prominent midrib. Flowers bright yellow, pentamerous, polygamus, in short crowded terminal panicles. Fruits ellipsoid, red.

Reproductive Cycle : April-May.

Materials Examined : HPU-EL-SKS-6672; Chhota Shimla (2,070m), Kanlog Cemetery (1,992m).

Propagation Notes : Succeeds in a well-drained open loam or in a peaty soil in a sunny position. Propagated by seeds.

Parts Used : Leaves. Inflorescence. Wood.

Folk Uses : Dried leaves used as **incense**. Wood used for **tool handles**. Inflorescence forms one of the ingredients of local wine, '**sur**'.

Active Constituents : Main ingredients are skimming, scopoletin (**bark**), lupenone, lupeol-β-sitosterol, skimmianine, scopoletin, umbelliferone, beyapten, laureoline, β-phellandrene, geraneol, myrcene, nerol, terpenes, azulene and traces of menthene (**leaves**).

Uses in Literature : Recorded to be used earlier as air purifier, antiseptic, incense, insectiside; and for boils, flavouring perfumery and cosmetics, gastric pain, rheumatic pains, scabies, small pox, making hoe and axe handles **(Ambasta, 1986; Chaudhuri, 1993; Chauhan, 1999; Dam & Hajra, 1997; Jain, 1991; Jain *et al.* 1991; Kapur & Srivastava, 1996; Kirtikar & Basu, 1984; Nautiyal *et al.*, 2003; Negi *et al.*, 1999; Rana *et al.*, 1996; Rawat & Chowdhury, 1998; Samant *et al.*, 1996, 2001; Singh, 1991, 1999; Singh & Kumar 2000a; Sood *et al.*, 2011; Uphof, 1968; Watt, 1889-1896)**.

Smilax aspera L.

Syn.: *S. balearica* (Willk. *ex* A.DC.) Burnat & Barbey; *S. catalonica* Poir.; *S. conferta* Jord.; *S. fulgens* Wall.; *S. goetzeana* Engl.; *S. intricatissima* Jord.; *S. longipes* Gand.; *S. maculata* Roxb.; *S. mauritanica* Poir.; *S. nigra* Willd.; *S. rigida* Sol. *ex* Sm.; *S. saxicola* Gand.; *S. willkommii* Gand.

Family: Smilacaceae.

Vern.: Bragh Bel.

Distribution : Pakistan-Bhutan. India: 1,200-2,500m.

Description : Evergreen climbers with zig-zag, spiny stems, oval leathery leaves which are heart - shaped at the base and tapering to a slender point and have prickly margins. Flowers small, pale green and borne in clusters in axil of leaves. Berries bluish-black on ripening.

Reproductive Cycle : September-November.

Materials Examined : HPU-EL-SKS-6675; Advance Study (2,128m), Grand Hotel (2,105m), Kanlog Cemetery (1,990m).

Propagation Notes : Propagated by seeds in spring.

Parts Used : Aerial Parts. Roots.

Folk Uses : Aerial parts cooked as **vegetable** and also **pickled**. 1-2g root powder useful against **fever**, **joint pains** and **gastric ailments**.

Active Constituents : Cheif constituents are sarasapogenin, its pseudogenin, a new sapogenin-asperagenin, norcyloartanol, β-sitosterol, parillin (**roots**).

Uses in Literature : Used so far as blood purifier, sheaf binder in crop fields with the belief of good yield, substitute of Indian Sarsaparilla, twiners, vegetable; and for cordages, pickle, rheumatic-arthritis, skin eruptions, sores, tanning, and wounds **(Abraham, 1997; Ambasta, 1986; Bhattacharjee, 2001; Farooq, 2005; Jain, 1991; Hay *et al.*, 1983; Hosagoudar & Henry 1996c; Khare, 2004; Pande *et al.*, 2006; Prajapati *et al.*, 2006; Singh & Kumar, 2000a; Sood & Thakur, 2004; Sood *et al.*, 2012b; Viswanathan *et al.*, 2006)**.

Solanum laxum Spreng.

Syn.: *S. boerhaaviifolium* Sendtn.; *S. jasminoides* Paxton.

Family: Solanaceae

Vern.: Nightshade Jasmine.

English Names:

Eng. : Jasmine Nightshade, Potato Creeper, Potato Vine.

Distribution : Native of Brazil.

Description : Unarmed semievergreen, woody twining climbers with cylindrical, glabrous smooth branches. Leaves cauline, alternate, opposite in floral region, exstipulate, petiolate, simple, ovate or ovate-lanceolate with dentate margins. Flowers pedicellate, complete, drooping, solitary or in extra-axillary cymes, rhipidium, white or pale blue. Berries succulent with persistent enlarged calyx, black.

Reproductive Cycle : June-October.

Materials Examined : HPU-EL-SKS-6679; Advance Study (2,128m), Auckland House (2,257m), Chapslee (2,243m), Chhota Shimla (2,090m), Hainault (2,294m), Town Hall (2,250m), Wood Villa (2,108m).

Propagation Notes : Propagation by cutting. Climber prefers well-drained porous soil and partial shade.

Parts Used : Whole Plant. Berries.

Folk Uses : Grown as a climbing **hedge** on trellis; also used as a pretty **climber** for cool houses. Berries **poisonous**.

Active Constituents : These are solasonine and solamargine (**berries**), solasodine (**leaves**), diosgenin, cholesterol and β-sitosterol (**flower**).

Uses in Literature : Known earlier as ornamental climber and poisonous to fowls (berries) **(Ambasta, 1986; Farooq, 2005; Paliwal, 2009; Rastogi & Mehrotra, 1991)**.

Solanum pseudocapsicum L.

Syn.: *S. capsicastrum* Link *ex* Schauer; *S. diflorum* Vell.; *S. mexiae* Standl.; *S. plurifurcipilum* Bitter; *S. validum* Rusby.

Family: Solanaceae.

Vern.: Charoti Mirch.

English Names:

Eng. : Christmas Cherry, Christmas Jerusalem, Winter Cherry.

Distribution : Temperate Himalayas from Kashmir- Sikkim.

Description : Erect branched shrubs with thin-textured light green, non-hairy alternately arranged leaves having slightly wavy margins. Flowers star - shaped, white, deeply-lobed held in axillary cymes. Berries orange-yellow coloured.

Reproductive Cycle : June-October.

Material Examined : HPU-EL-SKS-6680; Cecil Hotel (2,090m), Spring Field (2,075m).

Propagation Notes : Propagated by seeds.

Part Used : Whole Plant.

Folk Use : Grown as an **ornamental** plant.

Active Constituents : These are solanocapsine, solacapine, isosolacapine, episolacapine and o-methylsolanocapsine (**aerial part**).

Uses in Literature : Fruits reported to be poisonous and used earlier for dilation of pupil, nausea and pain in abdomen **(Ambasta, 1986; Baburaj *et al.*, 2000; Keller *et al.*, 2011; Rastogi & Mehrotra, 1990, 1993)**.

Solanum virginianum L.

Syn.: *S. mairei* H. Lev.; *S. surattense* Burm.f.; *S. xanthocarpum* Schrad. & Wendl.

Family: Solanaceae.

Vern.: Kant, Kari, Kateli.

English, Hindi and Sanskrit Names:

Eng. : Yellow Berried Night Shade;
Hindi : Bhatkatiya, Kateli, Ringni;
Sans. : Kantakari.

Distribution : Throughout India, Sri Lanka-S.E. Asia.

Description : Diffuse prickly shrubs which are more or less rough with scattered, stellate hair and procumbent branched stems. Leaves glabrous, oblong, pinatifid leaves with nerves on the surface armed with numerous long straight pickles. Flowers blue, solitary or in small cymes. Berries glabrous, yellow, often blotched with green.

Reproductive Cycle : April-September.

Materials Examined : HPU-EL-SKS-6682; Chhota Shimla (2,090m), Kanlog Cemetery (1,990m), Oak Over Cemetery (2,125m).

Propagation Notes : Propagated by seeds.

Parts Used : Root. Stem. Fruit. Seed.

Folk Uses : 10-15ml decoction of root with long pepper and honey given in case of **cough**. Powdered stem and fruits used for checking **fever** and **chest affections** (1 tsp, twice a day till relief). Roasted seeds **eaten**. Paste of seeds applied as poultice to check **headache** and **mouth sores** of cattle.

Active Constituents : These are solasodine, solasonine, solamargine and β-solamargine (**fruit**).

Uses in Literature : Reported to be used as anthelmintic, antipyretic, edible (fruit), laxative; and for asthma, cough, eye-diseases, fever, gonorrhoea, hernia, jaundice, jaw-ache, magico-religious, pain, piles, pyorrhoea, rheumatism and toothache **(Ambasta, 1986; Aminuddin & Girach, 1991; Badhwar & Fernandez, 2011; Bhattacharjee, 2001; Chauhan, 1999; Dhiman, 2005; Islam & Jha, 2003; Khare, 2004; Kirtikar & Basu, 1984; Pande *et al.*, 2006; Rastogi & Mehrotra, 1991; Ringmichon *et al.*, 2011; Singh & Pandey, 1998)**.

Spartium junceum L.

Family: Fabaceae.

Vern.: Weavers Broom.

English Name:

Eng. : Spanish or Weaver Broom.

Distribution : Native of Mediterranean, Canary Island and S. Europe; also found in N. Africa, Anatolia, Ukraine and W. Syria between 2,000-3,000m.

Description : Erect hardy shrubs with stems upto 4m tall. Branches round, greyish - green, erect. Leaves simple, scanty or almost wanting, linear lanceolate, thick, glabrous above pubescent below, deciduous. Flowers fragrant, yellow in loose terminal racemes. Bracts and bracteoles minute, caducous. Pods narrow, linear compressed, 12-20 seeded; black and smooth as it ripens.

Reproductive Cycle : May-September.

Materials Examined : HPU-EL-SKS-6685; B.C.S. (2,023m), Convent of Jesus & Mary (2,104m), Corner House (2,155m), R.K.M.V. (2,255m).

Propagation Notes : Propagation by seeds in spring. Prefers sunny slopes, can be grown on well- drained soil including lime or chalk.

Part Used : Whole Plant.

Folk Uses : Planted in gardens as a showy, fragrant **ornamental** and **hedge**. Fibres obtained from the plant good for filling **mattresses** and **making paper mats** and **ropes**.

Active Constituents : These are rutin, isoquercitrin, spieroside, luteolin-7-glucoside, orientin, ononin, formononetin, genistin, 5-methylgenistein and vitexin (**aerial parts**), chrysin, its 7β-D-glucoside and 7-gentiobioside (**flower**).

Uses in Literature : Reported to be used as cardiotonic, cathartic, diuretic, emetic, fibre, laxative, ornamental, purgative; and for basketry, cordage, perfumery, mats, rope, thread **(Ambasta, 1986; Bean, 1981; Brickell, 1990; Chej, 1984; Chittendon, 1956; Hadfield, 1956; Huxley, 1992; Jaeger, 1961; Nautiyal *et al.*, 1997; Rastogi & Mehrotra, 1993; Uphof, 1968; Usher, 1974)**.

Spiraea canescens D. Don

Family: Rosaceae.

Vern.: Chakhu.

Distribution : Temperate Himalayas: 2,000-4,000m.

Description : Softly tomentose or pubescent deciduous shrubs possessing ribbed, reddish-brown twigs and simple, ovate, nearly sessile leaves which are toothed towards the tip. Flowers white, in small compound corymbs. Follicles hairy.

Reproductive Cycle : May-October.

Materials Examined : HPU-EL-SKS-6689; Chhota Shimla (2,110m), Kanlog Cemetery (1,990m).

Propagation Notes : Propagated by seeds.

Parts Used : Twigs. Seeds.

Folk Uses : Twigs used as **walking stick**. Seeds used for **flavouring curries**.

Uses in Literature : Known in India as fodder, fuel, sedative; and for walking sticks **(Ambasta, 1986; Bean, 1981; Gamble, 1972; Huxley, 1992; Kapur, 1996b; Kapur & Srivastava, 1996; Nautiyal *et al.*, 2003; Phillips & Rix, 1989; Singh, 2000a; Singh & Kumar, 2000a)**.

Spiraea cantoniensis Lour.

Family: Rosaceae.

Vern.: Spirea.

English Name:

Eng. : Double Reeves Spirea.

Distribution : Native of China; naturalized in Ibiza.

Description : Glabrous deciduous shrubs having alternate, lanceolate, deeply toothed leaves and white showy flowers in numerous small umbels. Fruit a follicle.

Reproductive Cycle : April-July.

Materials Examined : HPU-EL-SKS-6690; Chapslee (2,243m), Nav Bahar Cemetery (2,031m).

Propagation Notes : Propagated by seeds, cutting of half-ripe wood in July-August and division of suckers in early spring.

Part Used : Whole Plant.

Folk Use : Planted for **ornamental** purposes.

Uses in Literature : Known earlier as an ornamental **(Brickell, 2003; Collett, 1902; Randhawa, 1960)**.

Taxodium distichum (L.) Rich.

Syn.: *Cuprespinnata disticha* (L.) J. Nelson; *Cupressus disticha* L.; *Glyptostrobus columnaris* Carriere; *Schubertia disticha* (L.) Mirb.; *T. denudatum* Carriere; *T. knightii* K. Koch; *T. pyramidatum* Beissn.

Family: Taxodiaceae.

Vern.: Cypress.

English Names:

Eng. : Bald Cypress, Swamp Cypress.

Distribution : N. America-New Jersey to Florida and Texas.

Description : Deciduous trees with a light cinnamon - brown bark, slightly pendulous spreading branches and linear-lanceolate, 2-ranked, acute, thin and light green leaves. Cones rugose without mucros.

Reproductive Cycle : April-October.

Material Examined : HPU-EL-SKS-6693; Gorton Castle (2,094m).

Propagation Notes : Propagated by seeds. Plant thrives in any soil but grows faster in normal unflooded soils.

Parts Used : Whole Plant. Bark. Wood.

Folk Uses : Grown as an **ornamental** tree. Bark used for **cordages**. Wood used for **construction works**.

Uses in Literature : Described earlier as analgesic, anticancerous, diuretic, fence, water pipes, vats; and for cordages, construction, furniture, posts **(Ambasta, 1986; Bailey, 1957; Huxley, 1992; Rushforth, 1987; Sood *et al.*, 2005; Uphof, 1968; Usher, 1974; Warda, 1985)**.

Taxus baccata L.

Syn.: *Cephalotaxus adpressa* Beissn.; *C. brevifolia* Beissn.; *C. tardiva* Siebold *ex* Endl.; *T. adpressa* Carriere; *T. aurea* K. Koch; *T. expansa* K. Koch; *T. fastigiata* Lindl.; *T. parvifolia* Wender.; *Verataxus adpressa* (Carriere) Carriere.

Family: Taxaceae.

Vern.: Rakhaal, Thuno.

English, Hindi and Sanskrit Names:

Eng. : Common Yew, Yew;
Hindi : Thuno, Zirnubbirmi;
Sans. : Mandupami.

Distribution : S. E. Asia: 2,100-3,400m.

Description : Evergreen trees with a stout thick trunk and spreading branches forming a broad low head. Leaves gradually flattened, distichous, acuminate with recurved margin, shining above pale-yellowish brown or rusty red below and aromatic when dry. Male strobili stalked, globose, arising from axil of leaves on the underside of branchlets whilst female ones are solitary and axillary. Fruits red, fleshy.

Reproductive Cycle : May -November.

Material Examined : HPU-EL-SKS-6695; Advance Study (2,128m).

Propagation Notes : Propagated by seeds and cuttings. Young plant requires shelter and deep shade for growth.

Parts Used : Whole Plant. Leaves.Wood. Bark.

Folk Uses : Planted as an **ornamental**. 1-3g powdered leaves and bark given with cow's milk till cure for **cough**, **cold**, **bronchitis**, **breast**

cancer and **spasm**. Bark used as substitute of **tea**. Wood used as **incense** and also suitable for low grade **furniture**.

Active Constituents : Taxine A, taxagifin, tetrol (I), amentoflavone, mono and di-O-mrthufamentoflavones, sciadopitysin, ginkgetin, sequoiaflavone, taxinine, hydrocyanic acid, formic acid, reducing sugars, resins, tannins, epherdrine, a glucoside, taxiphyllin, ferredoxin, acidyrterone and β-sitosterol (**leaves**) have been isolated.

Uses in Literature : Known so far as anticancerous, antispasmodic, aphrodisiac, aromatic, beauty and shaving creams, beverage, carminative, cheap grade pencils, dentrifices, diuretic, edible (aril, bark and leaves as a substitute for tea), emmenagogue, expectorant, fish poison, fodder, hair colour, incense, lithotropic, psychoactive, ornamental, sacred tree, sedative, stomachic, veneers, water-wheel; for abortions, asthma, back of combs, bronchitis, bone fracture (vet), carrying poles, carvings and layings, cold and cough, convulsions, diarrhoea, dyeing, epilepsy, flooring, furniture, gates, giddiness, knife handles, headache, hysteria, indigestion, irregular and depressal action of heart, nervousness, ploughs, severe biliousness, strength and vigour, tumours, turnery **(Agarwal, 2003; Ambasta, 1986; Aswal, 1996; Bhattacharjee, 2001; Chandra, 1997; Chauhan, 1999; Chevallier, 1996; Dhiman, 2003, 2005; Duke *et al.*, 2002; Dwivedi, 2007; Harborne & Baxter, 2001; Jain, 1991; Jain *et al.*, 1994; Jha & Kumar, 2003; Kaul, 1997; Khare, 2004; Kirn *et al.*, 2000; Kirtikar & Basu, 1984; Kumar & Rao, 2001; Lal *et al.*, 1996; Lata *et al.*, 2008; Nautiyal *et al.*, 1997, 2003; Negi & Gaur, 1991; Negi & Pant, 1990; Pande *et al.*, 2006; Prakash, 1999, 2001; Prajapati *et al.*, 2006; Rana *et al.*, 1996; Roy *et al.*, 1998; Said, 1997; Samant *et al.*, 1996; Sarin, 1996; Sharma, 2003; Sharma & Sood, 1997; Singh, 1991, 1999, 2001; Singh & Kumar, 2000a; Sinha & Sinha, 2001; Sood *et al.*, 2009b, 2011; Uphof, 1968; Warda, 1985; Watt, 1889-1896)**.

Thuja occidentalis L.

Syn.: *Cupressus arborvitae* O. Targ. Tozz.; *Juniperus ericoides* Mast.; *Retinispora devriesiana* Mast.; *R. pygmaea* Beissn.; *T. bodmeri* Beissn.; *T. caucasica* Gordon; *T. compacta* Standish *ex* Gordon; *T. odorata* Marshall; *T. sibirica* Gordon; *T. tatarica* Gordon.

Family: Cupressaceae.

Vern.: Thuja.

English Names:

Eng. : American Arborvitae, Eastern White Cedar.

Distribution : Native of N.E. America.

Description : Symmetrical, cone-shaped evergreen coniferous plants having short horizontal branches. Branchlets flattened, held in fan-like horizontal planes. Leaves scaly, closely attached to compressed branchlets. Foliage fragrant, dull yellowish - green, turning bronze in winter. Cones small, 1.3 cm long, egg-shaped, green first turning reddish brown when mature in autumn. Seeds winged.

Reproductive Cycle : April-October.

Materials Examined : HPU-EL-SKS-6696; Chhota Shimla (2,090m), Grand Hotel (2,105m), Naldehra (2,127m).

Propagation Notes : Propagated from semi-ripe heeled cuttings taken from the current season's growth and seeds.

Part Used : Whole Plant.

Folk Uses : Planted as **hedge** and for **ornamental** purposes.

Uses in Literature : Planted in gardens and boundary of parks as a shrub or specimen tree and used so far as anthelmintic, anti-inflammatory, antiseptic, aromatic, astringent, anti-cancer, anti-viral, diaphoretic, diuretic, emmenagogue, essential oil, expectorant, rubefacient; and for bronchitis, menstruational problems, pain, respiratory diseases **(Agarwal, 2003; Ambasta, 1986; Bean, 1981; Bhattacharjee, 2001; Brickell, 1990; Chej, 1984; Harborne & Baxter, 2001; Hill, 1951; Holtom & Hylton, 1979; Huxley, 1992; Paliwal, 2009; Prajapati *et al.*, 2006; Rastogi & Mehrotra, 1990, 1993; Rushforth, 1987; Sheat, 1948; Uphof, 1968; Usher, 1974; Warda, 1985)**.

Thuja orientalis L.

Syn.: *Biota coraeana* Siebold *ex* Gordon; *B. dumosa* Carriere; *B. excelsa* Gordon; *B. falcata* Carriere; *B. glauca* Carriere; *B. gracilifolia* Knight; *B. japonica* Siebold *ex* Gordon; *B. macrocarpa* Gordon; *B. orientalis* (L.) Endl.; *B. pendula* (Thunb.) Endl.; *Cupressus filiformis* Beissn.; *C. pendula* Thunb.; *Platycladus orientalis* (L.) Franco; *P. stricta* Spach.; *T. acuta* Moench; *T. australis* Ten.; *T. chengii* Borderes and Gaussen; *T. decora* Salisb.

Family: Cupressaceae.

Vern.: Morpankh.

English and Hindi Names:

Eng. : Chinese Arbovitae, Chinese Thuja, Oriental Arbovitae;

Hindi : Mayur Pankh, Morpankhi.

Distribution : Native of China; grown in Indian gardens.

Description : A small-sized evergreen conical tree with spreading and ascending branches. Branchlets thin. Leaves opposite, rhombic-ovate, acute, appressed, minute, bright green. Strobiles ovoid, fleshy, glaucous green at first turning reddish brown and formed of 6-8 thick scales that bear downcurved hooks protecting 1-2 wingless seeds.

Reproductive Cycle : July-October.

Materials Examined : HPU-EL-SKS-6705; Advance Study (2,128m), Bemloe (2,013m), Chhota Shimla (2,090m), Grand Hotel (2,105m), Hainault (2,294m), Summerhill (2,080m).

Propagation Notes : Prefers a moist loamy soil. Propagated by seeds requiring short cold stratification and cuttings of half-ripe wood of 5 - 8cm with a heel in July/August in a shaded frame.

Parts Used : Whole Plant. Leaves. Seeds.

Folk Uses : Grown as an **ornamental** for its dense foliage. Poultice of leaves or its seeds useful for healing open **wounds**. Powdered seeds (10g with honey, daily for one year) prescribed for **diabetes**.

Active Constituents : These are palmitic (5.28), stearic (7.30), C-18 acids (36.29), C-20 acids (6.10) and linoleic acid (44.6%) (**seed oil**), β-cuparenone (**twig**), α-thujene, (+) sabinene, (+) camphene, cedrol, cedrenol and α- and β-pinenes (**berries**), apigenin, cupressuflavone, quercitrin, kaempferol-7-O-glucoside, myricetin – 3 – O – α - L - rhamnoside, quercetin – 7 – O – rhamnoside, deoxypodophyllotoxin, ethyl p - methoxy – trans - cinnamate, nonacosanol, quercitrin, condensed tannins (**leaves**).

Uses in Literature : Described in literature for barrels and casks, dysentery, furniture, house building and scurvy; and as antiasthmatic, antifungal, anti-inflammatory, antipyretic, antitussive, expectorant, diuretic, fencing posts, psychoactive, tannin and tonic **(Ambasta, 1986; Benthall, 1946; Bhattacharjee, 2001; Farooq, 2005; Jain *et al.*, 1994; Paliwal, 2009; Pande *et al.*, 2006; Prajapati *et al.*, 2006; Singh & Prakash, 1996; Sood *et al.*, 2013; Warda, 1985).**

Thujopsis dolabrata Siebold & Zucc.

Syn.: *Libocedrus dolabrata* (L.f.) J. Nelson; *Platycladus dolabrata* (L.f.) Spach; *Thuja dolabrata* L.f.; *Thujopsis atrovirens* Lavallee; *Libocedrus dolabrata* (L.f.) J. Nelson.

Family: Cupressaceae.

Vern.: Thujopsis.

Distribution : Native of Japan.

Description : Trees upto 30m tall. Twigs irregularly whorled or scattered, wide spreading or nodding; branchlets flat and horizontally arranged. Leaves glossy and dark green above with distinct white marking beneath, lateral leaves with somewhat incurved tips. Cones broadly ovate with 6-10 scales, much thickened at apex and terminating in an obtuse, triangular tip. Seeds oblong or circular, compressed, both sides narrowly winged.

Reproductive Cycle : April-October.

Materials Examined : HPU-EL-SKS-6707; Advance Study (2,128m), Auckland House (2,257m), Chhota Shimla (2,083m), Convent of Jesus & Mary (2,104m).

Propagation Notes : Propagated by seeds and cuttings of almost ripe wood.

Part Used : Whole Plant.

Folk Use : Grown as an **ornamental**.

Uses in Literature : Reported to be used for building purposes, interior finish of houses, bridges, cabinet work, filling between boards to prevent leaking of water, wooden pipes; and as match-cord **(Chittendon, 1956; Huxley, 1992; Rushforth, 1987; Sheat, 1948; Uphof, 1968; Usher, 1974; Warda, 1985)**.

Trachelospermum jasminoides Lem.

Syn.: *Parechites adnascens* Hance; *P. bowringii* Hance; *P. thunbergii* A. Gray; *Rhynchospermum jasminoides* Lindl.; *T. bowringii* (Hance) Hemsl.; *T. divaricatum* Kanitz.

Family: Apocyanaceae.

Vern.: Star Jasmine.

English Names:

Eng. : Star Jasmine, Traders Compass.

Distribution : E. Asia - China, Japan, Korea.

Description : Evergreen woody lianas growing to 30m high. Leaves opposite, oval - lanceolate with an entire margin and an acuminate apex. Flowers fragrant, white with a tube - like corolla opening out into five petal-like lobes. Follicles slender, containing numerous seeds.

Reproductive Cycle : July-October.

Material Examined : HPU-EL-SKS-6709; Advance Study (2,128m).

Propagation Notes : Requires a well-drained moderately retentive soil in full sun or with part-day shade. Propagated by seeds, cuttings of half-ripe wood of 6 - 8cm with a heel in August and layering in summer.

Part Used : Whole Plant.

Folk Uses : Grown for **ornamental** purposes (**climbing vine, groundcover, fragrant** potted plant) on terraces and patios.

Uses in Literature : Used so far as anodyne, antibacterial, antirheumatic, antispasmodic, depurative, emmenagogue, febrifuge, light climber for limited space, tonic and vasodilator **(Bean, 1981; Brickell, 1990; Davis, 1990; Duke & Ayensu, 1985; Paliwal, 2009; Phillips & Rix, 1998; Randhawa, 1960; Sheat, 1948; Thomas, 1992)**.

***Tropaeolum majus* Linn.**

Syn.: *Cardamindum majus* Moench; *T. elatum* Salisb.; *T. hybridum* L.

Family: Tropaeolaceae.

Vern.: Tindu Phool.

English Names:

Eng. : Garden Nastrutium, Indian Cress, Nasturtium.

Distribution : S. America - Peru; naturalized in parts of Europe.

Description : Sprawling, tuberous - rooted climbing annuals possessing large-sized leaves borne on long petioles and divided into 5-narrow pointed lobes. Flowers orange, red or yellow in colour with a nectar-filled spur at the base. Capsule about 2 cm broad, 3-segmented, each with a single large seed.

Reproductive Cycle : May-November.

Materials Examined : HPU-EL-SKS-6715; Advance Study (2,128m), Chhota Shimla (2,090m), Holly Lodge (2,278m), Sidhowal Lodge (2,254m).

Propagation Notes : Prefers a rich light well-drained soil in full sun or partial shade. Propagated by seeds which germinates within 2 weeks and subsequently planted out in late spring or early summer.

Parts Used : Whole Plant. Flower.

Folk Uses : Commonly planted as a favourite **climbing** plant for its exceedingly orange coloured beautiful flowers. Flowers used for **offerings**.

Active Constituents : Plant contains glucocyanates, isoquercitrin, spilanthol, myrosin, oxalic acid and vit. C.

Uses in Literature : Known to be useful as antibacterial, antibiotic, antiseptic, cystitis, depurative, diuretic, expectorant, laxative, ornamental climber, pickle, salad (leaves, petiole, seed), stimulant, paint, tonic varnish; and for kidney inflammation, urinary infections **(Ambasta, 1986; Baburaj *et al.*, 2000; Bhattacharjee, 2001; Chej, 1982; Davis, 1990; Duke & Ayensu, 1985; Hedrick, 1972; Holtom & Hylton, 1979; Paliwal, 2009; Phillips, 1954; Pool, 1941; Prajapati *et al.*, 2006; Rastogi & Mehrotra, 1990; Sood *et al.*, 2013; Usher, 1974).**

Tsuga dumosa Eichl.

Syn.: *Abies brunoniana* Lindl.; *A. dumosa* (D. Don) Mirb.; *Pinus brunoniana* Wall; *P. dumosa* D. Don; *Tsuga brunoniana* (Wall.) Carriere; *T. calcarea* Downie; *T. wardii* Downie; *T. yunnanensis* (Franch.) E. Pritz.

Family: Pinaceae.

Vern.: Thingia.

English Names:

Eng. : Himalaya Hemlock, Indian Hemlock-Fir.

Distribution : Native of E. Himalayas.

Description : Trees having broad, irregular-pyrimidal open crown and pinkish - grey-brown bark ridged with broad, shallow flaky fissures. Branches oblique or horizontal. Leaves spirally arranged, pointing forward on the branches and placed distantly from one another, obtuse or rounded ends; upper surface green, shiny while the undersides have 2 wide silvery stomatal bands. Male cones globose, solitarily arranged, axillary; female ones round-ovate in shape, terminal and slightly down-curved; many spiral scales with 2 ovules contained within each scale.

Reproductive Cycle : April-November.

Material Examined : HPU-EL-SKS-6717; Chapslee (2,243m).

Propagation Notes : Propagated by seeds.

Parts Used : Whole Plant. Bark.

Folk Uses : Grown as an **ornamental** plant. Bark used for **thatching**.

Uses in Literature : Reported to be used for bobbins, construction, packing cases, reels, shingles, thatching; and as incense and ornamental **(Alan, 1974; Ambasta, 1986; Chaudhuri, 1993; Mehra & Jain, 1976; Randhawa, 1960; Sood *et al.*, 2010; Warda, 1985)**.

Ulmus wallichiana Planch.

Family: Ulmaceae.

Vern.: Brera.

Hindi Names:

Hindi : Chambar Maya, Mored, Pabuna.

Distribution : E. Asia - Himalayas from Afghanistan - W. Nepal.

Description : Trees with alternate, shortly - stalked, ovate long-pointed leaves having unequal base, cordate or tapering and often doubly toothed. Flowers red-brown appearing before leaves, bisexual, clustered in racemes. Fruits a winged nut or samara.

Reproductive Cycle : March-May.

Materials Examined : HPU-EL-SKS-6725; Advance Study (2,128m), Kanlog Cemetery (1,990m).

Propagation Notes : Propagated by seeds.

Parts Used : Leaves. Wood.

Folk Uses : Leaves employed as **fodder**. Wood used for making **furniture** and as **fuel**.

Uses in Literature : Reported to be used in India as fibre, fodder, fuelwood; and for bone fracture (vet.), construction works, furniture, sports good, tannins and toothache **(Ambasta, 1986; Brandis, 1921; Dwivedi, 2007; Kapur, 1996b; Sharma, 1996; Singh, 2000b; Pande *et al.*, 2006)**.

Viburnum cotinifolium D. Don

Syn.: *V. polycarpum* Wall.

Family: Caprifoliaceae.

Vern.: Termela.

Distribution : Afghanistan-C. Nepal, Bhutan: 1,800-3,600m.

Description : A large deciduous shrub with stellately tomentose young shoots and cyme-branches. Leaves thick, cordate, crenate or nearly entire; upper surface green, nearly glabrous and lower grey, stellately tomentose. Flowers numerous, white, tinged with pink. Drupes oblong red, turning black when ripe.

Reproductive Cycle : April-September.

Material Examined : HPU-EL-SKS-6729; Advance Study (2,128m), Boileauganj Cemetery (1,988m), Kanlog Cemetery (1,990m).

Propagation Notes : Propagated naturally by seeds.

Parts Used : Branches. Fruit.

Folk Uses : Branches used as **fuel** and **fodder** for cattle. Fruits **edible**.

Active Constituents : Ursolic acid and uvaol isolated from **aerial parts**.

Uses in Literature : Known so far for ripe and dried edible fruits, haemorrhages, hepatic disorders, indigestion, menorrhagia and meterrhagia **(Ambasta, 1986; Badhwar & Fernandez, 2011; Bennet *et al.*, 1991; Dwivedi, 2007; Farooq, 2005; Jain, 1991; Kapur, 1996a, b; Nautiyal *et al.*, 2003; Negi & Gaur, 1991; Pande *et al.*, 2006; Roy *et al.*, 1998; Sood *et al.*, 2001, 2012b; Watt, 1889-1896)**.

Viburnum mullaha Buch.-Ham. *ex* D. Don

Syn.: *V. stellulatum* Wall.

Family: Caprifoliaceae.

Vern.: Dhaab.

Distribution : Temperate Himalayas: 1,500-3,000m.

Description : Shrubs with ovate-lanceolate, sharply toothed leaves having upper glabrous surface and lower ones stellately hairy. Flowers numerous, white. Drupes bright red, shining and oblong.

Reproductive Cycle : May-July.

Materials Examined : HPU-EL-SKS-6731; Advance Study (2,128m), Boileauganj Cemetery (1,982m), Kanlog Cemetery (1,990m).

Propagation Notes : Propagated by seeds.

Parts Used : Leaves. Branches. Fruits.

Folk Uses : Leaves lopped for **fodder**. Branches suitable for making **walking sticks**. Fruits **edible**; its juice good against **indigestion**.

Uses in Literature : Reported earlier for its edible fruits, walking sticks; and as uterine tonic (vet.) **(Ambasta, 1986; Badhwar & Fernandez, 2011; Chandra, 1996; Chaudhuri, 1993; Roy *et al.*, 1998; Watt, 1889-1896)**.

Wisteria sinensis Sweet

Syn.: *Glycine sinensis* Sims; *Millettia chinensis* Benth.; *Rehsonia sinensis* (Sims) DC; *W. chinensis* DC.; *W. praecox* Hand.-Mazz.

Family: Fabaceae.

Vern.: Wisteria.

English Name:

Eng. : Chinese Wisteria.

Distribution : Native of China; naturalized in hills of N.E. India and also grown in Indian gardens.

Description : Woody climbers having odd-pinnately compound, alternate, pale green leaves. Leaflets in 7-11 pairs, light coppery when young. Flowers blue-violet, showy, arranged on 30 cm long pendulous terminal racemes. Pods 10 cm long, flat, brown, velvety, bean-like, thick disk-like evenly spaced seeds.

Reproductive Cycle : April-May.

Materials Examined : HPU-EL-SKS-6735; Advance Study (2,128m), Auckland House (2,257m), Chhota Shimla (2,080m), Wood Villa (2,108m).

Propagation Notes : Plant succeeds in a good loamy soil in a sunny south or S.W. facing position, sheltered from cold winds and from early morning sun on frosty mornings. Propagated by seeds, basal cuttings of side-shoots in early to mid summer in a frame and layering in spring.

Part Used : Whole Plant.

Folk Use : Planted as an **ornamental climber** due to its beautiful blue coloured hanging flowers.

Uses in Literature : So far useful as cardiac, diuretic, edible (flower, seed), fibre, ornamental climber and substitute of tea (leaves) **(Ambasta, 1986; Bean, 1981; Brickell, 1990; Duke & Ayensu, 1985; Kunkel, 1984; Paliwal, 2009; Phillips & Rix, 1989; Pool, 1941; Randhawa, 1960; Rastogi & Mehrotra, 1993, 1995b; Sood *et al.*, 2013; Thomas, 1992)**.

Xanthium strumarium L.

Syn.: *X. abyssinicum* Wallr.; *X. cuvescens* Millsp. & Sherff; *X. varians* Greene.

Family: Asteraceae.

Vern.: Banokra, Gheenjadu.

English, Hindi and Sanskrit Names:

Eng. : Burweed, Cocklebur; Duch bur, Louse-bur;

Hindi : Chota Dhatura, Chota Gokhru;
Sans. : Arishta.

Distribution : Native of America; distributed throughout India upto 2,500m.

Description : Erect shrubs possessing rough, long - stalked, cordate, lobed, alternate, toothed and triangular leaves. Heads 1-sexual, male and female on the same plant, combined in axillary or terminal clusters. Flowers numerous, white or green. Achenes obovoid, thick, enclosed in hard involucral bracts.

Reproductive Cycle : April-November.

Material Examined : HPU-EL-SKS-6739; Boileauganj Cemetery (1,998m).

Propagation Notes : Propagated by seeds and cutting.

Parts Used : Root. Leaves. Fruit. Seeds.

Folk Uses : 2-5ml of root extr. prescribed on empty stomach to cure **ulcer**, **boils** and **abscess**. Paste of leaves and seeds applied to check **swelling** and **inflammation**. Fruits considered **diuretic**, **coolant**, and also given in **small-pox**.

Active Constituents : Reported to possess a new polyphenol isolated and characterised as 1,3,5-tri-O-caffeoylquinic acid along with 3,5-di-O-caffeoylquinic acid, stigamasterol, strumaroside **(fruit)**, heptacosanol, sesquiterpene lactones, stigmasterol, β-sitosterol and its glucoside, 3,4-dihydroxycinnamic acid, KCL, KNO3 and K2SO4 **(root)**.

Uses in Literature : Described earlier as alexiteric, alternative, anthelmintic, antipyretic, appetizer, astringent, demulcent, diuretic, diaphoretic, laxative, sedative, stimulant, styptic, suppurative, tonic and tanning materials (flowers); and for bowel complaints, biliousness, diarrhoea, digestive, dyeing dysentery, epilepsy, fever, haemorrhages, herpes, leucoderma, leucorrhoea, liver complaints, menorrhagia, salivation, skin diseases, ulcers and vesicular eruptions **(Agarwal, 2003; Ambasta, 1986; Badhwar & Fernandez, 2011; Bhattacharjee, 2001; Borthakur *et al.*, 1999; Chauhan, 1999; Dhiman, 2005; Farooq 2005; Gogoi & Das, 2003; Hajra & Baishya, 1997; Kaushik & Dhiman, 2000; Kirtikar & Basu 1984; Meena *et al.*, 2003; Pande *et al.*, 2006; Parrotta, 2001; Prajapati *et al.*, 2006; Reddi *et al.*, 2005; Sharma 2004; Singh & Pandey, 1998; Singh & Srivastava, 2000; Singh *et al.*, 1996a, b; Srivastava *et al.*, 2003a; Uphof, 1968)**.

Yucca aloifolia L.

Syn.: *Dracaena lenneana* Regel; *Sarcoyucca aloifolia* (L.) Linding; *Y. arcuata* Haw.; *Y. conspicua* Haw.; *Y. draconis* L.; *Y. purpurea* Baker; *Y. serrulata* Haw.; *Y. tenuifolia* Haw.; *Y. tricolor* Baker; *Y. yucatana* Engelm.

Family: Agavaceae.

Vern.: Rambaan.

English and Hindi Names:

Eng. : Spanish Bayonet;
Hindi : Bhalu Ghas.

Distribution : S.E. America - N. Carolina to Florida; naturalized in Europe.

Description : A shrub with subterranean rhizome and erect, thick, cylindrical stem covered with leaves in younger stage. Leaves dagger - like, arranged in a crowded rosette, glaucous green in colour and ending in a stiff sharp spine. White flowers borne on column - like inflorescence arising from the center of plant. Fruits brownish, indehiscent, 5–8 cm long, obovate, seeds black, thickened.

Reproductive Cycle : July-August.

Materials Examined : HPU-EL-SKS-6745; Bemloe (2,010m), Grand Hotel (2,108m), Kennedy House (2,098m).

Propagation Notes : Thrives preferably in a sandy loam. Propagated by seeds, offsets, stem or rhizome cutting.

Parts Used : Whole Plant. Leaves. Flowers. Fruit.

Folk Uses : Planted for **ornamental** and **hedge** purposes and flowers and fruits **edible**. **Fibres** obtained from leaves used for making **ropes**, **baskets** and **mats**. Leaf splints used as **brush**.

Active Constituents : These are tigogenin (5α, 25R-spirostan-3β-ol), tigogenin, gitogenin, chlorogenin and its diacetate, smilagenin, hecogenin (**leaves**), cis-9-nonadecene, cis-8-heptadecene, nonadecane, heptadecane and 2,6,10-trimethyl-6,10-epoxy-2,7,11-dodecatrien-1-ol (**essential oil of flower**) and aloifoline (**flower**).

Uses in Literature : Useful in India as edible (flowers, fruit, stem), fibre for making ropes, baskets and mats (leaves), brush, planting in lawn, purgative, and substitute for soap (rhizome) **(Ambasta, 1986; Bean, 1981; Beri, 1992; Chopra *et al.*, 1986; Chittendon, 1956; Huxley, 1992; Moreman, 1998; Paliwal, 2009; Pool, 1941; Rastogi & Mehrotra, 1993; Uphof, 1968)**.

Zanthoxylum armatum DC.

Syn.: *Fagara armata* Thunb.; *Z. alatum* Roxb.; *Z. arenosum* Reeder & S.Y. Cheo; *Z. hostile* Wall.

Family: Rutaceae.

Vern.: Tirmir.

English, Hindi and Sanskrit Names:

Eng. : Winged Prickly Ash;
Hindi : Darmar, Nipalidhanya, Tejbal, Tejphal, Tezmal, Timbur, Timur, Tumru, Tun;
Sans. : Andhaka, Dviha, Gandhalu, Mahamuni, Sanuja, Saurabh, Sauraja, Sphutal, Sugandhi, Tikshnapatra, Tumbaru.

Distribution : Kashmir-S.W. China; S.E. Asia: 1,100-2,500m.

Description : Shrubs having stems and branches armed with long prickles. Leaves odd-pinnate; stalks winged. Stipular spines 2 at base. Leaflets opposite, sessile, lanceolate. Flowers small, yellow, dioecious, sparingly branched in axillary panicles. Drupes pale-red globose, small, ultimately splitting in 2 valves. Seeds solitary in a fruit, globose and shining black.

Reproductive Cycle : April-June.

Material Examined : HPU-EL-SKS-6756; B.C.S. (2,022m), Naldehra (2,127m).

Propagation Notes : Prefers a good deep well-drained moisture retentive soil in full sun or semi-shade. Propagated naturally by seeds.

Parts Used : Leaves. Branches. Root. Fruit. Bark.

Folk Uses : Young leaves and fruits **edible**, used for **'chutney'** preparation and flavouring **curries**. Also, these are chewed against **mouth ulcers**, **foul breath**; and as **blood purifier**. Branches used as **toothbrush**, **walking sticks**; and hung on main door to **ward off** evil spirits. Roots form one of the constituents of **local wine**, '**sur**'. 1-3g powdered bark with honey good against **cold** and **cough**.

Active Constituents : These are spilantho 2, 6, 8-decatrienoic isobutylamide (**roots**), tambulin (3,5-dihydroxy-7,8,4-trimethoxyflavone), tambulol (glucosyl-7,4-dimethyl ether of gossypetin), linalool, limonene, methyl cinnamate, lineole, trans-cinnamic acid, arginine, glycine, histidine, threonine, tyrosine, aspartic acid, glutamic acid, nevadensin, umbelliferone, β-sitosterol and its glucoside (**fruits**), β-sitosterol, pinoresionol dimethyl ether and a mixture of alkaloids (**aerial parts**).

Uses in Literature : Reported earlier as an anthelmintic, antiseptic, aromatic, astringent, carminative, condiment, deodorant, disinfectant, edible seeds, leaves, fruits, febrifuge, flavouring agent in 'paan', hair lotions, stimulant, stomachic, tonic in fever; and also for boils, cholera, clubs, cough, dental diseases, diarrhoea, dyspepsia, eczema, fencing, headache, gum troubles, indigestion, intoxicating fish, itching, leucoderma, local wine ('sur'), mouth blisters, serving furs, paralysis, pestles, piles, purification of water, religious ceremonies, rheumatism, round worms, scabies, smallpox, snake bite, soap making, toothache, warding off house flies, and wood for house building **(Agarwal, 2003; Ahmed & Borthakur, 2005; Ambasta, 1986; Ansari, 1991; Bhattacharjee, 2001; Borthakur & Sarma, 1996; Chakraborty, 2003; Chandra, 1997; Chauhan, 1999; Chopra *et al.*, 1956; Chowdhery, 1996; Das & Agarwal, 1991; Dey & Bahadur, 1973; Duke *et al.*, 2002; Girach, 2000; Harborne & Baxter, 2001; Jain, 1991; Kapur & Singh, 1996; Khare, 2004; Kirtikar & Basu, 1984; Lal *et al.*, 1996; Lalramnghinglova, 2000, 2002; Lata *et al.*, 2008; Manandhar, 1996; Mao, 1993; Mittre, 1997; Negi & Gaur, 1991; Negi *et al.*, 1999; Pal *et al.*, 2000; Pande *et al.*, 2006; Panthi & Chaudhary, 2003; Prajapati *et al.*, 2006; Rana *et al.*, 1996; Ranjan, 2000b; Ravi Shankar & Henry, 1992; Rawat & Chowdhury, 1998; Saklani & Jain, 1994; Samant *et al.*, 2001; Sharma, 2003; Singh, 1991, 1999; Singh & Kumar, 2000a; Sinha & Sinha, 2001; Siwakoti & Varma, 1996; Sood & Thakur, 2004; Sood *et al.* 2009b, 2012b, 2013; Uphof, 1968; Verma & Chauhan, 2006; Watt, 1889-1896)**.

Chapter 5
EPILOGUE

In overall perspective, the available plant diversity at heritage sites unravels 212 species (medicinal - 74 species; edible - 76 species; ornamental - 111 species; fodder - 50 species; timber - 47 species; fibre - 8 species; fencing - 43 species; avenue - 7 species; fuel -36 species) belonging to 129 genera under 70 families (**Figs. 8, 20; Table IV**). These include 98 trees, 96 shrubs, 12 climbers, 4 lianas and 2 sub-shrubs (**Fig. 6; Tables VII-XI**) of which 26 species belong to Gymnosperms (*Abies pindrow*, *Afrocarpus gracilior*, *Araucaria bidwillii*, *Araucaria columnaris*, *Araucaria cunninghamii*, *Cedrus deodara*, *Cryptomeria japonica*, *Cunninghamia lanceolata*, *Cupressus cashmeriana*, *Cupressus torulosa*, *Cycas revoluta*, *Ginkgo biloba*, *Juniperus chinensis*, *Juniperus horizontalis*, *Juniperus procera*, *Juniperus pseudosabina*, *Juniperus squamata*, *Picea smithiana*, *Pinus roxburghii*, *Pinus wallichiana*, *Taxodium distichum*, *Taxus baccata*, *Thuja occidentalis*, *Thuja orientalis*, *Thujopsis dolabrata*, *Tsuga dumosa*), 13 species to Monocots (*Agave americana*, *Arundo donax*, *Asparagus aethiopicus*, *Asparagus densiflorus*, *Asparagus racemosus*, *Asparagus setaceus*, *Asparagus virgatus*,

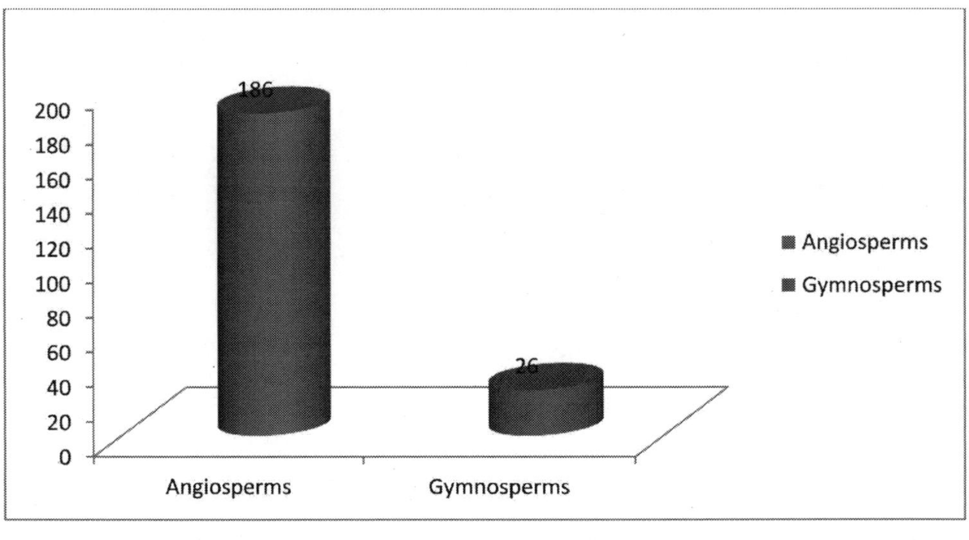

Fig. 5. Histogram Showing Total Number of Plants under Various Divisions of Plant Kingdom.

Chamaerops humilis, *Drepanostachyum falcatum*, *Musa paradisiaca*, *Rhapis excelsa*, *Smilax aspera*, *Yucca aloifolia*) and 173 species to Dicots (103 genera) inclusive of five plants of garden origin (*Abutilon* X *hybridum*, *Forsythia* X *intermedia*, *Fuchsia* X *hybrida*, *Ligustrum* X *vicaryi*, *Magnolia* X *soulangeana*) constituting bulk share of having 81.60% representation of the total diversity (**Figs. 5, 8; Tables III-V**). The plant diversity is predominantly woody with trees accounting for 46.22% of the represented diversity (**Fig. 7; Table IV**). The proportion of species belonging to Monocots to Dicots is 1: 13.30, of genera 1: 11.44 and of families is 1: 9.33 (**Fig. 9**). Evidently, the Dicotyledonous families are over 9 times larger than Monocotyledonous ones, the genera more than 11 times the Monocots and the species more than 13 times the Monocots **(Fig. 8; Table III**). The overall diversity can be grouped into 72 ethnobotanically important wild species, 119 cultivated ones and 21 growing as wild and under cultivation (**Table VI**). Phenologically, a large majority of collected taxa flowers and fruits during spring (33.96 %) and summer seasons (27.35%) (**Figs. 10, 11**). Unlike Hooker (1872-1897), the ratio of the total number of genera to species is low 1: 1.64 as compared to India. Good (1964) also observed the same and stated that within the same floral region, flora of smaller areas and remote islands have lower genus-species ratio as ecological diversity tends to be smaller.

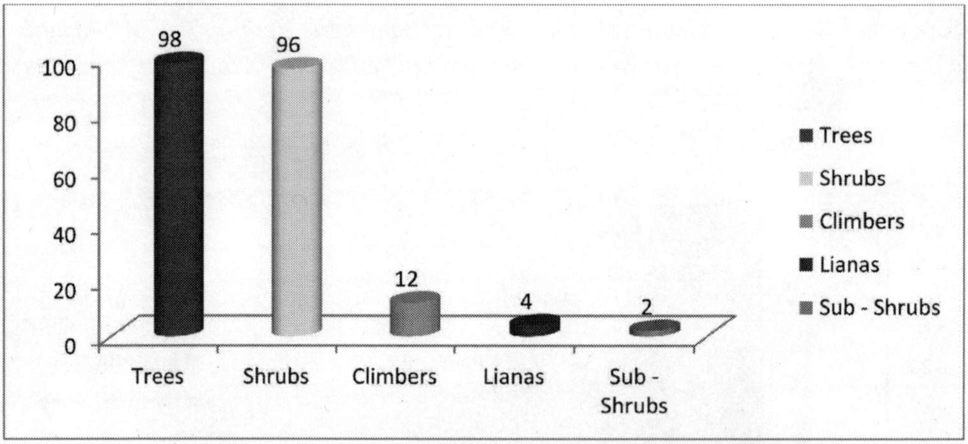

Fig. 6. Histogram Reflecting Quantitative Number of Trees, Shrubs, Sub-Shrubs, Climbers and Lianas of Heritage Sites.

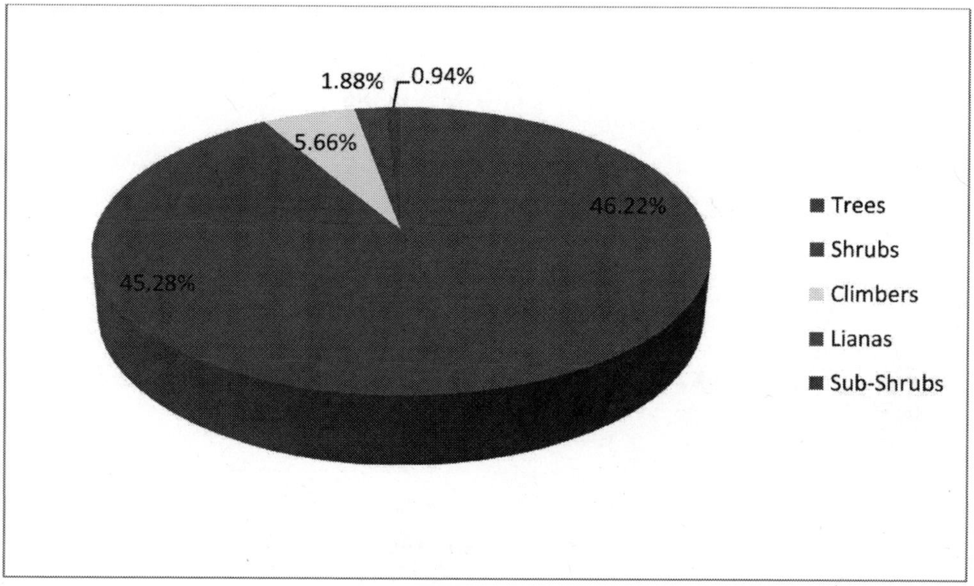

Fig. 7. Habit-Wise Classification of Plant Species of Heritage Sites in Shimla Hills.

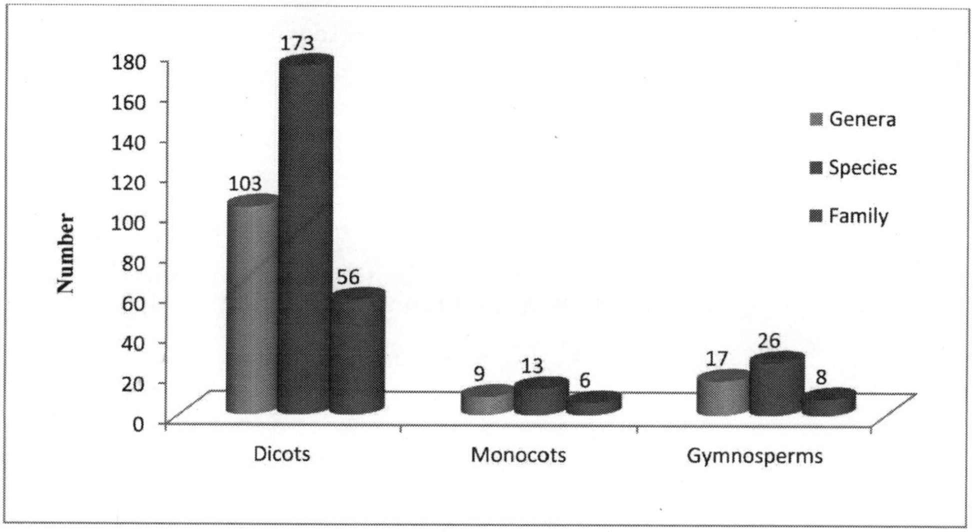

Fig. 8. Total Number of Species, Genera and Families under Various Divisions of Plant Kingdom.

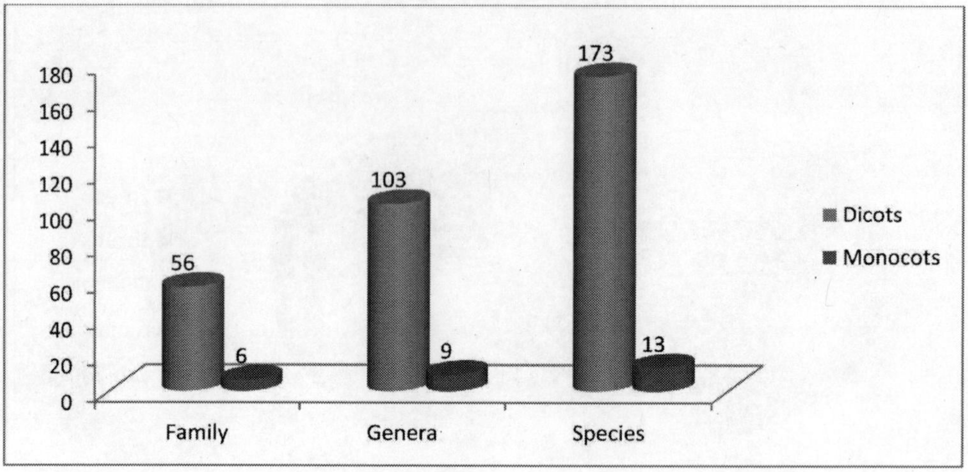

Fig. 9. Histogram Showing the Comparative Number of Genera, Species and Families of Dicots and Monocots.

TABLE III

Families, Genera and Species of Plants of Heritage Sites in Shimla Hills under Various Divisions of Plant Kingdom

S.No.	Division	Families	Genus	Species
1.	Dicots	56	103	173
2.	Monocots	6	9	13
3.	Gymnosperms	8	17	26
	Total	70	129	212

TABLE IV

Relative Percentage of Plants Belonging to Different Groups

Growth Form	Angiosperms		Gymnosperms	Total	Percentage (%)
	Dicots	Monocots			
Trees	71	2	25	98	46.22
Shrubs	85	10	1	96	45.28
Climbers	11	1	-	12	5.66
Lianas	4	-	-	4	1.88
Sub-Shrubs	2	-	-	2	0.94
Total	173	13	26	212	100

TABLE V
Plants of Garden Origin

S. No.	Botanical Name	Family	Origin
1.	*Abutilon* X *hybridum*	Malvaceae	Garden Origin
2.	*Forsythia* X *intermedia*	Oleaceae	Garden Origin
3.	*Fuchsia* X *hybrida*	Onagraceae	Garden Origin
4.	*Ligustrum* X *vicaryi*	Oleaceae	Garden Origin
5.	*Magnolia* X *soulangeana*	Magnoliaceae	Garden Origin

TABLE VI
Plant Resource Categories of Heritage Sites

S. No.	Cultivated	Wild	Wild as Well as Cultivated
1.	*Abutilon* X *hybridum*	*Abies pindrow*	*Agave americana*
2.	*Abutilon megapotamicum*	*Acer oblongum*	*Arundo donax*
3.	*Acacia dealbata*	*Aesculus indica*	*Asparagus racemosus*
4.	*Acer negundo*	*Berberis aristata*	*Buddleja asiatica*
5.	*Acer palmatum*	*Berberis lycium*	*Clematis montana*
6.	*Actinidia chinensis*	*Buddleja crispa*	*Colquhounia coccinea*
7.	*Afrocarpus gracilior*	*Cedrus deodara*	*Drepanostachyum falcatum*
8.	*Aloysia citriodora*	*Celtis australis*	*Erythrina arborescens*
9.	*Amygdalus mira*	*Cerasus cerasoides*	*Ficus neriifolia*
10.	*Araucaria bidwillii*	*Clematis buchananiana*	*Ficus religiosa*
11.	*Araucaria columnaris*	*Coriaria nepalensis*	*Hedera helix*
12.	*Araucaria cunninghamii*	*Cornus capitata*	*Hypericum oblongifolium*
13.	*Asparagus aethiopicus*	*Cornus macrophylla*	*Jasminum humile*
14.	*Asparagus setaceus*	*Daphne papyracea*	*Morus alba*
15.	*Asparagus virgatus*	*Datura stramonium*	*Morus serrata*
16.	*Asparagus densiflorus*	*Cotoneaster bacillaris*	*Opuntia dillenii*
17.	*Bougainvillea glabra*	*Debregeasia saeneb*	*Punica granatum*
18.	*Buddleja davidii*	*Desmodium elegans*	*Pyrus pyrifolia*
19.	*Buxus wallichiana*	*Desmodium multiflorum*	*Taxus baccata*
20.	*Callistemon lanceolatus*	*Elaeagnus umbellata*	*Yucca aloifolia*
21.	*Camellia japonica*	*Euonymus fimbriatus*	*Zanthoxylum armatum*
22.	*Campsis radicans*	*Euonymus lucidus*	
23.	*Cestrum elegans*	*Euonymus tingens*	

S. No.	Cultivated	Wild	Wild as Well as Cultivated
24.	*Cestrum nocturnum*	*Ficus auriculata*	
25.	*Cestrum parqui*	*Ficus palmata*	
26.	*Chamaerops humilis*	*Ficus sarmentosa*	
27.	*Citrus limon*	*Ficus subincisa*	
28.	*Citrus maxima*	*Girardinia diversifolia*	
29.	*Cryptomeria japonica*	*Hedera nepalensis*	
30.	*Cunninghamia lanceolata*	*Himalrandia tetrasperma*	
31.	*Cupressus cashmeriana*	*Hypericum patulum*	
32.	*Cupressus torulosa*	*Ilex dipyrena*	
33.	*Cycas revoluta*	*Indigofera dosua*	
34.	*Diospyros kaki*	*Inula cappa*	
35.	*Diospyros lotus*	*Lonicera quinquelocularis*	
36.	*Duranta erecta*	*Leea asiatica*	
37.	*Eriobotrya japonica*	*Myrica esculenta*	
38.	*Eucalyptus globulus*	*Lyonia ovalifolia*	
39.	*Euonymus japonicus*	*Myrsine Africana*	
40.	*Fagus sylvatica*	*Padus cornuta*	
41.	*Ficus benjamina*	*Parthenocissus semicordata*	
42.	*Ficus elastica*	*Picea smithiana*	
43.	*Forsythia ovata*	*Pinus roxburghii*	
44.	*Forsythia* X *intermedia*	*Pinus wallichiana*	
45.	*Fuchsia magellanica*	*Pistacia khinjuk*	
46.	*Fuchsia* X *hybrida*	*Prinsepia utilis*	
47.	*Ginkgo biloba*	*Pyracantha crenulata*	
48.	*Grevillea robusta*	*Quercus floribunda*	
49.	*Hedera canariensis*	*Quercus incana*	
50.	*Hedera colchica*	*Quercus oblongata*	
51.	*Hibiscus rosa-sinensis*	*Rhamnus persica*	
52.	*Hibiscus syriacus*	*Rhamnus virgatus*	
53.	*Hydrangea macrophylla*	*Pyrus pashia*	
54.	*Hypericum hookerianum*	*Rhododendron arboreum*	
55.	*Ipomoea purpurea*	*Ricinus communis*	
56.	*Jacaranda acutifolia*	*Rosa moschata*	
57.	*Jasminum grandiflorum*	*Rubia cordifolia*	
58.	*Jasminum mesnyi*	*Rubus ellipticus*	

S. No.	Cultivated	Wild	Wild as Well as Cultivated
59.	*Jasminum multiflorum*	*Rubus biflorus*	
60.	*Juglans regia*	*Rubus macilentus*	
61.	*Juniperus chinensis*	*Rubus niveus*	
62.	*Juniperus horizontalis*	*Rubus paniculatus*	
63.	*Juniperus procera*	*Salix pycnostachya* var. *oxycarpa*	
64.	*Juniperus pseudosabina*	*Sarcococca pruniformis*	
65.	*Juniperus squamata*	*Skimmia laureola*	
66.	*Kerria japonica*	*Smilax aspera*	
67.	*Lagerstroemia indica*	*Solanum virginianum*	
68.	*Ligustrum* X *vicaryi*	*Spiraea canescens*	
69.	*Ligustrum lucidum*	*Ulmus wallichiana*	
70.	*Liriodendron tulipifera*	*Viburnum cotinifolium*	
71.	*Lonicera japonica*	*Viburnum mullaha*	
72.	*Magnolia grandiflora*	*Xanthium strumarium*	
73.	*Magnolia obovata*		
74.	*Magnolia* X *soulangeana*		
75.	*Malus pumila*		
76.	*Malvaviscus arboreus* var. *mexicanus*		
77.	*Morus nigra*		
78.	*Musa paradisiaca*		
79.	*Myrtus communis*		
80.	*Nerium oleander*		
81.	*Paulownia tomentosa*		
82.	*Platanus orientalis*		
83.	*Populus ciliata*		
84.	*Populus nigra*		
85.	*Prunus amygdalus*		
86.	*Prunus armeniaca*		
87.	*Prunus domestica*		
88.	*Prunus laurocerasus*		
89.	*Prunus persica*		
90.	*Prunus salicina*		
91.	*Pyrus communis*		
92.	*Quercus robur*		
93.	*Rhapis excelsa*		

S. No.	Cultivated	Wild	Wild as Well as Cultivated
94.	*Rhododendron simsii* var. *vittatum*		
95.	*Robinia pseudoacacia*		
96.	*Rosa alba*		
97.	*Rosa banksiae*		
98.	*Rosa chinensis*		
99.	*Rosa damascena*		
100.	*Rosa jacquemontii*		
101.	*Rosa multiflora*		
102.	*Salix alba*		
103.	*Salix babylonica*		
104.	*Salvia leucantha*		
105.	*Salvia splendens*		
106.	*Sambucus nigra*		
107.	*Schefflera arboricola*		
108.	*Solanum laxum*		
109.	*Solanum pseudocapsicum*		
110.	*Spartium junceum*		
111.	*Spiraea cantoniensis*		
112.	*Taxodium distichum*		
113.	*Thuja occidentalis*		
114.	*Thuja orientalis*		
115.	*Thujopsis dolabrata*		
116.	*Trachelospermum jasminoides*		
117.	*Tropaeolum majus*		
118.	*Tsuga dumosa*		
119.	*Wisteria sinensis*		

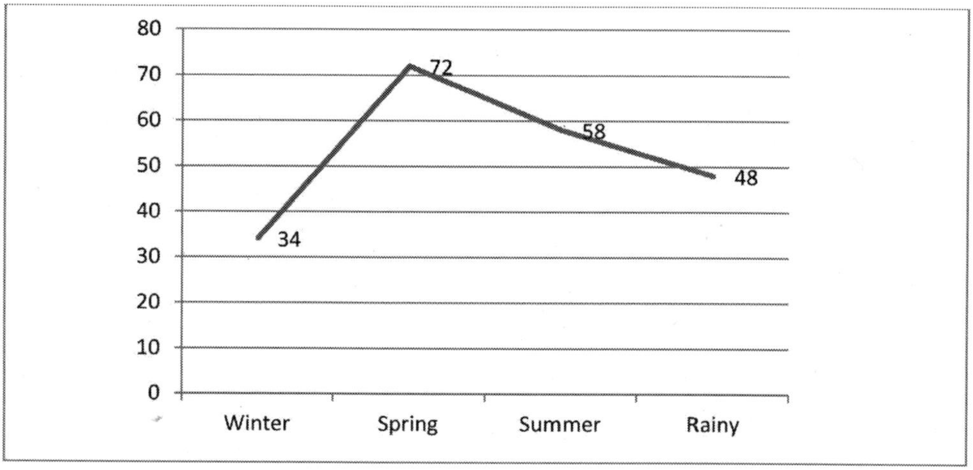

Fig. 10. Flowering and Fruiting Seasons of Plants.

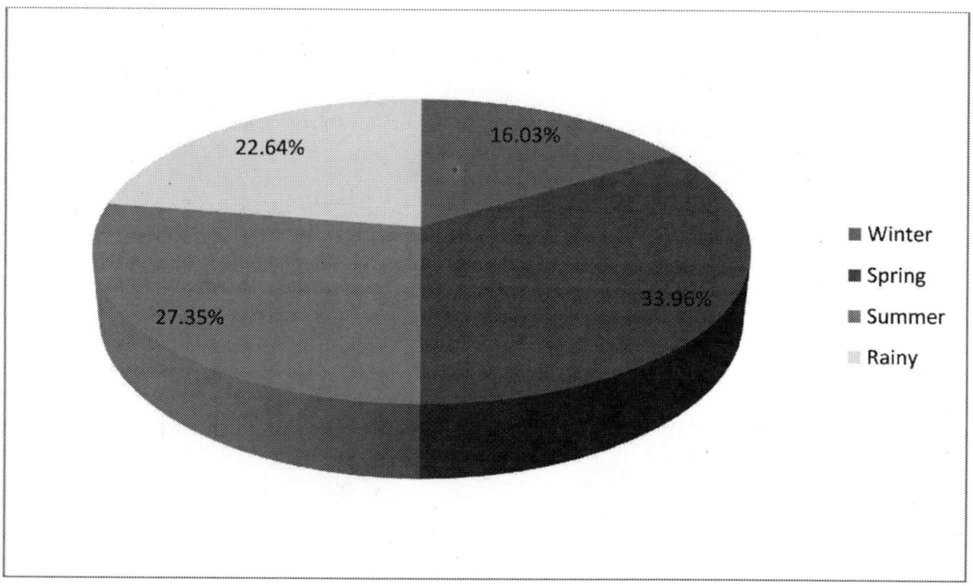

Fig. 11. Pie Diagram Showing Relative Percentage of Flowering and Fruiting Seasons of Plants.

TABLE VII
Trees of Heritage Sites

S. No.	Botanical Name	Local Name/s
1.	*Abies pindrow*	Fir, Tosh
2.	*Acacia dealbata*	Khair
3.	*Acer negundo*	Maple
4.	*Acer oblongum*	Padang
5.	*Acer palmatum*	Japanese Maple
6.	*Aesculus indica*	Kanhor, Khanor
7.	*Afrocarpus gracilior*	Podocarpus
8.	*Amygdalus mira*	Vaimi
9.	*Araucaria bidwillii*	Bunya Araucaria
10.	*Araucaria columnaris*	Araucaria
11.	*Araucaria cunninghamii*	Hoop pine
12.	*Callistemon lanceolatus*	Bottle Brush
13.	*Cedrus deodara*	Deaar, Devdar
14.	*Celtis australis*	Khirak
15.	*Cerasus cerasoides*	Pajja
16.	*Chamaerops humilis*	Fan Palm, Khajuri
17.	*Citrus limon*	Galgal
18.	*Citrus maxima*	Chakotra
19.	*Cornus capitata*	Teelahan
20.	*Cornus macrophylla*	Nang
21.	*Cryptomeria japonica*	Cryptomeria
22.	*Cunninghamia lanceolata*	Ghadhal, Thanela
23.	*Cupressus cashmeriana*	Kashmiri Cypress
24.	*Cupressus torulosa*	Saru
25.	*Cycas revoluta*	Cycas
26.	*Diospyros kaki*	Japani-phal
27.	*Diospyros lotus*	Barain
28.	*Eriobotrya japonica*	Lokat
29.	*Erythrina arborescens*	Pariyara
30.	*Eucalyptus globulus*	Eucalyptus, Safeda
31.	*Euonymus fimbriatus*	Euonymus
32.	*Euonymus japonicus*	Japani Euonymus
33.	*Euonymus lucidus*	Bharmela
34.	*Euonymus tingens*	Marmakoul
35.	*Fagus sylvatica*	Rang Badalane Bala Plant

S. No.	Botanical Name	Local Name/s
36.	*Ficus auriculata*	Timbal
37.	*Ficus benjamina*	Kabra
38.	*Ficus elastica*	Rubber Plant
39.	*Ficus neriifolia*	Dudhia
40.	*Ficus palmata*	Fegura
41.	*Ficus religiosa*	Peepal
42.	*Ficus sarmentosa*	Phaegdi
43.	*Ficus subincisa*	Ban Feguda
44.	*Ginkgo biloba*	Balkumari
45.	*Grevillea robusta*	Silver Fir
46.	*Ilex dipyrena*	Charkha, Holly Tree
47.	*Jacaranda acutifolia*	Jackranda, Nili Gulmohra
48.	*Juglans regia*	Akhrot
49.	*Juniperus chinensis*	Chinese Juniperus
50.	*Juniperus procera*	African Juniper
51.	*Juniperus pseudosabina*	Juniper
52.	*Juniperus squamata*	Thanara
53.	*Liriodendron tulipifera*	Tulip Tree
54.	*Lyonia ovalifolia*	Bhelar
55.	*Magnolia grandiflora*	Magnolia
56.	*Magnolia obovata*	Japani Magnolia
57.	*Magnolia* X *soulangeana*	Pink Magnolia
58.	*Malus pumila*	Seb
59.	*Morus alba*	Toot
60.	*Morus nigra*	Toot
61.	*Morus serrata*	Kimu
62.	*Musa paradisiaca*	Kela
63.	*Myrica esculenta*	Kaphal
64.	*Padus cornuta*	Jangli Jamun, Krun
65.	*Paulownia tomentosa*	Videshi Phool
66.	*Picea smithiana*	Rai
67.	*Pinus roxburghii*	Chil
68.	*Pinus wallichiana*	Kail
69.	*Pistacia khinjuk*	Kakad
70.	*Platanus orientalis*	Chinar
71.	*Populus ciliata*	Popular
72.	*Populus nigra*	Poplar
73.	*Prunus amygdalus*	Badam

S. No.	Botanical Name	Local Name/s
74.	*Prunus armeniaca*	Chuli, Khumani
75.	*Prunus domestica*	Aloo Bukhara
76.	*Prunus laurocerasus*	Cherry-laurel
77.	*Prunus persica*	Adu
78.	*Prunus salicina*	Shakarpara
79.	*Pyrus communis*	Nashpati
80.	*Pyrus pashia*	Kainth
81.	*Pyrus pyrifolia*	Zarainth
82.	*Quercus floribunda*	Moru
83.	*Quercus incana*	Banj
84.	*Quercus oblongata*	Ban
85.	*Quercus robur*	English Oak
86.	*Rhamnus persica*	Thalta
87.	*Rhododendron arboreum*	Braah, Brass
88.	*Robinia pseudoacacia*	Rabinia
89.	*Salix alba*	Benusa
90.	*Salix babylonica*	Majnu
91.	*Salix pycnostachya* var. *oxycarpa*	Changma Beeuins
92.	*Taxodium distichum*	Cypress
93.	*Taxus baccata*	Thuno
94.	*Thuja occidentalis*	Thuja
95.	*Thuja orientalis*	Morpankh
96.	*Thujopsis dolabrata*	Thujopsis
97.	*Tsuga dumosa*	Thingia
98.	*Ulmus wallichiana*	Brera

TABLE VIII
Shrubs of Heritage Sites in Shimla Hills

S. No.	Botanical Name	Local Name/s
1.	*Abutilon* X *hybridum*	Chinese Lantern
2.	*Abutilon megapotamicum*	Lantern Flower
3.	*Agave americana*	Keora
4.	*Aloysia citriodora*	Verbena
5.	*Arundo donax*	Nal
6.	*Asparagus aethiopicus*	Sprengeri Asparagus
7.	*Asparagus densiflorus*	Asparagus
8.	*Asparagus racemosus*	Satavar, Sanspai

S. No.	Botanical Name	Local Name/s
9.	*Asparagus setaceus*	Asparagus Fern
10.	*Asparagus virgatus*	Asparagus
11.	*Berberis aristata*	Kashmal
12.	*Berberis lycium*	Kashmal
13.	*Bougainvillea glabra*	Boganvillea
14.	*Buddleja asiatica*	Buddliya, Bana
15.	*Buddleja crispa*	Newarpati
16.	*Buddleja davidii*	Buddleya
17.	*Buxus wallichiana*	Shamshad
18.	*Camellia japonica*	Camellia
19.	*Campsis radicans*	Shehnai Phool
20.	*Cestrum elegans*	Raat -ki-Chameli
21.	*Cestrum nocturnum*	Rat-ki -Rani
22.	*Cestrum parqui*	Rat-ki -Rani
23.	*Colquhounia coccinea*	Bhekuyi
24.	*Coriaria nepalensis*	Kachog
25.	*Cotoneaster bacillaris*	Simbli
26.	*Daphne papyracea*	Chamar-Chumb, Raabal
27.	*Datura stramonium*	Dhatura
28.	*Debregeasia saeneb*	Syaru
29.	*Desmodium elegans*	Chamla
30.	*Desmodium multiflorum*	Martoi
31.	*Drepanostachyum falcatum*	Banji
32.	*Duranta erecta*	Neel Kanthi
33.	*Elaeagnus umbellata*	Ghain
34.	*Forsythia* X *intermedia*	Forsythia
35.	*Forsythia ovata*	Forsythia
36.	*Fuchsia* X *hybrida*	Doll Flower
37.	*Fuchsia magellanica*	Doll Flower
38.	*Girardinia diversifolia*	Bicchu Buti
39.	*Hibiscus rosa-sinensis*	China Rose
40.	*Hibiscus syriacus*	Brahm Phool
41.	*Himalrandia tetrasperma*	Hazara-kikra
42.	*Hydrangea macrophylla*	Hydrangea
43.	*Hypericum hookerianum*	Phiunli
44.	*Hypericum oblongifolium*	Lyoon
45.	*Hypericum patulum*	Phulanu
46.	*Indigofera dosua*	Kathee

S. No.	Botanical Name	Local Name/s
47.	*Inula cappa*	Basanti
48.	*Jasminum grandiflorum*	Chameli
49.	*Jasminum humile*	Peeli Chameli
50.	*Jasminum mesnyi*	Chameli
51.	*Jasminum multiflorum*	Chameli
52.	*Juniperus horizontalis*	Juniper
53.	*Kerria japonica*	Japani Rose
54.	*Lagerstroemia indica*	Har-Shingar
55.	*Leea asiatica*	Gan-Gola
56.	*Ligustrum lucidum*	Pataal Ped
57.	*Ligustrum* X *vicaryi*	Privet
58.	*Lonicera quinquelocularis*	Chaunsh
59.	*Malvaviscus arboreus var. mexicanus*	Gudhal
60.	*Myrsine africana*	Berang
61.	*Myrtus communis*	Myrtle
62.	*Nerium oleander*	Kaner
63.	*Opuntia dillenii*	Vilayati Suri
64.	*Prinsepia utilis*	Bekhal
65.	*Punica granatum*	Anar
66.	*Pyracantha crenulata*	Gingaru
67.	*Rhododendron simsii var. vittatum*	Ajalea
68.	*Rhamnus virgatus*	Reushi
69.	*Rhapis excelsa*	Lady Palm
70.	*Ricinus communis*	Arind
71.	*Rosa alba*	Gulab
72.	*Rosa banksiae*	Gulabri
73.	*Rosa chinensis*	Gulab
74.	*Rosa damascena*	Gulabadi
75.	*Rosa jacquemontii*	Gulab
76.	*Rosa moschata*	Kuja
77.	*Rosa multiflora*	Gulabari
78.	*Rubus biflorus*	Dher
79.	*Rubus ellipticus*	Akhe
80.	*Rubus macilentus*	Heere
81.	*Rubus niveus*	Klenchu
82.	*Rubus paniculatus*	Anchu
83.	*Sambucus nigra*	Salja, Sambucus
84.	*Sarcococca pruniformis*	Raabal

S. No.	Botanical Name	Local Name/s
85.	*Schefflera arboricola*	Umbrella Plant
86.	*Skimmia laureola*	Patti Dhup
87.	*Solanum pseudocapsicum*	Charoti Mirch
88.	*Solanum virginianum*	Kant
89.	*Spartium junceum*	Weavers Broom
90.	*Spiraea canescens*	Chakhu
91.	*Spiraea cantoniensis*	Spirea
92.	*Viburnum cotinifolium*	Junglee Dakh, Ricchnabu
93.	*Viburnum mullaha*	Dhaab
94.	*Xanthium strumarium*	Banokra, Gheenjadu
95.	*Yucca aloifolia*	Rambaan
96.	*Zanthoxylum armatum*	Tirmir

TABLE IX
Climbers of Heritage Sites

S.No.	Botanical Name	Local Name/s
1.	*Clematis buchananiana*	Dhudki Bel
2.	*Clematis montana*	Kaunia-bali
3.	*Hedera canariensis*	Banbat Kari
4.	*Hedera colchica*	Kiraya
5.	*Hedera helix*	Dakari
6.	*Ipomoea purpurea*	Gonani
7.	*Lonicera japonica*	Lushai
8.	*Parthenocissus semicordata*	Bhambti Phlankur
9.	*Rubia cordifolia*	Manjith
10.	*Smilax aspera*	Bragh Bel
11.	*Solanum laxum*	Nightshade Jasmine
12.	*Tropaeolum majus*	Tindu Phool

TABLE X
Lianas of Heritage Sites

S.No.	Botanical Name	Local Name/s
1.	*Actinidia chinensis*	Kiwi
2.	*Trachelospermum jasminoides*	Star Jasmine
3.	*Wisteria sinensis*	Wisteria
4.	*Hedera nepalensis*	Berbel

TABLE XI
Sub-Shrubs of Heritage Sites

S.No.	Botanical Name	Local Name/s
1.	*Salvia leucantha*	Velvet Salvia
2.	*Salvia splendens*	Salvia

TABLE XII
Fruit Types of Angiospermic Plants

S. No	Name of Species	Fruit Type
1.	*Abutilon* X *hybridum*	Capsule
2.	*Abutilon megapotamicum*	Capsule
3.	*Acacia dealbata*	Pod
4.	*Acer negundo*	Samara
5.	*Acer oblongum*	Samara
6.	*Acer palmatum*	Samara
7.	*Actinidia chinensis*	Berry
8.	*Aesculus indica*	Capsule
9.	*Agave americana*	Capsule
10.	*Aloysia citriodora*	Nutlets
11.	*Amygdalus mira*	Drupe
12.	*Arundo donax*	Caryopsis
13.	*Asparagus aethiopicus*	Berry
14.	*Asparagus densiflorus*	Berry
15.	*Asparagus racemosus*	Berry
16.	*Asparagus setaceus*	Berry
17.	*Asparagus virgatus*	Berry
18.	*Berberis aristata*	Berry
19.	*Berberis lycium*	Berry
20.	*Bougainvillea glabra*	Achene
21.	*Buddleja asiatica*	Capsule
22.	*Buddleja crispa*	Capsule
23.	*Buddleja davidii*	Capsule
24.	*Buxus wallichiana*	Capsule
25.	*Callistemon lanceolatus*	Capsule
26.	*Camellia japonica*	Capsule
27.	*Campsis radicans*	Pod
28.	*Celtis australis*	Drupe

S. No	Name of Species	Fruit Type
29.	*Cerasus cerasoides*	Drupe
30.	*Cestrum elegans*	Berry
31.	*Cestrum nocturnum*	Berry
32.	*Cestrum parqui*	Berry
33.	*Chamaerops humilis*	Drupes
34.	*Citrus limon*	Hesperidium
35.	*Citrus maxima*	Hesperidium
36.	*Clematis buchananiana*	Achene
37.	*Clematis montana*	Achene
38.	*Colquhounia coccinea*	Nutlets
39.	*Coriaria nepalensis*	Achene
40.	*Cornus capitata*	Drupe
41.	*Cornus macrophylla*	Drupe
42.	*Cotoneaster bacillaris*	Nutlets
43.	*Daphne papyracea*	Drupe
44.	*Datura stramonium*	Capsule
45.	*Debregeasia saeneb*	Drupe
46.	*Desmodium elegans*	Pod
47.	*Desmodium multiflorum*	Pod
48.	*Diospyros kaki*	Berry
49.	*Diospyros lotus*	Berry
50.	*Drepanostachyum falcatum*	Caryopsis
51.	*Duranta erecta*	Berrry
52.	*Elaeagnus umbellata*	Drupe
53.	*Eriobotrya japonica*	Pome
54.	*Erythrina arborescens*	Pod
55.	*Eucalyptus globulus*	Capsule
56.	*Euonymus fimbriatus*	Capsule
57.	*Euonymus japonicus*	Capsule
58.	*Euonymus lucidus*	Capsule
59.	*Euonymus tingens*	Capsule
60.	*Fagus sylvatica*	Nut
61.	*Ficus auriculata*	Fig
62.	*Ficus benjamina*	Fig
63.	*Ficus elastica*	Fig
64.	*Ficus neriifolia*	Fig
65.	*Ficus palmata*	Fig

S. No	Name of Species	Fruit Type
66.	*Ficus religiosa*	Fig
67.	*Ficus sarmentosa*	Fig
68.	*Ficus subincisa*	Fig
69.	*Forsythia* X *intermedia*	Capsule
70.	*Forsythia ovata*	Capsule
71.	*Fuchsia* X *hybrida*	Berry
72.	*Fuchsia magellanica*	Berry
73.	*Girardinia diversifolia*	Achene
74.	*Grevillea robusta*	Follicle
75.	*Hedera canariensis*	Berry
76.	*Hedera colchica*	Berry
77.	*Hedera helix*	Berry
78.	*Hedera nepalensis*	Berry
79.	*Hibiscus rosa-sinensis*	Capsule
80.	*Hibiscus syriacus*	Capsule
81.	*Himalrandia tetrasperma*	Berry
82.	*Hydrangea macrophylla*	Capsule
83.	*Hypericum hookerianum*	Capsule
84.	*Hypericum oblongifolium*	Capsule
85.	*Hypericum patulum*	Capsule
86.	*Ilex dipyrena*	Drupe
87.	*Indigofera dosua*	Pod
88.	*Inula cappa*	Achene
89.	*Ipomoea purpurea*	Capsule
90.	*Jacaranda acutifolia*	Capsule
91.	*Jasminum grandiflorum*	Berry
92.	*Jasminum humile*	Berry
93.	*Jasminum mesnyi*	Berry
94.	*Jasminum multiflorum*	Berry
95.	*Juglans regia*	Tryma
96.	*Kerria japonica*	Achenes
97.	*Lagerstroemia indica*	Capsule
98.	*Leea asiatica*	Berry
99.	*Ligustrum lucidum*	Berry
100.	*Ligustrum* X *vicaryi*	Berry
101.	*Liriodendron tulipifera*	Etaerio of Samaras
102.	*Lonicera japonica*	Berry

S. No	Name of Species	Fruit Type
103.	*Lonicera quinquelocularis*	Berry
104.	*Lyonia ovalifolia*	Capsule
105.	*Magnolia grandiflora*	Follicle
106.	*Magnolia obovata*	Follicle
107.	*Magnolia* X *soulangeana*	Follicle
108.	*Malus pumila*	Pome
109.	*Malvaviscus arboreus* var. *mexicanus*	Berry
110.	*Morus alba*	Sorosis
111.	*Morus nigra*	Sorosis
112.	*Morus serrata*	Sorosis
113.	*Musa paradisiaca*	Berry
114.	*Myrica esculenta*	Drupe
115.	*Myrsine africana*	Berry
116.	*Myrtus communis*	Berry
117.	*Nerium oleander*	Follicle
118.	*Opuntia dillenii*	Berry
119.	*Padus cornuta*	Drupe
120.	*Parthenocissus semicordata*	Berry
121.	*Paulownia tomentosa*	Capsule
122.	*Pistacia khinjuk*	Drupe
123.	*Platanus orientalis*	Achene
124.	*Populus ciliata*	Capsule
125.	*Populus nigra*	Capsule
126.	*Prinsepia utilis*	Drupe
127.	*Prunus amygdalus*	Drupe
128.	*Prunus armeniaca*	Drupe
129.	*Prunus domestica*	Drupe
130.	*Prunus laurocerasus*	Drupe
131.	*Prunus persica*	Drupe
132.	*Prunus salicina*	Drupe
133.	*Punica granatum*	Balastua
134.	*Pyracantha crenulata*	Nutlets
135.	*Pyrus communis*	Pome
136.	*Pyrus pashia*	Pome
137.	*Pyrus pyrifolia*	Pome
138.	*Quercus floribunda*	Acorn
139.	*Quercus incana*	Acorn

S. No	Name of Species	Fruit Type
140.	*Quercus oblongata*	Acorn
141.	*Quercus robur*	Acorn
142.	*Rhamnus persica*	Drupe
143.	*Rhamnus virgatus*	Drupe
144.	*Rhapis excelsa*	Drupe
145.	*Rhododendron arboreum*	Capsule
146.	*Rhododendron simsii* var. *vittatum*	Capsule
147.	*Ricinus communis*	Capsule
148.	*Robinia pseudoacacia*	Pod
149.	*Rosa alba*	Hip
150.	*Rosa banksiae*	Hip
151.	*Rosa chinensis*	Hip
152.	*Rosa damascena*	Hip
153.	*Rosa jacquemontii*	Hip
154.	*Rosa moschata*	Hip
155.	*Rosa multiflora*	Hip
156.	*Rubia cordifolia*	Berry
157.	*Rubus biflorus*	Etaerio of Drupes
158.	*Rubus ellipticus*	Etaerio of Drupes
159.	*Rubus macilentus*	Etaerio of Drupes
160.	*Rubus niveus*	Etaerio of Drupes
161.	*Rubus paniculatus*	Drupe
162.	*Salix alba*	Capsule
163.	*Salix babylonica*	Capsule
164.	*Salix pycnostachya* var. *oxycarpa*	Capsule
165.	*Salvia leucantha*	Carcerulus
166.	*Salvia splendens*	Carcerulus
167.	*Sambucus nigra*	Berry
168.	*Sarcococca pruniformis*	Drupe
169.	*Schefflera arboricola*	Berry
170.	*Skimmia laureola*	Berry
171.	*Smilax aspera*	Berry
172.	*Solanum laxum*	Berry
173.	*Solanum pseudocapsicum*	Berry
174.	*Solanum virginianum*	Berry
175.	*Spartium junceum*	Pod
176.	*Spiraea canescens*	Follicle

S. No	Name of Species	Fruit Type
177.	*Spiraea cantoniensis*	Follicle
178.	*Trachelospermum jasminoides*	Follicle
179.	*Tropaeolum majus*	Capsule
180.	*Ulmus wallichiana*	Samara
181.	*Viburnum cotinifolium*	Drupe
182.	*Viburnum mullaha*	Drupe
183.	*Wisteria sinensis*	Pod
184.	*Xanthium strumarium*	Achene
185.	*Yucca aloifolia*	Capsule
186.	*Zanthoxylum armatum*	Drupe

TABLE XIII
Predominant Fruit Types of Heritage Sites

S. No.	Fruit Types	No. of Species	%age of Total Species
1.	Berries	44	23.65
2.	Capsules	39	20.96
3.	Drupes	28	15.05
4.	Pods	9	4.83
5.	Achenes	9	4.83
6.	Follicles	8	4.30
7.	Syconis	8	4.30
8.	Hip	7	3.76
9.	Nut/Nutlets	7	3.76
10.	Pome	5	2.68
11.	Acorns	4	2.15
12.	Etaerio of Drupes	4	2.15
13.	Samara	4	2.15
14.	Sorosis	3	1.61
15.	Caryopsis	2	1.07
16.	Hesperidium	2	1.07
17.	Balausta	1	0.53
18.	Etaerio of Samara	1	0.53
19.	Tryma	1	0.53

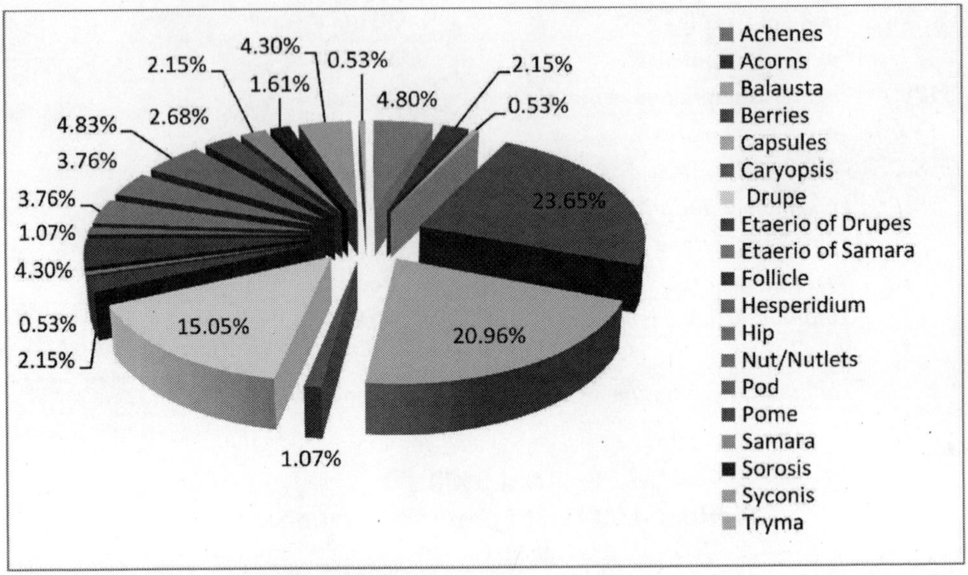

Fig. 12. Relative Percentage of Fruit Types

With respect to species richness, the most dominant families of flowering plants pertain to Rosaceae (32 species, 14 genera; 15.09%), Moraceae (11 species, 2 genera; 5.18%), Cupressaceae (10 species, 4 genera; 4.71%), Fabaceae (8 species, 7 genera; 3.77%), Oleaceae (8 species, 3 genera; 3.77%), Solanaceae (7 species, 3 genera; 3.30%), Pinaceae (6 species, 5 genera; 2.83%), Araliaceae (5 species, 2 genera; 2.35%), Asparagaceae (5 species,1 genus; 2.35%), Caprifoliaceae (5 species, 3 genera; 2.35%), Fagaceae (5 species, 2 genera; 2.35%), Malvaceae (5 species, 3 genera; 2.35%), Salicaceae (5 species, 2 genera; 2.35%), Celastraceae (4 species, 1 genus; 1.88%), Magnoliaceae (4 species, 2 genera; 1.88%), Rutaceae (4 species, 3 genera; 1.88%), Aceraceae (3 species, 1 genus; 1.41%), Araucariaceae (3 species, 1 genus; 1.41%), Buddlejaceae (3 species, 1 genus; 1.41%), Clusiaceae (3 species, 1 genus; 1.41%), Ericaceae (3 species, 2 genera; 1.41%), Lamiaceae (3 species, 2 genera; 1.41%), Myrtaceae (3 species, 3 genera; 1.41%), Taxodiaceae (3 species, 3 genera; 1.41%) etc. Moreover, these 24 families account for about 64.49% of overall recorded species diversity from the study area; of these the predominant Rosaceae amongst Dicots contribute to 15.09%, predominant Asparagaceae amongst Monocots to 2.35% and Cupressaceae amongst Gymnosperms to 4.71 % **(Figs. 13-16; Tables XIV-XVI)**. Amongst the predominant genera, these are *Ficus* (8 species), *Rosa* (7 species), *Prunus* (6 species), *Rubus* (5 species), *Euonymus*, *Hedera*, *Jasminum*, *Quercus* (4 species each), *Acer*, *Buddleja*, *Cestrum*, *Hypericum*, *Magnolia*, *Morus*, *Pyrus*, *Salix*, *Solanum* (3 species each), *Abutilon*,

Berberis, *Citrus*, *Clematis*, *Cornus*, *Desmodium*, *Diospyros*, *Forsythia*, *Fuchsia*, *Hibiscus*, *Ligustrum*, *Lonicera*, *Populus*, *Rhamnus*, *Rhododendron*, *Salvia*, *Spiraea*, *Viburnum* (2 species each) for Dicots (**Fig. 17; Table XVII**); *Asparagus* (5 species), *Agave*, *Arundo*, *Chamaerops*, *Drepanostachyum*, *Musa*, *Rhapis*, *Smilax*, *Yucca* (1 species each) for monocots (**Fig. 18; Table XVIII**) and *Juniperus* (5 species), *Araucaria* (3 species), *Cupressus*, *Pinus*, *Thuja* (2 species each), *Abies*, *Cedrus*, *Cryptomeria*, *Cycas*, *Ginkgo* (1 species each) for Gymnosperms (**Fig. 19; Table XIX**). Of the fruit types, berries (23.65 %) are the most common type, followed by capsules (20.96%), drupes (15.05%), pods (4.83%), achenes (4.83%), follicles (4.30%) and syconis (4.30%) (**Fig. 12**; **Tables XII, XIII**). There could be no two opinions that the heritage sites harbor rich plant diversity of which the dicotyledonous taxa predominate. Another fascinating feature of the present study is that 3 exotic primitive angiospermic taxa, *Liriodendron tulipifera*, *Magnolia grandiflora*, *Magnolia obovata* (Magnoliaceae) and indigenous *Myrica esculenta* (Myricaceae) (Rao, 1994) have also been represented in the courtyard of heritage sites.

TABLE XIV

Relative Percentage of Predominant Dicot Families

Family	Species	Relative Percentage
Rosaceae	32	15.09
Moraceae	11	5.18
Fabaceae	8	3.77
Oleaceae	8	3.77
Solanaceae	7	3.30
Araliaceae	5	2.35
Caprifoliaceae	5	2.35
Fagaceae	5	2.35
Malvaceae	5	2.35
Salicaceae	5	2.35
Celastraceae	4	1.88
Magnoliaceae	4	1.88
Rutaceae	4	1.88
Aceraceae	3	1.41
Buddlejaceae	3	1.41
Clusiaceae	3	1.41
Ericaceae	3	1.41
Lamiaceae	3	1.41
Myrtaceae	3	1.41

TABLE XV
Relative Percentage of Predominant Monocot Families

Family	Species	Relative Percentage
Asparagaceae	5	2.35
Agavaceae	2	0.94
Arecaceae	2	0.94
Poaceae	2	0.94
Musaceae	1	0.47
Smilacaceae	1	0.47

TABLE XVI
Relative Percentage of Predominant Gymnospermous Families

Family	Species	Relative Percentage
Cupressaceae	10	4.71
Pinaceae	6	2.83
Araucariaceae	3	1.41
Taxodiaceae	3	1.41
Cycadaceae	1	0.47
Ginkgoaceae	1	0.47
Podocarpaceae	1	0.47
Taxaceae	1	0.47

TABLE XVII
Predominant Dicot Genera of Heritage Sites

Genus	Family	No. of Species
Ficus	Moraceae	8
Rosa	Rosaceae	7
Prunus	Rosaceae	6
Rubus	Rosaceae	5
Euonymus	Celastraceae	4
Hedera	Araliaceae	4
Jasminum	Oleaceae	4
Quercus	Fagaceae	4
Acer	Aceraceae	3
Buddleja	Buddlejaceae	3
Cestrum	Solanaceae	3
Hypericum	Clusiaceae	3

Genus	Family	No. of Species
Magnolia	Magnoliaceae	3
Morus	Moraceae	3
Pyrus	Rosaceae	3
Salix	Salicaceae	3
Solanum	Solanaceae	3
Abutilon	Malvaceae	2
Berberis	Berberidaceae	2
Citrus	Rutaceae	2
Clematis	Ranunculaceae	2
Cornus	Cornaceae	2
Desmodium	Fabaceae	2
Diospyros	Ebenaceae	2
Forsythia	Oleaceae	2
Fuchsia	Onagraceae	2
Hibiscus	Malvaceae	2
Ligustrum	Oleaceae	2
Lonicera	Caprifoliaceae	2
Populus	Salicaceae	2
Rhamnus	Rhamnaceae	2
Rhododendron	Ericaceae	2
Salvia	Lamiaceae	2
Spiraea	Rosaceae	2
Viburnum	Caprifoliaceae	2

TABLE XVIII
Predominant Monocot Genera of Heritage Sites

Genus	Family	No. of Species
Asparagus	Asparagaceae	5
Agave	Agavaceae	1
Arundo	Poaceae	1
Chamaerops	Arecaceae	1
Drepanostachyum	Poaceae	1
Musa	Musaceae	1
Rhapis	Arecaceae	1
Smilax	Smilacaceae	1
Yucca	Agavaceae	1

TABLE XIX
Predominant Gymnospermous Genera of Heritage Sites

Genus	Family	No. of Species
Juniperus	Cupressaceae	5
Araucaria	Araucariaceae	3
Cupressus	Cupressaceae	2
Pinus	Pinaceae	2
Thuja	Cupressaceae	2
Abies	Pinaceae	1
Cedrus	Pinaceae	1
Cryptomeria	Taxodiaceae	1
Cycas	Cycadaceae	1
Ginkgo	Ginkgoaceae	1

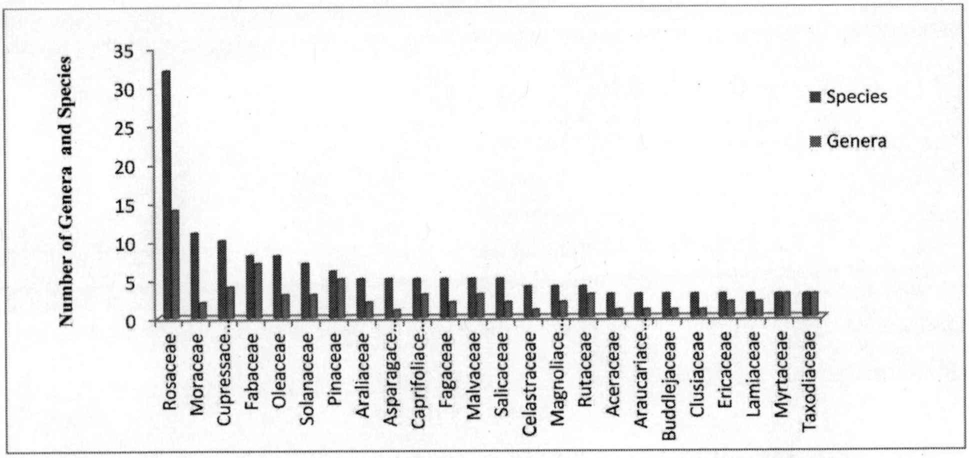

Fig. 13. Total Number of Genera and Species under Predominant Families of Heritage Sites.

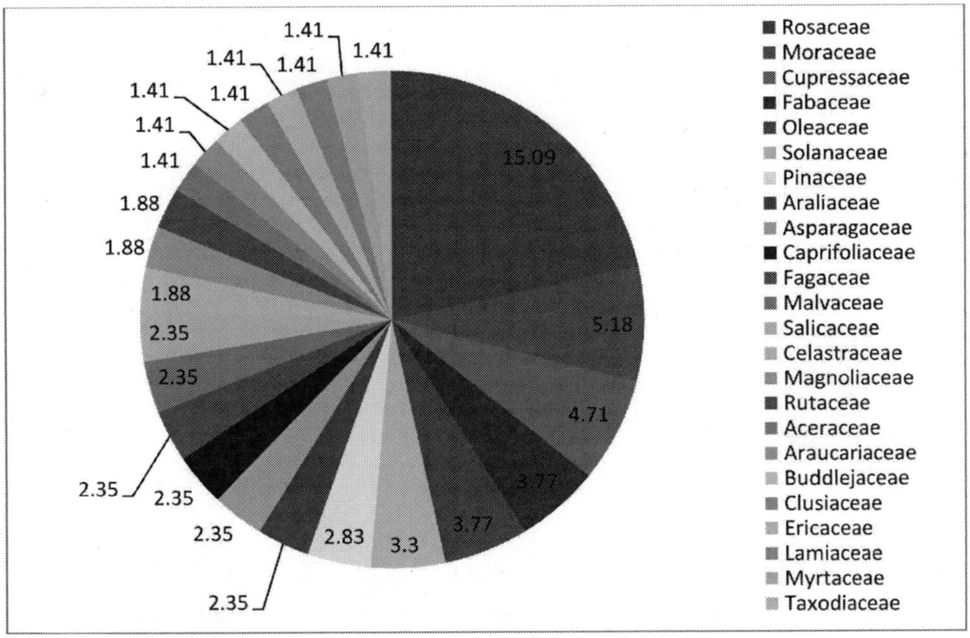

Fig. 14. Relative Percentage of Taxa Belonging to Predominant Families of Heritage Sites.

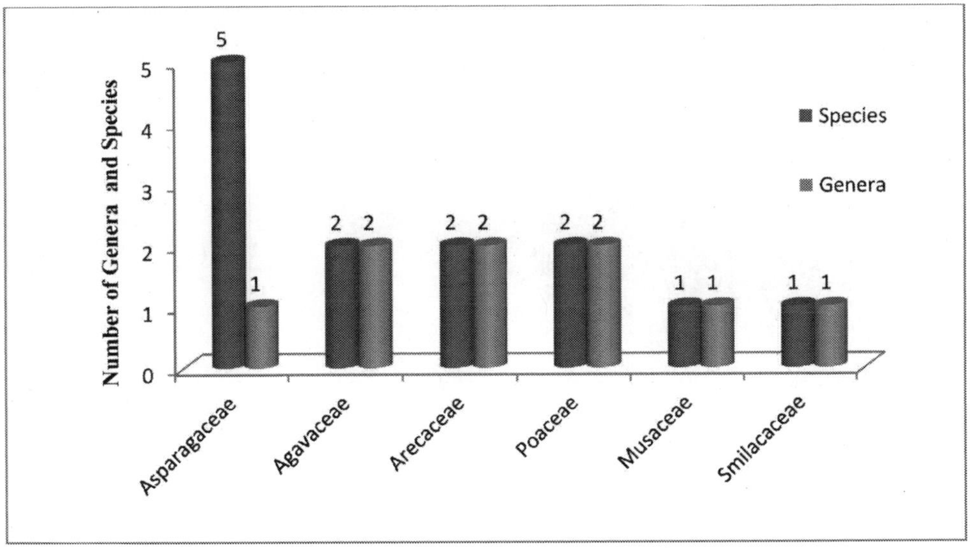

Fig. 15. Predominant Monocot Families of Heritage Sites.

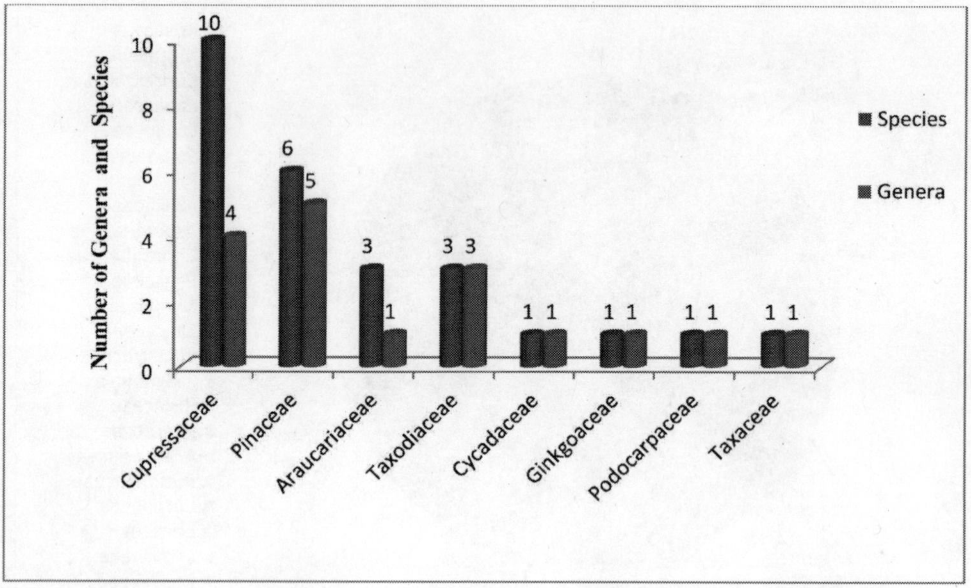

Fig. 16. Predominant Gymnospermous Families of Heritage Sites.

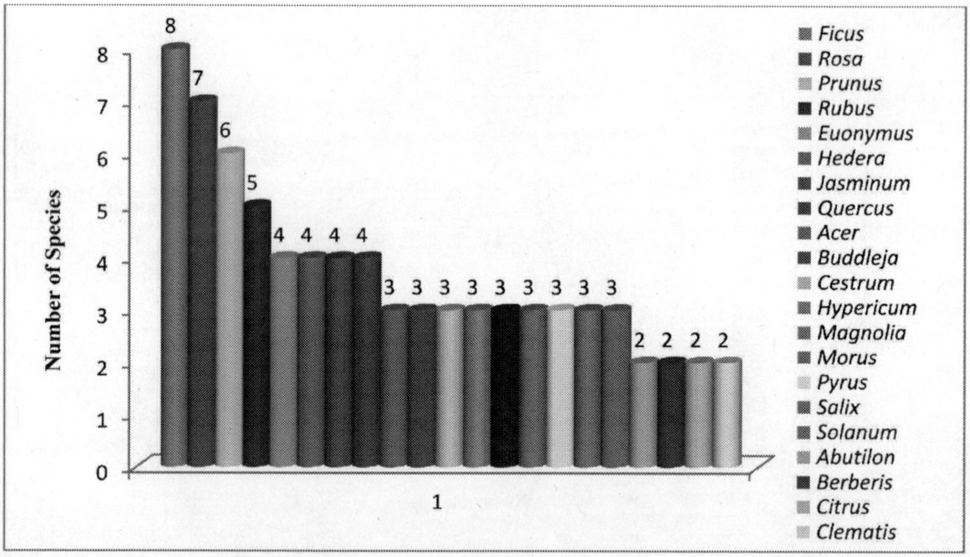

Fig. 17. Predominant Dicot Genera of Heritage Sites.

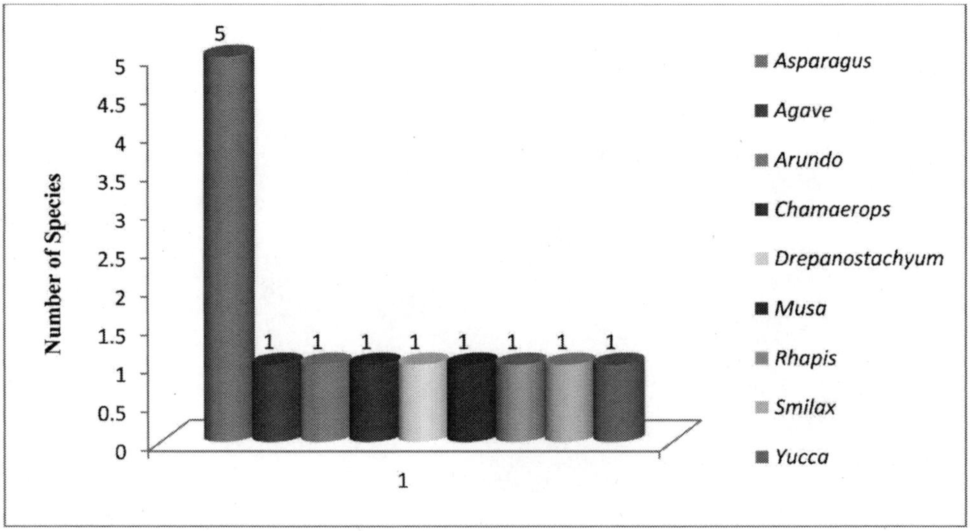

Fig. 18. Predominant Monocotyledonous Genera of Heritage Sites.

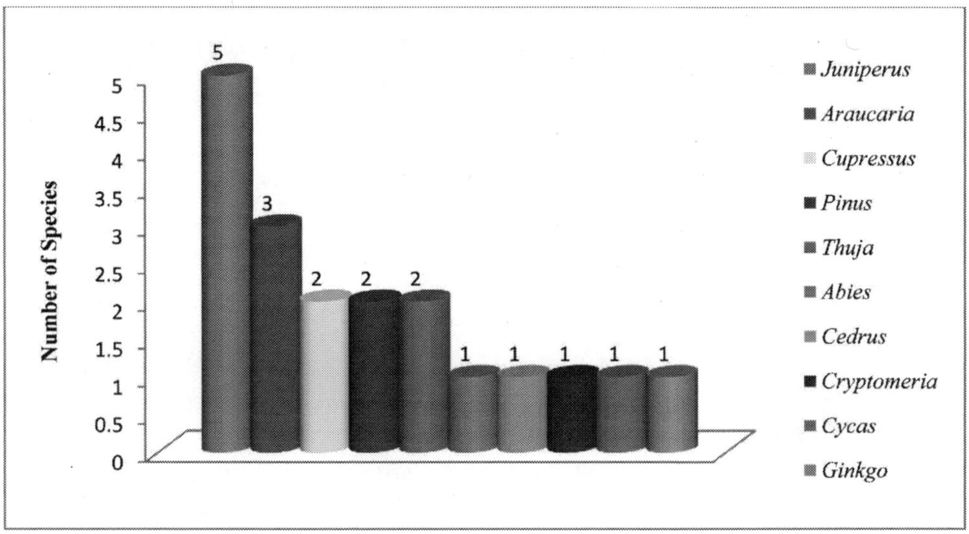

Fig. 19. Predominant Gymnospermous Genera of Heritage Sites.

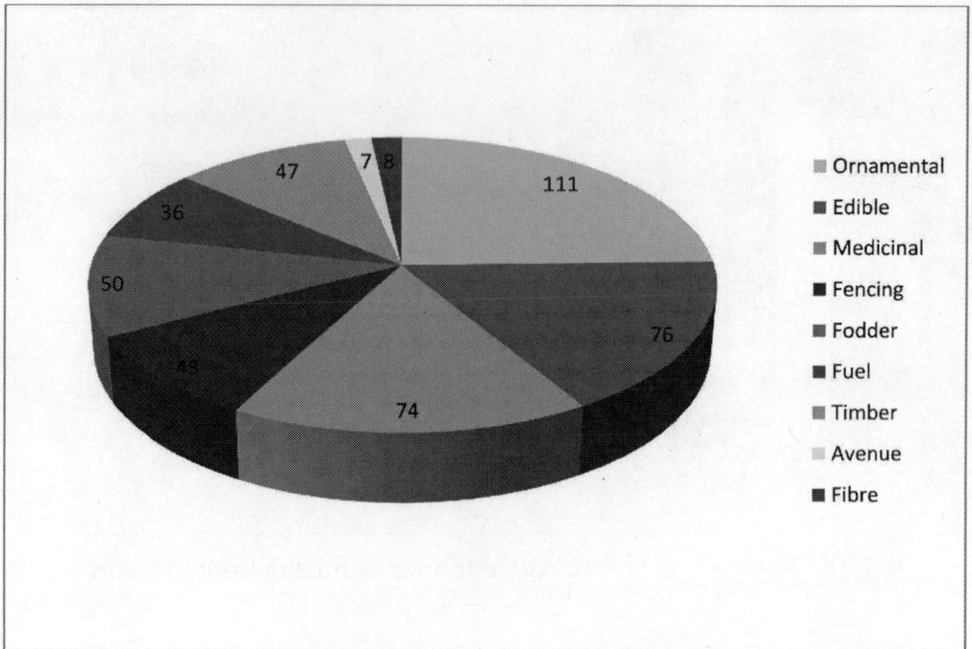

Fig. 20. Pie Diagram Showing Relative Number of Plants of Heritage Sites Used for Various Purposes.

ETHNOBOTANICAL ASPECTS OF FLORA

As many as 74 species belonging to 58 genera and 40 families, and covering 172 prescriptions for 63 human ailments, have been recorded presently **(Tables XX, XXII; Appendix III**). Of these, 48 occur in wild state; 26 species such as *Aloysia citriodora*, *Citrus limon*, *Citrus maxima*, *Cunninghamia lanceolata*, *Eriobotrya japonica*, *Eucalyptus globulus*, *Ginkgo biloba*, *Hibiscus rosa-sinensis*, *Jacaranda acutifolia*, *Jasminum grandiflorum*, *Jasminum multiflorum*, *Juniperus squamata*, *Lagerstroemia indica*, *Nerium oleander*, *Populus ciliata*, *Prunus amygdalus*, *Prunus persica*, *Pyrus communis*, *Robinia pseudoacacia*, *Rosa chinensis*, *Rosa damascena*, *Rosa multiflora*, *Salix alba*, *Salix babylonica*, *Sambucus nigra* and *Thuja orientalis* are grown (**Table XXI**) whereas *Agave americana*, *Buddleja asiatica*, *Ficus religiosa*, *Hedera helix*, *Hypericum oblongifolium*, *Jasminum humile*, *Opuntia dillenii*, *Punica granatum*, *Taxus baccata* and *Zanthoxylum armatum* occur in both wild as well as cultivated state. For curing various ailments, leaves have been predominantly employed (27 times), followed by fruits (16 times), flowers (13 times), roots (11 times), bark (9 times), seeds (8 times), whole plants (4 times), latex (2 times), twigs (3 times), wood (3 times), aerial parts, gall, rhizome, root-bark, stem (once each) (**Fig. 21**). Plants occurring at heritage sites are used for various ailments such as **skin diseases** (15 species: *Buddleja asiatica*, *Girardinia diversifolia*, *Hedera nepalensis*, *Jasminum grandiflorum*, *Jasminum humile*, *Juniperus squamata*, *Lyonia ovalifolia*, *Nerium oleander*, *Prunus amygdalus*, *Rhododendron arboreum*, *Ricinus communis*, *Rubia cordifolia*, *Salix alba*, *Salix babylonica*, *Xanthium strumarium*), **dental problems** (10 species: *Cerasus cerasoides*, *Desmodium multiflorum*, *Ficus religiosa*, *Jasminum grandiflorum*, *Jasminum humile*, *Juglans regia*, *Ricinus communis*, *Robinia pseudoacacia*, *Salix pycnostachya* var. *oxycarpa*, *Zanthoxylum armatum*), **wounds** (9 species: *Cunninghamia lanceolata*, *Hypericum oblongifolium*, *Hypericum patulum*, *Jacaranda acutifolia*, *Jasminum humile*, *Lagerstroemia indica*, *Leea asiatica*, *Rosa multiflora*, *Thuja orientalis*), **cough** (7 species: *Ficus palmata*, *Ficus religiosa*, *Pistacia khinjuk*, *Punica granatum*, *Solanum virginianum*, *Taxus baccata*, *Zanthoxylum armatum*), **suppuration of boils** (7 species: *Daphne papyracea*, *Hibiscus rosa-sinensis*, *Jasminum multiflorum*, *Juniperus squamata*, *Prinsepia utilis*, *Ricinus communis*, *Sarcococca pruniformis*), **mouth sores/ulcers** (6 species: *Eucalyptus globulus*, *Ficus religiosa*, *Jasminum grandiflorum*, *Pyrus pashia*, *Solanum virginianum*, *Zanthoxylum armatum*), **constipation** (5 species: *Euonymus tingens*, *Ficus palmata*, *Prunus amygdalus*, *Ricinus communis*, *Rosa damascena*), **hair care** (5 species: *Datura stramonium*, *Hedera nepalensis*, *Hibiscus rosa-sinensis*, *Pyrus pashia*, *Ricinus communis*), **pains and aches** (5 species: *Opuntia dillenii*, *Pistacia khinjuk*, *Prunus amygdalus*, *Sarcococca pruniformis*, *Smilax aspera*), **refrigerant** (5 species: *Citrus maxima*, *Malvaviscus arboreus* var. *mexicanus*, *Punica granatum*, *Rhododendron arboreum*, *Rubus niveus*), **rheumatic pains** (5 species: *Citrus limon*, *Girardinia diversifolia*, *Prinsepia utilis*, *Prunus persica*, *Ricinus communis*), **anthelmintic** (4 species: *Aesculus indica*, *Myrsine africana*, *Ricinus communis*, *Rubia cordifolia*), **blood purifier** (4 species: *Myrica esculenta*, *Myrsine africana*, *Populus ciliata*, *Zanthoxylum armatum*), **boils** (4 species: *Datura stramonium*, *Ficus auriculata*, *Hypericum oblongifolium*, *Hypericum patulum*), **headaches** (4 species: *Girardinia diversifolia*, *Nerium oleander*, *Salix pycnostachya* var. *oxycarpa*, *Solanum virginianum*), **antidote** (3 species: *Datura stramonium*, *Ficus religiosa*, *Rubus niveus*), **asthma** (3 species: *Datura stramonium*, *Opuntia dillenii*, *Pistacia khinjuk*), **astringent** (3 species: *Pyrus communis*, *Rhododendron arboreum*, *Rosa moschata*), **burns** (3 species: *Cunninghamia lanceolata*, *Prunus amygdalus*, *Sambucus nigra*), **cuts** (3 species: *Hypericum patulum*, *Lagerstroemia indica*, *Leea asiatica*), **fever** (3 species: *Indigofera dosua*, *Smilax aspera*, *Solanum virginianum*), **laxative** (3 species: *Myrsine africana*, *Pyrus communis*, *Rubia cordifolia*), **respiratory problems**

(3 species: *Eucalyptus globulus*, *Pistacia khinjuk*, *Taxus baccata*), **swelling** (3 species: *Girardinia diversifolia*, *Rubia cordifolia*, *Xanthium strumarium*), **ulcer** (3 species: *Cunninghamia lanceolata*, *Jasminum humile*, *Xanthium strumarium*), **bone fracture** (2 species: *Agave americana*, *Cerasus cerasoides*), **cold** (2 species: *Taxus baccata*, *Zanthoxylum armatum*), **coolant** (2 species: *Pyrus pashia*, *Xanthium strumarium*), **dysentery** (2 species: *Asparagus racemosus*, *Pyracantha crenulata*), **eye diseases** (2 species: *Berberis aristata*, *Euonymus tingens*), **gastro-intestinal disorders** (2 species: *Inula cappa*, *Smilax aspera*), **inflammation** (2 species: *Nerium oleander*, *Xanthium strumarium*), **menstruation complaints** (2 species: *Asparagus racemosus*, *Rubia cordifolia*), **nose bleeding** (2 species: *Pyrus pashia*, *Rhododendron arboreum*), **piles** (2 species: *Jasminum grandiflorum*, *Punica granatum*), **scalds** (2 species: *Cunninghamia lanceolata*, *Sambucus nigra*), **aphrodisiac** (1 species: *Prunus amygdalus*) **arthritis** (1 species: *Aesculus indica*), **breast cancer** (1 species: *Taxus baccata*), **cardiovascular disorders** (1 species: *Ginkgo biloba*), **carminative** (1 species: *Myrica esculenta*), **catarrh** (1 species: *Sambucus nigra*), **chest infection** (1 species: *Solanum virginianum*), **diabetes** (1 species: *Thuja orientalis*), **digestion (vet.)** (1 species: *Myrsine africana*), **diuretic** (1 species: *Xanthium strumarium*), **emollient** (1 species: *Salix babylonica*), **gargle** (1 species: *Sambucus nigra*), **gonorrohea** (1 species: *Sarcococca pruniformis*), **heart complaints** (1 species: *Rhododendron arboreum*), **indigestion** (1 species: *Viburnum mullaha*), **injuries** (1 species: *Opuntia dillenii*), **leucorrhoea** (1 species: *Rosa damascena*), **miscarriage** (1 species: *Asparagus racemosus*), **mouth odour** (1 species: *Zanthoxylum armatum*), **mouth wash** (1 species: *Sambucus nigra*), **nail disorders** (1 species: *Ilex dipyrena*), **smallpox** (1 species: *Xanthium strumarium*), **sores** (1 species: *Datura stramonium*), **spasm** (1 species: *Taxus baccata*), **sprain** (1 species: *Clematis buchananiana*), **tonsil** (1 species: *Desmodium elegans*), **vomiting** (1 species: *Citrus maxima*) (**Appendix III**). From the view point of plants employed for cure, high used families are Rosaceae (8 genera, 14 species), Fabaceae (3 genera, 4 species), Salicaceae (2 genera, 4 species), Moraceae (1 genus, 3 species), Oleaceae (1 genus, 3 species), Rutaceae (2 genera, 3 species), Araliaceae (1 genus, 2 species), Asteraceae (2 genera, 2 species), Caprifoliaceae (2genera, 2 species), Clusiaceae (1 genus, 2 species), Cupressaceae (2 genera, 2 species), Ericaceae (2 genera, 2 species), Lythraceae (2 genera, 2 species), Solanaceae (2 genera, 2 species) and the remaining families such as Agavaceae, Anacardiaceae, Apocyanaceae, Aquifoliaceae, Berberidaceae, Bignoniaceae, Buddlejaceae, Buxaceae, Cactaceae, Celastraceae, Euphorbiaceae, Ginkgoaceae, Hippocastanaceae, Leeaceae, Malvaceae, Myricaceae, Myrsinaceae, Myrtaceae, Ranunculaceae, Rubiaceae, Smilacaceae, Taxaceae, Taxodiaceae, Thymelaeaceae, Urticaceae and Verbenaceae represented by one species only (**Table XXII**).

Undeniably, the above medicinal plant diversity offers great potential for pharmaceutical industries and serve as an excellent source for a variety of lead molecules.

TABLE XX
Medicinal Plants of Heritage Sites

S. No.	Botanical Name	Family
1.	*Aesculus indica*	Hippocastanaceae
2.	*Agave americana**	Agavaceae
3.	*Aloysia citriodora*	Verbenaceae
4.	*Berberis aristata*	Berberidaceae
5.	*Buddleja asiatica*	Buddlejaceae
6.	*Cerasus cerasoides**	Rosaceae
7.	*Citrus limon**	Rutaceae
8.	*Citrus maxima*	Rutaceae
9.	*Clematis buchananiana**	Ranunculaceae
10.	*Cunninghamia lanceolata**	Taxodiaceae
11.	*Daphne papyracea**	Thymelaeaceae
12.	*Datura stramonium**	Solanaceae
13.	*Desmodium elegans**	Fabaceae
14.	*Desmodium multiflorum**	Fabaceae
15.	*Eucalyptus globulus*	Myrtaceae
16.	*Eriobotrya japonica**	Rosaceae
17.	*Euonymus tingens*	Celastraceae
18.	*Ficus auriculata**	Moraceae
19.	*Ficus palmata**	Moraceae
20.	*Ficus religiosa**	Moraceae
21.	*Ginkgo biloba*	Ginkgoaceae
22.	*Girardinia diversifolia*	Urticaceae
23.	*Hedera helix*	Araliaceae
24.	*Hedera nepalensis*	Araliaceae
25.	*Hibiscus rosa-sinensis**	Malvaceae
26.	*Hypericum oblongifolium*	Clusiaceae
27.	*Hypericum patulum**	Clusiaceae
28.	*Ilex dipyrena**	Aquifoliaceae
29.	*Indigofera dosua**	Fabaceae
30.	*Inula cappa*	Asteraceae
31.	*Jacaranda acutifolia**	Bignoniaceae
32.	*Jasminum grandiflorum*	Oleaceae

S. No.	Botanical Name	Family
33.	*Jasminum humile*	Oleaceae
34.	*Jasminum multiflorum*	Oleaceae
35.	*Juniperus squamata*	Cupressaceae
36.	*Lagerstroemia indica**	Lythraceae
37.	*Leea asiatica*	Leeaceae
38.	*Lyonia ovalifolia*	Ericaceae
39.	*Malvaviscus arboreus* var. *mexicanus**	Malvaceae
40.	*Myrica esculenta*	Myricaceae
41.	*Myrsine africana*	Myrsinaceae
42.	*Nerium oleander**	Apocyanaceae
43.	*Opuntia dillenii*	Cactaceae
44.	*Pistacia khinjuk*	Anacardiaceae
45.	*Populus ciliata*	Salicaceae
46.	*Prinsepia utilis*	Rosaceae
47.	*Prunus amygdalus*	Rosaceae
48.	*Prunus persica*	Rosaceae
49.	*Punica granatum**	Lythraceae
50.	*Pyracantha crenulata**	Rosaceae
51.	*Pyrus communis**	Rosaceae
52.	*Pyrus pashia**	Rosaceae
53.	*Robinia pesudoacacia**	Fabaceae
54.	*Rhododendron arboreum**	Ericaceae
55.	*Ricinus communis**	Euphorbiaceae
56.	*Rosa chinensis*	Rosaceae
57.	*Rosa damascena**	Rosaceae
58.	*Rosa moschata**	Rosaceae
59.	*Rosa multiflora*	Rosaceae
60.	*Rubia cordifolia**	Rubiaceae
61.	*Rubus ellipticus**	Rosaceae
62.	*Rubus niveus**	Rosaceae
63.	*Salix alba**	Salicaceae
64.	*Salix babylonica**	Salicaceae
65.	*Salix pycnostachya* var. *oxycarpa**	Salicaceae
66.	*Sambucus nigra*	Caprifoliaceae
67.	*Sarcococca pruniformis**	Buxaceae
68.	*Smilax aspera**	Smilacaceae
69.	*Solanum virginianum**	Solanaceae
70.	*Taxus baccata*	Taxaceae

S. No.	Botanical Name	Family
71.	*Thuja orientalis**	Cupressaceae
72.	*Viburnum mullaha**	Caprifoliaceae
73.	*Xanthium strumarium**	Asteraceae
74.	*Zanthoxylum armatum**	Rutaceae

***Additional Uses**

TABLE XXI
Cultivated Plants of Heritage Sites Used for Medicinal Purposes

S. No.	Botanical Name	Family
1.	*Aloysia citriodora*	Verbenaceae
2.	*Citrus limon**	Rutaceae
3.	*Citrus maxima*	Rutaceae
4.	*Cunninghamia lanceolata**	Taxodiaceae
5.	*Eriobotrya japonica**	Rosaceae
6.	*Eucalyptus globulus*	Myrtaceae
7.	*Ginkgo biloba*	Ginkgoaceae
8.	*Hibiscus rosa-sinensis**	Malvaceae
9.	*Jacaranda acutifolia**	Bignoniaceae
10.	*Jasminum grandiflorum*	Oleaceae
11.	*Jasminum multiflorum*	Oleaceae
12.	*Juniperus squamata*	Cupressaceae
13.	*Lagerstroemia indica**	Lythraceae
14.	*Nerium oleander**	Apocyanaceae
15.	*Populus ciliata*	Salicaceae
16.	*Prunus amygdalus*	Rosaceae
17.	*Prunus persica*	Rosaceae
18.	*Pyrus communis**	Rosaceae
19.	*Robinia pesudoacacia**	Fabaceae
20.	*Rosa chinensis*	Rosaceae
21.	*Rosa damascena**	Rosaceae
22.	*Rosa multiflora*	Rosaceae
23.	*Salix alba**	Salicaceae
24.	*Salix babylonica**	Salicaceae
25.	*Sambucus nigra*	Caprifoliaceae
26.	*Thuja orientalis**	Cupressaceae

***Additional Uses**

TABLE XXII
Medicinally Important Families of Heritage Sites

S. No.	Botanical Name	Family	Genera/ Species
1.	Agavaceae	*Agave americana*	1/1
2.	Anacardiaceae	*Pistacia khinjuk*	1/1
3.	Apocyanaceae	*Nerium oleander*	1/1
4.	Aquifoliaceae	*Ilex dipyrena*	1/1
5.	Araliaceae	*Hedera helix, Hedera nepalensis*	1/2
6.	Asteraceae	*Inula cappa, Xanthium strumarium*	2/2
7.	Berberidaceae	*Berberis aristata*	1/1
8.	Bignoniaceae	*Jacaranda acutifolia*	1/1
9.	Buddlejaceae	*Buddleja asiatica*	1/1
10.	Buxaceae	*Sarcococca pruniformis*	1/1
11.	Cactaceae	*Opuntia dillenii*	1/1
12.	Caprifoliaceae	*Sambucus nigra, Viburnum mullaha*	2/2
13.	Celastraceae	*Euonymus tingens*	1/1
14.	Clusiaceae	*Hypericum oblongifolium, Hypericum patulum*	1/2
15.	Cupressaceae	*Juniperus squamata, Thuja orientalis*	2/2
16.	Ericaceae	*Lyonia ovalifolia, Rhododendron arboreum*	2/2
17.	Euphorbiaceae	*Ricinus communis*	1/1
18.	Fabaceae	*Desmodium elegans, Desmodium multiflorum, Indigofera dosua, Robinia pesudoacacia*	3/4
19.	Ginkgoaceae	*Ginkgo biloba*	1/1
20.	Hippocastanaceae	*Aesculus indica*	1/1
21.	Leeaceae	*Leea asiatica*	1/1
22.	Lythraceae	*Lagerstroemia indica*, *Punica granatum*	2/2
23.	Malvaceae	*Hibiscus rosa-sinensis*, *Malvaviscus arboreus* var. *mexicanus*	2/2
24.	Moraceae	*Ficus auriculata*, *Ficus palmata*, *Ficus religiosa*	1/3
25.	Myricaceae	*Myrica esculenta*	1/1
26.	Myrsinaceae	*Myrsine africana*	1/1
27.	Myrtaceae	*Eucalyptus globulus*	1/1
28.	Oleaceae	*Jasminum grandiflorum*, *Jasminum humile*, *Jasminum multiflorum*	1/3
29.	Ranunculaceae	*Clematis buchananiana*	1/1

S. No.	Botanical Name	Family	Genera/ Species
30.	Rosaceae	*Cerasus cerasoides, Eriobotrya japonica, Prinsepia utilis, Prunus amygdalus, Prunus persica, Pyracantha crenulata, Pyrus communis, Pyrus pashia, Rosa chinensis, Rosa damascena, Rosa moschata, Rosa multiflora, Rubus ellipticus, Rubus niveus*	8/14
31.	Rubiaceae	*Rubia cordifolia*	1/1
32.	Rutaceae	*Citrus limon, Citrus maxima, Zanthoxylum armatum*	2/3
33.	Salicaceae	*Populus ciliata, Salix alba, Salix babylonica, Salix pycnostachya* var. *oxycarpa*	2/4
34.	Smilacaceae	*Smilax aspera*	1/1
35.	Solanaceae	*Datura stramonium, Solanum virginianum*	2/2
36.	Taxaceae	*Taxus baccata*	1/1
37.	Taxodiaceae	*Cunninghamia lanceolata*	1/1
38.	Thymelaeaceae	*Daphne papyracea*	1/1
39.	Urticaceae	*Girardinia diversifolia*	1/1
40.	Verbenaceae	*Aloysia citriodora*	1/1

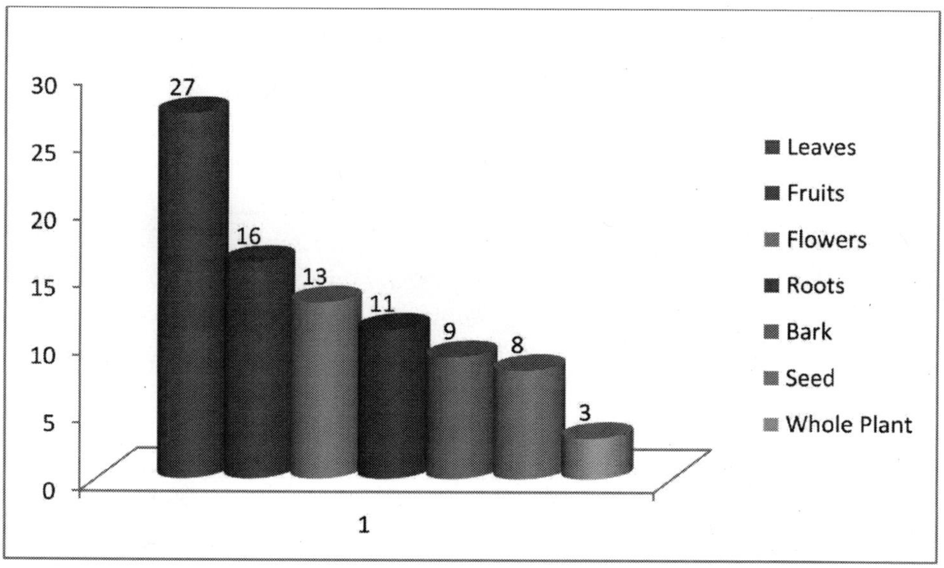

Fig. 21. Histogram Showing Relative Number of Plant Parts Used for Medicinal Purposes.

From the view-point of forages, altogether 50 species under 37 genera of 25 families have been recorded (**Tables XXIII, XXIV**). Species richness predominantly belongs to Moraceae (2 genera, 7 species), Rosaceae (6 genera, 7 species), Fabaceae (3 genera, 4 species), Caprifoliaceae (2 genera, 3 species), Fagaceae (1 genus, 3 species), Salicaceae (2 genera, 3 species), Berberidaceae (1 genus, 2 species), Rhamnaceae (1 genus, 2 species), Ulmaceae (2 genera, 2 species) and Urticaceae (2 genera, 2 species) etc. High used genera commonly employed for fodder purposes are *Ficus* (4), *Morus* (3), *Quercus* (3), *Berberis (*2), *Desmodium* (2), *Populus* (2), *Rhamnus* (2), *Rubus* (2) and *Viburnum* (2), etc (**Table XXIV**). Of these, 6 taxa like *Buxus wallichiana* (Buxaceae), *Morus nigra* (Moraceae), *Populus ciliata* (Salicaceae), *Populus nigra* (Salicaceae), *Robinia pseudoacacia* (Fabaceae) and *Salix babylonica* (Salicaceae) are cultivated for forages. As many as 8 plants are employed for cordages (**Table XXVII**). Of the native ornamentals comprising 21 species, Moraceae (3 species), Oleaceae (3 species), Apocynaceae (2 species), Clusiaceae (2 species), Cupressaceae (2 species) and Rosaceae (2 species) constitute its predominant families **(Tables LXXIV, LXXV)** whereas the high used genera are: *Ficus*, *Jasminum*, *Cupressus*, *Hypericum* and *Rosa* (**Table LXXVI**).

Regarding diversity for timber purposes, 47 species belonging to 38 genera under 29 families have been employed (**Table XXV**). These are *Abies pindrow*, *Acer oblongum*, *Aesculus indica*, *Araucaria cunninghamii, Buxus wallichiana*, *Cedrus deodara*, *Celtis australis*, *Cornus macrophylla*, *Cotoneaster bacillaris*, *Cryptomeria japonica*, *Cunninghamia lanceolata*, *Cupressus cashmeriana*, *Cupressus torulosa*, *Diospyros kaki*, *Drepanostachyum falcatum*, *Erythrina arborescens*, *Eucalyptus globulus*, *Euonymus fimbriatus*, *Euonymus tingens*, *Fagus sylvatica*, *Grevillea robusta*, *Ilex dipyrena*, *Jacaranda acutifolia*, *Juglans regia*, *Lonicera quinquelocularis*, *Morus alba*, *Morus nigra*, *Morus serrata*, *Picea smithiana*, *Pinus roxburghii*, *Pinus wallichiana*, *Pistacia khinjuk*, *Populus ciliata*, *Populus nigra*, *Pyrus pashia*, *Quercus floribunda*, *Quercus incana*, *Quercus oblongata*, *Rhamnus persica*, *Rhamnus virgatus*, *Rhododendron arboreum*, *Robinia pseudoacacia*, *Salix babylonica*, *Skimmia laureola*, *Taxodium distichum*, *Taxus baccata*, *Ulmus wallichiana* (**Table XXV**). High used families for this purpose are Pinaceae (5 species), Fagaceae (3 species), Moraceae (3 species), Salicaceae (3 species), Taxodiaceae (3 species) etc (**Table XXVI)**. Moreover, wood is also being used for agricultural implements (14 species; *Buxus wallichiana*, *Celtis australis*, *Cornus macrophylla*, *Cotoneaster bacillaris*, *Drepanostachyum falcatum*, *Fagus sylvatica*, *Pyrus pashia*, *Quercus floribunda*, *Quercus incana*, *Quercus oblongata*, *Rhamnus persica*, *Rhamnus virgatus*, *Rhododendron arboreum*, *Robinia pseudoacacia*), building works (26 species;

Abies pindrow, *Acer oblongum*, *Aesculus indica*, *Cedrus deodara*, *Cornus macrophylla*, *Cryptomeria japonica*, *Cunninghamia lanceolata*, *Cupressus cashmeriana*, *Cupressus torulosa*, *Erythrina arborescens*, *Eucalyptus globulus*, *Euonymus fimbriatus*, *Euonymus tingens*, *Fagus sylvatica*, *Jacaranda acutifolia*, *Picea smithiana*, *Pinus roxburghii*, *Pinus wallichiana*, *Pistacia khinjuk*, *Populus ciliata*, *Populus nigra*, *Quercus floribunda*, *Quercus incana*, *Quercus oblongata*, *Salix babylonica*, *Taxodium distichum*), carving (1 species; *Euonymus fimbriatus*), crate (1 species; *Picea smithiana*), furniture (25 species; *Abies pindrow*, *Araucaria cunninghamii*, *Cedrus deodara*, *Cornus macrophylla*, *Cryptomeria japonica*, *Cunninghamia lanceolata*, *Diospyros kaki*, *Fagus sylvatica*, *Grevillea robusta*, *Ilex dipyrena*, *Jacaranda acutifolia*, *Juglans regia*, *Morus alba*, *Morus nigra*, *Morus serrata*, *Pinus roxburghii*, *Pinus wallichiana*, *Pistacia khinjuk*, *Populus nigra*, *Quercus oblongata*, *Robinia pseudoacacia*, *Salix babylonica*, *Taxodium distichum*, *Ulmus wallichiana*, *Taxus baccata*), house interiors (2 species; *Abies pindrow*, *Picea smithiana*), plywood (2 species; *Grevillea robusta*, *Populus nigra*), roofing (2 species; *Arundo donax*, *Picea smithiana*), tool handles (5 species; *Lonicera quinquelocularis*, *Jacaranda acutifolia*, *Quercus incana*, *Rhododendron arboreum*, *Skimmia laureola*) and utensils (1 species; *Cryptomeria japonica*) (**Table XXIX**). Other aspects of material culture with plants used against each respective category have also been listed in **Table XXIX**. Likewise, use of 36 species (31 genera, 22 families) with predominance of Fabaceae, Rosaceae (4 species each), Fagaceae, Moraceae, Pinaceae (3 species each), Berberidaceae, Caprifoliaceae (2 species each), Aceraceae, Celastraceae, Coriariaceae, Cornaceae, Cupressaceae, Elaeagnaceae, Ericaceae, Lamiaceae, Myrsinaceae, Myrtaceae, Rhamnaceae, Salicaceae, Ulmaceae, Urticaceae and Verbenaceae (1 species each) for fuelwood have also been highlighted (**Table XXVIII**). Availability of such a varied plant genetic resources is important to support the present day crop improvement programmes to meet the needs and aspiration of future generations.

TABLE XXIII
Fodder Plants of Heritage Sites

S. No.	Botanical Name	Family
1.	*Aesculus indica*	Hippocastanaceae
2.	*Berberis aristata*	Berberidaceae
3.	*Berberis lycium*	Berberidaceae
4.	*Buddleja crispa*	Buddlejaceae
5.	*Buxus wallichiana*	Buxaceae
6.	*Celtis australis*	Ulmaceae

S. No.	Botanical Name	Family
7.	*Cerasus cerasoides**	Rosaceae
8.	*Colquhounia coccinea*	Lamiaceae
9.	*Cornus macrophylla*	Cornaceae
10.	*Cotoneaster bacillaris* *	Rosaceae
11.	*Debregeasia saeneb*	Urticaceae
12.	*Desmodium elegans*	Fabaceae
13.	*Desmodium multiflorum**	Fabaceae
14.	*Drepanostachyum falcatum*	Poaceae
15.	*Ficus auriculata*	Moraceae
16.	*Ficus palmata*	Moraceae
17.	*Ficus sarmentosa*	Moraceae
18.	*Ficus subincisa*	Moraceae
19.	*Girardinia diversifolia*	Urticaceae
20.	*Hedera helix*	Araliaceae
21.	*Himalrandia tetrasperma*	Rubiaceae
22.	*Hypericum oblongifolium**	Clusiaceae
23.	*Ilex dipyrena*	Aquifoliaceae
24.	*Indigofera dosua*	Fabaceae
25.	*Leea asiatica**	Leeaceae
26.	*Lonicera quinquelocularis*	Caprifoliaceae
27.	*Morus alba*	Moraceae
28.	*Morus nigra*	Moraceae
29.	*Morus serrata*	Moraceae
30.	*Myrica esculenta**	Myricaceae
31.	*Myrsine africana*	Myrsinaceae
32.	*Padus cornuta**	Rosaceae
33.	*Parthenocissus semicordata*	Vitaceae
34.	*Pistacia khinjuk*	Anacardiaceae
35.	*Populus ciliata*	Salicaceae
36.	*Populus nigra*	Salicaceae
37.	*Pyrus pashia*	Rosaceae
38.	*Quercus floribunda*	Fagaceae
39.	*Quercus incana*	Fagaceae
40.	*Quercus oblongata*	Fagaceae
41.	*Rhamnus persica*	Rhamnaceae
42.	*Rhamnus virgatus*	Rhamnaceae
43.	*Robinia pseudoacacia**	Fabaceae
44.	*Rosa moschata*	Rosaceae

S. No.	Botanical Name	Family
45.	*Rubus ellipticus*	Rosaceae
46.	*Rubus niveus*	Rosaceae
47.	*Salix babylonica*	Salicaceae
48.	*Ulmus wallichiana*	Ulmaceae
49.	*Viburnum cotinifolium**	Caprifoliaceae
50.	*Viburnum mullaha**	Caprifoliaceae

***Additional Uses**

TABLE XXIV
Families and Genera/Species Ratio of Fodder Plants

S. No.	Family	Botanical Name	Genus/ Species
1.	Anacardiaceae	*Pistacia khinjuk*	1/1
2.	Aquifoliaceae	*Ilex dipyrena*	1/1
3.	Araliaceae	*Hedera helix*	1/1
4.	Berberidaceae	*Berberis aristata, Berberis lycium*	1/2
5.	Buddlejaceae	*Buddleja crispa*	1/1
6.	Buxaceae	*Buxus wallichiana*	1/1
7.	Caprifoliaceae	*Lonicera quinquelocularis, Viburnum cotinifolium, Viburnum mullaha*	2/3
8.	Clusiaceae	*Hypericum oblongifolium*	1/1
9.	Cornaceae	*Cornus macrophylla*	1/1
10.	Fabaceae	*Desmodium elegans, Desmodium multiflorum, Indigofera dosua, Robinia pseudoacacia*	3/4
11.	Fagaceae	*Quercus floribunda, Quercus incana, Quercus oblongata*	1/3
12.	Hippocastanaceae	*Aesculus indica*	1/1
13.	Lamiaceae	*Colquhounia coccinea*	1/1
14.	Leeaceae	*Leea asiatica*	1/1
15.	Moraceae	*Ficus auriculata, Ficus palmata, Ficus sarmentosa, Ficus subincisa, Morus alba, Morus nigra, Morus serrata*	2/7
16.	Myricaceae	*Myrica esculenta*	1/1
17.	Myrsinaceae	*Myrsine africana*	1/1
18.	Poaceae	*Drepanostachyum falcatum*	1/1
19.	Rhamnaceae	*Rhamnus persica, Rhamnus virgatus*	1/2

S. No.	Family	Botanical Name	Genus/ Species
20.	Rosaceae	*Cerasus cerasoides*, *Cotoneaster bacillaris*, *Padus cornuta*, *Pyrus pashia*, *Rosa moschata*, *Rubus ellipticus*, *Rubus niveus*	6/7
21.	Rubiaceae	*Himalrandia tetrasperma*	1/1
22.	Salicaceae	*Populus ciliata*, *Populus nigra*, *Salix babylonica*	2/3
23.	Ulmaceae	*Celtis australis*, *Ulmus wallichiana*	2/2
24.	Urticaceae	*Debregeasia saeneb*, *Girardinia diversifolia*	2/2
25.	Vitaceae	*Parthenocissus semicordata*	1/1

TABLE XXV
Timber Plants of Heritage Sites

S. No.	Botanical Name	Family
1.	*Abies pindrow*	Pinaceae
2.	*Acer oblongum*	Aceraceae
3.	*Aesculus indica*	Hippocastanaceae
4.	*Araucaria cunninghamii*	Araucariaceae
5.	*Buxus wallichiana*	Buxaceae
6.	*Cedrus deodara*	Pinaceae
7.	*Celtis australis*	Ulmaceae
8.	*Cornus macrophylla**	Cornaceae
9.	*Cotoneaster bacillaris*	Rosaceae
10.	*Cryptomeria japonica*	Taxodiaceae
11.	*Cunninghamia lanceolata*	Taxodiaceae
12.	*Cupressus cashmeriana*	Cupressaceae
13.	*Cupressus torulosa*	Cupressaceae
14.	*Diospyros kaki*	Ebenaceae
15.	*Drepanostachyum falcatum*	Poaceae
16.	*Erythrina arborescens*	Fabaceae
17.	*Eucalyptus globulus*	Myrtaceae
18.	*Euonymus fimbriatus*	Celastraceae
19.	*Euonymus tingens*	Celastraceae
20.	*Fagus sylvatica*	Fagaceae
21.	*Grevillea robusta*	Proteaceae
22.	*Ilex dipyrena*	Aquifoliaceae
23.	*Jacaranda acutifolia*	Bignoniaceae

S. No.	Botanical Name	Family
24.	*Juglans regia*	Juglandaceae
25.	*Lonicera quinquelocularis*	Caprifoliaceae
26.	*Morus alba*	Moraceae
27.	*Morus nigra*	Moraceae
28.	*Morus serrata*	Moraceae
29.	*Picea smithiana*	Pinaceae
30.	*Pinus roxburghii*	Pinaceae
31.	*Pinus wallichiana*	Pinaceae
32.	*Pistacia khinjuk*	Anacardiaceae
33.	*Populus ciliata*	Salicaceae
34.	*Populus nigra*	Salicaceae
35.	*Pyrus pashia*	Rosaceae
36.	*Quercus floribunda*	Fagaceae
37.	*Quercus incana*	Fagaceae
38.	*Quercus oblongata*	Fagaceae
39.	*Rhamnus persica**	Rhamnaceae
40.	*Rhamnus virgatus**	Rhamnaceae
41.	*Rhododendron arboreum*	Ericaceae
42.	*Robinia pseudoacacia*	Fabaceae
43.	*Salix babylonica*	Salicaceae
44.	*Skimmia laureola*	Rutaceae
45.	*Taxodium distichum*	Taxodiaceae
46.	*Taxus baccata*	Taxaceae
47.	*Ulmus wallichiana*	Ulmaceae

***Additional Uses**

TABLE XXVI
Families and Genera/Species Ratio of Timber Plants

S. No.	Family	Botanical Name	Genus/ Species
1.	Aceraceae	*Acer oblongum*	1/1
2.	Anacardiaceae	*Pistacia khinjuk*	1/1
3.	Aquifoliaceae	*Ilex dipyrena*	1/1
4.	Araucariaceae	*Araucaria cunninghamii*	1/1
5.	Bignoniaceae	*Jacaranda acutifolia*	1/1
6.	Buxaceae	*Buxus wallichiana*	1/1
7.	Caprifoliaceae	*Lonicera quinquelocularis*	1/1

S. No.	Family	Botanical Name	Genus/ Species
8.	Celastraceae	*Euonymus fimbriatus, Euonymus tingens*	1/2
9.	Cornaceae	*Cornus macrophylla*	1/1
10.	Cupressaceae	*Cupressus cashmeriana, Cupressus torulosa*	1/2
11.	Ebenaceae	*Diospyros kaki*	1/1
12.	Ericaceae	*Rhododendron arboreum*	1/1
13.	Fabaceae	*Erythrina arborescens, Robinia pseudoacacia*	2/2
14.	Fagaceae	*Fagus sylvatica*	1/1
15.	Fagaceae	*Quercus floribunda, Quercus incana, Quercus oblongata*	1/3
16.	Hippocastanaceae	*Aesculus indica*	1/1
17.	Juglandaceae	*Juglans regia*	1/1
18.	Moraceae	*Morus alba, Morus nigra, Morus serrata*	1/3
19.	Myrtaceae	*Eucalyptus globulus*	1/1
20.	Pinaceae	*Abies pindrow, Cedrus deodara, Picea smithiana, Pinus roxburghii, Pinus wallichiana*	4/5
21.	Poaceae	*Drepanostachyum falcatum*	1/1
22.	Proteaceae	*Grevillea robusta*	1/1
23.	Rhamnaceae	*Rhamnus persica, Rhamnus virgatus*	1/2
24.	Rosaceae	*Cotoneaster bacillaris, Pyrus pashia*	2/2
25.	Rutaceae	*Skimmia laureola*	1/1
26.	Salicaceae	*Populus ciliata, Populus nigra, Salix babylonica*	2/3
27.	Taxaceae	*Taxus baccata*	1/1
28.	Taxodiaceae	*Cryptomeria japonica, Cunninghamia lanceolata, Taxodium distichum*	3/3
29.	Ulmaceae	*Celtis australis, Ulmus wallichiana*	2/2

TABLE XXVII
Fibre Plants of Heritage Sites

S. No.	Botanical Name	Family
1.	*Agave americana*	Agavaceae
2.	*Chamaerops humilis*	Arecaceae
3.	*Clematis buchananiana**	Ranunculaceae
4.	*Debregeasia saeneb*	Urticaceae

S. No.	Botanical Name	Family
5.	*Girardinia diversifolia*	Urticaceae
6.	*Spartium junceum*	Fabaceae
7.	*Taxodium distichum*	Taxodiaceae
8.	*Yucca aloifolia*	Agavaceae

***Additional Uses**

TABLE XXVIII
Fuel Wood Plants of Heritage Sites

S. No.	Family	Botanical Name	Genus/ Species
1.	Aceraceae	*Acer oblongum*	1/1
2.	Berberidaceae	*Berberis aristata*, *Berberis lycium*	1/2
3.	Caprifoliaceae	*Lonicera quinquelocularis*, *Viburnum cotinifolium**	2/2
4.	Celastraceae	*Euonymus tingens*	1/1
5.	Coriariaceae	*Coriaria nepalensis*	1/1
6.	Cornaceae	*Cornus capitata**	1/1
7.	Cupressaceae	*Juniperus squamata*	1/1
8.	Elaeagnaceae	*Elaeagnus umbellata*	1/1
9.	Ericaceae	*Rhododendron arboreum*	1/1
10.	Fabaceae	*Acacia dealbata*, *Desmodium multiflorum**, *Erythrina arborescens**, *Robinia pseudoacacia*	4/4
11.	Fagaceae	*Quercus floribunda*, *Quercus incana*, *Quercus oblongata*	1/3
12.	Lamiaceae	*Colquhounia coccinea*	1/1
13.	Moraceae	*Morus alba*, *Morus nigra*, *Morus serrata*	1/3
14.	Myrsinaceae	*Myrsine africana*	1/1
15.	Myrtaceae	*Eucalyptus globulus*	1/1
16.	Pinaceae	*Abies pindrow*, *Picea smithiana*, *Pinus roxburghii*	3/3
17.	Rhamnaceae	*Rhamnus persica**	1/1
18.	Rosaceae	*Cerasus cerasoides*, *Padus cornuta**, *Prunus domestica**, *Pyrus pashia*	4/4
19.	Salicaceae	*Populus nigra*	1/1
20.	Ulmaceae	*Ulmus wallichiana*	1/1
21.	Urticaceae	*Debregeasia saeneb*	1/1
22.	Verbenaceae	*Duranta erecta**	1/1

***Additional Uses**

TABLE XXIX
Plants Used for Material Culture

Use	Species	Genera/ Species
Agricultural Implements	*Buxus wallichiana, Celtis australis, Cornus macrophylla**, *Cotoneaster bacillaris, Drepanostachyum falcatum, Fagus sylvatica, Pyrus pashia, Quercus floribunda, Quercus incana, Quercus oblongata, Rhamnus persica**, *Rhamnus virgatus**, *Rhododendron arboreum, Robinia pseudoacacia*	11/14
Baskets	*Arundo donax, Drepanostachyum falcatum, Salix alba**, *Salix babylonica, Yucca aloifolia*	4/5
Brooms	*Arundo donax**, *Drepanostachyum falcatum, Myrsine Africana**, *Pinus roxburghii**	4/4
Brush	*Agave americana, Yucca aloifolia*	2/2
Building Works	*Abies pindrow, Acer oblongum, Aesculus indica, Cedrus deodara, Cornus macrophylla, Cryptomeria japonica, Cunninghamia lanceolata, Cupressus cashmeriana, Cupressus torulosa, Erythrina arborescens, Eucalyptus globulus, Euonymus fimbriatus, Euonymus tingens, Fagus sylvatica, Jacaranda acutifolia, Picea smithiana, Pinus roxburghii, Pinus wallichiana, Pistacia khinjuk, Populus ciliata, Populus nigra, Quercus floribunda, Quercus incana, Quercus oblongata, Salix babylonica, Taxodium distichum*	20/26
Carving	*Euonymus fimbriatus*	1/1
Cattle Shed	*Drepanostachyum falcatum*	1/1
Cordage	*Agave americana, Clematis buchananiana**, *Debregeasia saeneb, Taxodium distichum*	4/4
Crate	*Picea smithiana*	1/1
Decorative Purposes	*Cycas revoluta**, *Inula cappa**, *Picea smithiana** *Pinus roxburghii**	4/4
Dye	*Juglans regia, Rubia cordifolia*	2/2
Fibre	*Agave americana, Chamaerops humilis, Clematis buchananiana**, *Debregeasia saeneb, Girardinia diversifolia, Spartium junceum, Taxodium distichum, Yucca aloifolia*	8/8
Field Umbrella	*Drepanostachyum falcatum*	1/1

Use	Species	Genera/ Species
Furniture	*Abies pindrow, Araucaria cunninghamii, Cedrus deodara, Cornus macrophylla, Cryptomeria japonica, Cunninghamia lanceolata, Diospyros kaki, Fagus sylvatica, Grevillea robusta, Ilex dipyrena, Jacaranda acutifolia, Juglans regia, Morus alba, Morus nigra, Morus serrata, Pinus roxburghii, Pinus wallichiana, Pistacia khinjuk, Populus nigra, Quercus oblongata, Robinia pseudoacacia, Salix babylonica, Taxodium distichum, Ulmus wallichiana, Taxus baccata*	22/25
Green Manure	*Grevillea robusta*	1/1
House Interiors	*Abies pindrow, Picea smithiana*	2/2
Match-stick	*Populus nigra*	1/1
Mats	*Arundo donax, Drepanostachyum falcatum, Spartium junceum,Yucca aloifolia*	4/4
Pen (Kalam)	*Arundo donax**, *Cerasus cerasoides**, *Drepanostachyum falcatum*	3/3
Plywood	*Grevillea robusta, Populus nigra*	2/2
Pulp	*Picea smithiana*	1/1
Roofing	*Arundo donax, Picea smithiana*	2/2
Rope	*Agave americana, Debregeasia saeneb, Spartium junceum, Yucca aloifolia*	4/4
Tool Handles	*Lonicera quinquelocularis, Jacaranda acutifolia, Quercus incana, Rhododendron arboreum, Skimmia laureola*	5/5
Toothbrush (Scouring Teeth)	*Cerasus cerasoides**, *Citrus limon, Desmodium multiflorum, Juglans regia, Ricinus communis, Robinia pseudoacacia**, *Zanthoxylum armatum*	7/7
Utensils	*Cryptomeria japonica**	1/1
Walking Sticks	*Arundo donax**, *Buddleja asiatica, Buddleja crispa**, *Cerasus cerasoides Cornus macrophylla**, *Cotoneaster bacillaris, Drepanostachyum falcatum**, *Himalrandia tetrasperma, Lonicera quinquelocularis**, *Pyracantha crenulata**, *Pyrus pashia, Quercus floribunda, Spiraea canescens, Viburnum mullaha, Zanthoxylum armatum**	14/15
Water Trough	*Abies pindrow, Picea smithiana*	2/2
Welcome Gate	*Cerasus cerasoides**, *Drepanostachyum falcatum**, *Rhododendron arboreum, Sarcococca pruniformis**	4/4

***Additional Use**

For raising fence, 43 species, viz., *Abutilon* X *hybridum*, *Afrocarpus gracilior*, *Agave americana*, *Arundo donax*, *Bougainvillea glabra*, *Buddleja asiatica*, *Buxus wallichiana*, *Clematis montana*, *Cupressus torulosa*, *Drepanostachyum falcatum*, *Duranta erecta*, *Elaeagnus umbellata*, *Erythrina arborescens*, *Euonymus japonicus*, *Ficus neriifolia*, *Forsythia* X *intermedia*, *Hedera canariensis*, *Hedera colchica*, *Hedera helix*, *Hibiscus syriacus*, *Hydrangea macrophylla*, *Hypericum hookerianum*, *Juniperus chinensis*, *Ligustrum lucidum*, *Ligustrum* X *vicaryi*, *Lonicera japonica*, *Myrsine africana*, *Myrtus communis*, *Nerium oleander*, *Opuntia dillenii*, *Populus ciliata*, *Populus nigra*, *Prinsepia utilis*, *Pyracantha crenulata*, *Rosa banksiae*, *Rosa moschata*, *Rosa multiflora*, *Rubus niveus*, *Sambucus nigra*, *Solanum laxum*, *Spartium junceum*, *Thuja occidentalis* and *Yucca aloifolia* are being used (**Table XXX**). These species, due to spines, thorns and profuse branching are preferred for fencing to counter any damage to the growing vegetation by external agent. In order to add to the importance and beauty of the heritage sites, 7 species, viz. *Acacia dealbata*, *Eucalyptus globulus*, *Grevillea robusta*, *Jacaranda acutifolia*, *Populus ciliata*, *Populus nigra and Robinia pseudoacacia* (**Table XXXI**) are used for lining avenues. Such a data have great significance in evolving a meaningful sustainable rural development plan.

TABLE XXX
Hedge Plants of Heritage Sites

S. No.	Botanical Name	Family
1.	*Abutilon* X *hybridum**	Malvaceae
2.	*Afrocarpus gracilior**	Podocarpaceae
3.	*Agave americana*	Agavaceae
4.	*Arundo donax**	Poaceae
5.	*Bougainvillea glabra*	Nyctaginaceae
6.	*Buddleja asiatica**	Buddlejaceae
7.	*Buxus wallichiana*	Buxaceae
8.	*Clematis montana*	Ranunculaceae
9.	*Cupressus torulosa*	Cupressaceae
10.	*Drepanostachyum falcatum*	Poaceae
11.	*Duranta erecta*	Verbenaceae
12.	*Elaeagnus umbellata*	Elaeagnaceae
13.	*Erythrina arborescens*	Fabaceae
14.	*Euonymus japonicus*	Celastraceae
15.	*Ficus neriifolia*	Moraceae
16.	*Forsythia* X *intermedia*	Oleaceae
17.	*Hedera canariensis*	Araliaceae

S. No.	Botanical Name	Family
18.	*Hedera colchica*	Araliaceae
19.	*Hedera helix*	Araliaceae
20.	*Hibiscus syriacus*	Malvaceae
21.	*Hydrangea macrophylla*	Hydrangeaceae
22.	*Hypericum hookerianum*	Clusiaceae
23.	*Juniperus chinensis**	Cupressaceae
24.	*Ligustrum lucidum*	Oleaceae
25.	*Ligustrum* X *vicaryi*	Oleaceae
26.	*Lonicera japonica*	Caprifoliaceae
27.	*Myrsine africana**	Myrsinaceae
28.	*Myrtus communis*	Myrtaceae
29.	*Nerium oleander**	Apocyanaceae
30.	*Opuntia dillenii*	Cactaceae
31.	*Populus ciliata*	Salicaceae
32.	*Populus nigra*	Salicaceae
33.	*Prinsepia utilis**	Rosaceae
34.	*Pyracantha crenulata**	Rosaceae
35.	*Rosa banksiae*	Rosaceae
36.	*Rosa moschata**	Rosaceae
37.	*Rosa multiflora**	Rosaceae
38.	*Rubus niveus**	Rosaceae
39.	*Sambucus nigra**	Caprifoliaceae
40.	*Solanum laxum**	Solanaceae
41.	*Spartium junceum*	Fabaceae
42.	*Thuja occidentalis*	Cupressaceae
43.	*Yucca aloifolia*	Agavaceae

***Additional Uses**

TABLE XXXI
Avenue Plants of Heritage Sites

S. No.	Botanical Name	Family
1.	*Acacia dealbata*	Fabaceae
2.	*Eucalyptus globulus*	Myrtaceae
3.	*Grevillea robusta*	Proteaceae
4.	*Jacaranda acutifolia*	Bignoniaceae
5.	*Populus ciliata*	Salicaceae
6.	*Populus nigra*	Salicaceae
7.	*Robinia pseudoacacia*	Fabaceae

As a matter of fact, the present study highlights the use of 76 species belonging to 52 genera under 33 families for edible purposes (**Table XXXII**). Based on consumption pattern, these edibles can be categorised into various groups like **fruits** (54 species: *Actinidia chinensis*, *Amygdalus mira*, *Berberis aristata*, *Berberis lycium*, *Cerasus cerasoides*, *Citrus limon*, *Citrus maxima*, *Coriaria nepalensis*, *Cornus capitata*, *Daphne papyracea*, *Debregeasia saeneb*, *Diospyros kaki*, *Diospyros lotus*, *Elaeagnus umbellata*, *Eriobotrya japonica*, *Ficus auriculata*, *Ficus palmata*, *Ficus religiosa*, *Ficus sarmentosa*, *Ficus subincisa*, *Juglans regia*, *Leea asiatica*, *Malus pumila*, *Morus alba*, *Morus nigra*, *Morus serrata*, *Musa paradisiaca*, *Myrica esculenta*, *Myrsine africana*, *Opuntia dillenii*, *Padus cornuta*, *Parthenocissus semicordata*, *Pistacia khinjuk*, *Prinsepia utilis*, *Prunus amygdalus*, *Prunus armeniaca*, *Prunus domestica*, *Prunus persica*, *Prunus salicina*, *Punica granatum*, *Pyracantha crenulata*, *Pyrus communis*, *Pyrus pashia*, *Pyrus pyrifolia*, *Rhamnus persica*, *Rubus biflorus*, *Rubus ellipticus*, *Rubus macilentus*, *Rubus niveus*, *Rubus paniculatus*, *Viburnum cotinifolium*, *Viburnum mullaha*, *Yucca aloifolia*, *Zanthoxylum armatum*) **flowers/inflorescence** (11 species: *Abutilon* X *hybridum*, *Buddleja asiatica*, *Girardinia diversifolia*, *Hibiscus rosa-sinensis*, *Hibiscus syriacus*, *Indigofera dosua*, *Lonicera japonica*, *Rhododendron arboreum*, *Rosa moschata*, *Skimmia laureola*, *Yucca aloifolia*), **leaves** (6 speices: *Aloysia citriodora*, *Chamaerops humilis*, *Girardinia diversifolia*, *Morus alba*, *Rosa multiflora*, *Zanthoxylum armatum*), **tender stems** (4 species: *Berberis lycium*, *Drepanostachyum falcatum*, *Leea asiatica*, *Rosa moschata*), **seeds** (3 species: *Pinus roxburghii*, *Solanum virginianum*, *Spiraea canescens*), **petals** (2 species: *Rosa chinensis*, *Rosa damascena*), **aerial parts** (1 species: *Smilax aspera*), **gum** (1 species: *Acacia dealbata*), **penduncle** (1 species: *Agave americana*) **(Fig. 22)**. Nevertheless, the maximum diversity of edibles is represented by Rosaceae (12 genera, 25 species) followed by Moraceae (2 genera, 8 species), Rutaceae (3 genera, 4 species), Malvaceae (2 genera, 3 species), Caprifoliaceae (2 genera, 3 species), Agavaceae (2 genera, 2 species), Ebenaceae (1 genus, 2 species), Urticaceae (2 genera, 2 species) with top 5 families accounting for 56.57 % of the total edible taxa from the heritage sites (**Table XXXIII**). Among the genera, *Ficus*, *Prunus* and *Rubus* with 5 species each showed highest species diversity followed by *Rosa* (4 species), *Morus* (3 species), *Pyrus* (3 species), *Berberis* (2 species), *Citrus* (2 species), *Diospyros* (2 species), *Hibiscus* (2 species), *Viburnum* (2 species) and remaining *Abutilon* X *hybridum*, *Acacia dealbata*, *Actinidia chinensis*, *Agave americana*, *Aloysia citriodora*, *Amygdalus mira*, *Buddleja asiatica*, *Cerasus cerasoides*, *Chamaerops humilis*, *Coriaria nepalensis*, *Cornus capitata*, *Daphne papyracea*, *Debregeasia saeneb*, *Drepanostachyum falcatum*, *Elaeagnus umbellata*, *Eriobotrya japonica*, *Girardinia diversifolia*, *Indigofera* *dosua*, *Juglans regia*, *Leea asiatica*, *Lonicera japonica*, *Malus pumila*, *Musa paradisiaca*, *Myrica esculenta*, *Myrsine africana*, *Opuntia dillenii*, *Padus cornuta*, *Parthenocissus semicordata*, *Pinus roxburghii*, *Pistacia khinjuk*, *Prinsepia utilis*, *Punica granatum*, *Pyracantha crenulata*, *Rhamnus persica*, *Rhododendron arboreum*, *Skimmia laureola*, *Smilax aspera*, *Solanum virginianum*, *Spiraea canescens*, *Yucca aloifolia* and *Zanthoxylum armatum* represented by one species each **(Table XXXIII)**. Of the various plant parts, fruits top the list for edibility in the region (**Fig. 22**). It is quite obvious that the data generated would be of immense use to horticulturists and others in evolving improved varieties of edibles for meeting the food demand of growing population as well as to overcome the sufferings due to the deficiency of nutritional constituents.

TABLE XXXII
Edible Plants of Heritage Sites

S. No.	Botanical Name	Family	Parts used
1.	*Abutilon* X *hybridum*	Malvaceae	Flower
2.	*Acacia dealbata*	Fabaceae	Gum
3.	*Actinidia chinensis*	Actinidiaceae	Fruit
4.	*Agave americana*	Agavaceae	Peduncle
5.	*Aloysia citriodora*	Verbenaceae	Leaves
6.	*Amygdalus mira*	Rosaceae	Fruit
7.	*Berberis aristata*	Berberidaceae	Fruit
8.	*Berberis lycium*	Berberidaceae	Fruit, Tender Shoot
9.	*Buddleja asiatica*	Buddlejaceae	Flower
10.	*Cerasus cerasoides*	Rosaceae	Fruit
11.	*Chamaerops humilis*	Arecaceae	Leaves Bud
12.	*Citrus limon*	Rutaceae	Fruit
13.	*Citrus maxima*	Rutaceae	Fruit
14.	*Coriaria nepalensis*	Coriariaceae	Fruit
15.	*Cornus capitata*	Cornaceae	Fruit
16.	*Daphne papyracea*	Thymelaeaceae	Fruit
17.	*Debregeasia saeneb*	Urticaceae	Fruit
18.	*Diospyros kaki*	Ebenaceae	Fruit
19.	*Diospyros lotus*	Ebenaceae	Fruit
20.	*Drepanostachyum falcatum*	Poaceae	Young Stem
21.	*Elaeagnus umbellata*	Elaeagnaceae	Fruit
22.	*Eriobotrya japonica*	Rosaceae	Fruit
23.	*Ficus auriculata*	Moraceae	Fruit

S. No.	Botanical Name	Family	Parts used
24.	*Ficus palmata*	Moraceae	Fruit
25.	*Ficus religiosa*	Moraceae	Fruit
26.	*Ficus sarmentosa*	Moraceae	Fruit
27.	*Ficus subincisa*	Moraceae	Fruit
28.	*Girardinia diversifolia*	Urticaceae	Young Leaves, Inflorescence
29.	*Hibiscus rosa-sinensis*	Malvaceae	Flower
30.	*Hibiscus syriacus*	Malvaceae	Flower
31.	*Indigofera dosua*	Fabaceae	Flower
32.	*Juglans regia*	Juglandaceae	Fruit
33.	*Leea asiatica*	Leeaceae	Fruit, Juvenile Stem
34.	*Lonicera japonica*	Caprifoliaceae	Flower
35.	*Malus pumila*	Rosaceae	Fruit
36.	*Morus alba*	Moraceae	Fruit, Tender Leaves
37.	*Morus nigra*	Moraceae	Fruit
38.	*Morus serrata*	Moraceae	Fruit
39.	*Musa paradisiaca*	Musaceae	Fruit
40.	*Myrica esculenta*	Myricaceae	Fruit
41.	*Myrsine africana*	Myrsinaceae	Fruit
42.	*Opuntia dillenii*	Cactaceae	Fruit
43.	*Padus cornuta*	Rosaceae	Fruit
44.	*Parthenocissus semicordata*	Vitaceae	Fruit
45.	*Pinus roxburghii*	Pinaceae	Seeds
46.	*Pistacia khinjuk*	Anacardiaceae	Fruit
47.	*Prinsepia utilis*	Rosaceae	Fruit
48.	*Prunus amygdalus*	Rosaceae	Fruit
49.	*Prunus armeniaca*	Rosaceae	Fruit
50.	*Prunus domestica*	Rosaceae	Fruit
51.	*Prunus persica*	Rosaceae	Fruit
52.	*Prunus salicina*	Rosaceae	Fruit
53.	*Punica granatum*	Lythraceae	Fruit
54.	*Pyracantha crenulata*	Rosaceae	Fruit
55.	*Pyrus communis*	Rosaceae	Fruit
56.	*Pyrus pashia*	Rosaceae	Fruit
57.	*Pyrus pyrifolia*	Rosaceae	Fruit
58.	*Rhamnus persica*	Rhamnaceae	Fruit
59.	*Rhododendron arboreum*	Ericaceae	Flower
60.	*Rosa chinensis*	Rosaceae	Petals

S. No.	Botanical Name	Family	Parts used
61.	*Rosa damascena*	Rosaceae	Petals
62.	*Rosa moschata*	Rosaceae	Flower, Tender Stem
63.	*Rosa multiflora*	Rosaceae	Tender Leaves
64.	*Rubus biflorus*	Rosaceae	Fruit
65.	*Rubus ellipticus*	Rosaceae	Fruit
66.	*Rubus macilentus*	Rosaceae	Fruit
67.	*Rubus niveus*	Rosaceae	Fruit
68.	*Rubus paniculatus*	Rosaceae	Fruit
69.	*Skimmia laureola*	Rutaceae	Inflorescence
70.	*Smilax aspera*	Smilacaceae	Aerial Plant Parts
71.	*Solanum virginianum*	Solanaceae	Seeds
72.	*Spiraea canescens*	Rosaceae	Seeds
73.	*Viburnum cotinifolium*	Caprifoliaceae	Fruit
74.	*Viburnum mullaha*	Caprifoliaceae	Fruit
75.	*Yucca aloifolia*	Agavaceae	Flowers, Fruit
76.	*Zanthoxylum armatum*	Rutaceae	Young Leaves, Fruits

TABLE XXXIII
Families and Ratio of Genera/Species of Edible Plants

S. No.	Family	Botanical Name	Genus/ Species
1.	Actinidiaceae	*Actinidia chinensis*	1/1
2.	Agavaceae	*Agave americana, Yucca aloifolia*	2/2
3.	Anacardiaceae	*Pistacia khinjuk*	1/1
4.	Arecaceae	*Chamaerops humilis*	1/1
5.	Berberidaceae	*Berberis aristata, Berberis lycium*	1/2
6.	Buddlejaceae	*Buddleja asiatica*	1/1
7.	Cactaceae	*Opuntia dillenii*	1/1
8.	Caprifoliaceae	*Lonicera japonica, Viburnum cotinifolium, Viburnum mullaha*	2/3
9.	Coriariaceae	*Coriaria nepalensis*	1/1
10.	Cornaceae	*Cornus capitata*	1/1
11.	Ebenaceae	*Diospyros kaki, Diospyros lotus*	1/2
12.	Elaeagnaceae	*Elaeagnus umbellata*	1/1
13.	Ericaceae	*Rhododendron arboreum*	1/1
14.	Fabaceae	*Acacia dealbata, Indigofera dosua*	2/2
15.	Juglandaceae	*Juglans regia*	1/1

S. No.	Family	Botanical Name	Genus/ Species
16.	Leeaceae	*Leea asiatica*	1/1
17.	Lythraceae	*Punica granatum*	1/1
18.	Malvaceae	*Abutilon X hybridum, Hibiscus rosa-sinensis, Hibiscus syriacus*	2/3
19.	Moraceae	*Ficus auriculata, Ficus palmata, Ficus religiosa, Ficus sarmentosa, Ficus subincisa, Morus alba, Morus nigra, Morus serrata*	2/8
20.	Musaceae	*Musa paradisiaca*	1/1
21.	Myricaceae	*Myrica esculenta*	1/1
22.	Myrsinaceae	*Myrsine africana*	1/1
23.	Pinaceae	*Pinus roxburghii*	1/1
24.	Poaceae	*Drepanostachyum falcatum*	1/1
25.	Rhamnaceae	*Rhamnus persica*	1/1
26.	Rosaceae	*Amygdalus mira, Cerasus cerasoides, Eriobotrya japonica, Malus pumila, Padus cornuta, Prinsepia utilis, Prunus amygdalus, Prunus armeniaca, Prunus domestica, Prunus persica, Prunus salicina, Pyracantha crenulata, Pyrus communis, Pyrus pashia, Pyrus pyrifolia, Rosa chinensis, Rosa damascena, Rosa moschata, Rosa multiflora, Rubus biflorus, Rubus ellipticus, Rubus macilentus, Rubus niveus, Rubus paniculatus, Spiraea canescens*	12/25
27.	Rutaceae	*Citrus limon, Citrus maxima, Skimmia laureola, Zanthoxylum armatum*	3/4
28.	Smilacaceae	*Smilax aspera*	1/1
29.	Solanaceae	*Solanum virginianum*	1/1
30.	Thymelaeaceae	*Daphne papyracea*	1/1
31.	Urticaceae	*Debregeasia saeneb, Girardinia diversifolia*	2/2
32.	Verbenaceae	*Aloysia citriodora*	1/1
33.	Vitaceae	*Parthenocissus semicordata*	1/1

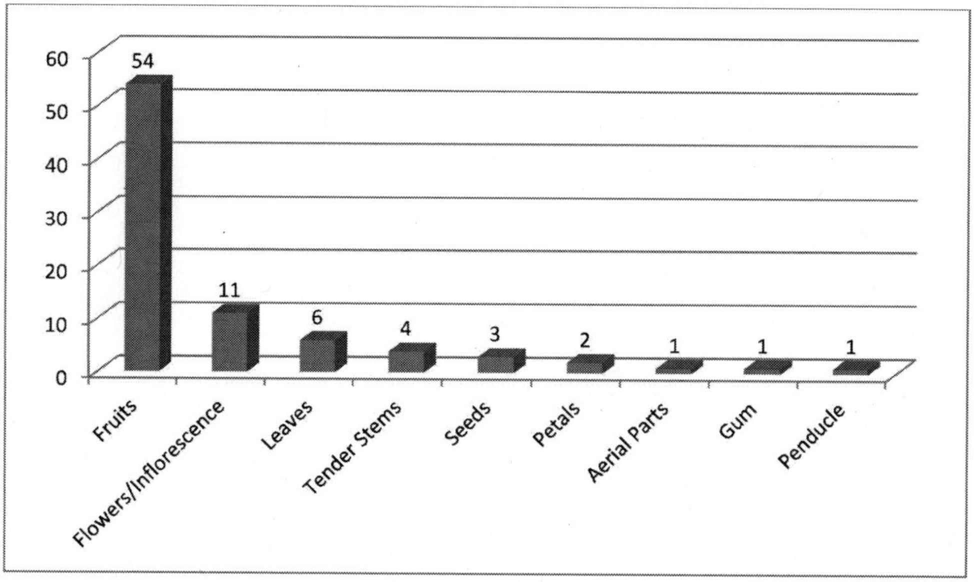

Fig. 22. Histogram Showing Relative Number of Plant Parts Used for Edible Purposes.

Of the information recorded presently, 154 additional uses of 96 plants have been reported from the study area (**Table XXXIV, Appendix IV**). It is relevant to mention here that additional ethnobotanical uses (marked with asterisk in various tables) (**Tables XX, XXI, XXIII, XXV, XXVII, XXVIII, XXIX, XXX)** have been recorded for the first time from the heritage sites of the study area. All these species hold great potential in shaping the welfare of mankind.

TABLE XXXIV
Plants with Additional Uses

S. No.	Name of the Species
1.	*Abutilon* X *hybridum*
2.	*Acer negundo*
3.	*Afrocarpus gracilior*
4.	*Agave americana*
5.	*Aloysia citriodora*
6.	*Arundo donax*
7.	*Buddleja asiatica*
8.	*Buddleja crispa*

S. No.	Name of the Species
9.	*Cerasus cerasoides*
10.	*Citrus limon*
11.	*Clematis buchananiana*
12.	*Colquhounia coccinea*
13.	*Cornus capitata*
14.	*Cornus macrophylla*
15.	*Cotoneaster bacillaris*
16.	*Cryptomeria japonica*
17.	*Cunninghamia lanceolata*
18.	*Cycas revoluta*
19.	*Daphne papyracea*
20.	*Datura stramonium*
21.	*Desmodium elegans*
22.	*Desmodium multiflorum*
23.	*Diospyros lotus*
24.	*Drepanostachyum falcatum*
25.	*Duranta erecta*
26.	*Eriobotrya japonica*
27.	*Erythrina arborescens*
28.	*Erythrina arborescens*
29.	*Euonymous fimbriatus*
30.	*Euonymus lucidus*
31.	*Fagus sylvatica*
32.	*Ficus auriculata*
33.	*Ficus palmata*
34.	*Ficus reliogiosa*
35.	*Hedera helix*
36.	*Hedera nepalensis*
37.	*Hibiscus rosa-sinensis*
38.	*Himalrandia tetrasperma*
39.	*Hypericum oblongifolium*
40.	*Hypericum patulum*
41.	*Ilex dipyrena*
42.	*Indigofera dosua*
43.	*Inula cappa*
44.	*Jacaranda acutifolia*
45.	*Jasminum grandiflorum*

S. No.	Name of the Species
46.	*Jasminum humile*
47.	*Juniperus chinensis*
48.	*Lagerstroemia indica*
49.	*Leea asiatica*
50.	*Lonicera quinquelocularis*
51.	*Malvaviscus arboreus* var. *mexicanus*
52.	*Musa paradisiaca*
53.	*Myrica esculenta*
54.	*Myrsine africana*
55.	*Nerium oleander*
56.	*Opuntia dillenii*
57.	*Padus cornuta*
58.	*Picea smithiana*
59.	*Pinus roxburghii*
60.	*Prinsepia utilis*
61.	*Prunus domestica*
62.	*Punica granatum*
63.	*Pyracantha crenulata*
64.	*Pyrus communis*
65.	*Pyrus pashia*
66.	*Rhamnus persica*
67.	*Rhamnus virgatus*
68.	*Rhododendron arboreum*
69.	*Ricinus communis*
70.	*Robinia pseudoacacia*
71.	*Rosa alba*
72.	*Rosa chinensis*
73.	*Rosa damascena*
74.	*Rosa moschata*
75.	*Rosa multiflora*
76.	*Rubia cordifolia*
77.	*Rubus ellipticus*
78.	*Rubus niveus*
79.	*Salix alba*
80.	*Salix babylonica*
81.	*Salix pycnostachya* var. *oxycarpa*
82.	*Sambucus nigra*

S. No.	Name of the Species
83.	*Sarcococca pruniformis*
84.	*Skimmia laureola*
85.	*Smilax aspera*
86.	*Solanum laxum*
87.	*Solanum virginianum*
88.	*Taxodium distichum*
89.	*Taxus baccata*
90.	*Thuja orientalis*
91.	*Tropaeolum majus*
92.	*Viburnum cotinifolium*
93.	*Viburnum mullaha*
94.	*Xanthium strumarium*
95.	*Yucca aloifolia*
96.	*Zanthoxylum armatum*

Based on analysis of data for Total Importance Values (TIV) of plant diversity at heritage sites in Shimla hills (**Table XXXV**), it is concluded that *Pyrus pashia* (Rosaceae) tops the list with highest economic potential value of 55%, followed by *Rhododendron arboreum* (47%), *Agave americana, Drepanostachyum falcatum*, (42.5% each), *Ficus religiosa, Pistacia khinjuk, Rosa damascena, Taxus baccata*, (40 % each), *Grevillea robusta, Hedera helix, Jacaranda acutifolia, Juglans regia, Pinus roxburghii, Populus ciliata, Robinia pseudoacacia, Rosa chinensis, Yucca aloifolia, Zanthoxylum armatum* (37.5% each), *Hibiscus rosa-sinensis, Morus alba, Morus nigra, Morus serrata, Opuntia dillenii and Salix babylonica* (35% each) etc. Undoubtedly, the resulted information on the economic potential of the germplasm material/s of the area can be profitably made use of for the welfare of mankind.

TABLE XXXV
Total Importance Value (TIV) of Plants of Heritage Sites

Plant Name	Ornamental	Hedge	Avenue	Medicine	Edible	Timber/ Furniture	Fuel/Fibre/ Fodder	Others	TIV %
Abies pindrow	1	-	-	-	-	4	3	-	20
Abutilon X hybridum	5	2	-	-	2	-	-	-	22.5
Abutilon megapotamicum	5	-	-	-	-	-	-	-	12.5
Acacia dealbata	1	1	3	-	2	-	2	2	27.5
Acer negundo	5	-	-	-	-	-	-	-	12.5
Acer oblongum	-	-	-	-	-	2	3	-	12.5
Acer palmatum	5	-	-	-	-	-	-	-	12.5
Actinidia chinensis	-	-	-	-	5	-	-	-	12.5
Aesculus indica	1	-	-	4	-	3	3	-	27.5
Afrocarpus gracilior	5	2	-	-	-	-	-	-	17.5
Agave americana	3	3	-	2	3	-	4	2	42.5
Aloysia citriodora	3	-	-	3	2	-	-	-	20
Amygdalus mira	-	-	-	-	4	-	-	2	15
Araucaria bidwillii	5	-	-	-	-	-	-	2	17.5
Araucaria columnaris	5	-	-	-	-	-	-	-	12.5
Araucaria cunninghamii	5	-	-	-	-	2	-	-	17.5
Arundo donax	-	3	-	-	-	-	-	5	20
Asparagus aethiopicus	5	-	-	-	-	-	-	-	12.5
Asparagus densiflorus	5	-	-	-	-	-	-	-	12.5
Asparagus racemosus	-	-	-	5	-	-	-	-	12.5
Asparagus setaceus	5	-	-	-	-	-	-	-	12.5
Asparagus virgatus	5	-	-	-	-	-	-	-	12.5

Plant Name	Ornamental	Hedge	Avenue	Medicine	Edible	Timber/ Furniture	Fuel/Fibre/ Fodder	Others	TIV %
Berberis aristata	-	-	-	3	5	-	3	2	32.5
Berberis lycium	-	-	-	1	5	-	3	2	27.5
Bougainvillea glabra	5	5	-	-	-	-	-	-	25
Buddleja asiatica	3	3	-	2	2	-	-	2	30
Buddleja crispa	-	-	-	-	2	-	3	2	17.5
Buddleja davidii	5	-	-	-	-	-	-	-	12.5
Buxus wallichiana	5	3	-	-	-	-	-	3	27.5
Callistemon lanceolatus	5	-	-	-	-	-	-	2	17.5
Camellia japonica	5	-	-	-	-	-	-	2	17.5
Campsis radicans	5	-	-	-	-	-	-	-	12.5
Cedrus deodara	1	-	-	-	-	5	-	-	15
Celtis australis	-	-	-	-	-	-	4	2	15
Cerasus cerasoides	-	-	-	2	3	-	3	5	32.5
Cestrum elegans	5	-	-	-	-	-	-	-	12.5
Cestrum nocturnum	5	-	-	-	-	-	-	2	17.5
Cestrum parqui	5	-	-	-	-	-	-	-	12.5
Chamaerops humilis	5	-	-	-	2	-	3	3	32.5
Citrus limon	-	-	-	3	5	-	-	3	27.5
Citrus maxima	-	-	-	3	5	-	-	-	20
Clematis buchananiana	-	-	-	2	-	-	3	3	20
Clematis montana	4	3	-	-	-	-	-	-	17.5
Colquhounia coccinea	4	-	-	-	-	-	3	-	17.5
Coriaria nepalensis	-	-	-	-	2	-	3	-	12.5
Cornus capitata	-	-	-	-	4	-	3	2	22.5

Plant Name	Ornamental	Hedge	Avenue	Medicine	Edible	Timber/ Furniture	Fuel/Fibre/ Fodder	Others	TIV %
Cornus macrophylla	-	-	-	-	-	3	4	3	25
Cotoneaster bacillaris	-	-	-	-	-	3	2	3	20
Cryptomeria japonica	5	-	-	-	-	3	-	3	27.5
Cunninghamia lanceolata	5	-	-	3	-	3	-	-	27.5
Cupressus cashmeriana	5	-	-	-	-	-	-	2	17.5
Cupressus torulosa	5	3	-	-	-	-	-	3	27.5
Cycas revoluta	5	-	-	-	-	-	-	2	17.5
Daphne papyracea	-	-	-	3	2	-	-	-	12.5
Datura stramonium	-	-	-	4	-	-	-	4	20
Debregeasia saeneb	-	-	-	-	4	-	5	3	30
Desmodium elegans	-	-	-	3	-	-	3	-	15
Desmodium multiflorum	-	-	-	-	-	-	3	2	12.5
Diospyros kaki	1	-	-	-	5	4	-	-	25
Diospyros lotus	4	-	-	-	4	-	-	-	20
Drepanostachyum falcatum	1	4	-	-	3	-	4	5	42.5
Duranta erecta	3	5	-	-	-	-	3	-	27.5
Elaeagnus umbellata	-	3	-	-	4	-	2	-	22.5
Eriobotrya japonica	-	-	-	2	5	-	-	-	17.5
Erythrina arborescens	3	2	-	-	-	2	2	-	22.5
Eucalyptus globulus	-	-	4	3	-	3	3	-	32.5
Euonymus fimbriatus	-	-	-	-	-	-	-	3	7.5
Euonymus japonicus	2	5	-	-	-	-	-	-	17.5
Euonymus lucidus	3	-	-	-	-	-	-	-	7.5
Euonymus tingens	-	-	-	3	-	-	3	2	20

Plant Name	Ornamental	Hedge	Avenue	Medicine	Edible	Timber/ Furniture	Fuel/Fibre/ Fodder	Others	TIV %
Fagus sylvatica	4	-	-	-	-	2	-	2	20
Ficus auriculata	-	-	-	3	5	-	4	-	30
Ficus benjamina	5	-	-	-	-	-	-	-	12.5
Ficus elastica	5	-	-	-	-	-	-	-	12.5
Ficus neriifolia	-	3	-	-	-	-	3	-	15
Ficus palmata	-	-	-	4	5	-	4	-	32.5
Ficus religiosa	3	-	-	5	3	-	-	5	40
Ficus sarmentosa	-	-	-	-	4	-	3	-	17.5
Ficus subincisa	-	-	-	-	3	-	2	-	12.5
Forsythia X *intermedia*	5	5	-	-	-	-	-	-	25
Forsythia ovata	5	-	-	-	-	-	-	-	12.5
Fuchsia X *hybrida*	5	-	-	-	-	-	-	-	12.5
Fuchsia magellanica	5	-	-	-	-	-	-	-	12.5
Ginkgo biloba	5	-	-	4	-	-	-	-	22.5
Girardinia diversifolia	-	-	-	4	4	-	4	-	30
Grevillea robusta	5	4	-	-	-	2	-	4	37.5
Hedera canariensis	5	3	-	-	-	-	-	-	20
Hedera colchica	5	3	-	-	-	-	-	-	20
Hedera helix	3	3	-	4	-	-	3	2	37.5
Hedera nepalensis	-	-	-	3	-	-	-	-	7.5
Hibiscus rosa-sinensis	5	-	-	3	3	-	-	3	35
Hibiscus syriacus	3	5	-	-	2	-	-	-	25
Himalrandia tetrasperma	-	-	-	-	-	-	3	3	15
Hydrangea macrophylla	5	3	-	-	-	-	-	-	20

Plant Name	Ornamental	Hedge	Avenue	Medicine	Edible	Timber/ Furniture	Fuel/Fibre/ Fodder	Others	TIV %
Hypericum hookerianum	3	2	-	-	-	-	-	-	12.5
Hypericum oblongifolium	3	-	-	3	-	-	2	2	25
Hypericum patulum	-	-	-	4	-	-	-	-	10
Ilex dipyrena	-	-	-	2	-	2	3	-	17.5
Indigofera dosua	-	-	-	2	4	-	3	-	17.5
Inula cappa	-	-	-	3	-	-	-	2	12.5
Ipomoea purpurea	4	-	-	-	-	-	-	-	10
Jacaranda acutifolia	3	-	5	2	-	3	-	2	37.5
Jasminum grandiflorum	4	-	-	4	-	-	-	3	27.5
Jasminum humile	4	-	-	4	-	-	-	3	27.5
Jasminum mesnyi	5	-	-	-	-	-	-	-	12.5
Jasminum multiflorum	4	-	-	3	-	-	-	3	25
Juglans regia	-	-	-	-	5	5	-	5	37.5
Juniperus chinensis	5	4	-	-	-	-	-	-	22.5
Juniperus horizontalis	5	-	-	-	-	-	-	3	20
Juniperus procera	5	-	-	-	-	-	-	-	12.5
Juniperus pseudosabina	5	-	-	-	-	-	-	-	12.5
Juniperus squamata	5	-	-	2	-	-	3	2	30
Kerria japonica	5	-	-	-	-	-	-	-	12.5
Lagerstroemia indica	5	-	-	2	-	-	-	-	17.5
Leea asiatica	-	-	-	2	3	-	2	-	17.5
Ligustrum lucidum	4	3	-	-	-	-	-	-	17.5
Ligustrum X *vicaryi*	-	5	-	-	-	-	-	-	12.5
Liriodendron tulipifera	5	-	-	-	-	-	-	-	12.5

Plant Name	Ornamental	Hedge	Avenue	Medicine	Edible	Timber/ Furniture	Fuel/Fibre/ Fodder	Others	TIV %
Lonicera japonica	5	3	-	-	3	-	-	-	27.5
Lonicera quinquelocularis	-	-	-	-	-	-	3	3	15
Lyonia ovalifolia	-	-	-	3	-	-	-	2	12.5
Magnolia grandiflora	5	-	-	-	-	-	-	-	12.5
Magnolia obovata	5	-	-	-	-	-	-	-	12.5
Magnolia X *soulangeana*	5	-	-	-	-	-	-	-	12.5
Malus pumila	-	-	-	-	5	-	-	-	12.5
Malvaviscus arboreus var. *mexicanus*	5	-	-	3	-	-	-	2	25
Morus alba	-	-	-	-	5	5	4	-	35
Morus nigra	-	-	-	-	5	5	4	-	35
Morus serrata	-	-	-	-	5	5	4	-	35
Musa paradisiaca	-	-	-	-	5	-	-	3	20
Myrica esculenta	-	-	-	3	5		3	-	27.5
Myrsine africana	-	3	-	3	2	-	3	2	32.5
Myrtus communis	5	3	-	-	-	-	-	-	20
Nerium oleander	5	3	-	3	-	-	-	-	27.5
Opuntia dillenii	3	3	-	4	4	-	-	-	35
Padus cornuta	-	-	-	-	3	-	3	-	15
Parthenocissus semicordata	-	-	-	-	3	-	3	-	15
Paulownia tomentosa	5	-	-	-	-	-	-	-	12.5
Picea smithiana	-	-	-	-	-	5	3	3	27.5
Pinus roxburghii	-	-	-	-	3	5	4	3	37.5
Pinus wallichiana	-	-	-	-	-	5	-	2	17.5
Pistacia khinjuk	-	-	-	5	3	5	3	-	40

Plant Name	Ornamental	Hedge	Avenue	Medicine	Edible	Timber/ Furniture	Fuel/Fibre/ Fodder	Others	TIV %
Platanus orientalis	5	-	-	-	-	-	-	-	12.5
Populus ciliata	-	4	-	3	-	4	4	-	37.5
Populus nigra	-	4	-	-	-	3	3	3	32.5
Prinsepia utilis	-	3	-	3	3	-	-	4	32.5
Prunus amygdalus	-	-	-	4	5	-	-	-	22.5
Prunus armeniaca	-	-	-	-	5	-	-	3	20
Prunus domestica	-	-	-	-	5	-	3	-	20
Prunus laurocerasus	4	-	-	-	-	-	-	2	15
Prunus persica	-	-	-	2	5	-	-	2	22.5
Prunus salicina	-	-	-	-	5	-	-	-	12.5
Punica granatum	-	-	-	4	5	-	-	3	30
Pyracantha crenulata	-	3	-	3	3	-	-	2	27.5
Pyrus communis	-	-	-	2	5	-	-	-	17.5
Pyrus pashia	-	-	-	5	5	2	5	5	55
Pyrus pyrifolia	-	-	-	-	5	1	-	-	15
Quercus floribunda	-	-	-	-	-	4	4	4	30
Quercus incana	-	-	-	-	-	4	4	4	30
Quercus oblongata	-	-	-	-	-	4	4	4	30
Quercus robur	5	-	-	-	-	-	-	-	12.5
Rhamnus persica	-	-	-	-	3	-	3	2	20
Rhamnus virgatus	-	-	-	-	-	-	2	3	12.5
Rhapis excelsa	5	-	-	-	-	-	-	-	12.5
Rhododendron arboreum	1	-	-	5	5	-	4	4	47
Rhododendron simsii var. *vittatum*	5	-	-	-	-	-	-	-	12.5

Plant Name	Ornamental	Hedge	Avenue	Medicine	Edible	Timber/ Furniture	Fuel/Fibre/ Fodder	Others	TIV %
Ricinus communis	-	-	-	5	-	-	-	2	17.5
Robinia pseudoacacia	-	-	4	2	-	4	3	2	37.5
Rosa alba	5	-	-	-	-	-	-	2	17.5
Rosa banksiae	5	3	-	-	-	-	-	-	20
Rosa chinensis	5	-	-	3	5	-	-	2	37.5
Rosa damascena	5	-	-	4	5	-	-	2	40
Rosa jacquemontii	5	-	-	-	-	-	-	-	12.5
Rosa moschata	2	3	-	3	3	-	2	-	32.5
Rosa multiflora	4	3	-	3	3	-	-	-	32.5
Rubia cordifolia	-	-	-	4	-	-	-	2	15
Rubus biflorus	-	-	-	-	5	-	-	-	12.5
Rubus ellipticus	-	-	-	2	5	-	3	-	25
Rubus macilentus	-	-	-	-	5	-	-	-	12.5
Rubus niveus	-	2	-	3	5	-	2	-	30
Rubus paniculatus	-	-	-	-	5	-	-	-	12.5
Salix alba	4	-	-	3	-	-	-	3	25
Salix babylonica	5	-	-	3	-	-	3	3	35
Salix pycnostachya var. *oxycarpa*	-	-	-	3	-	-	-	-	7.5
Salvia leucantha	5	-	-	-	-	-	-	-	12.5
Salvia splendens	5	-	-	-	-	-	-	-	12.5
Sambucus nigra	5	3	-	4	-	-	-	-	20
Sarcococca pruniformis	-	-	-	3	-	-	-	2	12.5
Schefflera arboricola	5	-	-	-	-	-	-	-	12.5
Skimmia laureola	-	-	-	-	3	-	-	3	15

Plant Name	Ornamental	Hedge	Avenue	Medicine	Edible	Timber/ Furniture	Fuel/Fibre/ Fodder	Others	TIV %
Smilax aspera	-	-	-	3	3	-	-	-	15
Solanum laxum	5	4	-	-	-	-	-	2	27.5
Solanum pseudocapsicum	4	-	-	-	-	-	-	-	10
Solanum virginianum	-	-	-	5	3	-	-	-	20
Spartium junceum	5	4	-	-	-	-	3	-	30
Spiraea canescens	-	-	-	-	3	-	-	2	12.5
Spiraea cantoniensis	5	-	-	-	-	-	-	-	12.5
Taxodium distichum	4	-	-	-	-	2	2	-	20
Taxus baccata	4	-	-	5	2	3	-	2	40
Thuja occidentalis	5	3	-	-	-	-	-	-	20
Thuja orientalis	5	-	-	3	-	-	-	-	20
Thujopsis dolabrata	5	-	-	-	-	-	-	-	12.5
Trachelospermum jasminoides	5	-	-	-	-	-	-	-	12.5
Tropaeolum majus	5	-	-	-	-	-	-	2	17.5
Tsuga dumosa	4	-	-	-	-	-	-	2	15
Ulmus wallichiana	-	-	-	-	-	4	5	-	22.5
Viburnum cotinifolium	-	-	-	-	4	-	3	-	17.5
Viburnum mullaha	-	-	-	3	4	-	3	2	30
Wisteria sinensis	5	-	-	-	-	-	-	-	12.5
Xanthium strumarium	-	-	-	4	-	-	-	-	10
Yucca aloifolia	4	4	-	-	2	-	3	2	37.5
Zanthoxylum armatum	-	-	-	5	5	-	-	5	37.5

Evidently from **Table XXXVI**, it is indicated that 69 species have been found to occur in considerable extent (++), 97 as abundant (+++) and 46 rare (*Abies pindrow*, *Abutilon* X *hybridum*, *Acer negundo*, *Acer palmatum*, *Actinidia chinensis*, *Aloysia citriodora*, *Amygdalus mira*, *Araucaria bidwillii*, *Araucaria cunninghamii*, *Asparagus setaceus*, *Asparagus virgatus*, *Bougainvillea glabra*, *Cestrum parqui*, *Citrus maxima*, *Cotoneaster bacillaris*, *Cycas revoluta*, *Diospyros kaki*, *Diospyros lotus*, *Duranta erecta*, *Erythrina arborescens*, *Euonymus fimbriatus*, *Euonymus lucidus*, *Fagus sylvatica*, *Ficus benjamina*, *Ficus religiosa*, *Forsythia ovata*, *Ginkgo biloba*, *Hedera canariensis*, *Hypericum hookerianum*, *Liriodendron tulipifera*, *Magnolia obovata*, *Magnolia* X *soulangeana*, *Morus nigra*, *Musa paradisiaca*, *Myrtus communis*, *Prunus laurocerasus*, *Quercus robur*, *Rhapis excelsa*, *Rhododendron simsii* var. *vittatum*, *Rosa jacquemontii*, *Rosa multiflora*, *Taxodium distichum*, *Taxus baccata*, *Trachelospermum jasminoides*, *Tsuga dumosa*, *Ulmus wallichiana*) in the study area. Among these, *Ginkgo biloba* is listed as endangered; *Cupressus cashmeriana* and *Ulmus wallichiana* as vulnerable; *Cryptomeria japonica* as near *Ulmus wallichiana* as vulnerable; *Cryptomeria japonica* as near threatened and 23 plants such as *Abies pindrow*, *Afrocarpus gracilior*, *Araucaria bidwillii*, *Araucaria columnaris*, *Araucaria cunninghamii*, *Cunninghamia lanceolata*, *Cycas revoluta*, *Desmodium elegans*, *Diospyros lotus*, *Juniperus chinensis*, *Juniperus horizontalis*, *Juniperus procera*, *Juniperus pseudosabina*, *Juniperus squamata*, *Picea smithiana*, *Platanus orientalis*, *Populus nigra*, *Quercus robur*, *Taxodium distichum*, *Taxus baccata*, *Thuja occidentalis*, *Thujopsis dolabrata* and *Tsuga dumosa* at low risk/least concern as per IUCN notified categories (**Table XXXVII; IUCN, 1978, 1995, 2013)**. In order to improve upon their threat status, policies regarding their conservation and cultivation be encouraged and strictly enforced with participation of local people, NGOs and forest department. In addition, biotechnological methods like cell and tissue cultures can provide new means for conservation and propagation of these invaluable plants. Concurrently, efforts should also be made to introduce them in the chain of experimental botanic gardens of different regional circles for their *ex situ* conservation, multiplication and scientific study.

TABLE XXXVI
Observed Density and Availability (ODA) of Plants at Heritage Sites

S. No.	Botanical Name	Local Name/s	Observed Density and Availability
1.	*Abies pindrow*	Fir, Tosh	+
2.	*Abutilon X hybridum*	Chinese Lantern	+
3.	*Abutilon megapotamicum*	Lantern Flower	++
4.	*Acacia dealbata*	Khair	++
5.	*Acer negundo*	Maple	+
6.	*Acer oblongum*	Padang	++
7.	*Acer palmatum*	Japanese Maple	+
8.	*Actinidia chinensis*	Kiwi	+
9.	*Aesculus indica*	Kanhor, Khanor	+++
10.	*Afrocarpus gracilior*	Podocarpus	++
11.	*Agave americana*	Keora	+++
12.	*Aloysia citriodora*	Verbena	+
13.	*Amygdalus mira*	Vaimi	+
14.	*Araucaria bidwillii*	Bunya Araucaria	+
15.	*Araucaria columnaris*	Araucaria	+++
16.	*Araucaria cunninghamii*	Hoop pine	+
17.	*Arundo donax*	Nal	++
18.	*Asparagus aethiopicus*	Sprengeri Asparagus	++
19.	*Asparagus densiflorus*	Asparagus	++
20.	*Asparagus racemosus*	Satavar, Sanspai	++
21.	*Asparagus setaceus*	Asparagus Fern	+
22.	*Asparagus virgatus*	Asparagus	+
23.	*Berberis aristata*	Kashmal	+++
24.	*Berberis lycium*	Kashmal	+++
25.	*Bougainvillea glabra*	Boganvillea	+
26.	*Buddleja asiatica*	Buddliya, Bana	++
27.	*Buddleja crispa*	Newarpati	+++
28.	*Buddleja davidii*	Buddleya	++
29.	*Buxus wallichiana*	Shamshad	++
30.	*Callistemon lanceolatus*	Bottle Brush	+++
31.	*Camellia japonica*	Camellia	++
32.	*Campsis radicans*	Shehnai Phool	++
33.	*Cedrus deodara*	Deaar, Devdar	+++
34.	*Celtis australis*	Khirak	+++
35.	*Cerasus cerasoides*	Pajja	++

S. No.	Botanical Name	Local Name/s	Observed Density and Availability
36.	*Cestrum elegans*	Raat- ki- Chameli	++
37.	*Cestrum nocturnum*	Rat-ki- Rani	++
38.	*Cestrum parqui*	Rat-ki- Rani	+
39.	*Chamaerops humilis*	Fan Palm, Khajuri	+++
40.	*Citrus limon*	Galgal	++
41.	*Citrus maxima*	Chakotra	+
42.	*Clematis buchananiana*	Dhudki Bel	+++
43.	*Clematis montana*	Kaunia-Bali	++
44.	*Colquhounia coccinea*	Bhekuyi	++
45.	*Coriaria nepalensis*	Kachog	++
46.	*Cornus capitata*	Teelahan	++
47.	*Cornus macrophylla*	Nang	+++
48.	*Cotoneaster bacillaris*	Simbli	+
49.	*Cryptomeria japonica*	Cryptomeria	++
50.	*Cunninghamia lanceolata*	Ghadhal, Thanela	++
51.	*Cupressus cashmeriana*	Kashmiri Cypress	++
52.	*Cupressus torulosa*	Saru	+++
53.	*Cycas revoluta*	Cycas	+
54.	*Daphne papyracea*	Chamar-Chumb, Raabal	+++
55.	*Datura stramonium*	Dhatura	++
56.	*Debregeasia saeneb*	Syaru	+++
57.	*Desmodium elegans*	Chamla	++
58.	*Desmodium multiflorum*	Martoi	++
59.	*Diospyros kaki*	Japani-Phal	+
60.	*Diospyros lotus*	Barain	+
61.	*Drepanostachyum falcatum*	Banji	+++
62.	*Duranta erecta*	Neel Kanthi	+
63.	*Elaeagnus umbellata*	Ghain	++
64.	*Eriobotrya japonica*	Lokat	+++
65.	*Erythrina arborescens*	Pariyara	+
66.	*Eucalyptus globulus*	Eucalyptus, Safeda	++
67.	*Euonymus fimbriatus*	Euonymus	+
68.	*Euonymus japonicus*	Japani Euonymus	++
69.	*Euonymus lucidus*	Bharmela	+
70.	*Euonymus tingens*	Marmakoul	+++
71.	*Fagus sylvatica*	Rang Badalane Bala Plant	+
72.	*Ficus auriculata*	Timbal	++

S. No.	Botanical Name	Local Name/s	Observed Density and Availability
73.	*Ficus benjamina*	Kabra	+
74.	*Ficus elastica*	Rubber Plant	++
75.	*Ficus neriifolia*	Dudhia	++
76.	*Ficus palmata*	Fegura	+++
77.	*Ficus religiosa*	Peepal	+
78.	*Ficus sarmentosa*	Phaegdi	++
79.	*Ficus subincisa*	Ban Feguda	++
80.	*Forsythia* X *intermedia*	Forsythia	++
81.	*Forsythia ovata*	Forsythia	+
82.	*Fuchsia* X *hybrida*	Doll Flower	++
83.	*Fuchsia magellanica*	Doll Flower	++
84.	*Ginkgo biloba*	Balkumari	+
85.	*Girardinia diversifolia*	Bicchu Buti	+++
86.	*Grevillea robusta*	Silver Fir	++
87.	*Hedera canariensis*	Banbat Kari	+
88.	*Hedera colchica*	Kiraya	++
89.	*Hedera helix*	Dakari	+++
90.	*Hedera nepalensis*	Berbel	+++
91.	*Hibiscus rosa-sinensis*	China Rose	++
92.	*Hibiscus syriacus*	Brahm Phool	+++
93.	*Himalrandia tetrasperma*	Hazara-Kikra	++
94.	*Hydrangea macrophylla*	Hydrangea	+++
95.	*Hypericum hookerianum*	Phiunli	+
96.	*Hypericum oblongifolium*	Lyoon	+++
97.	*Hypericum patulum*	Phulanu	++
98.	*Ilex dipyrena*	Charkha, Holly Tree	++
99.	*Indigofera dosua*	Kathee	++
100.	*Inula cappa*	Basanti	++
101.	*Ipomoea purpurea*	Gonani	++
102.	*Jacaranda acutifolia*	Jackranda, Nili Gulmohra	++
103.	*Jasminum grandiflorum*	Chameli	++
104.	*Jasminum humile*	Peeli Chameli	+++
105.	*Jasminum mesnyi*	Chameli	++
106.	*Jasminum multiflorum*	Chameli	++
107.	*Juglans regia*	Akhrot	+++
108.	*Juniperus chinensis*	Chinese Juniperus	+++
109.	*Juniperus horizontalis*	Juniper	++

S. No.	Botanical Name	Local Name/s	Observed Density and Availability
110.	*Juniperus procera*	African Juniper	++
111.	*Juniperus pseudosabina*	Juniper	++
112.	*Juniperus squamata*	Thanara	++
113.	*Kerria japonica*	Japani Rose	++
114.	*Lagerstroemia indica*	Har-Shingar	+++
115.	*Leea asiatica*	Gan-Gola	++
116.	*Ligustrum lucidum*	Pataal Ped	++
117.	*Ligustrum* X *vicaryi*	Privet	++
118.	*Liriodendron tulipifera*	Tulip Tree	+
119.	*Lonicera japonica*	Lushai	++
120.	*Lonicera quinquelocularis*	Chaunsh	++
121.	*Lyonia ovalifolia*	Bhelar	++
122.	*Magnolia grandiflora*	Magnolia	++
123.	*Magnolia obovata*	Japani Magnolia	+
124.	*Magnolia* X *soulangeana*	Pink Magnolia	+
125.	*Malus pumila*	Seb	++
126.	*Malvaviscus arboreus* var. *mexicanus*	Gudhal	++
127.	*Morus alba*	Toot	+++
128.	*Morus nigra*	Toot	+
129.	*Morus serrata*	Kimu	+++
130.	*Musa paradisiaca*	Kela	+
131.	*Myrica esculenta*	Kaphal	+++
132.	*Myrsine africana*	Berang	+++
133.	*Myrtus communis*	Myrtle	+
134.	*Nerium oleander*	Kaner	++
135.	*Opuntia dillenii*	Vilayati Suri	++
136.	*Padus cornuta*	Jangli Jamun, Krun	++
137.	*Parthenocissus semicordata*	Bhambti Phlankur	+++
138.	*Paulownia tomentosa*	Videshi Phool	++
139.	*Picea smithiana*	Rai	++
140.	*Pinus roxburghii*	Chil	+++
141.	*Pinus wallichiana*	Kail	+++
142.	*Pistacia khinjuk*	Kakad	+++
143.	*Platanus orientalis*	Chinar	++
144.	*Populus ciliata*	Popular	++
145.	*Populus nigra*	Poplar	++

S. No.	Botanical Name	Local Name/s	Observed Density and Availability
146.	*Prinsepia utilis*	Bekhal	+++
147.	*Prunus amygdalus*	Badam	++
148.	*Prunus armeniaca*	Chuli, Khumani	+++
149.	*Prunus domestica*	Aloo Bukhara	+++
150.	*Prunus laurocerasus*	Cherry-Laurel	+
151.	*Prunus persica*	Adu	+++
152.	*Prunus salicina*	Shakarpara	++
153.	*Punica granatum*	Anar	+++
154.	*Pyracantha crenulata*	Gingaru	+++
155.	*Pyrus communis*	Nashpati	+++
156.	*Pyrus pashia*	Kainth	+++
157.	*Pyrus pyrifolia*	Zarainth	++
158.	*Quercus floribunda*	Moru	++
159.	*Quercus incana*	Banj	++
160.	*Quercus oblongata*	Ban	+++
161.	*Quercus robur*	English Oak	+
162.	*Rhamnus persica*	Thalta	++
163.	*Rhamnus virgatus*	Reushi	+++
164.	*Rhapis excelsa*	Lady Palm	+
165.	*Rhododendron arboreum*	Braah, Brass	+++
166.	*Rhododendron simsii* var. *vittatum*	Ajalea	+
167.	*Ricinus communis*	Arind	++
168.	*Robinia pseudoacacia*	Rabinia	+++
169.	*Rosa alba*	Gulab	+++
170.	*Rosa banksiae*	Gulabri	+++
171.	*Rosa chinensis*	Gulab	+++
172.	*Rosa damascena*	Gulabadi	++
173.	*Rosa jacquemontii*	Gulab	+
174.	*Rosa moschata*	Kuja	+++
175.	*Rosa multiflora*	Gulabari	+
176.	*Rubia cordifolia*	Manjith	++
177.	*Rubus biflorus*	Dher	+++
178.	*Rubus ellipticus*	Akhe	+++
179.	*Rubus macilentus*	Heere	++
180.	*Rubus niveus*	Klenchu	+++
181.	*Rubus paniculatus*	Anchu	++

S. No.	Botanical Name	Local Name/s	Observed Density and Availability
182.	*Salix alba*	Benusa	++
183.	*Salix babylonica*	Majnu	++
184.	*Salix pycnostachya* var. *oxycarpa*	Changma Beeuins	++
185.	*Salvia leucantha*	Velvet Salvia	+++
186.	*Salvia splendens*	Salvia	+++
187.	*Sambucus nigra*	Salja, Sambucus	++
188.	*Sarcococca pruniformis*	Raabal	+++
189.	*Schefflera arboricola*	Umbrella Plant	++
190.	*Skimmia laureola*	Patti Dhup	++
191.	*Smilax aspera*	Bragh Bel	+++
192.	*Solanum laxum*	Nightshade Jasmine	+++
193.	*Solanum pseudocapsicum*	Charoti Mirch	++
194.	*Solanum virginianum*	Kant	+++
195.	*Spartium junceum*	Weavers Broom	++
196.	*Spiraea canescens*	Chakhu	++
197.	*Spiraea cantoniensis*	Spirea	++
198.	*Taxodium distichum*	Cypress	+
199.	*Taxus baccata*	Thuno	+
200.	*Thuja occidentalis*	Thuja	+++
201.	*Thuja orientalis*	Morpankh	+++
202.	*Thujopsis dolabrata*	Thujopsis	++
203.	*Trachelospermum jasminoides*	Star Jasmine	+
204.	*Tropaeolum majus*	Tindu Phool	+++
205.	*Tsuga dumosa*	Thingia	+
206.	*Ulmus wallichiana*	Brera	+
207.	*Viburnum cotinifolium*	Junglee Dakh, Ricchnabu	+++
208.	*Viburnum mullaha*	Dhaab	
209.	*Wisteria sinensis*	Wisteria	++
210.	*Xanthium strumarium*	Banokra, Gheenjadu	+++
211.	*Yucca aloifolia*	Rambaan	+++
212.	*Zanthoxylum armatum*	Tirmir	++

(+ Rare; ++ Considerable extent; +++ Abundant)

TABLE XXXVII
IUCN Status of Plants of Heritage Sites

Endangered	Vulnerable	Near Threatened	Low Risk/Least Concern
Ginkgo biloba	*Cupressus cashmeriana*	*Cryptomeria japonica*	*Abies pindrow*
	Ulmus wallichiana		*Afrocarpus gracilior*
			Araucaria bidwillii
			Araucaria columnaris
			Araucaria cunninghamii
			Cunninghamia lanceolata
			Cycas revoluta
			Desmodium elegans
			Diospyros lotus
			Juniperus chinensis
			Juniperus horizontalis
			Juniperus procera
			Juniperus pseudosabina
			Juniperus squamata
			Picea smithiana
			Platanus orientalis
			Populus nigra
			Quercus robur
			Taxodium distichum
			Taxus baccata
			Thuja occidentalis
			Thujopsis dolabrata
			Tsuga dumosa

IUCN Red Data List, 2013

PHYTOGEOGRAPHICAL ANALYSIS

Of the list of exotic (115 species) and native (92 species) plants, the exotics are more in occurrence than the natives at heritage sites in Shimla hills of H.P. (**Fig. 23; Tables XXXVIII, XXXIX**). Statistically, the ratio of exotic species has been estimated to be 55.55 %, while the representation for native species conforms to 44.44%. Habit-wise analysis of exotic/aliens collected presently indicates that the flora abounds in 55 trees **(Table XL)**, 47 shrubs **(Table XLI)**, 9 climbers **(Table XLII)**, 2 lianas **(Table XLIII)** and 2 sub-shrubs **(Table XLIV)**.

Amongst these, the trees predominate in the alien flora (**Fig. 24; Table XL**). Regarding taxonomic components, dicotyledonous taxa with 87 species grouped under 67 genera and 40 families constitute maximum (75.65%) proportion of aliens of the region **(Tables XLV, XLVI)**, followed by Gymnosperms (18 species) distributed in 11 genera under 7 families, viz., Araucariaceae (*Araucaria bidwillii*, *Araucaria columnaris*, *Araucaria cunninghamii*), Cupressaceae (*Juniperus chinensis*, *Juniperus horizontalis*, *Juniperus procera*, *Juniperus pseudosabina*, *Juniperus squamata*, *Thuja occidentalis*, *Thuja orientalis*, *Thujopsis dolabrata*), Cycadaceae (*Cycas revoluta*), Ginkgoaceae (*Ginkgo biloba*), Podocarpaceae (*Afrocarpus gracilior*), Taxaceae (*Taxus baccata*), Taxodiaceae (*Cryptomeria japonica*, *Cunninghamia lanceolata*, *Taxodium distichum*) **(Table XLVIII)**, and Monocots (10 species) distributed in 7 genera under 5 families, viz., Agavaceae (*Agave americana*, *Yucca aloifolia*), Arecaceae (*Chamaerops humilis*, *Rhapis excelsa*), Asparagaceae (*Asparagus aethiopicus*, *Asparagus densiflorus*, *Asparagus setaceus*, *Asparagus virgatus*), Poaceae (*Drepanostachyum falcatum*), Smilacaceae (*Smilax aspera*) (**Table XLVII**). Of the 52 alien families, Rosaceae (16 species) are at the top followed by Cupressaceae (8 species), Solanaceae (5 species), Araliaceae, Asparagaceae, Fabaceae, Malvaceae (4 species each), Araucariaceae, Bignoniaceae, Magnoliaceae, Moraceae, Oleaceae, Salicaceae, Taxodiaceae (3 species each), Aceraceae, Arecaceae, Agavaceae, Caprifoliaceae, Ebenaceae, Ericaceae, Fagaceae, Lamiaceae, Lythraceae, Myrtaceae, Verbenaceae (2 species each) whereas the remaining 27 families are with one species each **(Tables XLIX-LII)**. Relatively, the top ten alien families account for 46.95% representation. Likewise, the predominant genera with highest number of aliens for the heritage sites are *Juniperus*, *Prunus* (5 species each), *Asparagus*, *Rosa* (4 species each), *Araucaria*, *Cestrum*, *Hedera* (3 species each), *Acer*, *Diospyros*, *Hibiscus*, *Magnolia*, *Morus*, *Rhododendron*, *Salix*, *Salvia*, *Solanum*, *Thuja* (2 species each) and the remaining 68 genera are represented by one species each **(Table LIII)**. On the contrary, there is a preponderance of genera *Ipomoea*, *Euphorbia*, *Eucalyptus*, *Amaranthus* and *Solanum* amongst the alien species reported for H.P., India, Europe and China (Khuroo *et al*., 2012; Lambden *et al*., 2008, Wu *et al*., 2010).

TABLE XXXVIII
Exotic Plants of Heritage Sites

S. No.	Botanical Name	Family	Origin
1.	*Abutilon megapotamicum*	Malvaceae	Argentina, Brazil, Uruguay
2.	*Acacia dealbata*	Fabaceae	Australia, Tasmania
3.	*Acer negundo*	Aceraceae	N. America
4.	*Acer palmatum*	Aceraceae	Japan, N. Korea, China, Eastern Mongolia, S.E. Russia
5.	*Actinidia chinensis*	Actinidiaceae	China
6.	*Afrocarpus gracilior*	Podocarpaceae	Ethiopia, Kenya, Tanzania, Uganda
7.	*Agave americana*	Agavaceae	Mexican Highlands
8.	*Aloysia citriodora*	Verbenaceae	Argentina, Paraguay, Brazil, Uruguay, Chile, Bolivia, Peru
9.	*Amygdalus mira*	Rosaceae	W. Asia, China
10.	*Araucaria bidwillii*	Araucariaceae	S.E. Queensland
11.	*Araucaria columnaris*	Araucariaceae	New Caledonia
12.	*Araucaria cunninghamii*	Araucariaceae	S. Wales, Queensland, Papua New Guinea
13.	*Asparagus aethiopicus*	Asparagaceae	S. Africa
14.	*Asparagus densiflorus*	Asparagaceae	S. Africa
15.	*Asparagus setaceus*	Asparagaceae	S. Africa
16.	*Asparagus virgatus*	Asparagaceae	S.E. Africa
17.	*Bougainvillea glabra*	Nyctaginaceae	Brazil
18.	*Buddleja davidii*	Buddlejaceae	China, Japan
19.	*Callistemon lanceolatus*	Myrtaceae	Australia
20.	*Camellia japonica*	Theaceae	China, Taiwan, S. Korea, S. Japan
21.	*Campsis radicans*	Bignoniaceae	S.E. United States
22.	*Cestrum elegans*	Solanaceae	Mexico
23.	*Cestrum nocturnum*	Solanaceae	N. & S. America
24.	*Cestrum parqui*	Solanaceae	C. & S. America
25.	*Chamaerops humilis*	Arecaceae	Europe, N. Africa
26.	*Citrus maxima*	Rutaceae	Malaysia, Polynesia
27.	*Cryptomeria japonica*	Taxodiaceae	Japan
28.	*Cunninghamia lanceolata*	Taxodiaceae	China, Taiwan, Northern Vietnam, Laos
29.	*Cycas revoluta*	Cycadaceae	China, Japan
30.	*Diospyros kaki*	Ebenaceae	China, Turkey

S. No.	Botanical Name	Family	Origin
31.	*Diospyros lotus*	Ebenaceae	Armenia, Azerbaijan, China, Georgia, Iran, Korea, Nepal, Pakistan, Tajikistan, Turkey, Turkmenistan, Uzbekistan
32.	*Drepanostachyum falcatum*	Poaceae	Chinese-Japanese Region
33.	*Duranta erecta*	Verbenaceae	C. America
34.	*Elaeagnus umbellata*	Elaeagnaceae	Eurasia
35.	*Eriobotrya japonica*	Rosaceae	China, Japan
36.	*Eucalyptus globulus*	Myrtaceae	Australia
37.	*Euonymus japonicus*	Celastraceae	Japan, Korea, China
38.	*Fagus sylvatica*	Fagaceae	Europe
39.	*Ficus subincisa*	Moraceae	Tibet
40.	*Forsythia ovata*	Oleaceae	Korea
41.	*Fuchsia magellanica*	Onagraceae	Chile, Argentina
42.	*Ginkgo biloba*	Ginkgoaceae	China
43.	*Grevillea robusta*	Proteaceae	Queensland, New S. Wales
44.	*Hedera canariensis*	Araliaceae	Canary Island, N. Africa
45.	*Hedera colchica*	Araliaceae	S. E. Europe, Asia minor
46.	*Hedera helix*	Araliaceae	Europe, W. Asia
47.	*Hibiscus rosa-sinensis*	Malvaceae	China
48.	*Hibiscus syriacus*	Malvaceae	Syria
49.	*Hydrangea macrophylla*	Hydrangeaceae	Japan
50.	*Ipomoea purpurea*	Convolvulaceae	Tropical America
51.	*Jacaranda acutifolia*	Bignoniaceae	W.S. America, Peru
52.	*Jasminum mesnyi*	Oleaceae	China
53.	*Juglans regia*	Juglandaceae	Asia - Temperate-Afghanistan, Azerbaijan, Iran, Iraq, Turkey, Tajikistan, Turkmenistan, Asia-Tropical- Pakistan, Europe – Bulgaria, former Yugoslavia, Greece.
54.	*Juniperus chinensis*	Cupressaceae	China, Mangolia, Manchuria, Korea, Japan
55.	*Juniperus horizontalis*	Cupressaceae	N. America, United States in Alsaka
56.	*Juniperus procera*	Cupressaceae	E. Africa, S.W. Arabian Peninsula

S. No.	Botanical Name	Family	Origin
57.	*Juniperus pseudosabina*	Cupressaceae	C. Asia, Afghanistan, Tajikistan, Kazakhstan, W. China, W. Mongolia, South-Central Russia
58.	*Juniperus squamata*	Cupressaceae	Himalayas, China, N.E. Afghanistan
59.	*Kerria japonica*	Rosaceae	China, Japan, Korea
60.	*Lagerstroemia indica*	Lythraceae	China
61.	*Ligustrum lucidum*	Oleaceae	China
62.	*Liriodendron tulipifera*	Magnoliaceae	N. America, C. Florida, Louisiana
63.	*Lonicera japonica*	Caprifoliaceae	Japan, Korea
64.	*Magnolia grandiflora*	Magnoliaceae	S. E. United States (N. Carolina, Florida, Texas)
65.	*Magnolia obovata*	Magnoliaceae	Japan, Kurule Island of Russia
66.	*Malus pumila*	Rosaceae	Europe, W. Asia
67.	*Malvaviscus arboreus* var. *mexicanus*	Malvaceae	Mexico
68.	*Morus alba*	Moraceae	China, Europe, N. America
69.	*Morus nigra*	Moraceae	China
70.	*Myrsine africana*	Myrsinaceae	Macaronesia, Africa, S. Asia
71.	*Myrtus communis*	Myrtaceae	S. Europe, N. Africa
72.	*Opuntia dillenii*	Cactaceae	U.S.A. to Ecuador, Caribbean Territories.
73.	*Paulownia tomentosa*	Scrophulariaceae	C. & W. China
74.	*Platanus orientalis*	Platanaceae	Albania, Bulgaria, Former Yugoslavia, Greece, Italy
75.	*Populus nigra*	Salicaceae	Africa- Algeria, Morocco, Tunisia, Asia - temperate- Armenia, Azerbaijan, China, Georgia, Iran, Kazakhstan, Russian Federation (Ciscaucasia, Dagestan, Eastern Siberia, W. Siberia), Tajikistan, Turkey, EUROPE-Albania, Austria, Belarus, Belgium, Bulgaria, Czechoslovakia, Former Yugoslavia, France, Germany, Hungary, Italy, Netherlands, Poland, Romania, Russian Federation (European part), Spain, Switzerland, Ukraine

S. No.	Botanical Name	Family	Origin
76.	*Prunus amygdalus*	Rosaceae	S. Europe, Algeria
77.	*Prunus armeniaca*	Rosaceae	N. China, California, France, Turkey, Australia
78.	*Prunus domestica*	Rosaceae	Eurasia.
79.	*Prunus laurocerasus*	Rosaceae	S.W. Asia, S.E. Europe, Northern Iran
80.	*Prunus persica*	Rosaceae	U.S.A., Australia, Temperate Asia
81.	*Punica granatum*	Lythraceae	Persia, Afghanistan, Baluchistan
82.	*Pyrus communis*	Rosaceae	Eurasia
83.	*Quercus robur*	Fagaceae	Europe, N. Africa
84.	*Rhamnus persica*	Rhamnaceae	U.S.A.
85.	*Rhapis excelsa*	Arecaceae	S. China, Japan
86.	*Rhododendron arboreum*	Ericaceae	Eurasia, Europe, Armenia
87.	*Rhododendron simsii* var. *vittatum*	Ericaceae	China
88.	*Ricinus communis*	Euphorbiaceae	N. America
89.	*Robinia pseudoacacia*	Fabaceae	Canada, U.S.A., Sicilia
90.	*Rosa alba*	Rosaceae	China, Japan
91.	*Rosa banksiae*	Rosaceae	China
92.	*Rosa chinensis*	Rosaceae	China
93.	*Rosa multiflora*	Rosaceae	Japan, China
94.	*Rubia cordifolia*	Rubiaceae	Africa, Tropical Asia, Australia
95.	*Rubus biflorus*	Rosaceae	China
96.	*Salix alba*	Salicaceae	W. Asia & Europe.
97.	*Salix babylonica*	Salicaceae	China.
98.	*Salvia leucantha*	Lamiaceae	Central & E. Mexico
99.	*Salvia splendens*	Lamiaceae	Mexico, California, Brazil
100.	*Sambucus nigra*	Caprifoliaceae	Europe
101.	*Schefflera arboricola*	Araliaceae	China & Taiwan
102.	*Smilax aspera*	Smilacaceae	Turkey, Europe
103.	*Solanum laxum*	Solanaceae	Brazil
104.	*Solanum pseudocapsicum*	Solanaceae	C. Europe
105.	*Spartium junceum*	Fabaceae	S. Europe, S.W. Asia, N.W. Africa
106.	*Spiraea cantoniensis*	Rosaceae	China
107.	*Taxodium distichum*	Taxodiaceae	Southern U.S.A
108.	*Taxus baccata*	Taxaceae	Eurasia, N. Africa, Iran, Caucasus Origin

S. No.	Botanical Name	Family	Origin
109.	*Thuja occidentalis*	Cupressaceae	N.E. U.S.A., S.E. Canada
110.	*Thuja orientalis*	Cupressaceae	Mediterranean, China, S. Korea
111.	*Thujopsis dolabrata*	Cupressaceae	Japan
112.	*Tropaeolum majus*	Tropaeolaceae	Mexico, S. America
113.	*Wisteria sinensis*	Fabaceae	China
114.	*Xanthium strumarium*	Asteraceae	N. America
115.	*Yucca aloifolia*	Agavaceae	Mexico, W. Indies, Southern United State

TABLE XXXIX
Indigenous Plants of Heritage Sites

S. No.	Botanical Name	Family	Origin
1.	*Abies pindrow*	Pinaceae	W. Himalayas, Afghanistan, Pakistan, India, C. Nepal
2.	*Acer oblongum*	Aceraceae	Bhutan, India, Nepal, Pakistan
3.	*Aesculus indica*	Hippocastanaceae	India
4.	*Arundo donax*	Poaceae	E. & S. Asia
5.	*Asparagus racemosus*	Asparagaceae	India
6.	*Berberis aristata*	Berberidaceae	India, China, Caribbean Territories
7.	*Berberis lycium*	Berberidaceae	India, Nepal, Pakistan
8.	*Buddleja asiatica*	Buddlejaceae	India, Nepal, Bangladesh, China, Taiwan, Burma, Thailand, Vietnam, Malaysia, New Guinea, Philippines
9.	*Buddleja crispa*	Buddlejaceae	Afghanistan, Bhutan, N. India, Nepal, Pakistan, China
10.	*Buxus wallichiana*	Buxaceae	E. Asia, Himalayas, Afghanistan, C. Nepal
11.	*Cedrus deodara*	Pinaceae	Himalayas, Afghanistan, China, India, Pakistan
12.	*Celtis australis*	Ulmaceae	N. India, Temperate Himalayas
13.	*Cerasus cerasoides*	Rosaceae	India
14.	*Citrus limon*	Rutaceae	India, China, Australia, New Zealand, Mexico, U.S.A., Italy, Sicily, France, Spain
15.	*Clematis buchananiana*	Ranunculaceae	E. Asia, Himalayas, Tibet, Myanmar, W. China

S. No.	Botanical Name	Family	Origin
16.	*Clematis montana*	Ranunculaceae	China, Himalayas
17.	*Colquhounia coccinea*	Lamiaceae	China Nepal, Bhutan, India, Myanmar, Thailand
18.	*Coriaria nepalensis*	Coriariaceae	E. Asia, Himalayas, Pakistan, S.W. China
19.	*Cornus capitata*	Cornaceae	China, Indian Sub-continent, N. Myanmar
20.	*Cornus macrophylla*	Cornaceae	N. Pakistan, Nepal, N. India
21.	*Cotoneaster bacillaris*	Rosaceae	W. Himalayas, India, Pakistan
22.	*Cupressus cashmeriana*	Cupressaceae	E. Himalayas, Bhutan, N.E. India
23.	*Cupressus torulosa*	Cupressaceae	China, India-Himachal Pradesh, Uttar Pradesh
24.	*Daphne papyracea*	Thymelaeaceae	India-Himalayas
25.	*Datura stramonium*	Solanaceae	Caspian regions of Europe, N. America, Indian Subcontinent
26.	*Debregeasia saeneb*	Urticaceae	W. Asia- W. Himalayas
27.	*Desmodium elegans*	Fabaceae	Himalayas, China
28.	*Desmodium multiflorum*	Fabaceae	S.E. Asia, Himalayas, China
29.	*Erythrina arborescens*	Fabaceae	Central & E. Himalayas
30.	*Euonymus fimbriatus*	Celastraceae	E. Asia, Himalayas
31.	*Euonymus lucidus*	Celastraceae	N.E. Himalayas, Pakistan
32.	*Euonymus tingens*	Celastraceae	E. Asia, Himalayas
33.	*Ficus auriculata*	Moraceae	China, Bhutan, India, Nepal, Malaysia, Myanmar, Thailand
34.	*Ficus benjamina*	Moraceae	India, S. China, S.E. Asia, Malaysia, Philippine, N. Australia, Island of the S. Pacific
35.	*Ficus elastica*	Moraceae	Bhutan, India, Indonesia, Malaysia, Myanmar, Nepal
36.	*Ficus neriifolia*	Moraceae	S. Asia, India
37.	*Ficus palmata*	Moraceae	Afghanistan, Egypt, Ethiopia, Iran, India, Israel, Nepal, Oman, Pakistan, Somalia, Sudan
38.	*Ficus religiosa*	Moraceae	Bangladesh, India, Nepal, Pakistan, Thailand, Vietnam
39.	*Ficus sarmentosa*	Moraceae	E. Asia
40.	*Girardinia diversifolia*	Urticaceae	E. Asia, India, Nepal

S. No.	Botanical Name	Family	Origin
41.	*Hedera nepalensis*	Araliaceae	Nepal, Bhutan, Afghanistan, India, China, Laos, Myanmar, Thailand, Vietnam
42.	*Himalrandia tetrasperma*	Rubiaceae	Himalayas, Nepal, Bhutan, N. E. India
43.	*Hypericum hookerianum*	Clusiaceae	Bangladesh, Bhutan, India, Myanmar, Nepal, Thailand, Vietnam
44.	*Hypericum oblongifolium*	Clusiaceae	India- Himachal Pradesh, Jammu and Kashmir, Uttar Pradesh, Nepal, Pakistan.
45.	*Hypericum patalum*	Clusiaceae	E. Himalayas, China
46.	*Ilex dipyrena*	Aquifoliaceae	Nepal, Bhutan, India
47.	*Indigofera dosua*	Fabaceae	Bhutan, India, Indonesia, Laos, Myanmar, Nepal
48.	*Inula cappa*	Asteraceae	Asia, China, Bhutan, India
49.	*Jasminum grandiflorum*	Oleaceae	India
50.	*Jasminum humile*	Oleaceae	Afghanistan, Bhutan, China, India, Nepal, Pakistan, Tajikistan
51.	*Jasminum multiflorum*	Oleaceae	India
52.	*Leea asiatica*	Leeaceae	S.E. Asia
53.	*Lonicera quinquelocularis*	Caprifoliaceae	Himalayas
54.	*Lyonia ovalifolia*	Ericaceae	China, India, Japan, Bhutan
55.	*Morus serrata*	Moraceae	China, India- Himachal Pradesh, Jammu and Kashmir, Punjab, Uttar Pradesh, Nepal, Pakistan.
56.	*Musa paradisiaca*	Musaceae	Tropical Asia
57.	*Myrica esculenta*	Myricaceae	India
58.	*Nerium oleander*	Apocyanaceae	Africa- Algeria, Ethiopia, Libya, Morocco, Tunisia, Asia temperate -Afghanistan, China Cyprus, Iran, Iraq, Israel, Jordan, Lebanon, Oman, United Arab Emirates, Syria, Turkey, Asia Tropical- India, Nepal, Pakistan, Europe- Albania, Former Yugoslavia, Greece, Italy, Malta, France, Portugal, Spain
59.	*Padus cornuta*	Rosaceae	India

S. No.	Botanical Name	Family	Origin
60.	*Parthenocissus semicordata*	Vitaceae	Himalayas
61.	*Picea smithiana*	Pinaceae	W. Himalayas, N.E. Afghanistan, India, C. Nepal
62.	*Pinus roxburghii*	Pinaceae	Afghanistan, China, Pakistan, Bhutan, India, Myanmar, Nepal
63.	*Pinus wallichiana*	Pinaceae	Afghanistan, China, Pakistan, Bhutan, India, Myanmar, Nepal
64.	*Pistacia khinjuk*	Anacardiaceae	Afghanistan, India, Nepal, Pakistan, Myanmar
65.	*Populus ciliata*	Salicaceae	China, Bhutan, India, Nepal, Myanmar
66.	*Prinsepia utilis*	Rosaceae	Temperate Himalayas
67.	*Prunus salicina*	Rosaceae	S. China, S. E. Asia
68.	*Pyracantha crenulata*	Rosaceae	Asia
69.	*Pyrus pashia*	Rosaceae	Afghanistan, Bhutan, China, India, Nepal, Myanmar, Pakistan, Thailand, Vietnam
70.	*Pyrus* pyrifolia	Rosaceae	Asia Minor
71.	*Quercus floribunda*	Fagaceae	Afghanistan, India, Nepal, Pakistan
72.	*Quercus incana*	Fagaceae	India, Myanmar, Nepal, Pakistan, Thailand
73.	*Quercus oblongata*	Fagaceae	Himalayas, India
74.	*Rhamnus virgatus*	Rhamnaceae	E. Asia, Himalayas
75.	*Rosa damascena*	Rosaceae	India
76.	*Rosa jacquemontii*	Rosaceae	India
77.	*Rosa moschata*	Rosaceae	India
78.	*Rubus ellipticus*	Rosaceae	S. E. Asia, Himalayas
79.	*Rubus macilentus*	Rosaceae	Asia, China, India
80.	*Rubus niveus*	Rosaceae	Asia, China, India
81.	*Rubus paniculatus*	Rosaceae	Asia, China
82.	*Salix pycnostachya* var. *oxycarpa*	Salicaceae	India
83.	*Sarcococca pruniformis*	Buxaceae	W. Himalayas
84.	*Skimmia laureola*	Rutaceae	N. China, N. Himalayas
85.	*Solanum virginianum*	Solanaceae	India
86.	*Spiraea canescens*	Rosaceae	S.E. Asia, China, Japan

S. No.	Botanical Name	Family	Origin
87.	*Trachelospermum jasminoides*	Apocyanaceae	S.E. Asia, Japan, Korea, S. China, Vietnam
88.	*Tsuga dumosa*	Pinaceae	Bhutan, China, Tibet, Yunnan, India (Assam, Sikkim, Uttar Pradesh), Myanmar, Nepal
89.	*Ulmus wallichiana*	Ulmaceae	Afghanistan, N. Pakistan, N. India, W. Nepal
90.	*Viburnum cotinifolium*	Caprifoliaceae	Asia
91.	*Viburnum mullaha*	Caprifoliaceae	S. E. Asia, Himalayas
92.	*Zanthoxylum armatum*	Rutaceae	Bangladesh, Bhutan, China, India, Indonesia, Japan, Korea, Myanmar, Nepal, Pakistan, Taiwan, Thailand, Vietnam

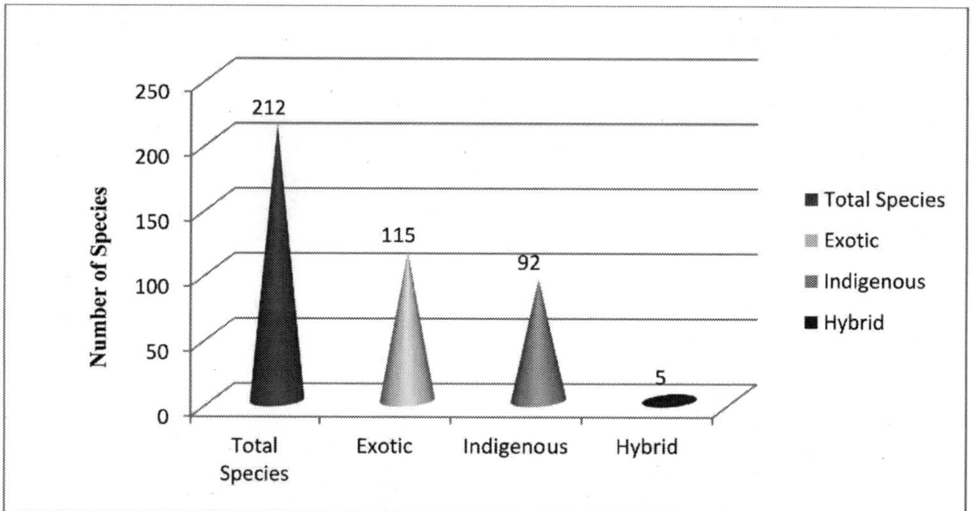

Fig. 23. Total Number of Exotic, Indigenous and Hybrid (Garden) Species of Heritage Sites.

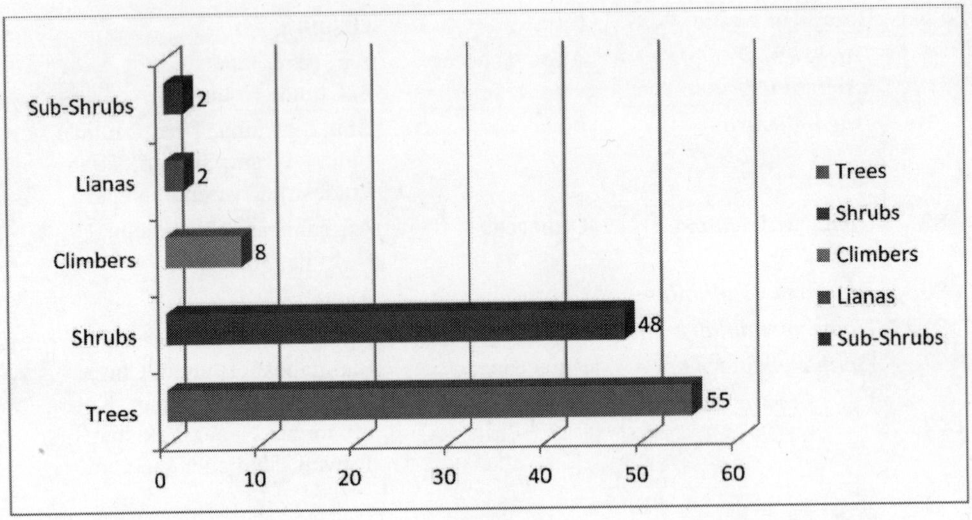

Fig. 24. Exotic Trees, Shrubs, Climbers, Lianas and Sub-Shrubs of Heritage Sites.

TABLE XL
Exotic Trees

S. No.	Botanical Name	Family
1.	*Acacia dealbata*	Fabaceae
2.	*Acer negundo*	Aceraceae
3.	*Acer palmatum*	Aceraceae
4.	*Afrocarpus gracilior*	Podocarpaceae
5.	*Amygdalus mira*	Rosaceae
6.	*Araucaria bidwillii*	Araucariaceae
7.	*Araucaria columnaris*	Araucariaceae
8.	*Araucaria cunninghamii*	Araucariaceae
9.	*Callistemon lanceolatus*	Myrtaceae
10.	*Chamaerops humilis*	Arecaceae
11.	*Citrus maxima*	Rutaceae
12.	*Cryptomeria japonica*	Taxodiaceae
13.	*Cunninghamia lanceolata*	Taxodiaceae
14.	*Cycas revoluta*	Cycadaceae
15.	*Diospyros kaki*	Ebenaceae
16.	*Diospyros lotus*	Ebenaceae
17.	*Eriobotrya japonica*	Rosaceae
18.	*Eucalyptus globulus*	Myrtaceae

S. No.	Botanical Name	Family
19.	*Euonymus japonicus*	Celastraceae
20.	*Fagus sylvatica*	Fagaceae
21.	*Ficus subincisa*	Moraceae
22.	*Ginkgo biloba*	Ginkgoaceae
23.	*Grevillea robusta*	Proteaceae
24.	*Jacaranda acutifolia*	Bignoniaceae
25.	*Juglans regia*	Juglandaceae
26.	*Juniperus chinensis*	Cupressaceae
27.	*Juniperus procera*	Cupressaceae
28.	*Juniperus pseudosabina*	Cupressaceae
29.	*Juniperus squamata*	Cupressaceae
30.	*Liriodendron tulipifera*	Magnoliaceae
31.	*Magnolia grandiflora*	Magnoliaceae
32.	*Magnolia obovata*	Magnoliaceae
33.	*Malus pumila*	Rosaceae
34.	*Morus alba*	Moraceae
35.	*Morus nigra*	Moraceae
36.	*Paulownia tomentosa*	Scrophulariaceae
37.	*Platanus orientalis*	Platanaceae
38.	*Populus nigra*	Salicaceae
39.	*Prunus amygdalus*	Rosaceae
40.	*Prunus armeniaca*	Rosaceae
41.	*Prunus domestica*	Rosaceae
42.	*Prunus laurocerasus*	Rosaceae
43.	*Prunus persica*	Rosaceae
44.	*Pyrus communis*	Rosaceae
45.	*Quercus robur*	Fagaceae
46.	*Rhamnus persica*	Rhamnaceae
47.	*Rhododendron arboreum*	Ericaceae
48.	*Robinia pseudoacacia*	Fabaceae
49.	*Salix alba*	Salicaceae
50.	*Salix babylonica*	Salicaceae
51.	*Taxodium distichum*	Taxodiaceae
52.	*Taxus baccata*	Taxaceae
53.	*Thuja occidentalis*	Cupressaceae
54.	*Thuja orientalis*	Cupressaceae
55.	*Thujopsis dolabrata*	Cupressaceae

TABLE XLI
Exotic Shrubs

S. No.	Botanical Name	Family
1.	*Abutilon megapotamicum*	Malvaceae
2.	*Agave americana*	Agavaceae
3.	*Aloysia citriodora*	Verbenaceae
4.	*Asparagus aethiopicus*	Asparagaceae
5.	*Asparagus densiflorus*	Asparagaceae
6.	*Asparagus setaceus*	Asparagaceae
7.	*Asparagus virgatus*	Asparagaceae
8.	*Bougainvillea glabra*	Nyctaginaceae
9.	*Buddleja davidii*	Buddlejaceae
10.	*Camellia japonica*	Theaceae
11.	*Campsis radicans*	Bignoniaceae
12.	*Cestrum elegans*	Solanaceae
13.	*Cestrum nocturnum*	Solanaceae
14.	*Cestrum parqui*	Solanaceae
15.	*Drepanostachyum falcatum*	Poaceae
16.	*Duranta erecta*	Verbenaceae
17.	*Elaeagnus umbellata*	Elaeagnaceae
18.	*Forsythia ovata*	Oleaceae
19.	*Fuchsia magellanica*	Onagraceae
20.	*Hibiscus rosa-sinensis*	Malvaceae
21.	*Hibiscus syriacus*	Malvaceae
22.	*Hydrangea macrophylla*	Hydrangeaceae
23.	*Jasminum mesnyi*	Oleaceae
24.	*Juniperus horizontalis*	Cupressaceae
25.	*Kerria japonica*	Rosaceae
26.	*Lagerstroemia indica*	Lythraceae
27.	*Ligustrum lucidum*	Oleaceae
28.	*Malvaviscus arboreus* var. *mexicanus*	Malvaceae
29.	*Myrsine africana*	Myrsinaceae
30.	*Myrtus communis*	Myrtaceae
31.	*Opuntia dillenii*	Cactaceae
32.	*Punica granatum*	Lythraceae
33.	*Rhapis excelsa*	Arecaceae
34.	*Rhododendron simsii* var. *vittatum*	Ericaceae
35.	*Ricinus communis*	Euphorbiaceae

S. No.	Botanical Name	Family
36.	*Rosa alba*	Rosaceae
37.	*Rosa banksiae*	Rosaceae
38.	*Rosa chinensis*	Rosaceae
39.	*Rosa multiflora*	Rosaceae
40.	*Rubus biflorus*	Rosaceae
41.	*Sambucus nigra*	Caprifoliaceae
42.	*Schefflera arboricola*	Araliaceae
43.	*Solanum pseudocapsicum*	Solanaceae
44.	*Spartium junceum*	Fabaceae
45.	*Spiraea cantoniensis*	Rosaceae
46.	*Xanthium strumarium*	Asteraceae
47.	*Yucca aloifolia*	Agavaceae

TABLE XLII
Exotic Climbers

S. No.	Botanical Name	Family
1.	*Hedera canariensis*	Araliaceae
2.	*Hedera colchica*	Araliaceae
3.	*Hedera helix*	Araliaceae
4.	*Ipomoea purpurea*	Convolvulaceae
5.	*Lonicera japonica*	Caprifoliaceae
6.	*Rubia cordifolia*	Rubiaceae
7.	*Smilax aspera*	Smilacaceae
8.	*Solanum laxum*	Solanaceae
9.	*Tropaeolum majus*	Tropaeolaceae

TABLE XLIII
Exotic Lianas

S. No.	Botanical Name	Family
1.	*Actinidia chinensis*	Actinidiaceae
2.	*Wisteria sinensis*	Fabaceae

TABLE XLIV
Exotic Sub-Shrubs

S. No.	Botanical Name	Family
1.	*Salvia leucantha*	Lamiaceae
2.	*Salvia splendens*	Lamiaceae

TABLE XLV
Taxonomic Components of Exotics

	No. of Families	No. of Genera	No. of Species
Dicots	40	67	87
Monocots	5	7	10
Gymnosperms	7	11	18
Total	52	85	115

TABLE XLVI
Exotic Dicots

S. No.	Botanical Name	Family	Origin
1.	*Abutilon megapotamicum*	Malvaceae	Argentina, Brazil, Uruguay
2.	*Acacia dealbata*	Fabaceae	Australia & Tasmania
3.	*Acer negundo*	Aceraceae	N. America
4.	*Acer palmatum*	Aceraceae	Japan, N. Korea, China, Eastern Mongolia and S.E. Russia
5.	*Actinidia chinensis*	Actinidiaceae	China
6.	*Aloysia citriodora*	Verbenaceae	Argentina, Paraguay, Brazil, Uruguay, Chile, Bolivia, Peru
7.	*Amygdalus mira*	Rosaceae	W. Asia, China
8.	*Bougainvillea glabra*	Nyctaginaceae	Brazil
9.	*Buddleja davidii*	Buddlejaceae	China, Japan
10.	*Callistemon lanceolatus*	Myrtaceae	Australia
11.	*Camellia japonica*	Theaceae	China, Taiwan, Southern Korea, Southern Japan
12.	*Campsis radicans*	Bignoniaceae	S.E. United States
13.	*Cestrum elegans*	Solanaceae	Mexico
14.	*Cestrum nocturnum*	Solanaceae	N. & S. America
15.	*Cestrum parqui*	Solanaceae	Central & S. America
16.	*Citrus maxima*	Rutaceae	Malaysia, Polynesia
17.	*Diospyros kaki*	Ebenaceae	China, Turkey

S. No.	Botanical Name	Family	Origin
18.	*Diospyros lotus*	Ebenaceae	Armenia, Azerbaijan, China, Georgia, Iran, Korea, Nepal, Pakistan, Tajikistan, Turkey, Turkmenistan, Uzbekistan
19.	*Duranta erecta*	Verbenaceae	C. America
20.	*Elaeagnus umbellata*	Elaeagnaceae	Euresia
21.	*Eriobotrya japonica*	Rosaceae	China, Japan
22.	*Eucalyptus globulus*	Myrtaceae	Australia
23.	*Euonymus japonicus*	Celastraceae	Japan, Korea, China
24.	*Fagus sylvatica*	Fagaceae	Europe
25.	*Ficus subincisa*	Moraceae	Tibet
26.	*Forsythia ovata*	Oleaceae	Korea
27.	*Fuchsia magellanica*	Onagraceae	Chile, Argentina
28.	*Grevillea robusta*	Proteaceae	Queensland, New S. Wales
29.	*Hedera canariensis*	Araliaceae	Canary Island, Northern Africa
30.	*Hedera colchica*	Araliaceae	S. E. Europe, Asia minor
31.	*Hedera helix*	Araliaceae	Europe, W. Asia
32.	*Hibiscus rosa-sinensis*	Malvaceae	China
33.	*Hibiscus syriacus*	Malvaceae	Syria
34.	*Hydrangea macrophylla*	Hydrangeaceae	Japan
35.	*Ipomoea purpurea*	Convolvulaceae	Tropical America
36.	*Jacaranda acutifolia*	Bignoniaceae	Western S. America, Peru
37.	*Jasminum mesnyi*	Oleaceae	China
38.	*Juglans regia*	Juglandaceae	Asia - Temperate - Afghanistan, Azerbaijan, Iran, Iraq, Turkey, Tajikistan, Turkmenistan, Asia-Tropical- Pakistan, Europe - Bulgaria, former Yugoslavia, Greece
39.	*Kerria japonica*	Rosaceae	China, Japan, Korea
40.	*Lagerstroemia indica*	Lythraceae	China
41.	*Ligustrum lucidum*	Oleaceae	China
42.	*Liriodendron tulipifera*	Magnoliaceae	N. America, Central Florida, Louisiana
43.	*Lonicera japonica*	Caprifoliaceae	Japan, Korea
44.	*Magnolia grandiflora*	Magnoliaceae	S.E. United States (N. Carolina, Florida and Texas)
45.	*Magnolia obovata*	Magnoliaceae	Japan, Kurule Island of Russia
46.	*Malus pumila*	Rosaceae	Europe, W. Asia

S. No.	Botanical Name	Family	Origin
47.	*Malvaviscus arboreus* var. *mexicanus*	Malvaceae	Mexico
48.	*Morus alba*	Moraceae	China, Europe, N. America
49.	*Morus nigra*	Moraceae	China.
50.	*Myrsine africana*	Myrsinaceae	Macaronesia, Africa, S. Asia
51.	*Myrtus communis*	Myrtaceae	S. Europe, N. Africa
52.	*Opuntia dillenii*	Cactaceae	U.S.A. to Ecuador, Caribbean Territories.
53.	*Paulownia tomentosa*	Scrophulariaceae	Central & W. China
54.	*Platanus orientalis*	Platanaceae	Albania, Bulgaria, Former Yugoslavia, Greece, Italy
55.	*Populus nigra*	Salicaceae	Africa- Algeria, Morocco, Tunisia; Asia- temperate- Armenia, Azerbaijan, China, Georgia, Iran, Kazakhstan, Russian Federation (Ciscaucasia, Dagestan, E. Siberia, W. Siberia), Tajikistan, Turkey; Europe -Albania, Austria, Belarus, Belgium, Bulgaria, Czechoslovakia, Former Yugoslavia, France, Germany, Hungary, Italy, Netherlands, Poland, Romania, Russian Federation (European part), Spain, Switzerland, Ukraine
56.	*Prunus amygdalus*	Rosaceae	S. Europe, Algeria
57.	*Prunus armeniaca*	Rosaceae	N. China, California, France, Turkey, Australia
58.	*Prunus domestica*	Rosaceae	Eurasia
59.	*Prunus laurocerasus*	Rosaceae	S.W. Asia, S.E. Europe, N. Iran
60.	*Prunus persica*	Rosaceae	U.S.A., Australia, Temperate Asia
61.	*Punica granatum*	Lythraceae	Persia, Afghanistan, Baluchistan
62.	*Pyrus communis*	Rosaceae	Eurasia
63.	*Quercus robur*	Fagaceae	Europe, N. Africa
64.	*Rhamnus persica*	Rhamnaceae	U.S.A.
65.	*Rhododendron arboreum*	Ericaceae	Eurasia, Europe, Armenia
66.	*Rhododendron simsii* var. *vittatum*	Ericaceae	China

S. No.	Botanical Name	Family	Origin
67.	*Ricinus communis*	Euphorbiaceae	N. America
68.	*Robinia pseudoacacia*	Fabaceae	Canada, U.S.A., Sicilia
69.	*Rosa alba*	Rosaceae	China, Japan
70.	*Rosa banksiae*	Rosaceae	China
71.	*Rosa chinensis*	Rosaceae	China
72.	*Rosa multiflora*	Rosaceae	Japan, China
73.	*Rubia cordifolia*	Rubiaceae	Africa, Tropical Asia, Australia
74.	*Rubus biflorus*	Rosaceae	China.
75.	*Salix alba*	Salicaceae	W. Asia, Europe.
76.	*Salix babylonica*	Salicaceae	China.
77.	*Salvia leucantha*	Lamiaceae	C. & E. Mexico
78.	*Salvia splendens*	Lamiaceae	Mexico, California, Brazil
79.	*Sambucus nigra*	Caprifoliaceae	Europe
80.	*Schefflera arboricola*	Araliaceae	China, Taiwan
81.	*Solanum laxum*	Solanaceae	Brazil
82.	*Solanum pseudocapsicum*	Solanaceae	C. Europe
83.	*Spartium junceum*	Fabaceae	S. Europe, S.W. Asia, N.W. Africa
84.	*Spiraea cantoniensis*	Rosaceae	China
85.	*Tropaeolum majus*	Tropaeolaceae	Mexico, S. America
86.	*Wisteria sinensis*	Fabaceae	China
87.	*Xanthium strumarium*	Asteraceae	N. America

TABLE XLVII
Exotic Monocots

S. No.	Botanical Name	Family	Origin
1.	*Agave americana*	Agavaceae	Mexican Highlands
2.	*Asparagus aethiopicus*	Asparagaceae	S. Africa
3.	*Asparagus densiflorus*	Asparagaceae	S. Africa
4.	*Asparagus setaceus*	Asparagaceae	S. Africa
5.	*Asparagus virgatus*	Asparagaceae	S.E. Africa
6.	*Chamaerops humilis*	Arecaceae	Europe, N. Africa
7.	*Drepanostachyum falcatum*	Poaceae	Chinese-Japanese Region
8.	*Rhapis excelsa*	Arecaceae	S. China, Japan
9.	*Smilax aspera*	Smilacaceae	Turkey, Europe
10.	Yucca aloifolia	Agavaceae	Mexico, W. Indies, S. United States

TABLE XLVIII
Exotic Gymnosperms

S. No.	Botanical Name	Family	Origin
1.	*Afrocarpus gracilior*	Podocarpaceae	Ethiopia, Kenya, Tanzania, Uganda
2.	*Araucaria bidwillii*	Araucariaceae	S.E. Queensland
3.	*Araucaria columnaris*	Araucariaceae	New Caledonia
4.	*Araucaria cunninghamii*	Araucariaceae	S. Wales, Queensland, Papua New Guinea
5.	*Cryptomeria japonica*	Taxodiaceae	Japan
6.	*Cunninghamia lanceolata*	Taxodiaceae	China, Taiwan, N. Vietnam, Laos
7.	*Cycas revoluta*	Cycadaceae	China, Japan
8.	*Ginkgo biloba*	Ginkgoaceae	China
9.	*Juniperus chinensis*	Cupressaceae	China, Mangolia, Manchuria, Korea, Japan
10.	*Juniperus horizontalis*	Cupressaceae	N. America, United States in Alsaka
11.	*Juniperus procera*	Cupressaceae	E. Africa, S.W. Arabian Peninsula
12.	*Juniperus pseudosabina*	Cupressaceae	C. Asia, Afghanistan, Tajikistan, Kazakhstan, W. China, W. Mongolia, South-Central Russia
13.	*Juniperus squamata*	Cupressaceae	Himalayas, China, N.E. Afghanistan
14.	*Taxodium distichum*	Taxodiaceae	Southern USA
15.	*Taxus baccata*	Taxaceae	Eurasia, N. Africa, Iran, Caucasus Origin
16.	*Thuja occidentalis*	Cupressaceae	N.E. USA, S.E. Canada
17.	*Thuja orientalis*	Cupressaceae	Mediterranean, China, S. Korea
18.	*Thujopsis dolabrata*	Cupressaceae	Japan

TABLE XLIX
Exotic Families

S. No.	Family	Species
1.	Rosaceae	16
2.	Cupressaceae	8
3.	Solanaceae	5
4.	Araliaceae	4
5.	Asparagaceae	4
6.	Fabaceae	4
7.	Malvaceae	4
8.	Araucariaceae	3
9.	Bignoniaceae	3
10.	Magnoliaceae	3
11.	Moraceae	3
12.	Oleaceae	3
13.	Salicaceae	3
14.	Taxodiaceae	3
15.	Aceraceae	2
16.	Arecaceae	2
17.	Agavaceae	2
18.	Caprifoliaceae	2
19.	Ebenaceae	2
20.	Ericaceae	2
21.	Fagaceae	2
22.	Lamiaceae	2
23.	Lythraceae	2
24.	Myrtaceae	2
25.	Verbenaceae	2
26.	Actinidiaceae	1
27.	Asteraceae	1
28.	Buddlejaceae	1
29.	Cactaceae	1
30.	Celasteraceae	1
31.	Convolvulaceae	1
32.	Cycadaceae	1
33.	Elaeganaceae	1
34.	Euphorbiaceae	1
35.	Ginkgoaceae	1
36.	Hydrangaceae	1

S. No.	Family	Species
37.	Juglandaceae	1
38.	Myrsinaceae	1
39.	Nyctaginaceae	1
40.	Onagraceae	1
41.	Platanaceae	1
42.	Poaceae	1
43.	Podocarpaceae	1
44.	Proteaceae	1
45.	Rhamnaceae	1
46.	Rubiaceae	1
47.	Rutaceae	1
48.	Scrophulariaceae	1
49.	Smilacaceae	1
50.	Taxaceae	1
51.	Theaceae	1
52.	Tropaelaceae	1

TABLE L
Exotic Dicotyledonous Families with Corresponding Genera/Species

S. No.	Family	Botanical Name	Genera/ Species
1.	Aceraceae	*Acer negundo*, *Acer palmatum*	1/2
2.	Actinidiaceae	*Actinidia chinensis*	1/1
3.	Araliaceae	*Hedera canariensis*, *Hedera colchica*, *Hedera helix*, *Schefflera arboricola*	2/4
4.	Asteraceae	*Xanthium strumarium*	1/1
5.	Bignoniaceae	*Campsis radicans*, *Jacaranda acutifolia*	2/2
6.	Buddlejaceae	*Buddleja davidii*	1/1
7.	Cactaceae	*Opuntia dillenii*	1/1
8.	Caprifoliaceae	*Lonicera japonica*, *Sambucus nigra*	2/2
9.	Celastraceae	*Euonymus japonicus*	1/1
10.	Convolvulaceae	*Ipomoea purpurea*	1/1
11.	Ebenaceae	*Diospyros kaki*, *Diospyros lotus*	1/2
12.	Elaeagnaceae	*Elaeagnus umbellata*	1/1
13.	Ericaceae	*Rhododendron arboreum*, *Rhododendron simsii* var. *vittatum*	1/2
14.	Euphorbiaceae	*Ricinus communis*	1/1
15.	Fabaceae	*Acacia dealbata*, *Robinia pseudoacacia*, *Spartium junceum*, *Wisteria sinensis*	4/4

S. No.	Family	Botanical Name	Genera/ Species
16.	Fagaceae	*Fagus sylvatica, Quercus robur*	2/2
17.	Hydrangeaceae	*Hydrangea macrophylla*	1/1
18.	Juglandaceae	*Juglans regia*	1/1
19.	Lamiaceae	*Salvia leucantha, Salvia splendens*	1/2
20.	Lythraceae	*Lagerstroemia indica, Punica granatum*	2/2
21.	Magnoliaceae	*Liriodendron tulipifera, Magnolia grandiflora, Magnolia obovata*	2/3
22.	Malvaceae	*Abutilon megapotamicum, Hibiscus rosa-sinensis, Hibiscus syriacus, Malvaviscus arboreus* var. *mexicanus*	3/4
23.	Moraceae	*Ficus subincisa, Morus alba Morus nigra*	2/3
24.	Myrsinaceae	*Myrsine africana*	1/1
25.	Myrtaceae	*Callistemon lanceolatus, Eucalyptus globulus, Myrtus communis*	3/3
26.	Nyctaginaceae	*Bougainvillea glabra*	1/1
27.	Oleaceae	*Forsythia ovata, Jasminum mesnyi, Ligustrum lucidum*	3/3
28.	Onagraceae	*Fuchsia magellanica*	1/1
29.	Platanaceae	*Platanus orientalis*	1/1
30.	Proteaceae	*Grevillea robusta*	1/1
31.	Rhamnaceae	*Rhamnus persica*	1/1
32.	Rosaceae	*Amygdalus mira, Eriobotrya japonica, Kerria japonica, Malus pumila, Prunus amygdalus, Prunus armeniaca, Prunus domestica, Prunus laurocerasus, Prunus persica, Pyrus communis, Rosa alba, Rosa banksiae, Rosa chinensis, Rosa multiflora, Rubus biflorus, Spiraea cantoniensis*	9/16
33.	Rubiaceae	*Rubia cordifolia*	1/1
34.	Rutaceae	*Citrus maxima*	1/1
35.	Salicaceae	*Populus nigra, Salix alba, Salix babylonica*	2/3
36.	Scrophulariaceae	*Paulownia tomentosa*	1/1
37.	Solanaceae	*Cestrum elegans, Cestrum nocturnum, Cestrum parqui Solanum* laxum, *Solanum pseudocapsicum*	2/5
38.	Theaceae	*Camellia japonica*	1/1
39.	Tropaeolaceae	*Tropaeolum majus*	1/1
40.	Verbenaceae	Aloysia citriodora, *Duranta erecta*	2/2

TABLE LI
Exotic Monocotyledonous Families with Corresponding Genera/Species

S. No.	Family	Species	Genera/ Species
1.	Agavaceae	*Agave americana*, *Yucca aloifolia*	2/2
2.	Arecaceae	*Chamaerops humilis*, *Rhapis excelsa*	2/2
3.	Asparagaceae	*Asparagus aethiopicus*, *Asparagus densiflorus*, *Asparagus setaceus*, *Asparagus virgatus*	1/4
4.	Poaceae	*Drepanostachyum falcatum*	1/1
5.	Smilacaceae	*Smilax aspera*	1/1

TABLE LII
Exotic Gymnospermous Families with Corresponding Genera/Species

S. No.	Family	Species	Genera/ Species
1.	Araucariaceae	*Araucaria bidwillii*, *Araucaria columnaris*, *Araucaria cunninghamii*	1/3
2.	Cupressaceae	*Juniperus chinensis*, *Juniperus horizontalis*, *Juniperus procera*, *Juniperus pseudosabina*, *Juniperus squamata*, *Thuja occidentalis*, *Thuja orientalis*, *Thujopsis dolabrata*	3/8
3.	Cycadaceae	*Cycas revoluta*	1/1
4.	Ginkgoaceae	*Ginkgo biloba*	1/1
5.	Podocarpaceae	*Afrocarpus gracilior*	1/1
6.	Taxaceae	*Taxus baccata*	1/1
7.	Taxodiaceae	*Cryptomeria japonica*, *Cunninghamia lanceolata*, *Taxodium distichum*	3/3

TABLE LIII
Predominant Exotic Genera

Genera	Species	Genera	Species
Juniperus	5	*Hibiscus*	2
Prunus	5	*Magnolia*	2
Asparagus	4	*Morus*	2
Rosa	4	*Rhododendron*	2
Araucaria	3	*Salix*	2
Cestrum	3	*Salvia*	2
Hedera	3	*Solanum*	2
Acer	2	*Thuja*	2
Diospyros	2		

So far as the nativity status of exotic aliens of heritage sites is concerned (**Fig. 25; Tables LIV-LVI**), most of the species owe their origin to Asia (Afghanistan, Bangladesh, Bhutan, China, Japan, Myanmar, Nepal, Pakistan, Sri-Lanka, Thailand, etc.) (40%), followed by Central, North and South American Continent (Brazil, Caribbean Territories, Chile, Mexico, Venezuela, United States, etc.) (25.21%), Common Origin (Asia, America, Africa, Europe, Australia) (19.13%), African Continent (Algeria, Egypt, Ethiopia, Ghana, Kenya, South Africa, Tunisia, Zimbabwe, etc.) (6.08%), Europe (Austria, Belgium, Bulgaria, France, Germany, Italy, Romania, Spain, Ukraine, etc.) (5.21 %) and Australia (4.34 %). In contrast, majority of the aliens have been introduced from Europe in Kashmir (38%; Khuroo *et al.*, 2007), South America in H.P. (22.94%; Jaryan *et al.*, 2012) and America to the Whole India (55%; Nayar, 1977). Transportation corridors due to increasing global trade have been recognized as an important pathway for unintentional translocation of alien species and homogenization of earth's biota (Hodkinson & Thompson, 1997; Perrings *et al.*, 2005; Sharma & Rogharbanshi, 2009; Valtonen *et al.*, 2006) and consequently resulting in altering the composition of biodiversity in different ecosystems (Mack *et al.*, 2000; Vitousek *et al.*, 1996).

TABLE LIV
Native Status of Exotic Plants

S. No.	Continent/Country	Species	Genera	Families
1.	Asia (Afghanistan, Bangladesh, Bhutan, China, Japan, Myanmar, Nepal, Pakistan, Sri-Lanka, Thailand, etc.)	46	39	28
2.	Central, North and South American Continent (Brazil, Caribbean Territories, Chile, Mexico, Venezuela, United States, etc.)	29	26	20
3.	Australian	5	5	4
4.	African Continent (Algeria, Egypt, Ethiopia, Ghana, Kenya, South Africa, Tunisia, Zimbabwe, etc.)	7	4	4
5.	Europe (Austria, Belgium, Bulgaria, France, Germany, Italy, Romania, Spain, Ukraine, etc.)	6	6	6
6.	Common Origin (Asia, America, Africa, Europe, Australia)	22	17	14

TABLE LV
Origin and Habit of Exotic Plants

Categories	Sub-categories	No. of Species	%age of Species
Origin	Asian	46	40
	American (Central, South & North)	29	25.21
	African	7	6.08
	European	6	5.21
	Australian	5	4.34
		22	19.13
Habit	Trees	55	47.82
	Shrubs	47	40.86
	Climbers	9	7.82
	Lianas	2	1.73
	Sub-Shrubs	2	1.73

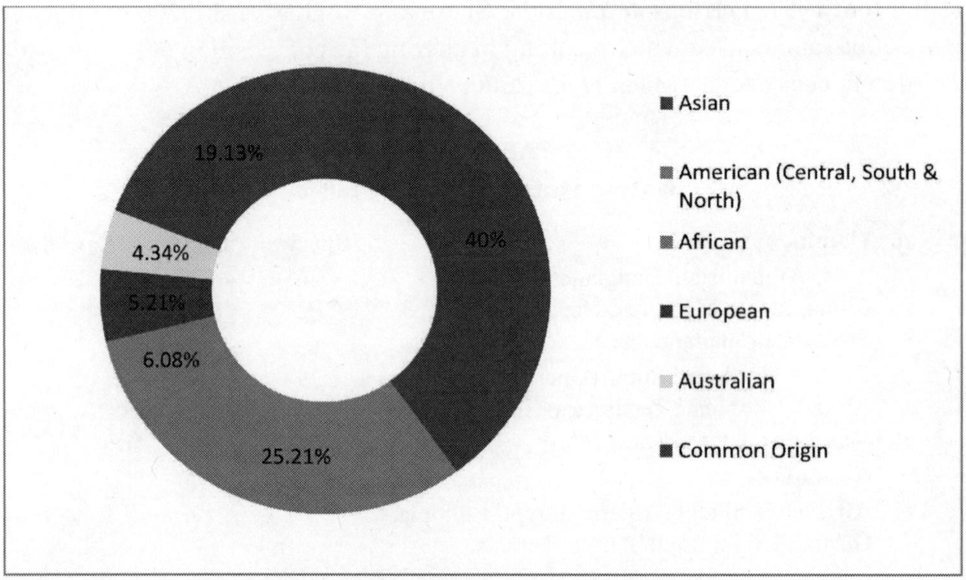

Fig. 25. Doughnut Showing Native Status of Plant Species of Heritage Sites.

TABLE LVI
Affinities of Exotic Plants

Asia	America	Australia	African	Europe	Common origin
Acer palmatum	*Abutilon megapotamicum*	*Acacia dealbata*	*Afrocarpus gracilior*	*Araucaria columnaris*	*Araucaria cunninghamii*
Actinidia chinensis	*Acer negundo*	*Araucaria bidwillii*	*Asparagus aethiopicus*	*Fagus sylvatica*	*Chamaerops humilis*
Amygdalus mira	*Agave americana*	*Callistemon lanceolatus*	*Asparagus densiflorus*	*Platanus orientalis*	*Elaeagnus umbellata*
Buddleja davidii	*Aloysia citriodora*	*Eucalyptus globulus*	*Asparagus setaceus*	*Sambucus nigra*	*Hedera colchica*
Camellia japonica	*Bougainvillea glabra*	*Grevillea robusta*	*Asparagus virgatus*	*Smilax aspera*	*Hedera helix*
Citrus maxima	*Campsis radicans*		*Hedera canariensis*	*Solanum pseudocapsicum*	*Malus pumila*
Cryptomeria japonica	*Cestrum elegans*		*Juniperus procera*		*Morus alba*
Cunninghamia lanceolata	*Cestrum nocturnum*				*Myrsine africana*
Cycas revoluta	*Cestrum parqui*				*Myrtus communis*
Diospyros kaki	*Duranta erecta*				*Populus nigra*
Diospyros lotus	*Fuchsia magellanica*				*Prunus amygdalus*
Drepanostachyum falcatum	*Ipomoea purpurea*				*Prunus armeniaca*
Eriobotrya japonica	*Jacaranda acutifolia*				*Prunus domestica*
Euonymus japonicus	*Juniperus horizontalis*				*Prunus laurocerasus*
Ficus subincisa	*Liriodendron tulipifera*				*Prunus persica*

Asia	America	Australia	African	Europe	Common origin
Forsythia ovata	*Magnolia grandiflora*				*Pyrus communis*
Ginkgo biloba	*Malvaviscus arboreus* var. *mexicanus*				*Quercus robur*
Hibiscus rosa-sinensis	*Opuntia dillenii*				*Rhododendron arboreum*
Hibiscus syriacus	*Rhamnus persica*				*Rubia cordifolia*
Hydrangea macrophylla	*Ricinus communis*				*Salix alba*
Jasminum mesnyi	*Robinia pseudoacacia*				*Spartium junceum*
Juglans regia	*Salvia leucantha*				*Taxus baccata*
Juniperus chinensis	*Salvia splendens*				
Juniperus pseudosabina	*Solanum laxum*				
Juniperus squamata	*Taxodium distichum*				
Kerria japonica	*Thuja occidentalis*				
Lagerstroemia indica	*Tropaeolum majus*				
Ligustrum lucidum	*Xanthium strumarium*				
Lonicera japonica	*Yucca aloifolia*				
Magnolia obovata					
Morus nigra					
Paulownia tomentosa					
Punica granatum					
Rhapis excelsa					
Rhododendron simsii var. *vittatum*					

Asia	America	Australia	African	Europe	Common origin
Rosa alba					
Rosa banksiae					
Rosa chinensis					
Rosa multiflora					
Rubus biflorus					
Salix babylonica					
Schefflera arboricola					
Spiraea cantoniensis					
Thuja orientalis					
Thujopsis dolabrata					
Wisteria sinensis					

On tracing the reasons for introduction, it is inferred that exotic species have been introduced intentionally at the heritage sites for various purposes such as ornamentals (85 species), fencing (22 species), avenue (5 species), edible (35 species), medicinal (28 species), fuelwood (13 species), fibre (5 species), seed-oil (2 species), sacred (6 species), incense (4 species), fodder (10 species), material culture (21 species) and horticultural (15 species). Enumerative details of various species for the aforesaid categories are enlisted in **Tables LVII-LXX**. With regard to the origin of these alien ornamental species, it is evident that introduction pathway/s predominantly radiate from Asia and America (**Table LXXI**). Comparatively, introduction from other continents like Australia, Africa and Europe are bare minimal (**Table LXXI**). Of the 40 alien rich ornamental families, Cupressaceae (3 genera, 8 species) are at the top followed by Rosaceae (4 genera, 7 species), Solanaceae (2 genera, 5 species), Araliaceae (2 genera, 4 species), Asparagaceae (1 genus, 4 species), Malvaceae (3 genera, 4 species), Araucariaceae (1 genus, 3 species), Magnoliaceae (2 genera, 3 species), Oleaceae (3 genera, 3 species), Taxodiaceae (3 genera, 3 species), Aceraceae (1 genus, 2 species), Agavaceae (2 genera, 2 species), Arecaceae (2 genera, 2 species), Bignoniaceae (2 genera, 2 species), Caprifoliaceae (2 genera, 2 species), Fabaceae (2 genera, 2 species), Fagaceae (2 genera, 2 species), Lamiaceae (1 genera, 2 species), Myrtaceae (2 genera, 2 species), Salicaceae (1 genus, 2 species), Verbenaceae (2 genera, 2 species) etc. (**Table LXXII**). Of the exotic/alien ornamental genera, predominant ones are *Juniperus* (5 species), *Asparagus* (4 species), *Rosa* (4 species), *Araucaria* (3 species), *Cestrum* (3 species), *Hedera* (3 species), etc. (**Table LXXIII**). The present assessment also unravels that ornamentals truly had been a major pathway of alien species introduction into the region. The ornamental exotic plants were introduced intentionally for landscaping gardens, houses, parks, hospitals, public places, cemeteries and socio-cultural spaces. Similar observations have been made earlier by Mack & Erneberg (2002), Reichard &White (2001), Silva & Smith (2004) and Wu *et al.* (2004a, b). Interestingly, of the 85 ornamental aliens reported from the heritage sites (**Table LXX**), 66 are reported for the first time from the region (present study) (see Jaryan *et al.*, 2012). No doubt, the present study will surely serve as a data base of alien ornamental flora of the region.

TABLE LVII
Exotic Hedge Plants

Botanical Name	Family	Botanical Name	Family
Afrocarpus gracilior	Podocarpaceae	*Myrtus communis*	Myrtaceae
Agave americana	Agavaceae	*Opuntia dillenii*	Cactaceae
Bougainvillea glabra	Nyctaginaceae	*Rosa banksiae*	Rosaceae
Drepanostachyum falcatum	Poaceae	*Rosa chinensis*	Rosaceae
Duranta erecta	Verbenaceae	*Rosa multiflora*	Rosaceae
Elaeagnus umbellata	Elaeagnaceae	*Sambucus nigra*	Caprifoliaceae
Euonymus japonicus	Celastraceae	*Solanum laxum*	Solanaceae
Hibiscus syriacus	Malvaceae	*Spartium junceum*	Fabaceae
Hydrangea macrophylla	Hydrangeaceae	*Thuja occidentalis*	Cupressaceae
Juniperus chinensis	Cupressaceae	*Yucca aloifolia*	Agavaceae
Ligustrum lucidum	Oleaceae		
Myrsine africana	Myrsinaceae		

TABLE LVIII
Exotic Avenue Plants

Botanical Name	Family	Botanical Name	Family
Acacia dealbata	Fabaceae	*Populus nigra*	Salicaceae
Eucalyptus globulus	Myrtaceae	*Robinia pseudoacacia*	Fabaceae
Jacaranda acutifolia	Bignoniaceae		

TABLE LIX
Exotic Edible Plants

Botanical Name	Family	Botanical Name	Family
Abutilon X *hybridum*,	Malvaceae	*Lonicera japonica*	Caprifoliaceae
Acacia dealbata	Fabaceae	*Malus pumila*	Rosaceae
Actinidia chinensis	Actinidiaceae	*Morus alba*	Moraceae
Agave americana	Agavaceae	*Morus nigra*	Moraceae
Aloysia citriodora	Verbenaceae	*Opuntia dillenii*	Cactaceae
Amygdalus mira	Rosaceae	*Prunus amygdalus*	Rosaceae
Chamaerops humilis	Arecaceae	*Prunus armeniaca*	Rosaceae
Citrus maxima	Rutaceae	*Prunus domestica*	Rosaceae

Botanical Name	Family	Botanical Name	Family
Diospyros kaki	Ebenaceae	*Prunus persica*	Rosaceae
Diospyros lotus	Ebenaceae	*Punica granatum*	Punicaceae
Drepanostachyum falcatum	Poaaceae	*Pyrus communis*	Rosaceae
Elaeagnus umbellata	Elaeagnaceae	*Rhamnus persica*	Rhamnaceae
Eriobotrya japonica	Rosaceae	*Rhododendron arboreum*	Ericaceae
Ficus subincisa	Moraceae	*Rosa multiflora*	Rosaceae
Hedera helix	Araliaceae	*Rubus biflorus*	Rosaceae
Hibiscus rosa-sinensis	Malvaceae	*Smilax aspera*	Smilacaceae
Hibiscus syriacus	Malvaceae	*Yucca aloifolia*	Agavaceae
Juglans regia	Juglandaceae		

TABLE LX
Exotic Medicinal Plants

Botanical Name	Family	Botanical Name	Family
Agave americana	Agavaceae	*Punica granatum*	Punicaceae
Citrus maxima	Rutaceae	*Pyrus communis*	Rosaceae
Cunninghamia lanceolata	Taxodiaceae	*Robinia pesudoacacia*	Fabaceae
Eucalyptus globulus	Myrtaceae	*Rhododendron arboreum*	Ericaceae
Ginkgo biloba	Ginkgoaceae	*Ricinus communis*	Euphorbiaceae
Hedera helix	Araliaceae	*Rosa chinensis*	Rosaceae
Jacaranda acutifolia	Bignoniaceae	*Rosa multiflora*	Rosaceae
Juniperus squamata	Cupressaceae	*Rubia cordifolia*	Rubiaceae
Lagerstroemia indica	Lythraceae	*Salix alba*	Salicaceae
Malvaviscus arboreus var. *mexicanus*	Malvaceae	*Salix babylonica*	Salicaceae
Myrsine africana	Myrsinaceae	*Sambucus nigra*	Caprifoliaceae
Opuntia dillenii	Cactaceae	*Smilax aspera*	Smilacaceae
Prunus amygdalus	Rosaceae	*Taxus baccata*	Taxaceae
Prunus persica	Rosaceae	*Xanthium strumarium*	Asteraceae

TABLE LXI
Exotic Fuel Wood Plants

Botanical Name	Family	Botanical Name	Family
Acacia dealbata	Fabaceae	*Myrsine africana*	Myrsinaceae
Duranta erecta	Verbenaceae	*Populus nigra*	Salicaceae
Elaeagnus umbellata	Elaeagnaceae	*Prunus domestica*	Rosaceae
Eucalyptus globulus	Myrtaceae	*Rhamnus persica*	Rhamnaceae
Juniperus squamata	Cupressaceae	*Rhododendron arboreum*	Ericaceae
Morus alba	Moraceae	*Robinia pseudoacacia*	Fabaceae
Morus nigra	Moraceae		

TABLE LXII
Exotic Fibre Yielding Plants

Botanical Name	Family	Botanical Name	Family
Agave americana	Agavaceae	*Taxodium distichum*	Taxodiaceae
Chamaerops humilis	Arecaceae	*Yucca aloifolia*	Agavaceae
Spartium junceum	Fabaceae		

TABLE LXIII
Exotic Oil Yielding Plants

Botanical Name	Family
Amygdalus mira	Rosaceae
Prunus armeniaca	Rosaceae

TABLE LXIV
Exotic Sacred Plants

Botanical Name	Family	Botanical Name	Family
Cestrum nocturnum	Solanaceae	*Punica granatum*	Punicaceae
Hibiscus rosa-sinensis	Malvaceae	*Rosa alba*	Rosaceae
Malvaviscus arboreus var. *mexicanus*	Malvaceae	*Rosa chinensis*	Rosaceae

TABLE LXV
Exotic Plants Used as Incense

Botanical Name	Family	Botanical Name	Family
Cryptomeria japonica	Taxodiaceae	*Juniperus squamata*	Cupressaceae
Juniperus horizontalis	Cupressaceae	*Taxus baccata*	Taxaceae

TABLE LXVI
Exotic Fodder Plants

Botanical Name	Family	Botanical Name	Family
Drepanostachyum falcatum	Poaceae	*Myrsine africana*	Myrsinaceae
Ficus subincisa	Moraceae	*Populus nigra*	Salicaceae
Hedera helix	Araliaceae	*Rhamnus persica*	Rhamnaceae
Morus alba	Moraceae	*Robinia pseudoacacia*	Fabaceae
Morus nigra	Moraceae	*Salix babylonica*	Salicaceae

TABLE LXVII
Exotic Plants for Material Culture

Use	Species
Agricultural Implements	*Drepanostachyum falcatum*, *Rhamnus persica*, *Rhododendron arboreum*
Baskets	*Drepanostachyum falcatum*, *Salix alba*, *Salix babylonica*
Brooms	*Drepanostachyum falcatum*, *Myrsine africana*
Cattle Shed	*Drepanostachyum falcatum*
Dye	*Juglans regia*
Field's Umbrella	*Drepanostachyum falcatum*
Furniture	*Araucaria cunninghamii, Cryptomeria japonica, Cunninghamia lanceolata, Diospyros kaki, Fagus sylvatica, Grevillea robusta, Jacaranda acutifolia, Juglans regia, Morus alba, Morus nigra, Populus ciliata, Robinia pseudoacacia, Salix babylonica, Taxodium distichum, Taxus baccata*
Green Manure	*Grevillea robusta*
Mats	*Drepanostachyum falcatum*
Pen (Kalam)	*Drepanostachyum falcatum*
Tool Handles	*Rhododendron arboreum*
Tooth brush	*Robinia pseudoacacia*
Utensil	*Cryptomeria japonica*
Welcome Gate	*Drepanostachyum falcatum*

TABLE LXVIII
Exotic Horticultural Plants

Botanical Name	Family	Botanical Name	Family
Actinidia chinensis	Actinidiaceae	*Morus nigra*	Moraceae
Amygdalus mira	Rosaceae	*Prunus amygdalus*	Rosaceae
Citrus maxima	Rutaceae	*Prunus armeniaca*	Rosaceae
Diospyros kaki	Ebenaceae	*Prunus domestica*	Rosaceae
Eriobotrya japonica	Rosaceae	*Prunus persica*	Rosaceae
Juglans regia	Juglandaceae	*Punica granatum*	Punicaceae
Malus pumila	Rosaceae	*Pyrus communis*	Rosaceae
Morus alba	Moraceae		

TABLE LXIX
Exotic Plants for Curing Various Ailments

Ailment	Plant Species
Anthelmintic	*Myrsine africana, Ricinus communis*
Asthma	*Opuntia dillenii*
Blood Purifier	*Myrsine africana*
Boils	*Xanthium strumarium*
Breast Cancer	*Taxus baccata*
Bronchitis	*Taxus baccata*
Burning Sensation of Soles	*Rhododendron arboreum*
Burns	*Cunninghamia lanceolata, Sambucus nigra*
Cardiovascular Disorders	*Ginkgo biloba*
Catarrh	*Sambucus nigra*
Cold	*Taxus baccata*
Constipation	*Prunus amygdalus, Ricinus communis*
Cough	*Punica granatum, Taxus baccata*
Cuts	*Lagerstroemia indica*
Digestive Disorders	*Rosa chinensis*
Fever	*Smilax aspera*
Fracture	*Agave americana*
Gastric Ailments	*Smilax aspera*
Headache	*Nerium oleander*
Inflammation	*Xanthium strumarium*
Joint Pain	*Prunus amygdalus, Smilax aspera*
Laxative	*Myrsine africana, Pyrus communis*
Menstruational Problems	*Rubia cordifolia*

Ailment	Plant Species
Mouth Ulcer	*Eucalyptus globulus*
Nose Bleeding	*Rhododendron arboreum*
Piles	*Punica granatum*
Respiratory Infections	*Eucalyptus globulus*
Rheumatic Pain	*Prunus persica*, *Ricinus communis*
Scabies	*Nerium oleander*
Scalds	*Cunninghamia lanceolata*, *Sambucus nigra*
Skin Diseases	*Juniperus squamata*, *Prunus amygdalus*, *Ricinus communis*, *Rubia cordifolia*, *Salix alba*, *Salix babylonica*
Smallpox	*Xanthium strumarium*
Spasm	*Taxus baccata*
Suppuration of Furuncles	*Ricinus communis*
Swellings	*Nerium oleander*, *Rubia cordifolia*, *Xanthium strumarium*
Throat Infections	*Sambucus nigra*
Ulcer	*Cunninghamia lanceolata*, *Rosa multiflora*, *Xanthium strumarium*
Vomiting	*Citrus maxima*
Wounds	*Cunninghamia lanceolata*, *Jacaranda acutifolia*, *Lagerstroemia indica*, *Rosa multiflora*

TABLE LXX
Exotic Ornamentals of Heritage Sites

S. No.	Botanical Name	Family	Origin
1.	*Abutilon megapotamicum**	Malvaceae	Argentina, Brazil, Uruguay
2.	*Acer negundo*	Aceraceae	N. America
3.	*Acer palmatum**	Aceraceae	Japan, N. Korea, China, Eastern Mongolia, S.E. Russia
4.	*Afrocarpus gracilior**	Podocarpaceae	Ethiopia, Kenya, Tanzania, Uganda
5.	*Agave americana**	Agavaceae	Mexican Highlands
6.	*Aloysia citriodora* *	Verbenaceae	Argentina, Paraguay, Brazil, Uruguay, Chile, Bolivia, Peru
7.	*Araucaria bidwillii**	Araucariaceae	S.E. Queensland
8.	*Araucaria columnaris**	Araucariaceae	New Caledonia
9.	*Araucaria cunninghamii**	Araucariaceae	S. Wales, Queensland, Papua New Guinea

S. No.	Botanical Name	Family	Origin
10.	*Asparagus aethiopicus**	Asparagaceae	S. Africa
11.	*Asparagus densiflorus**	Asparagaceae	S. Africa
12.	*Asparagus setaceus**	Asparagaceae	S. Africa
13.	*Asparagus virgatus**	Asparagaceae	S.E. Africa
14.	*Bougainvillea glabra**	Nyctaginaceae	Brazil
15.	*Buddleja davidii*	Buddlejaceae	China, Japan
16.	*Callistemon lanceolatus**	Myrtaceae	Australia
17.	*Camellia japonica**	Theaceae	China, Taiwan, S. Korea, S. Japan
18.	*Campsis radicans* *	Bignoniaceae	S. E. United States
19.	*Cestrum elegans**	Solanaceae	Mexico
20.	*Cestrum nocturnum*	Solanaceae	N. & S. America
21.	*Cestrum parqui**	Solanaceae	C. & S. America
22.	*Chamaerops humilis**	Arecaceae	Europe, N. Africa
23.	*Cryptomeria japonica**	Taxodiaceae	Japan
24.	*Cunninghamia lanceolata*	Taxodiaceae	China, Taiwan, Northern Vietnam, Laos
25.	*Cycas revoluta**	Cycadaceae	China, Japan
26.	*Diospyros lotus**	Ebenaceae	Armenia, Azerbaijan, China, Georgia, Iran, Korea, Nepal, Pakistan, Tajikistan, Turkey, Turkmenistan, Uzbekistan
27.	*Duranta erecta**	Verbenaceae	C. America
28.	*Euonymus japonicus**	Celastraceae	Japan, Korea, China
29.	*Fagus sylvatica**	Fagaceae	Europe
30.	*Forsythia ovata* *	Oleaceae	Korea
31.	*Fuchsia magellanica**	Onagraceae	Chile, Argentina
32.	*Ginkgo biloba*	Ginkgoaceae	China
33.	*Grevillea robusta**	Proteaceae	Queensland, New S. Wales
34.	*Hedera canariensis**	Araliaceae	Canary Island, N. Africa
35.	*Hedera colchica**	Araliaceae	S. E. Europe, Asia minor
36.	*Hedera helix**	Araliaceae	Europe, W. Asia
37.	*Hibiscus rosa-sinensis**	Malvaceae	China
38.	*Hibiscus syriacus*	Malvaceae	Syria
39.	*Hydrangea macrophylla**	Hydrangeaceae	Japan
40.	*Ipomoea purpurea*	Convolvulaceae	Tropical America
41.	*Jacaranda acutifolia**	Bignoniaceae	W.S. America, Peru
42.	*Jasminum mesnyi**	Oleaceae	China

S. No.	Botanical Name	Family	Origin
43.	*Juniperus chinensis**	Cupressaceae	China, Mangolia, Manchuria, Korea, Japan
44.	*Juniperus horizontalis**	Cupressaceae	N. America, United States in Alsaka
45.	*Juniperus procera**	Cupressaceae	E. Africa, S.W. Arabian Peninsula
46.	*Juniperus pseudosabina* *	Cupressaceae	C. Asia, Afghanistan, Tajikistan, Kazakhstan, W. China, Western Mongolia, South-Central Russia
47.	*Juniperus squamata**	Cupressaceae	Himalayas, China, N.E. Afghanistan
48.	*Kerria japonica*	Rosaceae	China, Japan, Korea
49.	*Lagerstroemia indica**	Lythraceae	China
50.	*Ligustrum lucidum*	Oleaceae	China
51.	*Liriodendron tulipifera**	Magnoliaceae	N. America, C. Florida, Louisiana
52.	*Lonicera japonica**	Caprifoliaceae	Japan, Korea
53.	*Magnolia grandiflora**	Magnoliaceae	S.E. United States (N. Carolina, Florida, Texas)
54.	*Magnolia obovata**	Magnoliaceae	Japan, Kurule Island of Russia
55.	*Malvaviscus arboreus* var. *mexicanus**	Malvaceae	Mexico
56.	*Myrtus communis**	Myrtaceae	S. Europe, N. Africa
57.	*Opuntia dillenii*	Cactaceae	U.S.A. to Ecuador, Caribbean Territories.
58.	*Paulownia tomentosa* *	Scrophulariaceae	C. & W. China
59.	*Platanus orientalis**	Platanaceae	Albania, Bulgaria, Former Yugoslavia, Greece, Italy
60.	*Prunus laurocerasus**	Rosaceae	S.W. Asia, S.E. Europe, N. Iran
61.	*Quercus robur* *	Fagaceae	Europe, N. Africa
62.	*Rhapis excelsa*	Arecaceae	S. China, Japan
63.	*Rhododendron simsii* var. *vittatum**	Ericaceae	China
64.	*Rosa alba**	Rosaceae	China, Japan
65.	*Rosa banksiae**	Rosaceae	China
66.	*Rosa chinensis**	Rosaceae	China
67.	*Rosa multiflora**	Rosaceae	Japan, China
68.	*Salix alba*	Salicaceae	W. Asia & Europe.

S. No.	Botanical Name	Family	Origin
69.	*Salix babylonica*	Salicaceae	China.
70.	*Salvia leucantha*	Lamiaceae	C. & E. Mexico
71.	*Salvia splendens**	Lamiaceae	Mexico, California, Brazil
72.	*Sambucus nigra**	Caprifoliaceae	Europe
73.	*Schefflera arboricola*	Araliaceae	China, Taiwan
74.	*Solanum laxum* *	Solanaceae	Brazil
75.	*Solanum pseudocapsicum**	Solanaceae	C. Europe
76.	*Spartium junceum*	Fabaceae	S. Europe, S.W. Asia, N.W. Africa
77.	*Spiraea cantoniensis*	Rosaceae	China
78.	*Taxodium distichum*	Taxodiaceae	Southern U.S.A
79.	*Taxus baccata**	Taxaceae	Eurasia, N. Africa, Iran, Caucasus Origin
80.	*Thuja occidentalis**	Cupressaceae	N.E. U.S.A., S.E. Canada
81.	*Thuja orientalis**	Cupressaceae	Mediterranean, China, S. Korea
82.	*Thujopsis dolabrata**	Cupressaceae	Japan
83.	*Tropaeolum majus*	Tropaeolaceae	Mexico, S. America
84.	*Wisteria sinensis**	Fabaceae	China
85.	*Yucca aloifolia**	Agavaceae	Mexico, W. Indies, Southern United States

***New Ornamental Records**

TABLE LXXI
Native Status of Exotic Ornamental Plants

S.No.	Continent/Country	Species
1.	Asia (Afghanistan, Bangladesh, Bhutan, China, Japan, Myanmar, Nepal, Pakistan, Sri-Lanka, Thailand, etc.)	35
2.	Central, North and South American Continent (Brazil, Caribbean Territories, Chile, Mexico, Venezuela, United States, etc.)	25
3.	Australia	3
4.	African Continent (Algeria, Egypt, Ethiopia, Ghana, Kenya, South Africa, Tunisia, Zimbabwe, etc.)	7
5.	Europe (Austria, Belgium, Bulgaria, France, Germany, Italy, Romania, Spain, Ukraine, etc.)	5
6.	Common Origin (Asia, America, Africa, Europe, Australia)	10

TABLE LXXII
Exotic Ornamental Families

S. No.	Family	Botanical Names	Genera/ Species
1.	Aceraceae	*Acer negundo*, *Acer palmatum*	1/2
2.	Agavaceae	*Agave americana*, *Yucca aloifolia*	2/2
3.	Araliaceae	*Hedera canariensis*, *Hedera colchica*, *Hedera helix*, *Schefflera arboricola*	2/4
4.	Araucariaceae	*Araucaria bidwillii*, *Araucaria columnaris*, *Araucaria cunninghamii*	1/3
5.	Arecaceae	*Chamaerops humilis*, *Rhapis excelsa*	2/2
6.	Asparagaceae	*Asparagus aethiopicus*, *Asparagus densiflorus*, *Asparagus setaceus*, *Asparagus virgatus*	1/4
7.	Bignoniaceae	*Campsis radicans*, *Jacaranda acutifolia*	2/2
8.	Buddlejaceae	*Buddleja davidii*	1/1
9.	Cactaceae	*Opuntia dillenii*	1/1
10.	Caprifoliaceae	*Lonicera japonica*, *Sambucus nigra*	2/2
11.	Celastraceae	*Euonymous japonicus*	1/1
12.	Convolvulaceae	*Ipomoea purpurea*	1/1
13.	Cupressaceae	*Juniperus chinensis*, *Juniperus horizontalis*, *Juniperus procera*, *Juniperus pseudosabina*, *Juniperus squamata*, *Thuja occidentalis*, *Thuja orientalis*, *Thujopsis dolabrata*	3/8
14.	Cycadaceae	*Cycas revoluta*	1/1
15.	Ebenaceae	*Diospyros lotus*	1/1
16.	Ericaceae	*Rhododendron simsii* var. *vittatum*	1/1
17.	Fabaceae	*Spartium junceum*, *Wisteria sinensis*	2/2
18.	Fagaceae	*Fagus sylvatica*, *Quercus robur*	2/2
19.	Ginkgoaceae	*Ginkgo biloba*	1/1
20.	Hydrangeaceae	*Hydrangea macrophylla*	1/1
21.	Lamiaceae	*Salvia leucantha*, *Salvia splendens*	1/2
22.	Lythraceae	*Lagerstroemia indica*	1/1
23.	Magnoliaceae	*Liriodendron tulipifera*, *Magnolia grandiflora*, *Magnolia obovata*	2/3
24.	Malvaceae	*Abutilon megapotamicum*, *Hibiscus rosa-sinensis*, *Hibiscus syriacus*, *Malvaviscus arboreus* var. *mexicanus*	3/4
25.	Myrtaceae	*Callistemon lanceolatus*, *Myrtus communis*	2/2
26.	Nyctaginaceae	*Bougainvillea glabra*	1/1

S. No.	Family	Botanical Names	Genera/ Species
27.	Oleaceae	*Forsythia ovata, Jasminum mesnyi, Ligustrum lucidum*	3/3
28.	Onagraceae	*Fuchsia magellanica*	1/1
29.	Platanaceae	*Platanus orientalis*	1/1
30.	Podocarpaceae	*Afrocarpus gracilior*	1/1
31.	Proteaceae	*Grevillea robusta*	1/1
32.	Rosaceae	*Kerria japonica, Prunus laurocerasus, Rosa alba, Rosa banksiae, Rosa chinensis, Rosa multiflora, Spiraea cantoniensis*	4/7
33.	Salicaceae	*Salix alba, Salix babylonica*	1/2
34.	Scrophulariaceae	*Paulownia tomentosa*	1/1
35.	Solanaceae	*Cestrum elegans, Cestrum nocturnum, Cestrum parqui, Solanum laxum, Solanum pseudocapsicum*	2/5
36.	Taxaceae	*Taxus baccata*	1/1
37.	Taxodiaceae	*Cryptomeria japonica, Cunninghamia lanceolata, Taxodium distichum*	3/3
38.	Theaceae	*Camellia japonica*	1/1
39.	Tropaeolaceae	*Tropaeolum majus*	1/1
40.	Verbenaceae	*Aloysia citriodora, Duranta erecta*	2/2

TABLE LXXIII
Predominant Exotic Ornamental Genera

S. No.	Genera	No. of Species
1.	*Juniperus*	5
2.	*Asparagus*	4
3.	*Rosa*	4
4.	*Araucaria*	3
5.	*Cestrum*	3
6.	Hedera	3
7.	*Acer*	2
8.	*Hibiscus*	2
9.	*Magnolia*	2
10.	*Salix*	2
11.	*Salvia*	2
12.	*Solanum*	2
13.	*Thuja*	2

TABLE LXXIV
Indigenous Ornamental Plants

S. No.	Botanical Name	Family	Origin
1.	*Buddleja asiatica*	Buddlejaceae	India, Nepal, Bangladesh, China, Taiwan, Burma, Thailand, Vietnam, Malaysia, New Guinea, Philippines
2.	*Buxus wallichiana*	Buxaceae	E. Asia, Himalayas, Afghanistan, C. Nepal
3.	*Clematis montana*	Ranunculaceae	China, Himalayas
4.	*Colquhounia coccinea*	Lamiaceae	China, Nepal, Bhutan, India, Myanmar, Thailand
5.	*Cupressus cashmeriana*	Cupressaceae	E. Himalayas, Bhutan, N.E. India
6.	*Cupressus torulosa*	Cupressaceae	China, India-Himachal Pradesh, Uttar Pradesh
7.	*Erythrina arborescens*	Fabaceae	Central & E. Himalayas
8.	*Euonymus lucidus*	Celastraceae	N.E. Himalayas, Pakistan
9.	*Ficus benjamina*	Moraceae	India, S. China, S.E. Asia, Malaysia, Philippine, N. Australia, Island of the S. Pacific
10.	*Ficus elastica*	Moraceae	Bhutan, India, Indonesia, Malaysia, Myanmar, Nepal
11.	*Ficus religiosa*	Moraceae	Bangladesh, India, Nepal, Pakistan, Thailand, Vietnam
12.	*Hypericum hookerianum*	Clusiaceae	Bangladesh, Bhutan, India, Myanmar, Nepal, Thailand, Vietnam
13.	*Hypericum oblongifolium*	Clusiaceae	India- Himachal Pradesh, Jammu and Kashmir, Uttar Pradesh, Nepal, Pakistan.
14.	*Jasminum grandiflorum*	Oleaceae	India
15.	*Jasminum humile*	Oleaceae	Afghanistan, Bhutan, China, India, Nepal, Pakistan, Tajikistan
16.	*Jasminum multiflorum*	Oleaceae	India
17.	*Nerium oleander*	Apocyanaceae	Africa- Algeria, Ethiopia, Libya, Morocco, Tunisia, Asia temperate -Afghanistan, China Cyprus, Iran, Iraq, Israel, Jordan, Lebanon, Oman, United Arab Emirates, Syria, Turkey, Asia Tropical- India, Nepal, Pakistan, Europe- Albania, Former Yugoslavia, Greece, Italy, Malta, France, Portugal, Spain

S. No.	Botanical Name	Family	Origin
18.	*Rosa damascena*	Rosaceae	India
19.	*Rosa jacquemontii*	Rosaceae	India
20.	*Trachelospermum jasminoides*	Apocyanaceae	S.E. Asia, Japan, Korea, S. China, Vietnam
21.	*Tsuga dumosa*	Pinaceae	Bhutan, China, Tibet, Yunnan, India (Assam, Sikkim, Uttar Pradesh), Myanmar, Nepal

TABLE LXXV
Indigenous Ornamental Families

S. No.	Family	Botanical Name	Genera/ Species
1.	Apocyanaceae	*Nerium oleander*, *Trachelospermum jasminoides*	2/2
2.	Buddlejaceae	*Buddleja asiatica*	1/1
3.	Buxaceae	*Buxus wallichiana*	1/1
4.	Celastraceae	*Euonymus lucidus*	1/1
5.	Clusiaceae	*Hypericum hookerianum*, *Hypericum oblongifolium*	1/2
6.	Cupressaceae	*Cupressus cashmeriana*, *Cupressus torulosa*	1/2
7.	Fabaceae	*Erythrina arborescens*	1/1
8.	Lamiaceae	*Colquhounia coccinea*	1/1
9.	Moraceae	*Ficus benjamina*, *Ficus elastica*, *Ficus religiosa*	1/3
10.	Oleaceae	*Jasminum grandiflorum*, *Jasminum humile*, *Jasminum multiflorum*	1/3
11.	Pinaceae	*Tsuga dumosa*	1/1
12.	Ranunculaceae	*Clematis montana*	1/1
13.	Rosaceae	*Rosa damascena, Rosa jacquemontii*	1/2

TABLE LXXVI
Predominant Indigenous Ornamental Genera

S. No.	Genera	No. of Species
1.	*Ficus*	3
2.	*Jasminum*	3
3.	*Cupressus*	2
4.	*Hypericum*	2
5.	*Rosa*	2

Based on the catalogue of invasive alien flora of India (Reddy, 2008), 16 species namely *Robinia pseudoacacia*, *Spartium junceum*, *Wisteria sinensis* (Fabaceae), *Agave americana* (Agavaceae), *Arundo donax* (Poaceae), *Asparagus densiflorus* (Asparagaceae), *Cestrum nocturnum*, (Solanaceae), *Elaeagnus umbellata* (Elaeagnaceae), *Eucalyptus globulus* (Myrtaceae), *Hedera helix* (Araliaceae), *Lonicera japonica* (Caprifoliaceae), *Morus alba* (Moraceae), *Paulownia tomentosa* (Scrophulariaceae), *Ricinus communis* (Euphorbiaceae), *Rosa multiflora* (Rosaceae), *Xanthium strumarium* (Asteraceae) have been identified in the courtyards, cemeteries and socio-cultural spaces of the heritage sites of the study area but none enjoy the status of invasive aliens. In agreement to earlier report of Chandra Sekar (2012), the present work lends support to the earlier view-point that the family Fabaceae is the dominant invasive family of Indian Himalayan region. It is obvious that such a base line regional data when pooled together is likely to provide vital clues for elucidation of the probable causes and consequences of the phenomenon of invasion and can strengthen the policy efforts for their effective predictive systems, monitoring tools and management practices.

CONCLUDING REMARKS

From the above foregoing accounts, it is quite obvious that courtyards of heritage sites in Shimla Hills are rich in genetically diverse population of herbs, shrubs and trees of ornamental, medicinal and other economic virtues hailing from the indigenous and exotic resources. Though unintentional, many of these sites have played a key role in introduction, acclimatisation of exotic plants, and facilitating towards their ultimate conservation along with the prevalent native elements which no doubt are the products of millions of years of evolution. Due to such a varied richness of plant diversity, it is of paramount importance to recognize these as *ex situ* centres of conservation otherwise the priceless, heritable germplasm would be squandered and dissipitated to anthropogenic pressures for which our future generations would not forgive us. Without any further delay, the State Government should act in collaboration with local Biodiversity Board on a priority-wise basis and take adequate policy decision/s in this regard. At the same time, these enacted acts/ regulations must ensure that it must have full transparency, including the public right to information and public participation. Ultimately, this would go a long way in conservation of invaluable germplasm of varied economic virtues occurring at the heritage sites.

Chapter 6
BIBLIOGRAPHY

Abraham, Z. 1997. Further ethnobotanical study of Todas and Kotas of Nilgiris: 249-254. *In:* Jain, S.K. (*ed.*) *Contribution to Indian Ethnobotany.* Sci. Publ., Jodhpur, India.

Adams, R.P. 2004. *Junipers of the World.* Trafford Publication, Canada.

Agarwal, V.S. 2003. *Directory of Indian Economic Plants.* B.S.M.P.S., Dehradun.

Agate, E. 1998. *Hedging: A Practical Handbook.* British Trust for Conservation, London.

Aggarwal, A., Prakash, V. & Sharma, S. 2004. Ethnobotanical studies on some wild plants of lower hills. *Appl. Bot. Abs.* ***24***(2): 136.

Ahluwalia, K.S. 1952. Medicinal plants of Kangra valley. *Indian For.* ***78*** (4): 181-194.

Ahmed, A.A & Borthakur, S.K. 2005. *Ethnobotanical Wisdom of Khasis (Hynniew Treps) of Meghalaya.* B.S.M.P.S., Dehradun.

Alan, M. 1974. *Trees of Britain and Northern Europe.* Harper Collins Publ., London.

Ambasta, S.P. (*ed.*)1986. *The Useful Plants of India.* C.S.I.R., New Delhi.

Aminuddin & Girach, R.D. 1996. Native phytotherapy among the Paudi Bhuinya of Bonai Hills. *Ethnobotany* ***8***: 66-70.

Anonymous 1990. *Phytochemical Investigation of Certain Medicinal Plants Used in Ayurveda.* C.C.R.A.S., New Delhi.

Anonymous, 1948-76. *The Wealth of India (A Dictionary of India Raw Materials and Industrial Products),* Vol. I-VIII. C.S.I.R., New Delhi (India).

Anonymous, 2008. *History of Himachal Pradesh.* Himachal Pradesh India.Com.

Anonymous, 2012. *Census GIS House Hold.* Censusindiamaps.net.

Ansari, A.A. 1991. Ethnobotanical notes on some plants of Khirsu, Pauri Garhwal, U.P. *Ethnobotany* ***3***: 105-106.

Arora, J.S. 1993. *Introductory Ornamental Horticulture.* Kalyani Publ., Ludhiana.

Arora, R.K. 1997. Native food plants of tribals of northeast India: 137-152. *In:* Jain, S.K. (*ed.*) *Contributions to Indian Ethnobotany.* Sci. Publ., Jodhpur, India.

Arora, R.K. & Nayar, E.R. 1984. Wild relatives of crop plants in India. *Sci. Monogr.* *7*: 90.

Arora, R.K. & Pandey, A. 1996. *Wild Edible Plants of India- Diversity, Conservation and Use*. N.B.P.G.R, New Delhi.

Arora, R.K., Maheswari, M.L., Chandel, K.P.S. & Gupta, R. 1980. Mano (*Inula racemosa*): Little known aromatic plants of Lahaul valley, India. *Eco. Bot.* ***34***(2): 175-180.

Arya, K.R., Pande, P.C. & Prakesh, V. 1999. Ethnobotanical study on tribal areas of Almora district-II. *Ethnobotany* ***11***: 100-104.

Aslam, S., Ganaie, K.A., John A.Q. & Dar, G.H. 2010. Tracking the invasion pathway: assesment of α-diversity and invasiveness of alien ornamental plants of Srinagar (Kashmir, J&K), India. *Natl. & Sci.* ***8***(3): 79-95.

Asolkar, L.V., Kakkar, K.K. & Chakre, O.J. 1992. *Second Supplement to Glossary of Indian Medicinal Plants with Active Principles,* Part–I (A-K). C.S.I.R., New Delhi.

Aswal, B.S. 1996. Conservation of ethno-medicinal plants diversity of Garhwal Himalaya in India: 133-135. *In:* Jain, S.K. (*ed.*) *Ethnobiology in Human Welfare*. Deep Publ., New Delhi.

Aswal, B.S. & Mehrotra, B.N. 1994. *Flora of Lahaul-Spiti (A Cold Desert in North –West Himalayas).* B.S.M.P.S., Dehra Dun.

Athar, M. & Nasir, S.M. 2005. Taxonomic perspective of plant species yielding vegetable oils used in cosmetics and skin care products. *Afr. J. Biotech.* ***4*** (1): 36-44.

Atkinson, E.T. 1882. *The Himalayan District of North West Provinces of India.* Cosmo Publ., New Delhi, Reprinted (1973).

Babu, C.R. 1977. *Herbaceous Flora of Dehradun*. C.S.I.R., New Delhi.

Baburaj, D.S., Britto, S.J., Mathew, G.K. & Rajan, S. 2000. Cultivated medicinal plants useful in homoeopathy found in Nilgiri District, Tamil Nadu: 381-389. *In:* Maheshwari, J.K. *(ed.) Ethnobotany and Medicinal Plants of Indian Subcontinent*. Sci. Publ., Jodhpur (India).

Badhwar, S.R.L. & Fernandez, R.R. 2011. *Edible wild plants of the Himalayas*. Daya Publ. House, New Delhi, India.

Badola, H.K. 2001. Medicinal plant diversity in Himachal Pradesh: 407-430. In Samant, S.S., Dhar, U. & Palani, L.M.S. (*eds.*) *Himalayan Medicinal Plants: Potential and Prospects.* Himvikas Occasional Publ., No. 14, Gyanodya Prakashan, Nainital, India.

Badola, H.K. & Pal, M. 2002. Endangered medicinal plant species in Himachal Pradesh. *Curr. Sci.* ***83***: 797-798.

Bailey, L.H. 1957. *Manual of Cultivated Plants. Most Commonly Grown in the Continental United States and Canada.* The MacMillan Co., New York.

Bamber, C.J. 1916. *Plants of the Punjab.* Govt. Printing Press, Lahore. (Reprinted: 1976, B.S.M.P.S., Dehra Dun & Periodical Experts, Delhi).

Banerjee, A. 1996. Medicinal uses of some flowers by Santhals of West Bengal. *J. Econ. Taxon. Bot. Addl. Ser.* ***12***: 314-317.

Banerjee, D.K. & Pal, D.C. 1996. Plant used by the tribals of North Indian plains for hair and scalp protection: 187-188. *In*: Jain, S.K. (*ed.*) *Ethnobiology in Human Welfare*. Deep Publ., New Delhi.

Bean, W. 1981. *Trees and Shrubs Hardy in Great Britain. Vol. 1 - 4 and Supplement.* Murray Publ. House, London.

Bedi, R. 2000. *Floral Spectrum of India*. Pelican Creations Intl., Noida, India.

Beg, M. J., Beg, M. Z. & Ali, S. J. 2005. Folklore treatment of some common ailments in Eastern U.P.: 147-177. *In*: Khan, I.A. & Khanum, A. (*eds.*): *Role of Biotechnology in Medicinal and Aromatic Plants.* Ukaaz Publ., Hyderabad.

Beigh, Y.S., Nawchoo, I. A. & Muhammad, I. 2003. Plants in traditional medicine of Kashmir Himalaya: 99-104. *In:* Singh, V. & Jain, A. P. (*eds.*) *Ethnobotany and Medicinal Plants of India and Nepal*, Vol. I. Sci. Publ., Jodhpur, India.

Belal, A.E. & Springuel, I. 1996. Economic value of plant diversity in arid environments. *Nature & Res.* ***32***(1): 33-39.

Bennett, S.S.R. 1987. *Name Changes in Flowering Plants of India and Adjacent Regions.* Triseas Publ., Dehradun, India.

Bennet, S.S.R., Biswas, S., Chandra, V., Singh, P.J., Chandra, S. & Dayal, R. 1991. *Food From Forest.* I.C.F.R. E., Dehradun.

Benthall, A.P. 1946. *The Trees of Calcutta and its Neighbourhood.* Thacker Spink & Co., Ltd. Calcutta.

Beri, S.N. 1992. *The Houseplant Guidebook.* Hind Pocket Books Pvt., Ltd., New Delhi, India.

Berry, S. & Bradley, S. 2000. *Illustrated Encyclopedia*: *Garden Plants*. Anness Publ. Ltd., London.

Bhagat, R., Simpale, V.B. & Deshmukh, R.B. 2009. *Flora of Barmati*. Vedams Book Intl., Pune.

Bhalla, S. & Bhalla, T.C. 2006. Traditional food and beverages of Himachal Pradesh. *Indian J. Trad. Know.* ***6***(1): 17-24.

Bhandary, M.J., Chandrashekar, K.R. & Kaveriappa, K.M. 1996. Ethnobotany of Gowlis of Uttar Kannada district, Karnataka. *J. Econ. Taxon. Bot.* Addl. Ser. ***12***: 244-249.

Bhatt, D.C. & Mitaliya, K.D. 1999. Ethnomedical plants of Victoria Park (Reserved Forest) of Bhavnagar, Gujrat, India. *Ethnobotany* ***11***: 81-84.

Bhatt, D.C., Mitaliya, K.D. & Babaria, A.V. 2003. Studies on some ethnobotanical plants of Barda Hills from Saurashtra: 221-227. *In:* Singh, V. & Jain, A.P. *(eds) Ethnobotany and Medicinal Plants of India and Nepal.* Sci. Publ., Jodhpur.

Bhatt, D.C., Mitaliya, K.D. & Mehta, S.K. 2001. Some sacred plants of Gujrat and their medicinal uses. *Ethnobotany* ***13***: 146-149.

Bhattacharjee, S.K. 2001. *Handbook of Medicinal Plants*. Pointer Publ., Jaipur.

Bhattacharjee, S.K. 2006. *Advances in Ornamental Horticulture*, 6 Vols. Pointer Publ., Jaipur (Rajasthan), India.

Bhattacharya, A.K., Tiwari, P., Patra, K. & Sinha, V.K. 2005. Seasonal availability and consumption of non-wood forest products (NWFP) as food among Baiga tribe of Dindori district of Madhya Pradesh, India. *Ethnobotany* ***17***: 64-68.

Bhuyan, L.R. 2003. Some plants used as medicine by the Nishi tribe of Arunanchal Pradesh: a preliminary study: 447-450. *In:* Singh, V. & Jain, A.P. (*ed.*) *Ethnobotany and Medicinal Plants of India and Nepal,* Vol. I. Sci. Publ., Jodhpur (India).

Billore, K.V., Joseph, T.G. & Dave, S. K. 1998. Interesting folk remedies by the 'Lok Vaidyas' of Rajasthan for 'Swas Roga'. *Ethnobotany* ***10***: 42-45.

Bisht, A. & Jain, S.P. 2006. Review of ethnobotanical studies of genus *Rubus* (Rosaceae) from Western Himalaya. *Ethnobotany* ***18***: 127-130.

Blatter, E.J. & Hallberg, F. 1918. The flora of the Indian Desert (Jodhpur and Jaisalmer). *Proc. Nat. Inst. Sci. India* ***26***: 218-246.

Bole, P.V. & Vaghani, Y. 1986. *Field Guide to the Common Trees of India*. Oxford Univ. Press, U.K.

Booth, C.O. 1999. *Encyclopaedia of Garden Plants,* 2 Vols. Daya Publ. House, New Delhi.

Bor, N.L. & Raizada, M.B. 1999. *Some Beautiful Indian Climbers and Shrubs*. Oxford Univ. Press, U.K.

Bor, N.R. 1960. *The Grasses of Burma, Ceylon, India and Pakistan (Excluding Bambuseae)*. Intl. Book Distr., Dehra Dun.

Bora, H.R., Borthakur, S.K. & Hazarika, L.K. 2003. Use of plants in control of pests in Assam-an ethnobotanical approach: 956-963. *In:* Singh, V. & Jain, A.P. (*eds.*) *Ethnobotany and Medicinal Plants of India and Nepal,* Vol. II. Sci. Publ., Jodhpur (India).

Borthakur, S.K. 1993. Native phytotherapy for child and women diseases from Assam. *Ethnobotany* **5** (87): 87-91.

Borthakur, S.K. 1996a. Wild edible plants in the market of Assam, India. An ethnobotanical investigation: 31-34. *In:* Jain, S.K. (*ed.*) *Ethnobotany in Human Welfare*. Deep Publ., New Delhi.

Borthakur, S.K. 1996b. Postnatal care of women in traditional system in Assam. *Ethnobotany* ***8***: 51-55.

Borthakur, S.K. 1997. Studies in ethnobotany of the Karbis (Mikiris): plant masticatories and dyestuffs: 181-190. *In:* Jain, S.K. (*ed.*) *Contribution to Indian Ethnobotany.* Sci. Publ., Jodhpur, India.

Borthakur, S.K., Sarma, T.R., Nath, K., & Deka, P. 1999. The house gardens of Assam: a traditional Indian experiences of management and conservation of biodiversity –II. *Ethnobotany* ***11***:65-80.

Borthakur, S.K. & Sarma, U.K. 1996. Ethnoveterinary medicine with special reference to cattle prevalent among the Nepalis of Assam, India: 197-199. *In:* Jain, S.K. (*ed.*) *Ethnobiology in Human Welfare.* Deep Publ., New Delhi.

Bose, K.C. 1984. *Pharmacopoeia Indica*. B.S.M.P.S., Dehradun.

Bose, T.K. & Chowdhury, B. 1991. *Tropical Garden Plants in Colour.* Hort. & Allied Publ., Calcutta, India.

Bose, T.K., Pal, P., Chowdhury, B. & Sharma, S.P. (*eds.*) 2004. *Garden Plants in Colour: House Plants*. Naya Udyog, Kolkata, India.

Brandis, D. 1874. *The Forest Flora of the North-West and Central India*. London (Reprinted 1972: B.S.M.P.S., Dehra Dun).

Brandis, D. 1906. *Indian Trees*. B.S.M.P.S., Dehra Dun, India.

Brandis, D. 1921. *Indian Trees. An Account of Trees, Shrubs, Woody Climbers, Bamboos and Palms Indigenous or Commonly Cultivated in the British Indian Empire.* Constable & Comp. Ltd., London.

Brickell, C. 1990. *The RHS Gardener's Encyclopedia of Plants and Flowers.* Dorling Kindersley Publ. Ltd., London.

Brickell, C. 2003. *RHS A-Z Encyclpedia of Garden Plant (IIIrd edition).* Dorling Kindersley Publ. Ltd., London.

Brickell, C., Zuk, J.D. & Zuk, J. 1997. *The American Horticultural Society A-Z Encyclopedia of Garden Plants*. Am. Hort. Soc., USA.

Burkill, I.H. 1965. *Chapters on the History of Botany in India*. Govt. of India Press, Delhi.

Burman, J. 1769. *Flora Malabarica*. Amsterdam.

Burman, N.L. 1768. *Flora Indica*. Ludg, Batavia (Indonesia).

Caius, J.F. 1986. *The Medicinal and Poisonous Plants of India*. Sci. Publ., Jodhpur.

Cambessedes, J. 1844. Plantae Rarias quas in Indae Orientali collegit Victor Jacquemont: 1-56. *In*: Jacquemont, Victor (*ed.*). Voyage dans L' Indae, Paris.

Chakrabarty, T. & Balakrishnan, N.P. 2003. Ethnobotany of the Andeman & Nicobar Islands, India: 859-893. *In:* Singh, V. & Jain, A.P. (*eds.*) *Ethnobotany and Medicinal Plants of India and Nepal.* Sci. Publ., Jodhpur, India.

Chakraborty, P. 2003. Wild edible plants sold in the daily market in and around of Imphal, Manipur: 481-484. *In:* Singh, V. & Jain, A.P. (*eds.*) *Ethnobotany and Medicinal Plants of India and Nepal.* Sci. Publ., Jodhpur, India.

Champion, H.G. & Seth, S.K. 1968. *A Revised Survey of the Forest Types of India.* Govt. India Publ., Delhi.

Chandra, V. 1997. *Edible Plants of Forestry Origin.* I.C.F.R.E., Dehradun, India.

Chandrasekar, K. & Srivastava, S.K. 2003. Ethnomedicinal studies in Pin valley National Park, Lahaul-Spiti, Himachal Pradesh. *Ethnobotany* ***15***: 44-47.

Chandrasekar, K. & Srivastava, S.K. 2005. New reports on aphrodisiac plants from Pin Valley National Park, Himachal Pradesh. *Ethnobotany* ***17***: 189-190.

Chatterjee, A. & Pakrashi, S.C. 1997. *The Treatise on Indian Medicinal Plants.* C.S.I.R., New Delhi (India).

Chatterjee, D. 1939. Studies on the endemic flora of India and Burma. *J. Royal Asiat. Soc. Bengal (Sci.)* ***5***: 19-67.

Chatterjee, D. 1947. Influence of east Mediterranean region flora on that of India. *Sci. Cult.* **13**: 9-11.

Chaudhuri, A.B. 1993. *Forest Plants of Eastern India.* Ashish Publ. House, New Delhi.

Chauhan, N.S. 1974. Medicinal plants of Una forest division, Una district, H.P. lying in Shivalik range. *Nagarjun* ***27***: 31-39.

Chauhan, N.S. 1983. Some rare and interesting medicinal plants of H.P. All India Summer Inst. on Agroforestry in the Western Himalaya, July 22-August 7, 1983.

Chauhan, N.S. 1984. *Medicinal Wealth of Pabbar Valley in Himachal Pradesh*, M.Sc. Thesis, HPKVV, Palampur.

Chauhan, N.S. 1988. Ethnobotanical study of medicinal plants of Himachal Pradesh: 187-198. *In:* Kaushik, P. (*ed.*) *Indigenous Medicinal Plants*. Today & Tomorrow's Printer & Publ., New Delhi.

Chauhan, N.S. 1989a. Medicinal and aromatic plants of Himachal Pradesh and their role in social Forestry programme IIIrd Forestery Conference (proceeding) Vol. II: 89 (1), 89 (9): FRI, Dehra Dun.

Chauhan, N.S. 1989b. Potential of aromatic plants flora in Himachal Pradesh. *Indian Perf.* ***33***(2): 118-122.

Chauhan, N.S. 1989c. Endangered ayurvedic pharmacopoeial plant resources of H.P.: 199-205. *In*: Kaushik, P. (*ed.*) *Indiginous Medicinal Plants Including Microbes and Fungi.* Today & Tommorow's Printer & Publ., New Delhi.

Chauhan, N.S. 1990. Medicinal orchids of Himachal Pradesh. *J. Orchid Soc. India* ***4***(2): 99-105.

Chauhan, N.S. 1995. *Plant Resources of Economic Use in Himachal Pradesh.* Deptt. of Forest Products, Dr. Y.S. Parmar Univ. Hort. & For., Solan (H.P.).

Chauhan, N.S. 1996. Ethno-medicobotany of Pabbar valley in district Shimla, Himachal Pradesh (India): 30. *In: UHF-IUFRO Intl. Workshop in Prospects of Medicinal Plants,* Solan (H.P.), India.

Chauhan, N.S. 1999. *Medicinal and Aromatic Plants of Himachal Pradesh.* Indus Publ. Co., New Delhi.

Chauhan, N.S. 2003. Important medicinal and aromatic Plants of Himachal Pradesh. *Indian For.* ***129***(8): 979-998.

Chauhan, N.S. & Lakhanpal, T.N. 2000. Ecological and ethnobotanical studies on the flora of Spiti, Himachal Pradesh - a cold desert in western Himalaya, 293-300. *In:* Kohli, R.K., Singh, H.P., Vij, S.P., Dhir, K.K., Batish, D.R. & Khurana, D.R. (*eds*) *Man and Forests.* DNAES, ISTS, Bot. Dept. & C.N.S, Punjab Univ., Chandigarh.

Chauhan, V. & Chauhan, N.S. 1985. Ethnobotany of Trans-Giri area of Sirmaur District of Himachal Pradesh. *Bull. Medico. Ethno. Bot. Res.* ***9***(3/4): 106-122.

Chaurasia, O.P., Ahmed, Z. & Ballabh, B. 2007. *Ethnobotany and Plants of Trans-Himalaya.* Satish Serial Publ., New Delhi.

Chaurasia, O.P., Singh, B. & Ballabh, B. 2001. Potential aromatic floras of Himalayan cold district: Ladakh & Lahul-Spiti. *J. Econ. Taxon. Bot.* ***25***(1): 91-97.

Chej, R. 1984. *Encyclopaedia of Medicinal Plants.* MacDonald Publ., Hansley Industrial Court, U.S.A.

Chevallier, A. 1996. *The Encyclopedia of Medicinal Plants.* Dorling Kindersley Publ. Ltd., London.

Chhetri, D.R. 2005. Ethnomedicinal Plants of the Khangchendzonga National Park, Sikkim, India. *Ethnobotany* ***17***: 96-103.

Chhetri, R.B. 1994. Further observations on ethnomedicobotany of Khasi hills in Meghalaya, India. *Ethnobotany* ***6***: 33-36.

Chhetri, R.B. & Shrestha, R. 2004. Ethnobotany of some weeds of winter crops in Dhulikhel, Nepal. *Ethnobotany* ***16***:108-112.

Chittendon, F. 1956. *RHS Dictionary of Plants Plus Supplement.* Oxford Univ. Press, U.K.

Chopra, R.N., Badhwar, R.L. & Gosh, S. 1965. *Poisonous Plants of India.* 2 Vols., I.C.A.R., New Delhi.

Chopra, R.N., Nayar, S.L. & Chopra, I.C. 1956. *Glossary of Indian Medicinal Plants.* C.S.I.R., New Delhi.

Chopra, R.N., Nayar, S. L. & Chopra, I. C. 1986. *Glossary of Indian Medicinal Plants (Including the Supplement).* C.S.I.R., New Delhi.

Chopra, R.N., Nayar, S.L. & Varma, B.S 1969. *Supplement of Glossary of Indian Medicinal Plants*. C.S.I.R., New Delhi.

Chowdhery, H.J. 1996. Arunachal Pradesh, its people, ethnobotanical diversity and conservation: 323-328. *In:* Jain, S.K. (*ed.*) *Ethnobiology in Human Welfare*. Deep. Publ., New Delhi.

Chowdhery, H.J. 1998. *Orchid Flora of Arunachal Pradesh*. B.S.M.P.S., Dehra Dun, India.

Chowdhery, H.J. & Murti, S. K. 2000. *Plant Diversity and Conservation in India - An Overview*. B.S.M.P.S., Dehradun.

Chowdhery, H.J. & Wadhwa, B.M. 1984. *Flora of Himachal Pradesh Analysis*. Vol. I - III. B.S.I., Howrah.

Chowdhury, J.U., Alam, M.K., & Hassan, M.A. 1996. Some traditional folk formularies against dysentery and diarrhoea in Bangladesh. *J. Econ. Taxon. Bot.* Addl. Ser. ***12***: 20-23.

Chun-Lin, L. & Jieru, W. 1995. Ethnobotany of Jinuo nationality in Xishuangbanna, Southwest China: edible plants. *Ethnobotany 7*: 39-49.

Clarkes, C.B. 1898. Sub-areas of British India, illustrated by the detailed distribution of Cyperaceae in that Empire. *J. Linn. Soc. Bot. London **34**:* 1-146.

Collett, H. 1902. *Flora Simlensis*. Thacker Spink & Co. Calcutta & Shimla.

Conway, D. 1973. *The Magic of Herbs*. Jonathan, Co. Ltd., U.K.

Cooke, T. 1901-1908. *The Flora of the Presidency of Bombay,* Vols. 1-2. London. (Reprinted: 1958, Vols. 1-3. B.S.I., Calcutta).

Coventry, E.M. 1901. *A Catalogue of the Trees and Shrubs of Punjab*. Lahore.

Cowen, D.V.1950. *Flowering Trees and Shrubs in India*. Thacker & Co. Ltd., Bombay.

Dagar, H.S. 2003. Ethnobotanical evaluation of some endemic plants in Andaman and Nicobar Islands: 134-147. *In*: Khan, S.S. (*ed.*) *Vistas in Ethnobotany*, Vol.-I, *Indian J. Pure & Appl. Biol.* Naval Kishore Press, Lucknow.

Dagar, H.S. & Dagar, J.C. 1996. Ethnobotanical studies of the Nicobarese of Chowra island of Nicobar group of islands. *J. Econ. Taxon. Bot.* Addl. Ser. ***12***: 381-388.

Dagar, J.C. & Dagar, H.S. 1999. *Ethnobotany of Aborigines of Andaman-Nicobar Islands.* Survey of India, Port Blair, Andaman-Nicobar Islands, India.

Dalzell, N.A. 1858. *Catalogue of the Indigenous Flowering Plants of the Bombay Presidency*. Surat.

Dalzell, N.A. & Gibson, A. 1861. *The Bombay Flora*. Bombay Edu. Soc. Press, Mumbai, India.

Dam, D.P. & Hajra, P.K. 1997. Observations on ethnobotany of the Monpas of Kameng district, Arunachal Pradesh: 153-159. *In:* Jain, S.K. (*ed.*) *Contribution to Indian Ethnobotany.* Sci. Publ., Jodhpur, India.

Dang, R. 1998. *Flowers of the Western Himalayas*. Wilderness Films, India.

D'Rozario, A., Bera, S. & Mukherji, D. 1999. *A Handbook of Ethnobotany*. Kalyani Publ., New Delhi.

Das, D. & Agarwal, V.S. 1991. *Fruit Drug Plants of India*. Kalyani Publ., New Delhi.

Das, H.J., Panda, P.C. & Patnaik S. N. 1996. Traditional uses of wetland plants of Eastern Orissa: 306-312. *In:* Maheshwari, J.K. (*eds.*) *Ethnobotany in South Asia*. Sci. Publ., Jodhpur, India.

Dash, S.S., Maiti, A. & Rai, S.K. 2003. Traditional uses of plants among the urban population of Gangtok-Sikkim: 317-324. *In*: Singh, V. & Jain, A.P. (*eds.*) *Ethnobotany and Medicinal Plants of India and Nepal.* Sci. Publ., Jodhpur, India.

Dastur, J.F. 1970. *Medicinal Plants of India and Pakistan*. D.B. Taraporevala Sons & Co., India.

Datt, B.1996. Ethnobotanical resources of Chhatarpur district (Madhya Pradesh): 400-402. *In:* Jain, S.K. (*ed.*) *Ethnobiology in Human Welfare,* Deep Publ., New Delhi.

Davinson, J.A. & Miller, D.R. 1990. *Ornamental Plants.* Elsevier Sci. Publ., Amsterdam, Netherlands.

Davis. B. 1990. *Climbers and Wall Shrubs.* Viking Publ., New York.

Dean, J. 1964. *A Second Book of Wild Flowers*. A & C Black Ltd., London.

Deb, D.B. 1981. *The Flora of Tripura State,* Vol. I. Today and Tomorrow's Printer & Publ., New Delhi.

Deb, D.B. 1983. *The Flora of Tripura State,* Vols II. Today and Tomorrow's Printer & Publ., New Delhi.

Deokota, R. & Chhetri, R.B. 2007. Ethnobotanical study in Sunsari district of eastern Nepal. *Ethnobotany* ***19***: 67-72.

Deokota, R. & Chhetri, R.B. 2009. Traditional medication in Bhedetar of District Sunsari (Nepal). *Ethnobotany* ***21***: 112-115.

Deorani, S.C. & Naithani, H.B. 1995. *Orchids of Nagaland.* Oriental Enterprises, Dehra Dun, India.

Devi, Y.N. & Singh, P.K. 2008. Ethnobotany of Verbenaceae in Manipur. *Ethnobotany* ***20***: 111-114.

Dey, A.C. 1980. *Indian Medicinal Plants Used in Ayurvedic Preparations.* B.S.M.P.S., Dehradun.

Dey, K.L. & Bahadur, R. 1973. *The Indigenous Drugs of India*. Pama Primlane, New Delhi.

Dhaliwal, D.S. & Sharma, M. 1999. *Flora of Kullu District (Himachal Pradesh)*. B.S.M.P.S., Dehra Dun.

Dhar, U. & Kachroo, P. 1983. *Alpine Flora of Kashmir Himalaya*. Sci. Publ., Jodhpur, India.

Dhiman, A.K. 2003. *Scared Plants and Their Medicinal Uses*. Daya Publ. House, Delhi.

Dhiman, A.K. 2005. *Wild Medicinal Plants of India (with Ethnomedicinal Uses)*. B.S.M.P.S., Dehradun.

Dhiman, D.R. 1976. *Himachal Pradesh Ki Vanoshdhiya Sampada.* Imperial Printing Press, Dharamshala, H.P.

Dhore, M.M., Lachure, P.S. & Gawande, P.D., 2011. *Trees of Vidarbha (A Field Guide to Flowering Trees)*. Paygun Publ., Amravati.

Dirr, M.A. 2009. *Dirr's Hardy Trees and Shrubs-An Illustrated Encyclopedia.* Timber Press, Oregon.

Diwakar, P.G. & Ansari, A.A. 2003. Econimic importance of plant resources of Buldhana district of Maharashtra: 592-600. *In:* Singh, V. & Jain, A.P. (*eds.*) *Ethnobotany and Medicinal Plants of India and Nepal.* Sci. Publ., Jodhpur.

Dixit, R.D. & Kumar, R. 2003. Plants used by local people in human welfare: 53-59. *In*: Singh, V. & Jain, A.P. (*eds.*) *Ethnobotany and Medicinal Plants of India and Nepal.* Sci. Publ., Jodhpur, India.

Dobriyal, R.M., Singh, G.S., Rao, K.S. & Saxena, K.G. 1997. Medicinal plant resources in Chakinal watershed in north-western Himalaya. *J. Herbs, Spices & Medicinal Plants* ***5***: 15-27.

Donald, W.1950. The Forsythias. *Natl. Hort. Mag.* ***10***: 9-16.

Drury, H. 1864. *Handbook of Indian Flora,* Vols. I & III. Travancore Sircar Press, Travancore.

Duhoon, S.S., Kopper, M.N. & Chandra, U. 1996. Seabuckthorn (*Hippophae spp*) - a less known wonder plant of ethno-medico-botanical importance in cold desert of India. *J. Econ. Taxon. Bot. Addl. Ser.* ***12***: 43-45.

Duke, J.A. & Ayensu, E.S. 1985. *Medicinal plants of China*. Reference Publ., Inc, Algonac, Michigan.

Duke, J.A., Bogenschutz-Godwi, M.J., Judi duCellier & Duke, P.K. 2002. *Handbook of Medicinal Herbs*. CRC Press, New York.

Duthie, J.F. 1882. Plants found in Kumaon, Garhwal and the adjacent parts of Tibet by Capt. Richard Strachey and Mr. Winterbottom: 403-670. *In:* Atkinson, E.T.: *Gazetteer of North Western Provinces*. Vol. 10.

Duthie, J.F. 1903-1929. *Flora of the Upper Gangetic Plain and of the Adjacent Siwalik and Sub-Himalayan tracts,* Vols. 1-3. Superintendent Govt. Printing, Calcutta. (Reprinted: 1960, Vols. 1-2. B.S.I., Calcutta).

Duthie, J.F. 1906. *Catalogue of Plants of Kumaon and the Adjacent Parts of Gharwal and Tibet Based on the Collections Made by Strachey and Winterbottom During the Year 1846–1849 and on the Catalogue Originally Prepared in 1852 by Sir Richard Strachey.* Reeve London, UK. (Reprinted: 1974, B.S.M.P.S., Dehradun).

Dwivedi, A.P. 2007. *Forests: The Nonwood Resources.* Intl., Book Distr., Dehradun, India.

Dwivedi, S.N. 2003. Ethnobotanical studies and conservational strategies of wild and natural resources of Rewa district of Madhya Pradesh: 233-244. *In:* Singh, V. & Jain, A.P. (*eds.*) *Ethnobotany and Medicinal Plants of India and Nepal.* Sci. Publ., Jodhpur, India.

Dwivedi, S.N., Dwivedi, S. & Patel, P.C. 2007. Antimicrobial herbs used among the tribal and rural communities. *Ethnobotany* ***19***: 111-115.

Edgeworth, M.P. 1867. Flora of Banda. *J. Linn. Soc. London* ***9***: 304-326.

Farjon, A. 1994. *Cupressus cashmeriana. Kew Mag.* ***253***: 156-166.

Farjon, A. 2005. *Monograph of Cupressaceae and Sciadopitys.* Royal Bot. Gard., Kew.

Farooq, S. 2005. *555 Medicinal Plants: Field and Laboratory Manual.* Intl. Book Distr., Dehradun (India).

Faulks, R.J. 1958. *An Introduction to Ethnobotany.* Moredale Publ., Ltd., London.

Foster, S. & Duke, J.A. 1990. *A Field Guide to Medicinal Plants. Eastern and Central N. America.* Houghton Mifflin Co.

Fyson, P.F. 1915. *The Flora of Nilgiri and Pulney Hill Tops,* Vols. 1-3, Madras. (Reprinted 1975, B.S.M.P.S., Dehra Dun.).

Fyson, P.F. 1932. *The Flora of the South Indian Hill Stations,* Vols. I & II. Suptd. Govt. Press, Madras.

Gamble, J.S. 1898. The Hill Route from Mussoorie to Simla. *Indian For.* ***24***: 105-111.

Gamble, J.S. 1972. *A Manual of Indian Timbers.* B.S.M.P.S., Dehradun.

Gamble, J.S. & Fischer, C.E.C. 1915–1936. *Flora of the Presidency of Madras.* 3 Vols. West Newman & Co., Adlard & Son: London.

Ganai, K.A. & Nawachoo, I.A. 2003. Traditional treatment of toothache by the Gujjar and Bakerwal tribes of Kashmir in India: 105-107. *In*: Singh, V. & Jain, A.P. (*eds.*) *Ethnobotany and Medicinal Plants of India and Nepal.* Sci. Publ., Jodhpur, India.

Gaur, R.D. 1999. *Flora of District Garhwal-North-West Himalaya with Ethnobotanical Notes.* Transmedia Publ., Srinagar, Garhwal, India.

Gaur, R.D. & Singh, P.B. 1993. Ethno-medicinal plants of Mandi district, Himachal Pradesh. *Bull. Medico. Ethnobot. Res.* ***14***: 1-11.

Ghate, V.S. 1998. Plants in Patra-Pooja: notes on their identity and utilization. *Ethnobotany* ***10***: 6-15.

Ghosh, A., Maity, S. & Maity, M. 1996. Ethnomedicine in Bankura district, West Bengal. *J. Econ. Taxon. Bot.* Addl. Ser. ***12***: 318-320.

Gill, L.S., Idu, M. & Ogbor, D.N. 1997. Folk medicinal plants; practices and beliefs of the Bini People in Nigeria. *Ethnobotany* ***9***: 1-5.

Gill, L.S., Nyawuame, H.G.K., Esezobor, E.I., & Osagie, I.S. 1993. Nigerian folk medicine: practices and beliefs of Esan people. *Ethnobotany* ***5***: 129-142.

Girach, R.D. 1992. Medicinal plants used by Kondh tribe of district Phulbani (Orissa) in Eastern India. *Ethnobotany* ***4***: 53-66.

Girach, R.D. & Aminuddin 1995. Ethnomedicinal uses of plants among the tribals of Singhbhum district, Bihar, India. *Ethobotany 7*: 103-107.

Girach, R.D., Aminuddin, Ahmed, M., Brahmam, M. & Misra, M.K. 1996. Native phytotherapy among rural population of district Bhadrak, Orissa: 162-164. *In*: Jain, S.K. (*ed.*) *Ethnobiology in Human Welfare.* Deep Publ., New Delhi.

Goel, A.K. & Tripathi, S. 2009. Ethnobotanical spectrum of Indian flora: an overview during the past 20 years in the journal Ethnobotany. *Ethnobotany* ***21***: 131-134.

Gogoi, P. & Borthakur, S.K. 1991. Plants in religio-cultural beliefs of Tai Khamtis of Assam (India). *Ethnobotany* ***3***: 89-95.

Gogoi, R., Bokolial, D. & Hazarika, D. 2003. Preliminary observations on the medicinal plants of Chandrapur area of Kamrup district, Assam: 410-415. *In:* Singh, V. & Jain, A.P. (*eds*) *Ethnobotany and Medicinal Plants of India and Nepal.* Sci. Publ., Jodhpur.

Gogoi, R. & Das, M.K. 2003. Observation on some weed of medicinal importance in the Brahmaputra Valley of Assam: 434-441. *In:* Singh, V. & Jain, A.P. (*eds.*) *Ethnobotany and Medicinal Plants of India and Nepal,* Vol. I. Sci. Publ., Jodhpur, India.

Good, R. 1964. *The Geography of the Flowering Plants*. Longman Group Ltd., London.

Govaerts, R, 2001. How many species of seed plants are there? *Taxon* ***50***: 1085-1090.

Goyal, A.K & Arora, S. 2009. *India's Fourth National Report to the Convention on Biological Diversity*. M.O.E.F., G.O.I., New Delhi.

Grae, I. 1974. *Nature's Colors - Dyes from Plants.* MacMillan Publ. Co., New York.

Graf, A.B. 1980. *Exotica. Pictorial Cyclopedia of Exotic Plants from Tropical and Near-Tropic Regions.* Roehrs Co. INC., U.S.A.

Graf, A.B. 1981. *Tropica. Color Cyclopedia of Exotic Plants and Trees for Warm-region Horticulture – in Cool Climate the Summer Garden or Sheltered Indoors.* Roehrs Company INC., U.S.A.

Graham, J. 1839. *A Catalogue of the Plants Growing in Bombay and its Vicinity.* Bombay.

Grey-Wilson, C. & Matthews, V. 1983. *Gardening on Walls.* Collins Publ., Sidney.

Griffith, W. 1854. *Notulae und Plantas Asiaticas,* Vols. 1-4. Chrles A. Serrao Publ., Calcutta.

Grotenhermen, F. & Russo, E. (*eds.*) 2002. *Cannabis and Cannabinoids Pharmacology, Toxicology and Therapeutic Potential.* The Haworth Press, Inc., New York.

Guha Bakshi, D.N., Sensarma, P. & Pal, D.C. 1999. *A Lexicon of Medicinal Plants in India*, Vol. I. Naya Prokash, Calcutta (India).

Gupta, M.P. 1988. Present status and future prospects for the cultivation and collection of medicinal plants of Himachal Pradesh. *Indian For.* ***114*** (1): 19-25.

Gupta, R. 1961. Flora of Lam Dal. *Indian For.* ***87***(5): 316-324.

Gupta, R. 1964. Survey record of medicinal and aromatic plants of Chamba forest division, Himachal Pradesh. *Indian For.* ***90***: 454-468.

Gupta, R. 1971. Medicinal and aromatic plants of Bhandal ranges and Churah forest division, Chamba district, Himachal Pradesh. *J. Bomb. Nat. Hist. Soc.* ***68*** (3): 791-803.

Gupta, R. 1981. Plants in folk medicine of the Himalaya: 83-90. *In*: Jain, S.K. (*ed.*) *Glimpses of Indian Ethnobotany*. Oxford & IBH Publ. Co., New Delhi.

Gupta, R.K., 1957. Botanical exploration in erstwhile tehri Garhwal State II. *J. Bomb. Nat. Hist. Soc.* ***54***: 878-886.

Gupta, R.K., 1962. Botanical exploration in erstwhile tehri Garhwal State III. *J. Bomb. Nat. Hist. Soc.* ***59***: 486-512.

Gupta, S.M. 1989. Ethnobotanical aspects of art and literature: 59-68. *In*: Jain, S.K. (*ed.*) *Methods and Approaches in Ethnobotany*. Soc. Ethnobot., C.D.R.I., Lucknow.

Hadfield, M. 1956. *Everymans Wild Flowers and Trees*. J.M. Dent & Sons Ltd., London.

Haines, H.H. 1921-1925. *The Botany of Bihar and Orissa,* 6 Vols., London (Reprinted: 1961, 1-3 Vols. B.S.I., Calcutta).

Hajra, P.K. & Baishya, A.K. 1997. Ethnobotanical notes on the Miris (Mishings) of Assam plains: 161-168. *In:* Jain, S.K. (*ed.*) *Contribution to Indian Ethnobotany.* Sci. Publ., Jodhpur, India.

Hajra, P.K. & Das, B.K. 1982. Vegetation of Gangtok with special reference to exotic plants. *Indian For.* ***107***: 554-566.

Harborne, F.R.S. J. B. & Baxter, H. 2001. *Chemical Dictionary of Economic Plants.* John Wiley & Sons, England.

Haridasan, K. & Rao, R.R. 1985. *A Forest Flora of Meghalaya,* 2 Vols. B.S.M.P.S., Dehra Dun.

Haridasan, K., Sukla, G.P. & Deori, M.L. 2001 Cultivation prospects of medicinal plants of Arunanchal Pradesh: a review: 329-343. *In*: Samant, S.S., Dhar, U. & Palni, L.M.S. (*eds.*) *Himalayan Medicinal Plants-Potential and Prospects*. Gyanodaya Prakashan, Nainital (India).

Hay, R., McQuown, F.R., & Beckett, G.K. 1983. *The Dictionary of Indoor Plants in Colour*. Ebury Press & Michael Joseph Ltd., London.

Hazara, P.K. & Mudgal, V. 1997. *Plant Diversity Hot Spots in India - An Overview.* B.S.I., Calcutta, India.

Hedrick, U.P. 1972. *Sturtevant's Edible Plants of the World*. Dover Publ., New York.

Hembrom, P.P. 1996. Contact therapy practised by Mundas of Chotanagpur (Bihar). *Ethnobotany* ***8***: 36-39.

Henry, A.N., Chitra, V. & Balakrishnan, N.P. (*eds.*) 1989. *Flora of Tamil Nadu*, Series 1, Vol. 3. B.S.I., Coimbatore.

Henry, A.N., Hosagoudar, V.B. & Ravi Kumar, K. 1996. Ethno-medico-botany of the Southern Western Ghats of India: 173-180. *In:* Jain, S.K. (*ed.*) *Ethnobiology in Human Welfare*. Deep Publ., New Delhi.

Hill, A.F. 1951. *Economic Botany: A Textbook of Useful Plants and Plants Product*. Tata McGraw Publ. Co. Ltd., New Delhi.

Hodkinson, D.J. & Thompson, K. 1997. Plant dispersal: the role of man. *J. Appl. Ecol.* ***34***: 1484-1496.

Hohenacker, R.F. 1849. *Plants of Canara and Neighbouring areas of Maharashtra and Malabar*. *Flora* ***32***: 556-560.

Holtom, J. & Hylton, W. 1979. *Complete Guide to Herbs*. Rodale Press, New York.

Hooker, J.D. 1872-1897. *The Flora of British India,* Vols. I-VII. B.S.M.P.S., Dehra Dun.

Hooker, J.D., 1906. A Sketch of the Flora of British India. Imperial Gazetter of Inma, Oxford.

Hooker, J.D. & Thomson, T. 1855. *Flora Indica*. W. Pamplin, Fifth street. Soho. (Reprinted: 1976, B.S.M.P.S., Dehra Dun & Periodical Experts, Delhi).

Hosagoudar, V.B. & Henry, A.N. 1996a. Ethnobotany of Kadars, Malasars and Kuthuvans of the Anamalais in Coimbatore District, Tamil Nadu, India. *J. Econ. Taxon. Bot.* Addl. Ser. ***12***: 260-267.

Hosagoudar, V.B. & Henry, A.N. 1996b. Ethnobotany of Soligas in Biligiri Rangana Betta, Karnataka, Southern India. *J. Econ. Taxon. Bot.* Addl. Ser. ***12***: 228-243.

Hosagoudar, V.B. & Henry, A.N. 1996c. Ethnobotany of tribes Irular, Kurumban and Paniyan of Nilgiris in Tamil Nadu, Southern India. *J. Econ. Taxon. Bot.* Addl. Ser. ***12***: 272-283.

Huidrom, S.B.K. 1996. Plants used in medico-sexual purposes by Meitein community in Manipur state, India. *J. Econ. Taxon. Bot.* Addl. Ser. ***12***: 364-366.

Huxley, A. 1992. *The New RHS Dictionary of Gardening.* MacMillan Press, New York.

Hynniewta, T.M., Kataki, S.K. & Wadhwa, B.M. 2000. *Orchids of Nagaland.* B.S.I., Calcutta.

Ingalhalikar, S., 2007. *Flowers of Sahyadri.* Corolla Publ., Pune.

Islam, M.A. & Jha, R.M. 2003. Ethnopharmacology of tribals of Ranchi district in Jharkhand: 300-310. *In:* Singh, V. & Jain, A.P. (*eds.*) *Ethnobotany and Medicinal Plants of India and Nepal,* Vol. I. Sci. Publ., Jodhpur, India.

IUCN, 1978. *Categories, Objectives and Criteria for Protected Areas. Commission in National Parks and Protected Areas, Committee on Criteria and Nomenclature.* IUCN, Gland, Switzerland.

IUCN, 2013. *IUCN Red List of Threatened Species. Version 2013.1.* IUCN, Gland, Switzerland.

IUCN, 1995. *IUCN Guidelines for Re-introductions.* IUCN, Gland, Switzerland.

Jadhav, D. 2009. Ethnomedicinal plants used in leaf therapy in Ratnam district (Madhya Pradesh). *Ethnobotany* ***21***: 84-90.

Jaeger, P. 1961. *The Wonderful Life of Flower.* George, G. Harrap & Co., Ltd., London.

Jain, S.K. 1968. *Medicinal Plants.* Nat. Book Trust, New Delhi.

Jain, S.K. 1981. *Glimpses of Indian Ethnobotany.* Oxford & IBH Publ. Co., New Delhi.

Jain, S.K. 1987. Ethnobotany its concepts and relevance: 1-12. *Presid. Addr. X Bot. Conf.* Patna.

Jain, S.K. 1991. *Dictionary of Indian Folk Medicine and Ethnobotany.* Deep Publ., New Delhi.

Jain, S.K., Mudgal, V., Banerjee, D.K., Guha, A., Pal, D.C. & Das. D. 1984. *Bibliography of Ethnobotany.* B.S.I., Howrah.

Jain, S.K., Ranjan, V., Sikarwar, R.K.S., & Saklani, A. 1994. Botanical distribution of psychoactive plants of India. *Ethnobotany* ***6***: 65-75.

Jain, S.K. & Rao, R.R. (*eds*) 1977. *A Handbook of Field and Herbarium Methods.* Today's & Tomorrow's Printer & Publ., New Delhi.

Jain, S.K. & Saklani, A. 1992. Cross-cultural ethnobotanical studies in North East India. *Ethnobotany* ***4***: 25-38.

Jain, S.K., Sinha, B.K., & Gupta, R.C. 1991. *Notable Plants in Ethnomedicine of India*. Deep Publ., New Delhi.

Jain, S.K. & Srivastava, S. 1999. *Dictionary of Ethnoveterinary Plants of India*. Deep Publ., New Delhi.

Jain, S.K. & Srivastava, S. 2001. Indian ethnobotanical literature in last two decades - a graphic review and future directions. *Ethnobotany* ***13***: 1-8

Jain, S.P. & Puri, H.S. 1994. An ethno-medico-botanical survey of Parvati valley in Himachal Pradesh (India). *J. Econ. Taxon. Bot.* ***10***: 221-222.

Jaiswal, S., Singh, S.V., Singh, B. & Singh, H.N. 2004. Plant used for tissue healing of animals. *Nat. Product Rad.* ***3***(4): 284-292.

Jamir, N.S., Lanusunep & Konyak, C.H.M. 2011. Some less known ethnomedicinal plants used by Angami-Naga tribes in Kohima district, Nagaland (India). *Ethnobotany* ***23***:116-120.

Jaryan, V., Uniyal, S., Gupta, R.C. & Singh, R.D. 2012. Alien flora of Indian Himalayan state of Himachal Pradesh. *Environ. Monit. Assess.* ***185***(7): 6129-6153.

Jeyapardha, D., Murugesan, P., Rajeshwar, G. & Karunakaran, K. 2011. Ethnopharmacological application of medicinal plants for skin ailments and cosmetics used by various tribal communities in Kolli Hills of South India. *Intl. J. Pharm. Tech. Res.* ***11***: 102-104.

Jha, B.J. & Kumar, M. 2003. Herbal remedies for various diseases: 78-87. *In:* Singh, V. & Jain, A. P. (*eds.*) *Ethnobotany and Medicinal Plants of India and Nepal,* Vol. I. Sci. Publ., Jodhpur, India.

Jha, V. & Basak, P.K. 1994. Ethnobotanical aspects of Mithila Painting. *Ethnobotany* ***6***: 9-18.

Jha, V., Choudhary, U.N. & Saraswati, K.C. 1991. Botanical aspects of an ethnoveterinary prescription in Mithila, North Bihar (India). *Ethnobotany* ***3***: 101-104.

Jha, V., Mishra, S., Kargupta, A.N. & Jha, A. 1996. Leaves and flowers utilised as supplementary vegetables in Darbhanga (North Bihar) and their ethnobotanical significance: 395-402. *In:* Maheshwari, J.K. (*eds.*) *Ethnobotany in South Asia*. Sci. Publ., Jodhpur, India.

Joseph, J. 1986. *Flora of Nongpoh and Vicinity*. B.S.I., Calcutta.

Joseph, J. 1987. *Orchids of Nilgiris*. B.S.I., Calcutta.

Joshi, A. R. & Joshi, K. 2005. Piscicidal plants of Bagmati wasteland, Nepal: indigenous practices. *Ethnobotany* ***17***: 184-186.

Joshi, P. 1995. *Ethnobotany of the Primitive Tribes in Rajasthan.* Rupa Books Pvt. Ltd., Jaipur.

Kachroo, P., Sapru, B.L. & Dhar, U. 1977. *Flora of Ladakh.* B.S.M.P.S., Dehra Dun.

Kala, C.P. 2003. *Medicinal Plants of Indian Trans-Himalaya (Focus on Tibetan Use of Medicinal Resources).* B.S.M.P.S., Dehra Dun.

Kala, C.P. & Manjrekar, N. 2000. Ethno-medicobotany of Indian Trans-Himalaya: a case study from Spiti: 177-183. *In:* Maheshwari, J.K. (*ed.*) *Ethnobotany and Medicinal Plants of Indian Subcontinent.* Sci. Publ., Jodhpur, India.

Kalakoti, B.S. 2001. Nainital Hills- A transtitional zone for tropical, temperate and alpine plants and medicinal value for selected plants in pharmaceuticals: 583-591. *In*: Pandey, P.C. & Samant, S.S. (*eds.*) *Plant Diversity of The Himalaya.* Gyanodaya Prakashan, Nainital, India.

Kanjilal, P.C. 1933. *A Forest Flora of Pilibhit, Oudh, Gorakhpur and Bundelkhand.* Govt. Printing Press, Calcutta.

Kanjilal, U.N. 1901. *Forest Flora of the School Circle, North West Province.* Govt. Printing Press, Calcutta.

Kanjilal, U.N. 1911. *Forest Flora of the Siwalik and Jaunsar Forest Divisions of United Provinces of Agra and Oudh.* Govt. Printing Press, Calcutta.

Kanjilal, U.N. & Gupta, B.L. 1928. *Forest Flora of Chakrata, Dehradun and Saharanpur Forest Divisions, United Provinces of Agra and Oudh.* Govt. Printing Press, Calcutta.

Kanjilal, U.N., Kanjilal, P.C., Das, A., De, R.N. & Bor, N.L. 1934-1940. *Flora of Assam,* 5 Vols. Govt. Press, Shillong.

Kapoor, V.P. 2005. Herbal cosmetics for skin and hair care. *Nat. Product Radi.* ***4***: 306-314.

Kapur, S.K. 1985. Observations on the floristic composition of Kangra Valley (Himachal Pradesh). *New Bot.* ***12***: 157-165.

Kapur, S.K. 1993. Ethnomedico plants of Kangra valley (Himachal Pradesh). *J. Econ. Taxon. Bot.* ***17***(2): 395-408.

Kapur, S.K. 1996a. Traditionally important medicinal plants of Bhaderwah Hills – Jammu Province–III: 62-69. *In:* Maheshwari, J.K. (*eds.*) *Ethnobotany in South Asia.* Sci. Publ., Jodhpur, India.

Kapur, S.K. 1996b. Traditionally important medicinal plants of Bhaderwah Hills – Jammu Province–IV: 70-74. *In:* Maheshwari, J.K. (*eds.*) *Ethnobotany in South Asia.* Sci. Publ., Jodhpur, India.

Kapur, S.K. & Nanda, S. 1996. Traditionally important medicinal plants of Bhaderwah hills (Jammu Province)- Part II: 56-61. *In:* Maheshwari, J.K. (*eds.*) *Ethnobotany in South Asia.* Sci. Publ., Jodhpur, India.

Kapur, S.K., Nanda, S. & Srivastava, T.N. 1996. Ethnobotanical uses of RRL-Herbarium - III: 50-55. *In:* Maheshwari, J. K. (*eds.*) *Ethnobotany in South Asia*. Sci. Publ., Jodhpur, India.

Kapur, S.K. & Singh, P. 1996. Traditionally important medicinal plants of Udhampur district (Jammu Province) Part–I: 75-81. *In*: Maheshwari, J.K. (*eds.*) *Ethnobotany in South Asia*. Sci. Publ., Jodhpur, India.

Kapur, S.K. & Srivastava, T.N. 1996. Traditionally important medicinal plants of Udampur district (Jammu Province) Part–II: 82-88. *In*: Maheshwari, J. K. (*eds.*) *Ethnobotany in South Asia*. Sci. Publ., Jodhpur, India.

Karnick, C.R. 1994. *Pharmacopoeial Standards of Herbel Plants.* Sri Satguru Publ., Delhi.

Katewa, S.S. & Arora, A. 1997. Some plants in folk medicine of Udaipur district (Rajasthan). *Ethnobotany* ***9***: 48-51.

Katewa, S.S., Choudhary, B.L., Jain, A. & Takhar, H.K. 2001. Some plants in folk medicine of Rajsamand district (Rajasthan). *Ethnobotany* ***13***: 129-134.

Kaul, M.K. 1986. *Weed Flora of Kashmir Valley*. Sci. Publ., Jodhpur.

Kaul, M.K. 1997. *Medicinal Plants of Kashmir and Ladakh Temperate and Cold Arid Himalaya.* Indus Publ. Co., New Delhi.

Kaur, H. & Sharma, M. 2004. *Flora of Sirmaur. (Himachal Pradesh*). B.S.M.P.S., Dehra Dun.

Kaushik, P. 1988. *Indigenous Medicinal Plants* (*Including Microbes and Fungi*). Today & Tomorrow's Printer & Publ., New Delhi.

Kaushik, P. & Dhiman, A.K. 2000. *Medicinal Plants and Raw Drugs of India.* B.S.M.P.S., Dehra Dun (India).

Kayal, R.N., Saha, S. & Ghosal, P.P. 2003. A floristic account of Indian medicinal plants (originated, cultivated, naturalized or introduced) used in homoeopathic drug preparation: 60-67. *In:* Singh, V. & Jain, A.P. (*eds.*) *Ethnobotany and Medicinal Plants of India and Nepal,* Vol. I. Sci. Publ., Jodhpur, India.

Keller, H., Prance, G.T., Tressen, S.G. & Duarte. 2011. Ethnobotanical resources of multiple-use Guarani Reserve, Misiones (Argentina): their importance to indigenous communities of Caramelito and Taruma Poty-II. *Ethnobotany* ***23***: 33-50.

Khan, A.A. 2005. Ethnobotany in twenty first century in India. *Ethnobotany* ***17***: 71-78.

Khanna, K.K. 2002. Unreported ethnomedicinal uses of plants from the tribal and rural folklores of Gonda district, Uttar Pradesh. *Ethnobotany* ***14***: 52-56.

Khanna, K.K., Mudgal, V., Shukla, G. & Srivastava, P.K. 1996a. Unreported ethnomedicinal uses of plants from Mirzapur district, Uttar Pradesh. *J. Econ. Taxon. Bot.* Addl. Ser. ***12***: 112- 117.

Khanna, K.K., Shukla, G. & Mudgal, V. 1996b. New traditional medicinal uses of plants from Jalaun district, Uttar Pradesh. *J. Econ. Taxon. Bot.* Addl. Ser. ***12***: 108-111.

Khanna, K.K., Srivastava, P.K. & Mudgal, V. 1996c. Noteworthy medicinal plants uses from rural folklore of Raebareli district, Uttar Pradesh. *J. Econ. Taxon. Bot.* Addl. Ser. ***12***: 118-122.

Khare, C.P. 2004. *Encyclopedia of Indian Medicinal Plants, Rational Western Therapy, Ayurvedic and Other Traditional Usage, Botany*. Springer-Verlag, Berlin Heidelberg, New York.

Kharkongor, P. & Joseph, J. 1981. Folklore medicobotany of rural Khasi and Jaintia tribes Meghalaya: 124-136. *In*: Jain, S.K. (*ed.*) *Glimpses of Indian Ethnobotany*. Oxford & IBH Publ. Co., New Delhi.

Khuroo, A.A., Rashid, I., Zafar, R., Dhar, G. H., & Wafai, B. A. 2007. The alien flora of Kashmir Himalaya. *Biol. Invas.* ***9***: 269-292.

Khuroo, A.A., Reshi, Z. A., Malik, A. H., Weber, E., Rashid, I., & Dar, G. H. 2012. Alien flora of India: taxonomic composition, invasion status and biogeographic affiliations. *Biol. Invas.* ***14***: 99–113.

Kirn, H.S., Kapahi, B.K. & Srivastava, T.N. 2000. Taxo-ethnobotanical observations on the gymnosperms of Poonch district (J. & K State) India: 155-160: *In:* Maheshwari, J.K. (*eds.*) *Ethnobotany and Medicinal Plants of Indian Subcontinent.* Sci. Publ., Jodhpur, India.

Kirtikar, K.R. & Basu, B.D. 1984. *Indian Medicinal Plants*. Vol. I-IV. Lalit Mohan Basu, Allahabad, India.

Koelz, W.N. 1979. Notes on the ethnobotany of Lahaul, a province of the Punjab. *Quar. J. Crude Drug. Res.* ***17***: 1-56.

Krisa, B. 1982. *The Illustrated Encyclopedia of Flowering Plants*. Octopus Books Ltd., London.

Kshirsagar, R.D. & Singh, N.P. 2000. Less known ethnomedicinal uses of plants in Coorg district of Karnataka state, Sourthern India. *Ethnobotany* ***12***: 12-16.

Kshirsagar, R.D. & Singh, N.P. 2007. *Ethnobotany of Mysore and Coorg, Karnataka State*. B.S.M.P.S., Dehradun.

Kulkarni, D.K. & Kumbhojkar, M.S. 1993. Kitchen garden plants of Mahadeokali tribe in Maharastra. *Ethnobotany* ***5***: 119-127.

Kulkarni D.K. & Kumbhojkar, M.S. 2000. Plant Conservation in Maharashtra during 17th century. *Ethnobotany* ***12***: 29-38

Kumar, A., Tewari, D.D. & Pandey, Y.N. 2003. Ethno-phytotherapeutics among Tharus of Beerpur- Semra forest range of Balrampur district (U.P.): 839-844. *In:* Singh, V. & Jain, A.P. (*eds*) *Ethnobotany and Medicinal Plants of India and Nepal.* Sci. Publ., Jodhpur.

Kumar, K.K., Sasidharan, N., & Swarupanandan, K. 2000. Ethnobotanical studies on the hills tribes in the Shola forests of high ranges, Kerala, South India: 451-466. *In:* Maheshwari, J.K. *(ed.) Ethnobotany and Medicinal Plants of Indian Subcontinent*. Sci. Publ., Jodhpur (India).

Kumar, N. 2005. *Studies on the Distribution and Importance of Medicinal and Aromatic Plants of Nahan area, District Sirmaur (H.P.).* M.Sc. Thesis, Dept. Forest Products. Dr. Y.S. Parmar Univ. Hort. & For., Solan, H.P.

Kumar, S. 2001. *Flora of Haryana (Materials)*. B.S.M P.S., Dehra Dun.

Kumar, S. 2002. *The Medicinal Plants of North-East India*. Sci. Publ., Jodhpur, India.

Kumar, S., Goyal, S. & Parveen, F. 2004. Ethnoveterinary plants in Indian arid zone. *Ethnobotany **16***: 91-95.

Kumar, V. 2003. Wild edible plants of Surguja district of Chhttisgarh State, India: 272-282. *In:* Singh, V. & Jain, A.P. (*eds.*) *Ethnobotany and Medicinal Plants of India and Nepal*, Vol. I. Sci. Publ., Jodhpur (India).

Kumar, V. & Rao, R.R. 2001. Some plant beverages used in traditional medicine. *Ethnobotany **12***: 36-39.

Kunkel, G. 1984. *Plants for human consumption*. Koeltz Sci. Books Publ, Koenigstein, Germany.

Kurian, J.C. (*ed.*) 1995. *Plants That Heal*. Oriental Watchman Publ. House, Pune, India.

Kurz, S. 1870. *Report on the Vegetation of the Andaman Islands*. Govt. Printing Press, Calcutta.

Kurz, S. 1876. A sketch of the vegetation of the Nicobar Islands. *J. Asiat. Soc. Bengal **45***(2): 105-164.

Lal, B., Vats, S.K. Singh, R.D. & Gupta, A. K. 1996. Plants used as ethnomedicine and supplementary food by Gaddis of Himachal Pradesh, India: 384-387. *In*: Jain, S.K. (*ed.*) *Ethnobiology in Human Welfare*. Deep Publ., New Delhi.

Lalramnghinglova, H. 1996. Ethnobotany of Mizoram- A preliminary survey: 439-459. *In*: Maheshwari, J.K. (*eds.*) *Ethnobotany in South Asia*. Sci. Publ., Jodhpur, India.

Lalramnghinglova, H. 2000. Ethnobotanical and agroecological studies on genetic resources of food plants in Mizoram state: 637-644. *In*: Maheshwari, J.K. (*eds.*) *Ethnobotany and Medicinal Plants of Indian Subcontinent.* Sci. Publ., Jodhpur, India.

Lalramnghinglova, H. 2002. Ethnobotanical study on the edible plants of Mizoram. Ethnobotany ***14***: 23-33.

Lalramnghinglova, H. 2003. *Ethnobotanical Study on the Edible Plants of Mizoram*. B.S.M.P.S., Dehradun (India).

Lambdon, P.W., Pysek, P., Basnou, C., Hejda, M., Arianoutsou, M., Essl, F., Jarosık, V., Pergl, J., Winter, M., Anastasiu, P., Andriopoulos, P., Bazos, I., Brundu, G., Celesti-Grapow,L., Chassot, P., Delipetrou, P., Josefsson, M., Kark, S., Klotz, S., Kokkoris, Y., Kuhn, I., Marchante, H., Perglova,I., Pino, J., Vila, M., Zikos, A., Roy, D., & Hulme, P. 2008. Alien flora of Europe: species diversity, temporal trends, geographical patterns and research needs. *Preslia* ***80***: 101–149.

Lata, S., Maurya, A.K. & Jain, S.K. 2008. An overview of ethnobotanical studies on Uttarakhand state in north India. *Ethnobotany* ***20***: 25-40.

Llamas, K.A. 2003. *Tropical Flowering Plants: A Guide to Identification and Cultivation.* Timber Press, Oregon.

Lunardi, C. 1987. *Simon and Schuster's Guide to Shrubs and Vines and Other Small Ornamentals.* Simon & Schuster Inc., New York.

Mack, R. N., & Erneberg, M. 2002. The United States naturalized flora: largely the product of deliberate introductions. *Ann. Miss. Bot. Gdn.* ***89***: 176-189.

Mack, R.N., Simberloff, D., Lonsdale, W.M., Evans, H., Clout, M. & Bazzaz, F.A. 2000. Biotic invasions: causes, epidemiology, global consequences, and control. *Ecol. Appl.* ***10***: 689 -710.

Mahato, A.K. & Mahato, P. 1996. Ethnobotanical wealth of Chhotanagpur Plateau–IV. Some medicinal plants used against intestinal worms. *J. Econ. Taxon. Bot.* Addl. Ser. ***12***: 389-391.

Mahato, R.B. & Chaudhary, R.P. 2005. Ethnomedicinal plants of Palpa district, Nepal. *Ethnobotany* ***17***: 152-163.

Maheshwari, J.K. 1963. *The Flora of Delhi.* C.S.I.R., New Delhi.

Maheshwari, J.K. 1966. *Illustrations to the Flora of Delhi.* C.S.I.R., New Delhi.

Maheshwari, J.K. (*ed.*) 1992. *Ethnobotany in India.* Sci. Publ., Jodhpur.

Maheshwari, J.K. 1995. Ethnobotany in development and conservation of resources: 114-120. *In*: Jain, S.K. (*ed.*) *A Manual of Ethnobotany, 2nd ed.,* Sci. Publ., Jodhpur.

Maheshwari, J.K. & Paul, S.R. 1975. The exotic flora of Ranchi. *J. Bombay Nat. Hist. Soc.* *72*(1): 158-188.

Maheshwari, J.K. & Singh, J. P. 1984. Contribution to the ethnobotany of Bhoxa tribe of Bijnor and Pauri Garhwal districts, U.P. *J. Econ. Taxon. Bot.* ***5***: 251-259.

Maheshwari, J.K., Singh, K.K. & Saha, S. 1981. *The Ethnobotany of the Tharus of Kheri District, Uttar Pradesh.* N.B.R.I., Lucknow, India.

Maithani, G.P., Bahuguna, V.K., Negi, J.D.S. & Nautiyal, S. 1991. *Handbook of Some Important Himalayan Shrubs.* ICFRE-1, FRI, Dehra Dun, India.

Maliya, S.D. 2004. Some new or less-known folk medicine of district Bahraich, Uttar Pradesh, India-2. *Ethnobotany* ***16***: 113-115.

Manandhar, N.P. 1980a. *Medicinal Plants of Nepal Himalaya*. Ratna Pustak Bhandar, Kathmandu, Nepal.

Manandhar, N.P. 1980b. Some less known medicinal plants of Rasuwa district (Nepal). *Quart. J. Crude Drug Res*. ***18***(3): 147-151.

Manandhar, N.P. 1990. Folk-lore medicine of Chitwan district, Nepal. *Ethnobotany* ***2***: 31-38.

Manandhar, N.P. 1995. Ethnobotanical notes on unexploited wild food plants of Nepal. *J. Econ. Taxon. Bot. 7*: 95-101.

Manandhar, N.P. 1996. Traditional practice for oral health care in Nepal. *J. Econ. Taxon. Bot.* Addl. Ser. ***12***: 408-413.

Manandhar. N.P. 2002. *Plants and People of Nepal*. Timber Press, Oregon.

Manandhar, N.P. 2003. Ethnobotanical notes on the family *Leguminosae* in Nepal: 354-357. *In:* Singh, V. & Jain, A.P. (*eds.*) *Ethnobotany and Medicinal Plants of India and Nepal,* Vol. I. Sci. Publ., Jodhpur, India.

Mander, U. & Jongman, R.H.G. 1998. Human impact on rural landscapes in Central and Northern Europe. *Land. & Urban Plan.* ***41***: 149-154.

Manilal, K.S. 1988. Flora of Silent Valley, tropical rain forest of India. *J. Trop. For.* ***4***: 402.

Manilal, K.S. & Sivarajan, V.V. 1982. *Flora of Calicut*. B.S.M.P.S., Dehra Dun.

Manral, B. & Pande, P.C. 2004. Plant associated with the folk songs of Kumaon Himalaya: 195-207. *In*: Pande, P.C. & Joshi, N.K. (*eds.*) *Traditions and Traditional Knowledge of Central Himalayas.* B.S.M.P.S., Dehradun (India).

Mao, A.A. 1993. A preliminary report on the folklore botany of Mao Nagas of Manipur (India). *Ethnobotany* ***5***: 143-147.

Martin, M. 1972. *Introduction a l' Ethnobotanique du Cembodge.* Paris.

Martin, M., Mathias, E. & Censtance (*eds*) 2001. *Ethnoveterinary Medicine - An Annonated Bibliography of Community Animal Healthcare.* ITDG Publ., London.

Masih, S.K. 2003. Diversity of ethnomedicinal wealth in Amar Kantak plateau region (India): 204-214. *In*: *Vistas in Ethnobotany,* Vol.-I. *Indian J. Appl. & Pure Biol.* Naval Kishore Press, Lukhnow.

Matthew, K.M. 1969. Exotic flora of Kodaikanal and Palni hills. *Rec. Bot. Surv. India.* ***20***(1):1-241

Mccann, C. 1950. *Trees of India - A Popular Handbook.* D.B. Taraporevata Sons & Co., Bombay.

McMillan-Browse, P. 1993. *Palms for Cooler Climates.* Trebah Enterprises, Falmouth, U.K.

McNeely, J.A., Miller, K.R., Reid, W.V., Mittermeier, R.A., & Werner, T.B. 1990. *Conserving the World's Biological Diversity*. IUCN, World Res. Inst., Cons. Intl., WWF-US & the World Bank: Washington DC.

Meena, S.L., Sharma, K.C. & Gopalan, R. 2003. Ethnomedicinal plants of Karauli District, Rajasthan: 177-180. *In:* Singh, V. & Jain, A.P. (*eds.*) *Ethnobotany and Medicinal Plants of India and Nepal*, Vol. I. Sci. Publ., Jodhpur (India).

Megoneitso, K. & Rao, R.R. 1983. Ethnobotanical studies in Nagaland-4. Sixty two medicinal plants used by Angami Nagas. *J. Econ. Taxon. Bot. **4***: 167-172.

Mehra, P.N. & Jain, K.K. 1976. *Abies and Juniperus complexes in the E. Himalaya with observations on Larix griffithii Hook.f. and Tsuga dumosa Eichler*. Sree Saraswaty Press Ltd., Calcatta.

Mehtani, S. & Ingle, P.K. 1999. *Plants Food Flavours*. Nat. Inst. Sci. Comm. New Delhi.

Menninger, E.A. 1962. *Flowering Trees of the World (for Tropics and Warm Climates)*. Hearthside Press Inc., New York.

Menninger, E.A. 1970. *Flowering Vines of the World: An Encyclopedia of Climbing Plants*. Hearthside Press Inc., New York.

Miller Gault, S. 1976. *The Dictionary of Shrubs in Colour*. Rainbird Publ. Group Ltd., London & Peerage Books, London.

Mishra, D. K., Samanta, Godhuli & Mishra, T. K. 1996. Ethnomedicine of the tribe Kharia of Midnapore District, West Bengal. *J. Econ. Taxon. Bot.* Addl. Ser. ***12***: 329-331.

Mishra, M.K. & Gokhale, S.S. 2003. *Baseline Information on Medicinal Plants Conservation and Sustainable Utilization* (*Himachal Pradesh*). A project sponsored by UNDP/GEF, MOEF, Govt. of India.

Mishra, R.N. & Chandra, V. 1998. *Trees of Rajasthan*. B.S.M.P.S., Dehradun.

Mishra, T.K., Bhunia, A., Sahoo, T.K. & Roy, S.B. 1997. Ethnobotany vis-à-vis joint forest management movement of Midnapore (West Bengal), India. *Ethnobotany **9***: 52-55.

Mitaliya, K.D. & Bhatt, D.C. 2003. Ethnobotanical studies on wild edible plants of Saurashtra, Gujrat, India: 228-232. *In*: Singh, V. & Jain, A.P. (*eds.*) *Ethnobotany and Medicinal Plants of India and Nepal*. Sci. Publ., Jodhpur, India.

Mittre, V. 1997. Wild plants in Indian folk life - a historical perspective: 35-53. *In:* Jain, S.K. (*ed.*) *Contribution to Indian Ethnobotany*. Sci. Publ., Jodhpur, India.

Moerman, D. 1998. *Native American Ethnobotany*. Timber Press, Oregon.

Mohanty, R.B. & Padhy, S.N. 1996. Traditional phytotheraphy for diarrhoeal diseases in Ganjam and Phulbani districts of South Orissa, India. *Ethnobotany* ***8***: 60-65.

Mokat, D.N. & Deokule, S.S. 2004. Plants used as veterinary medicine in Ratnagiri district of Maharashtra. *Ethnobotany* ***16***: 131-135.

Molla, H.A. & Roy, B. 1985. Traditional uses of some medicinal plants by the Robha tribals in Jalpaiguri district, West Bengal. *J. Econ. Taxon. Bot.* *7*(3): 578-580.

Moorcroft, W. & Trebeck, G. 1841. *Travels in the Himalayan Provinces (1819-1835), Vols 1-2. Illustrated.* John Murray, London.

Mudgal, V., Pal, D.C., Kayal, R.N. & Saha, S. 1999. *Ethnobotany of Totopora.* B.S.M.P.S., Dehradun, India.

Mukherjee, A., Mukherjee, S. & Mukherjee, M. 1986. Less known medicinal plants of Rubiaceae in Darjeeling Hills, West Bengal (India). *Indian J. Appl. & Pure Bot.* ***1***: 86-91.

Murthy, E.N., Raju, V.S. & Reddy, C.S. 2007. Occurrence of exotic *Hyptis suaveolens. Curr. Sci.* ***93*** (9): 1203.

Murthy, S.R., Rani, S.S. & Pullaiah, T. 2003. Wild edible plants of Andhra Pradesh, India: 613-630. *In*: Singh, V. & Jain, A. P. (*eds.*) *Ethnobotany and Medicinal Plants of India and Nepal.* Sci. Publ., Jodhpur, India.

Murti, S.K. 2001. *Flora of Cold Deserts of Western Himalaya (Vol. 1) Monocotyledons*. B.S.I., Calcutta.

Nadanakunjidam, S. & Abirai, S. 2005. Comparative study of traditional medical knowledge of Pondicherry and Karaikal regions in Union Territory of Pondicherry. *Ethnobotany* ***17***: 112-117.

Nadeem, M., Kumar, A., Nandi, S.K. & Palni, M.S. 2001. Tissue culture of medicinal plants with particular reference to Kumaon Himalaya: 231-270. *In*: Samant, S.S., Dhar, U. & Palni, L.M.S. (*eds.*) *Himalayan Medicinal Plants - Potential and Prospects*. Gyanodaya Prakashan, Nainital (India).

Nair, N.C. 1977. *Flora of Bashahr Himalaya.* Intl. Biosci. Publ., Hissar.

Nath, A. & Maiti, G.G. 2003. Ethnobotany of Barak Vally (Southern Assam) with special reference to folk medicines: 964-971. *In*: Singh, V. & Jain, A.P. (*eds.*) *Ethnobotany and Medicinal Plants of India and Nepal.* Sci. Publ., Jodhpur, India.

Nautiyal, A.R., Nautiyal, M.C., & Purohit, A.N. 1997. *Harvesting Herbs-2000. Medicinal and Aromatic Plants - An Action Plan for Uttrakhand.* B.S.M.P.S., Dehradun.

Nautiyal, S. 2010. *Plant-Biodiversity Conservation in Academic Institutions: An Efficient Approach for Conserving Biodiversity Across Ecological Regions in India*. Inst. Soc. & Econ. Change (ISEC) Bangalore, India.

Nautiyal, S., Maikhuri, R.K., Rao, K.S. & Saxena, K.G. 2003. Ethnobotany of the Tolchha Bhotiya tribe of the buffer zone villages in Nanda Devi Biosphere Reserve, India: 119-142. *In:* Singh, V. & Jain, A. P. (*eds.*) *Ethnobotany and Medicinal Plants of India and Nepal,* Vol. I. Sci. Publ., Jodhpur, India.

Nayak, P.K., Nayak, R.K. & Choudhury, B.P. 2003. A check list of medicinal plants of Kalahandi district in Orissa: 519-532. *In*: Singh, V. & Jain, A.P. (*eds.*) *Ethnobotany and Medicinal Plants of India and Nepal.* Sci. Publ., Jodhpur, India.

Nayar, M.P. 1977. Changing patterns of the Indian flora. *Bull. Bot. Surv. Ind.* ***19***: 145-154.

Negi, K.S. & Gaur, R.D. 1991. A contribution to the edible wild fruits of Uttar Pradesh hills. *Bull. Bot. Surv. India* ***43***: 233-266.

Negi, K.S., Gaur, R.D. & Tiwari, J.K. 1999. Ethnobotanical notes on flora of the Har-Ki-Doon (District Uttar Kashi), Garhwal Himalaya, Uttar Pradesh, India. *Ethnobotany* ***11****:* 9-17.

Negi, K.S. & Pant, K.C. 1990. Notes on ethnobotany of Gangwal - a tribe of Garhwal Himalaya. *Ethnobotany* ***2***: 81-85.

Negi, K.S., Tiwari, J.K. & Gaur, R.D. 1985. Economic importance of some common trees in Garhwal Himalaya: An ethnobotanical study. *Indian J. For.* ***8***: 276-289.

Negi, K.S., Tiwari, J.K., Gaur, R.D. & Pant, K.C. 1993. Notes on ethnobotany of five districts of Garhwal Himalaya, Uttar Pradesh, India. *Ethnobotany* ***5***: 73-81.

Negi, P.S. & Hajra, P.K. 2007. Alien flora of Doon valley, North West Himalaya. *Curr. Sci.* ***92***(7): 968-978.

Negi, P.S. & Subramani, S.P. 2002. Ethnobotanical study in the village Chhitkul of Sangla valley, district Kinnaur, Himachal Pradesh. *J. Non-Timber For. Prod.* ***9***(3-4): 113-120.

Negi, P.S. & Subramani, S.P. 2006. Ethnobotanical study in the Naggar area of Kullu Valley, Himachal Pradesh. *J. Econ. Taxon. Bot.* ***30***(2): 349-358.

Negi, Y.S. & Bhalla, P. 2002. Collection and marketing of important medicinal and aromatic plants in tribal areas of Himachal Pradesh. *Indian For.* ***128***(6): 641-649.

Neuwinger, H.D. 2000. *African Traditional Medicine: A Dictionary of Plant Uses and Applications.* Medphrm. Sci. Pub. Sturtgart, Germany.

Nisha, V.M. & Sivadasan, M. 2007. Ethnodermatologically significant plants used by traditional healers of Wayanad district, Kerela. *Ethnobotany* ***19****:* 55-61.

Oakman, H. 1990. *Shrubs for Tropical and Subtropical Gardens*. Jacaranda Press, Auckland, NZ.

Oommachan, M. 1977. *The Flora of Bhopal (Angiosperm)*. J.K. Jain Brothers, Bhopal (India).

Osmaston, A.E. 1927. *Forest Flora for Kumaon*. Intl. Book Distr., Dehra Dun, India.

Padhy, S.N. & Dash, S.K. 2003. Ethnobotanical survey of insect repellent plant species in folklore and literature of Orissa State, India: 505-512. *In:* Singh, V. & Jain, A.P. (*eds.*) *Ethnobotany and Medicinal Plants of India and Nepal*. Sci. Publ., Jodhpur.

Pal, B.P. 1972. *Beautiful Climbers of India*. I.C.A.R., New Delhi, India.

Pal, B.P. & Krishnamurthi, S. 1967. *Flowering Shrubs*. I.C.A.R., New Delhi.

Pal, D.C., Guha, A. & Sen, R. 2000. Medicinal plants and plant products used in children diseases among aboriginal in India: 121-130. *In*: Maheshwari, J.K. (*eds.*) *Ethnobotany and Medicinal Plants of Indian Subcontinent*. Sci. Publ., Jodhpur, India.

Pal, D.C. & Mudgal, V. 1985. Conservational practices of tribals of Purulia and their relevance in Eco-development programmes. *Proc. Ecodevelopment Camp, Baghmundi (Purulia). School of Fundamental Research, Calcutta*: 37-42.

Paliwal, H.K. 2009. *Ornamental Gardening: A User's Companion*. Natl. Book Trust, New Delhi, India.

Panda, S. 1996. Plant use–a recent perspective from Eastern-Himalayas. *J. Econ. Taxon. Bot.* Addl. Ser. ***12***: 332-337.

Pande, P.C., Tiwari, L., & Pande, H.C. 2006. *Folk-Medicine and Aromatic Plants of Uttaranchal*. B.S.M.P.S., Dehradun.

Pandey, D.N. 1998. *Ethnoforestry* (*Local Knowledge for Sustainable Forestry and Livelihood Security*). Himanshu Publ., Udaipur, New Delhi.

Pandey, I.B. 2003. Some traditional herbal remedies used in and around Kanpur city (Uttar Pradesh), India. *Ethnobotany* ***15***: 129-131.

Pandey, P.C. & Joshi, J.S. 2001. Cultured plants of Kumaun Himalaya used for medicinal purposes: 117-124. *In*: Samant, S.S., Dhar, U. & Palni, M.S. (*eds.*) *Himalayan Medicinal Plants, Potential and Prospects*. Gyanodaya Prakasan, Nainital, India.

Pandey, R.P. & Parmar, P.J. 1994. The exotic flora of Rajasthan. *J. Econ. Taxon. Bot.* ***18***(1): 105-121.

Panthi, M.P. & Chaudhary, R.M. 2003. Ethnomedicinal plant resources of Arghakhanchi district. *Ethnobotany* ***15***: 71-86.

Parker, R.N. 1918. *A Forest Flora for the Punjab with Hazara and Delhi*. Govt. Press, Lahore.

Parkinson, C.E. 1923. *A Forest Flora of the Andaman Islands*. Govt. of India, Simla.

Parmar, C. & Kaushal, M. K. 1982. *Wild Fruits of the Sub-Himalayan Region*. Kalyani Publ., Ludhiana.

Paroda, R.S. 1996. India's national bureau of plant genetic resources leads nationwide system. *Diversity* ***12***(3): 43-46.

Parrotta, J.A. 2001. *Healing Plants of Peninsular India*. CABI Publ., New York.

Patil, D.A. 2008. Scientific plant names in an ethnobotanical perspective. *Ethnobotany* ***20***: 115-121.

Pawar, S. & Patil, D.A. 2004. Herbal folk remedies for urinogenital complaints in Jalgaon district of Maharashtra. *Ethnobotany* ***16***: 86-90.

Payee, G.D. (*ed.*) 2000. *Cultural Uses of Plants: A Guide to Learning About Ethnobotany*. Bot. Gdn. Press, New York.

Pennel, F.W. 1943. The Scrophulariaceae of the Western Himalayas. *Acad. Nat. Sci. Philad. Monogr.* ***5***: 1-163. (Reprinted: 1997, B.S.M.P.S., Dehra Dun).

Perrings, C., Dehnen-Schmntz, K., Tonja, J. & Williamson, M. 2005. How to manage biological invasions under globalization?. *Trends Ecol. Evol.* ***20***: 212-215.

Perry, F. & Hay, R. 1982. *Tropical and Subtropical Plants*. British Library Publ., London.

Perumal Samy, R., & Gopalakrishnakone, P. 2007. Current status of herbal and their future perspectives. *Nature Proc.* ***1***: 1176-1188.

Phillips, G.A.R. 1954. *The Book of Garden Flowers*. Frederick Warne & Co., Ltd. London & New York.

Phillips, R. & Rix, M. 1989. *Shrubs*. Pan Books, London.

Phillips, R. & Rix, M. 1998. *Conservatory and Indoor Plants* Vols. I & II. Pan Books, London.

Polunin, O. 1969. *A Field Guide to the Flowers of Europe*. Oxford Univ. Press, Melbourne & London.

Polunin, O. & Stainton, A. 1984. *Flowers of the Himalaya*. Oxford Univ. Press, Delhi.

Polunin, O. & Stainton, A. 1987. *Concise Flowers of the Himalaya*. Oxford Univ. Press, Delhi.

Pool, R.J. 1941. *Flowers and Flowering Plants. An Introduction of the Nature and Work of Flower and Classification of Flowering Plant*. McGraw-Hill Book Co., Inc., New York & London.

Prain, D. 1903. Flora of the Sunderbans. *Rec. Bot. Surv. India* ***2***: 231-370.

Prajapati, N.S., Purohit, S.S., Sharma, A.K. & Kumar, T. 2006. *A Handbook of Medicinal Plants- A Complete Source Book*. Agrobios, India.

Prakash, V. 1999. Indian medicinal plants: current status-1. *Ethnobotany* ***10***: 112-121.

Prakash, V. 2001. Indian medicinal plants: current status: 45-63. *In*: Samant, S.S., Dhar, U. & Palni, M.S. (*ed.*) *Himalayan Medicinal Plants, Potential and Prospects.* Gyanodaya Prakasan, Nainital, India.

Prakash, V., Aggarwal, A., Sharma, A. & Mehrotra, R.S. 2003. Ethnobotanical studies on some wild plants of district Hamirpur in Himachal Pradesh: 374. *In:* Prakash, P. (*ed.*) *Proc. Natl. Sem. Biochem. Sci.: Health & Environment Aspects*, Allied Publ., Pvt. Ltd., New Delhi.

Prasad, P., Nagendra, Singh, R.A.J.A. Narayanan, L.M. & Natarajan, C.R. 1996. Ethnobotany of the Kanikkars of South Tamil Nadu-1. *J. Econ. Taxon. Bot. Addl.* Ser. ***12***: 292-298.

Prashantkumar, P. & Vidyasagar, G.M. 2008. Traditional knowledge on medicinal plants used for the treatment of skin diseases in Bidar district, Karnataka. *Indian J. Trad. Know.* *7*(2): 273-276.

Pruthi, J.S. 1976. *Spices and Condiments.* Natl. Book Trust, India.

Pullaiah, T. (*ed.*) 2004. *Biodiversity in India*, Vols. 1-3. Regency Publ., New Delhi.

Purohit, K. & Samant, S.S. 1995. *Fodder Trees and Shrubs of Central Himalaya.* Gyanodaya Prakashan, Nainital (U.P.), India.

Quattrocchi, U. 2000. *CRC World Dictionary of Plant Names.* C.R.C. Press, Washington.

Raizada, M.B. & Saxena, H.O. 1978. *Flora of Mussoorie,* Vol. I. B.S.M.P.S., Dehra Dun.

Rajasekaran, A. & Singh, J. 2009. Ethnobotany of Indian horse chestnut (*Aesculus indica*) in Mandi district, Himachal Pradesh. *Indian. J. Trad. Know.* ***8***(2): 285-286.

Rajendran, A. & Henry, A. N. 1994. Plants used by the tribe Kadar in Anaimalai hills of Tamil Nadu. *Ethnobotany* ***6***: 19-24.

Rajendran, A., Rama Rao, N. & Henry, A.N. 1996. Hepatic stimulant plants of Andhra Pradesh, India. *J. Econ. Taxon. Bot. Addl. Seires* ***12***: 221-223.

Rajendran, A. & Sikarwar, R.L.S. 2003. Intra-cultural ethnobotanical studies on the tribe Kadars (Tamilnadu) and Sahariyas (Madhya Pradesh), India: 662-664. *In*: Singh, V. & Jain, A.P. (*eds.*) *Ethnobotany & Medicinal Plants of India and Nepal.* Sci. Publ., Jodhpur, India.

Rajendran, K., Balakrishnan, R. & Chandrasekaran, S. 2003. Common medicinal plants and their utilization by villagers in East Coast districts of Tamilnadu: 715-726. *In:* Singh, V. & Jain, A.P. (*eds.*) *Ethnobotany and Medicinal Plants of India and Nepal.* Sci. Publ., Jodhpur, India.

Rajendran, S.M. 2003. West Sikkim-a treasure house of medicinal plants and their utility by local tribals: 311-316. *In:* Singh, V. & Jain, A.P. (*eds.*) *Ethnobotany and Medicinal Plants of India and Nepal*, Vol. I. Sci. Publ., Jodhpur (India).

Raju, R.A. 2000. *Wild Plants of Indian Sub-Continent and Their Economic Use.* C.B.S. Publ. & Distr., New Delhi.

Rama Rao, M. 1914. *Flowering Plants of Travancore*. Govt. Press, Trivandrum, India.

Rama Rao, N. & Henry, A. N. (*ed.*) 1996. *Ethnobotany of Eastern Ghats in Andhra Pradesh, India*. B.S.I., Calcutta.

Ramaswamy, S.V. & Razi, B.A. 1973. *Flora of Banglore District*. Prasarenga. Mysore Univ., Mysore.

Rana, J.C. 2004. *Lesser Known Food Plants of Himachal Pradesh.* N.B.P.G.R. Reg. Stn., Phagli, Shimla (H.P.).

Rana, T.S., Datt, B. & Rao, R.R. 1996. Strategies of sustainable utilization of plant resources by the tribals of the Tons valley, Western Himalaya. *Ethnobotany **8***: 96-104.

Rana, T.S., Datt, B & Rao, R.R. 2003. *Flora of Tons Valley, Garhwal Himalayas (Uttranchal).* B.S.M.P.S, Dehra Dun.

Randhawa, G.S. 1973. *Ornamental Horticulture in India*. Today & Tomorrow's Printer & Publ., New Delhi.

Randhawa, M.S., 1947. A bio-aesthetic plan for India. *Curr. Sci. **16*** (2): 41-46.

Randhawa, M.S. 1957. *Flowering Trees in India*. I.C.A.R., New Delhi.

Randhawa, M.S. 1960. *Beautiful Trees and Gardens*. I.C.A.R., New Delhi.

Randhawa, M.S 1965. *Flowering Trees*. Natl. Book Trust, New Delhi, India.

Ranjan, P. 2000a. A contribution to some of the medicinal plants of Indo-Nepal border area adjoining the districts of Madhubani and Sitamarhi: 651-659. *In*: Maheshwari, J.K. (*ed.*) *Ethnobotany and Medicinal Plants of Indian Subcontinent*, Vol. I. Sci. Publ. Jodhpur (India).

Ranjan, P. 2000b. A contribution to some wild edible plants of Indo-Nepal Border area: 661-666. *In:* Maheshwari, J.K. (*ed.*) *Ethnobotany and Medicinal Plants of Indian Subcontinent.* Sci. Publ., Jodhpur.

Ranjan, P. 2003. A contribution to some ethnomedicinally important plants of Nepal: 345-353. *In:* Singh, V. & Jain, A.P. (*eds*) *Ethnobotany and Medicinal Plants of India and Nepal,* Vol. I. Sci. Publ., Jodhpur.

Rao, A.S. 1978. Inventory of West Himalayan Flora: 133-145. *In*: *Natl. Sem. Resour. Devl. Himalayan Region*. DST, New Delhi.

Rao, A.S. & Verma, D.M. 1982. *Cyperaceae of North East India*. B.S.I., Calcutta.

Rao, R.R. 1990. Ethnobiology in the study and conservation of fragile ecosystem: some issue from the Himalayan region. *Ethnobotany **2***: 45-55.

Rao, R.R. 1994. *Biodiversity in India (Floristic Aspects)*. B.S.M.P.S., Dehra Dun.

Rao, R.R. 1996. Traditional knowledge and sustainable development: key role of ethnobiologists. *Ethnobotany* ***8***: 14-24.

Rao, R.R. & Razi, B.A. 1974. *A Synoptic Flora of Mysore District*. Today & Tomorrow's Printers & Publ., New Delhi.

Rao, R.S. 1985-1986. *Flora of Goa, Diu, Daman and Nagarhaveli*. B.S.I., Calcutta.

Rastogi, M.A. 1960. Medicines from the wild. A case study of the Great Himalayan Park. *The Indian Magz. of Her People & Cult.*: 74-75.

Rastogi, R.P. & Mehrotra, B.N. 1990. *Compendium of Indian Medicinal Plants*, Vol. I. C.D.R.I. & N.I.S.C., New Delhi.

Rastogi, R.P. & Mehrotra, B.N. 1991. *Compendium of Indian Medicinal Plants*, Vol. II. C.D.R.I. & N.I.S.C., New Delhi.

Rastogi, R.P. & Mehrotra, B.N. 1993. *Compendium of Indian Medicinal Plants*, Vol. III. C.D.R.I. & N.I.S.C., New Delhi.

Rastogi, R.P. & Mehrotra, B.N. 1995a. *Compendium of Indian Medicinal Plants,* Vol. IV. C.D.R.I. & N.I.S.C., New Delhi.

Rastogi, R.P. & Mehrotra, B.N. 1995b. *Compendium of Indian Medicinal Plants*, Vol. V. C.D.R.I. & N.I.S.C., New Delhi.

Rau, M.A. 1961. A recent survey of medicinal plant resources of Lahoul valley (Punjab, India). *Pakistan J. Sci. & Res. (Intl. Symp.)* ***4***: 202-205.

Rau, M.A. 1974. Vegetation and Phytogeography of the Himalaya: 247-280. *In*: M.S. Mani, (*ed.*), *Ecology and Biogeography in India*. Junk Publ. The Hague.

Rau, M.A. 1975. *High Altitude Flowering Plants of Western Himalaya*. B.S.I., Calcutta.

Ravishankar, T. & Henry, A.N. 1992. Ethnobotany of Aditlabad District, Andhra Pradesh, India. *Ethnobotany* ***4***: 45-52.

Rawat, D.S., Kharwal, A.D. & Rawat, S. 2009. Ethnobotanical studies on dental hygiene in district Hamirpur, Himachal Pradesh, India. *Ethnobot. Leafl.* ***13***: 1434-1442.

Rawat, M.S. & Chowdhury, S. 1998. *Ethno-Medico-Botany of Arunachal Pradesh (Nishi & Apatani Tribes)*. B.S.M.P.S., Dehradun.

Razi, B.A. 1955. Some observations on plants of the south Indian hill tops and their distribution. *Proc. Nat. Inst. Sci. India* ***21***B: 79-89.

Reddi, T.V.V.S., Prasanthi, S. & Naidu, B.V.A.R. 2005. Medicinal and aromatic plants of India: 1-146. *In*: Khan, I.A. & Khanum, A. (*eds.*) *Role of Biotechnology in Medicinal and Aromatic Plants.* Ukaaz Publ., Hyderabad.

Reddy, C.S. 2008. Catalogue of invasive alien flora of India. *Life Sci.* ***5*** (2): 84-89.

Reddy, C.S., Bhanja, M.R. & Raju, V.S. 2000. *Cassia uniflora* Miller: a new record for Andhra Pradesh, India. *Indian J. For.* ***23*** (3): 324-325.

Reddy, C.S. & Raju, V.S. 2002. Additions to the weed flora of Andhra Pradesh. *J. Econ. Taxon. Bot.* ***26***: 195-198.

Reddy, R.S., Reddy, A. M. & Yasodamma, N. 2012. Exploration of wild ornamental flora of YSR districts, Andhra Pradesh, India. *Indian J. Fundamen. & Life Sci. (online)* ***2*** (1): 192-199.

Reddy, C.S. & Reddy, K.N. 2004. *Cassia rotundifolia* Pers. (Caesalpiniaceae): a new record for India. *J. Econ. Taxon. Bot.* ***28***: 73-74.

Redford, K.H. & Mansur, J.A. 1997. *Traditional People and Biodiversity Conservation in Large Tropical Landscapes.* America Verde Publ., Arlington, Virginia, USA.

Rehder, A. 1949. *Bibliography of Cultivated Trees and Shrubs Hardy in the Cooler Temperate Regions of the Northern Hemisphere.* Jamaica Plain, Mass.

Reichard, S.H., & White, P.S. 2001. Horticulture as a pathway of invasive plant introductions in the United States. *Biosci.* ***51***: 103–113.

Ringmichon, C.L., Shimpi, S.N. & Gopalkrishnan, B. 2011. Some noteworthy herbal remedies used by Naga tribes in Manipur. *Ethnobotany* ***23***: 82-85.

Riotte, L. 1978. *Companion Planting for Successful Gardening.* Garden Way, Vermont, U.S.A.

Roberts, C.A., Stallman, D. & Bieri, J.A. 2002. Modelling complex human-environment interactions: the grand canyon river trip simulator. *Ecol. Modl.* ***153***: 181-196.

Robson, N.K.B. 1970. Shrubby Asiatic *Hypericum* species in cultivation. *J. Roy. Hort. Soc.* ***95***: 490.

Roxburgh, W. 1795-1819. *Plants of the Coast of Coromandel,* Vols. 1-3. W. Bulmer & Co., London.

Roxburgh, W. 1814. *Hortus Bengalensis.* Mission Press, Serampore.

Roxburgh, W. 1832. *Flora Indica,* 3 Vols. W. Thacker & Co., Serampore.

Roy, B., Halder, A.C. & Pal, D.C. 1998. *Plants for Human Consumption in India.* B.S.I., Calcutta.

Roy, G.P., Shukla, B.K. & Bhaskar Datt. 1992. *Flora of Madhya Pradesh (Chhatapur and Damoh).* Amerind Pub. Co. Pvt. Ltd., New Delhi.

Royle, J.F. 1833-1840. *Illustrations of the Botany and Other Branches of the Natural History of the Himalayan Mountains and of the Flora of Cashmere,* 2 Vols. W.N. Allen & Co., London.

Rushforth. K. 1987. *Conifers.* Christopher Helm, Scotland.

Sagwal, S.S. & Gupta, N.K. 1988. Role of forestry extension in exploitation on medicinal and aromatic plants of Himachal Pradesh: 50-54. *In:* Khosla, P.K. & Sehgal, R.N. (*eds*) *Trends in Tree Sci.* I.S.T.S. Publ., Solan.

Sahni, K.C. 1998. *The Book of Indian Trees.* Bombay Nat. Hist. Soc., Oxford Univ. Press, New York.

Sahoo, A.K. & Mudgal, V. 1995. Less known ethnobotanical uses of plants of Phulbani district, Orissa, India. *Ethnobotany 7*: 63-67.

Sahoo, B.B. & Satapathy, K.B. 2009. Plants used by the tribals and rural folks for common ailments in Jaipur District (Orissa). *Ethnobotany* ***21***: 107-111.

Said, H.M. 1997. *Hamdard Pharmacopoeia of Eastern Medicine.* Sri Satguru Publ., Delhi.

Saikia, B., Doley, B., Rethy, P. & Gujurel, P.R. 2010. Traditional uses of flowers as food in Sonitpur district of Assam and Papumpare district of Arunachal Pradesh. *Ethnobotany* ***22***:121-124.

Saikia, N. & Nath, S.C. 2003. Traditional use of some plants in religio-cultural purposes by 'Koch' community in Assam, India: 997-1001. *In:* Singh, V. & Jain, A.P. (*eds.*) *Ethnobotany and Medicinal Plants of India and Nepal,* Vol. II. Sci. Publ., Jodhpur.

Saini, D.C. 1996a. Ethnobotany of Tharus of Basti district, Uttar Pradesh. *J. Econ. Taxon. Bot.* Addl. Ser. ***12***: 138-153.

Saini, V.K. 1996b. Plants in the welfare of tribal women and children in certain areas of Central India: 140-144. *In*: Jain, S.K. (*ed.*) *Ethnobiology in Human Welfare*. Deep Publ., New Delhi.

Saklani, S. & Jain, S.K. (*ed.*) 1994. *Cross-Cultural Ethnobotany of North-East India*. Deep Publ., New Delhi.

Saldanha, C.J. 1984. *Flora of Karnataka,* Vol. 1. Oxford & IBH Publishing Co., New Delhi, Bombay, Calcutta.

Saldanha, C.J. & Nicolson, D.H. (*eds.*) 1976. *Flora of Hassan District, Karnataka, India.* Amerind Publ. Co. Pvt. Ltd. New Delhi.

Samant, S.S., Dhar, U. & Rawal, R.S. 1996. Natural resource use by some natives within Nanda Devi Biosphere reserve in West Himalaya. *Ethnobotany* ***8***: 40-50.

Samant, S.S., Dhar, U. & Rawal, R.S. 2001. Diversity and distribution of wild edible plants of the Indian Himalaya: 421-482. *In*: Pande, P.C. & Samant, S.S. (*ed.*) *Plant diversity of the Himalaya.* Gyanodaya Prakashan, Nainital, India.

Sambamurty, A.V.S.S. 2002. *Dictionary of Angiosperm Families and Their Genera.* Asiatic, New Delhi.

Sane, H. & Ghate, V. 2006. Sacred conservation practices at species level through tree worship. *Ethnobotany* ***18***: 46-52.

Santapau, H. 1953. The Flora of Khandala on the Western Ghats of India. *Rec. Bot. Surv. India* ***16***: 1-396.

Santapau, H. 1958. *History of Botanical Research in India, Burma and Ceylon, Part II (Systematic Botany of Angiosperms)*. Indian Bot. Soc., New Delhi.

Santapau, H. 1962. The present state of taxonomic and floristic in India after independence. *Bull. Bot. Surv. India* ***4***: 209-216.

Santapau, H. & Henry, A.N. 1973. *A Dictionary of the Flowering Plants in India*. C.S.I.R., New Delhi.

Sanyal, D. 1994. *Vegetable Drugs of India*. B.S.M.P.S., Dehradun.

Sarin, Y.K. 1967. A survey of vegetable raw material resources of Lahul. *Indian For.* ***93***(7): 489-495.

Sarin, Y.K. 1996. *Illustrated Manual of Herbal Drugs Used in Ayurveda*. C.S.I.R. & I.C.M.R., New Delhi.

Sarin, Y.K. & Chopra, C.L. 1985. Medicinal and aromatic plant resources of Himachal Pradesh, present status and future prospects. *Workshop Sci. &Tech., Shimla*, Oct. 29-30, 1985.

Sarin, Y.K., Kapoor, L.D. & Chopra, I.C. 1963. *Physochlaina praelta* (Don) Miers: a hyoscyamine yielding plant important in Lahhul's economy. *Indian For.* ***89***: 610-611.

Sasi, R., Rajendran, A. & Aravindhan, V. 2010. Ethno-hair tonic percepts in Velliangiri Hills of South Western Ghats. *Ethnobotany* ***22***: 107-110.

Satapathy, K.B. & Brahmam, M. 1996. Some medicinal plants used by tribals of Sundargarh district, Orissa, India: 153-155. *In*: Jain, S.K. (*ed.*) *Ethnobiology in Human Welfare*, Deep Publ., New Delhi.

Satyal, P., Paudel, P., Lamichhane, B. & Setzer, W.N. 2012. Volatile constituents and biological activities of the leaf essential oil of *Jasminum mesnyi* growing in Nepal. *Jour. Chemi. & Pharma. Res.* ***4***(1): 437-439.

Saxena, A.P. & Vyas, K.M. 1981. Ethnobotanical records on infectious diseases from tribals of Banda district. *J. Econ. Taxon. Bot.* ***2***: 191-194.

Saxena, H.O. & Brahmam, M. 1994-1996. *The Flora of Orissa*, Vols. I-IV. Reg. Res. Lab., Bhubaneswar & Orissa Forest Devl. Corp. Ltd., Bhubaneswar.

Saxena, K.G. 1991. Biological invasion in the Indian sub-continent: review of invasion by plants: 53-73. *In*: Ramakrishnan, P.S. (*ed.*) *Ecology of Biological Invasion in the Tropics*. Intl.. Sci. Publ., New Delhi.

Schmidt-Vogt, D. 1998. Defining degradation: the impacts of Suidden on forests in Northern Thailand. *Mountain Res. Dev.* ***18***: 135-149.

Schultes, R.E. 1962. Tapping our heritage of ethnobotanical lore. *Econ. Bot.* ***8***: 224-227.

Schultes, R.E. 1983. Richard Spruce: an early ethnobotanist and explorer of the northwest Amazon and northern Andes. *J. Ethnobiol.* ***3***: 139-147.

Schultes, R.E. 1986. The reason for ethnobotanical conservation. *Bull. Bot. Surv. India* ***28***: 203-324.

Schultes, R.E. & Raffauf, R.F. 1990. *The Healing Forest.* Dioscorides Press, Portland, Oregon.

Seetharam, Y.N., Chalageri, G., Haleshi, C. & Vijay. 1999. Folk medicine and ethnomedicine of North-Eastern Karnataka. *Ethnobotany* ***11***:32-37.

Seitel, P. 2001. *Safeguarding Traditional Cultures: A Global Assessment. Centre for Folk Life and Cultural Heritage.* Smithsonian Inst., Washington DC.

Sekar, K.C. 2012. Invasive alien plants of Indian Himalayan region - diversity and implication. *Am. J. Plant Sci.* ***3***: 177–184.

Sen, S., Batra, A. & Ali, Z.D. 2002. Some herbal home remedies form Jaipur district, Rajasthan. *Ethnobotany* ***14***: 116-122.

Sen, S.K. & Behera, L.M. 2003. Ethnomedicinal plants used against skin diseases in Bargarh district in Orissa. *Ethnobotany* ***15***: 90-96.

Sensarma, P. 1991. Herbal veterinary medicines in an ancient-Sanskrit work – The Garuda Purana. *Ethnobotany* ***3***: 83-87.

Seth, M.K. 2004. Trees and their economic importance. *Bot. Rev.* ***69*** (4): 321-376.

Sethi, G.S. 2003. Potential application biotechnology in medicinal plants of Himachal Pradesh. *In: Proc. Natl.. Sem. Emerging Trends in Res. & Business Management of Medicinal & Aromatic Plants,* March 11-13, 2003. CCSHAH, Hissar, Haryana.

Sharma, B.D. 1984. *Exotic Flora of Allahabad.* B.S.I., Dehradun.

Sharma, B.D. & Chandel, K.P.S. 1996. Occurrence, distribution and diversity of soft fruits in N.W. and W. Himalayas and prospects of their conservation and utilization. *Indian J. Pl. Gen. Res.* ***9***(2): 237-240.

Sharma, B.D. & Rana, J.C. 1999. Traditional medicinal uses of plants of Himachal hills. *J. Econ. Taxon. Bot.* ***23***(1): 173-176.

Sharma, B.D. & Rana, J.C. 2000. Traditional medicinal uses of plants of Himachal Hills: 173-176. *In*: Maheshwari, J.K. (*eds.*) *Ethnobotany and Medicinal Plants of Indian Subcontinent.* Sci. Publ., Jodhpur, India.

Sharma, B.D. & Rana, J.C. 2005. *Plant Genetic Resources of Western Himalayas: Status and Prospects.* B.S.M.P.S., Dehradun, India.

Sharma, H.M., Devi, A.R., & Sharma, B.M. 2003. Ethnomedicinal uses of monocotyledonous plants by the Meitei of Manipur: 473-480. *In:* Singh, V. & Jain, A.P. (*eds.*) *Ethnobotany and Medicinal Plants of India and Nepal,* Vol. I. Sci. Publ., Jodhpur, India.

Sharma, H.M., Sharma, B.M. & Devi, A.R. 2000. Contributions to the edible fruits of Manipur: 615-623. *In*: Maheshwari, J.K. (*eds.*) *Ethnobotany and Medicinal Plants of Indian Subcontinent.* Sci. Publ., Jodhpur, India.

Sharma, K.R. & Sood, M. 1997. *Important Medicinal Plants of Himachal Pradesh*. Vol.-I & II. Dr. Y.S. Parmar Univ. Hort. & For., Solan, H.P.

Sharma, M.P. 1996. Distribution and uses of some important dendrological components of Garhwal-Kumaon Himalaya: 102-107. *In*: Maheshwari, J. K. (*eds.*) *Ethnobotany in South Asia*. Sci. Publ., Jodhpur, India.

Sharma, N.K. 2002. Ethno-medico-religious plants of Hadoti plateau (S.E. Rajasthan)-a preliminary survey: 394-411. *In:* Trivedi, P.C. *(ed.) Ethnobotany.* Aavishkar Publ. & Distributors, Jaipur.

Sharma, P.K. & Chauhan, N.S. 2000. Ethnobotanical studies of Gaddies - a tribal community of district Kangra, Himachal Pradesh: 301. *In:* Kohli, R.K., Singh, H.P., Vij, S.P., Dhir, K.K., Batish, D.R. & Khurana, D.K. (*eds*) *Man and Forests.* DNAES, IUFRO, ISTS, Bot. Dept. & CVS, Panjab Univ., Chandigarh.

Sharma, P.K., Chauhan, N.S. & Lal, B. 2004. Observation on the traditional phytotherapy among the inhabitants of the Parvati valley in Western Himalaya, India. *J. Ethnopharmacol.* ***92***: 167-176.

Sharma, P.K., Chauhan, N.S. & Lal, B. 2005. Studies on plant associated indigenous knowledge among the Malanis of Kullu district, Himachal Pradesh. *Indian J. Trad. Know.* ***4***(4): 403-408.

Sharma, P.K., Dhyani, S.K. & Shanker, V. 1979. Some useful and medicinal plants of the district Dehradun and Siwalik. *J. Sci. Res. Pl. Med.* ***1***: 17-43.

Sharma, P.S., & Raghubanshi, A. S. 2009. Plant invasions along roads: a case study from central highland, India. *Environ. Monitor. & Assess.* ***157***: 191-198.

Sharma, R. 2003. *Medicinal Plants of India - An Encyclopaedia.* Daya Publ. House, Delhi.

Sharma, S. 2002. *Ethnoveterinary Important Plants of Renuka* (District Sirmaur, H.P.). M.Phil. Thesis, Dept. of Biosci., H.P. Univ., Shimla, India.

Sharma, S., Gautam, A.K. & Bhadauria, R. 2009. Some important supplementary food plants and wild edible fungi of upper hilly region of district Shimla (Himachal Pradesh), India. *Ethnobot. Leafl.* ***13***: 1020-1028.

Sharma, U.K. 2004. *Medicinal Plants of Assam*. B.S.M.P.S., Dehradun.

Sheat, W.G. 1948. *Propagation of Trees, Shrubs and Conifers.* MacMillan Publ. Co., New York.

Shenghi, P. 1991. Conservation of biological diversity in temple-yards and holly hills by Dai ethnic minorities of China. *Ethnobotany* ***3***: 27-35.

Shepherd, F.W. 1974. *Hedges and Screens.* Royal Hort. Soc., London.

Shome, U., Rawat, A.K.S. & Mehrotra, S. 1996. Time tested household remedies: 96-100. *In*: Jain, S.K. (*ed.*) *Ethnobiology in Human Welfare*, Deep Publ., New Delhi.

Shrivastava, A.K. & Singh, M. 2003. Medicinal potential of grasses in Rajasthan: 170-176. *In:* Singh, V. & Jain, A. P. (*eds.*) *Ethnobotany and Medicinal Plants of India and Nepal*, Vol. I. Sci. Publ., Jodhpur, India.

Shukla, G., Verma, B.K. & Pandey, H.P. 2003. Folk-lore herbotherapy of Jalaun district, U.P., India: 143-147. *In*: Singh, V. & Jain, A.P. (*eds.*) *Ethnobotany and Medicinal Plants of India and Nepal.* Sci. Publ., Jodhpur, India.

Sikarwar, R.L.S. 1996. Ethnoveterinary herbal medicines in Morena district of Madhya Pradesh, India: 194-196. *In:* Jain, S.K. (*ed.*) *Ethnobiology in Human Welfare.* Deep Publ., New Delhi.

Sikarwar, R.L.S. & Kaushik, J.P. 1992. Some less known medicinal uses of trees among the Sahariyas of Morena district, M.P., India. *Ethnobotany* ***4***: 71-74.

Silva, L. & Smith, C.W., 2004. A characterization of the non-indigenous flora of the Azores Archipelago. *Biol. Invas*. ***6***: 193–204.

Singh, A.K. 2000a. A contribution to the ethnobotany of sub-Himalayan region of Eastern Uttar Pradesh: 237-246. *In:* Maheshwari, J.K. (*eds.*) *Ethnobotany and Medicinal Plants of Indian Subcontinent.* Sci. Publ., Jodhpur, India.

Singh, A.K., Singh, R.N. & Singh, S.K. 1987. Some ethnobotanical plants of Terai region of Gorakhpur district. *J. Econ. Taxon. Bot.* ***23***: 185-198.

Singh, B., Chaurasia, O.P. & Jadhav, K.L. 1996a. An ethnobotanical study of Indus valley (Ladakh): 92-101. *In:* Maheshwari, J.K. (*eds.*) *Ethnobotany in South Asia*. Sci. Publ., Jodhpur, India.

Singh, G. & Kachroo, P. 1976. *Forest Flora of Srinagar*. B.S.M.P.S., Dehra Dun & Periodical Expert, Delhi.

Singh G. & Kachroo P. 1983. Exotic trees and shrubs of Kashmir. *Indian For.****109***: 60-76.

Singh, G. & Misri, B. 1974. Some exotic ornamentals of Kashmir. *Indian J. Hort.* ***31***: 834.

Singh, G.S. 1990. Ethnobotanical studies of useful plants of the Kullu district in North-Western Himalaya, India. *J. Econ. Taxon. Bot.* ***23***: 185-188.

Singh, G.S. 2000b. Ethnobotanical study of useful plants of Kullu District in North Western Himalaya, India: 185-198. *In:* Maheshwari, J.K. (*eds.*) *Ethnobotany and Medicinal Plants of Indian Subcontinent.* Sci. Publ., Jodhpur, India.

Singh, G.S., Rao, K.S. & Saxena, K.G. 1998. Environment, ecological and socio-economic impact of introduced crops in Western Himalaya: a case study of Kullu valley. *J. Human. Ecol.* ***9***: 63-72.

Singh, H. 1988. Ethnobiological treatment of piles by Bhoxas of Uttar Pradesh. *Anc. Sci. Life* ***8***: 167-170.

Singh, H. & Sharma, M. 2006. *Flora of Chamba District (Himachal Pradesh*). B.S.M.P.S., Dehra Dun.

Singh, H., Kumar, B., Hyniewta, T.M. & Bora, P.J. 1997. Ethno-medico-botanical studies in Tripura, India. *Ethnobotany* ***9***: 56-58.

Singh, H.B. & Arora, R.K. 1978. *Wild Edible Plants of India.* New Delhi, I.C.A.R.

Singh, J., Bhuyan, T.C. & Ahmed, A. 1996b. Ethnobotanical studies on the Mishing tribes of Assam with special reference to food and medicinal plants–I: 350-356. *In:* Maheshwari, J.K. (*eds.*) *Ethnobotany in South Asia*. Sci. Publ., Jodhpur, India.

Singh, J.B. & Kachroo, P. 1994. *Forest Flora of Pir Panjal Range. (Northwestern Himalaya)*. B.S.M.P.S., Dehra Dun.

Singh, K.K. & Kumar, K. 2000a. *Ethnobotanical Wisdom of Gaddie Tribe of Western Himalayas.* B.S.M.P.S., Dehradun.

Singh, K.K. & Kumar, K. 2000b. Observations in ethnoveterinary medicine among the Gaddie tribe of Kangra valley, Himachal Pradesh. *Ethnobotany* ***12***: 42-44.

Singh, K.K. & Prakash, A. 1994. Indigenous phytotherapy among the Gond tribe of Uttar Pradesh, India. *Ethnobotany* ***6***: 37-41.

Singh, K.K. & Prakash, A. 1996. Observations on ethnobotany of the Kol tribe of Varanasi District, Uttar Pradesh. *J. Econ. Taxon. Bot.* Addl. Ser.***12***: 133-137.

Singh, K.K., Prakash, A. & Palvi, S.K. 2003a. Observations on some energy plants among the tribals of Madhya Pradesh, 291-296. *In:* Maheshwari, J.K. (*ed.*) *Ethnobotany and Medicinal Plants of Indian Subcontinent.* Sci. Publ., Jodhpur.

Singh, M.P., Srivastava, J.L. & Pandey, S.N. 2003b. *Indigenous Medicinal Plants, Social Forestry and Tribals.* Daya Publ. House, Delhi.

Singh, P.B. 1991. *Illustrated Field Guide to Commercially Important Medicinal and Aromatic Plants of Himachal Pradesh.* Soc. Herbal Med. & Himalayan Biodiversity, Mandi, Himachal Pradesh.

Singh, P.B. 1993. Medicinal plants of Ayurvedic importance from Mandi district of Himachal Pradesh. *Bull. of Medico-Ethnobot. Res.* ***14*** (3-4): 126-136.

Singh, P.B. 1996. Wild edible plants of Mandi district in N.W. Himalaya: 22-25. *In:* Jain, S.K. (*ed.*) *Ethnobotany in Human Welfare*. Deep Publ., New Delhi.

Singh, P.B. 1999. *Illustrated Field Guide to Commercially Important Medicinal and Aromatic Plants of Himachal Pradesh.* Azad Offset Printer, Chandigarh.

Singh, P.B. 2001. Medicinal plant diversity of Kangra district, Himachal Pradesh: a focus on folk, commercial and threatened medicinal plants: 185-196. *In*: Samant, S.S., Dhar, U. & Palni, M.S. (*eds*) *Himalayan Medicinal Plants, Potential And Prospects.* Gyanodya Prakashan, Nainital, India.

Singh, P.B. & Aswal, B.S. 1992. Medicinal plants of Himachal Pradesh used in Indian pharmaceutical industry. *Bull. Medico-Ethnobot. Res.* ***13***: 172-208.

Singh, R. 1969. *Fruits*. Natl. Book Trust, New Delhi, India.

Singh, S. & Tiagi, B. 1979. *Flora of North-East Rajasthan*. Kalyani Publ., New Delhi.

Singh, S.C. & Srivastava, G.N. 2000. Exotic medicinal plants of Lucknow district, U.P., India: 223-235. *In:* Maheshwari, J.K. (*ed.*) *Ethnobotany and Medicinal Plants of Indian Subcontinent*. Sci. Publ., Jodhpur (India).

Singh, S.K. & Rawat G. S. 2000. *Flora of Great Himalayan National Park – Himachal Pradesh*. B.S.M.P.S., Dehra Dun.

Singh, S.P., Tripathi, S. & Shukla, R.S. 2003c. Ethnomedicinal heritage for bioprospecting and drug development in North Eastern state of India: 384-395. *In:* Singh, V. & Jain, A.P. (*eds*) *Ethnobotany and Medicinal Plants of India and Nepal,* Vol. I. Sci. Publ., Jodhpur.

Singh, V & Chauhan, N.S. 2005. Traditional practices of herbal medicines in the Lahaul valleys, Himachal Himalayas. *Indian J. Trad. Know.* ***4***(2): 208-220.

Singh, V. & Pandey, R.P. 1998. *Ethnobotany of Rajasthan, India*. Sci. Publ., Jodhpur, India.

Singh, V.K. & Zaheer, A.A. 1996. Folk medicinal plants used for family planning in India: 184-186. *In*: Jain, S.K. (*ed.*) *Ethnobiology in Human Welfare*. Deep Publ., New Delhi.

Sinha, R.K. & Sinha, S. 2001. *Ethnobiology.* Surabhi Publ., Jaipur, India.

Sinha, S.C. 1996. Wild edible plants of Manipur, India: 42-47. *In:* Jain, S.K. (*ed.*) *Ethnobiology in Human Welfare*. Deep Publ., New Delhi.

Sivakumar, A., Subramaniam, M.S., Karunakaran, M. & Burkanududeen, A. 2003. Ethnobotany of Poliyars of Annamalai hills, Tamil Nadu: 679-865. *In:* Singh, V. & Jain, A.P. (*eds.*) *Ethnobotany and Medicinal Plants of India and Nepal*, Vol. II. Sci. Publ., Jodhpur (India).

Sivaperumal, R., Ramya, S., Ravi, A.V., Rajasekaran, C. & Jayakumararaja, R. 2009. Herbal remedies practiced by Malayali's to treat skin diseases. *Envioron. We Int. J. Sci. Tech.* ***4***: 35-44.

Sivarajan, V.V. & Balachandran, I. 1994. *Ayurvedic Drugs and Their Plant Sources*. Oxford & IBH Publ. Co., New Delhi.

Siwakoti, M. & Siwakoti, S. 2000. Ethnomedicinal uses of plants among the Satar tribe of Nepal: 99-108. *In:* Maheshwari, J.K. (*ed.*) *Ethnobotany and Medicinal Plants of Indian Subcontinent.* Sci. Publ., Jodhpur.

Siwakoti, M. & Varma, S.K. 1996. Medicinal plants of the Terai of Eastern Nepal: 423-438. *In:* Maheshwari, J.K. (*ed.*) *Ethnobotany in South Asia*. Sci. Publ., Jodhpur (India).

Sood, S.K., Gupta, P. & Kumar, S. 2010. *Flavouring and Fragrant Resources of India*. Sci. Publ., Jodhpur, India.

Sood, S.K. & Kaushal, B. 2008. *Fibre and Fodder Resources of India.* Satish Serial Publ. House, Azadpur, Delhi.

Sood, S.K., Kumar, S. & Lakhanpal, T. N 2013. *Sacred Plants for Sustenance of Mankind.* Daya Publ. House, New Delhi.

Sood, S.K., Kumar, S., Bassi, S.K. & Rana, J.C. 2012a. *Ethnobotany of the Heritage Region of Shiwalik Himalaya.* Anamika Publ. & Dist. (Pvt.) Ltd., New Delhi.

Sood, S.K., Nath, R., & Kalia, D.C. 2001. *Ethnobotany of Cold Desert Tribes of Lahaul-Spiti (N. W. Himalaya).* Deep Publ., New Delhi.

Sood, S.K., Parmar, S. & Lakhanpal, T.N. 2005. *Ethnic Plants of India. Used in Cancer Cure - A Compendium.* B.S.M.P.S., Dehra Dun.

Sood, S.K., Rawat, D., Kumar, S., & Rawat, S. 2012b. *Handbook of Wild Edible Plants.* Pointer Publ., Jaipur.

Sood, S.K., Rawat, S. & Rawat, D. 2009a. *Dye, Masticatory and Fumitory Resources of India.* Intl.. Book Distr., Dehradun.

Sood, S.K., Sharma, D., Kumar, S. & Lakhanpal, T.N. 2010. *Healing Herbs: Traditional Medications for Wounds, Sores and Bones.* Pointer Publ., Jaipur, Rajasthan, India.

Sood, S.K., Sharma, R. & Mattu, V.K. 2006. *Honey Plants of the North-West Himalaya.* Indus Publ. Co., New Delhi.

Sood, S.K., Sharma, S., Lakhanpal, T.N. & Kumar, S. 2011. *Indian Herbs for Purfumery and Beauty Care.* Daya Publ. House, Delhi.

Sood, S.K., Sharma, S.K., Kumar, N. & Kumar, H. 2009b. *Ethnobotanical Studies on Trees, Shrubs and Climbers of Himalaya.* Satish Serial Publ. House, Azadpur, Delhi.

Sood, S.K. & Suderhsna. 2008. *Root Drug Plants of Ethnic India.* Satish Serial Publ. House, Azadpur, Delhi.

Sood, S.K. & Thakur, S. 2004. *Ethnobotany of Rewalsar Himalaya (Dist. Mandi, Himachal Pradesh, India).* Deep Publ., New Delhi.

Srivastava, R.C. 1989. *Drug Plant Resources of Central India.* Today & Tomorrow's Printer & Publ., New Delhi.

Srivastava, R.C., Singh, V.P. & Singh, M.K. 2003a. Medicinal plants of Jaunpur district (U.P.): 148-159. *In:* Singh, V. & Jain, A.P. (*eds.*) *Ethnobotany and Medicinal Plants of India and Nepal,* Vol. I. Sci. Publ., Jodhpur, India.

Srivastava, S.C., Mitra, S. & Bandyopadhyay, S. 2003b. Some plants of ethnomedicinal importance from Darjeeling district, West Bengal: 972-977. *In*: Singh, V. & Jain, A.P. (*eds.*) *Ethnobotany and Medicinal Plants of India and Nepal,* Vol. II. Sci. Publ., Jodhpur.

Srivastava, S.K. & Chandrasekar, K. 2004. Ethnomedicine of the Pin Valley National Park. Himachal Pradesh: plant used in treatment of dysentery. *Ethnobotany* ***16***: 62-63.

Srivastava, S.K., Mehrotra, S., Srivastava, G.K. & Chauhan, D.K. 2003c. Folk uses of herbs from Mahabaleshwar region of Sahyadri Hills: 857-863. *In*: Singh, V. & Jain, A.P. (*eds*) *Ethnobotany and Medicinal Plants of India and Nepal.* Sci. Publ., Jodhpur.

Srivastava, T.N. 1976. *Flora Gorakhpurensis*. Today & Tommorrow's Printer & Publs., New Delhi.

Stainton, A. 1988. *Flowers of the Himalaya, A Supplement.* Oxford Univ. Press, Delhi.

Stewart, J.L. 1868. *Forest and Fuel Plantations of the Punjab.* Govt. Press, Lahore.

Stewart, R.R. 1917. The Flora of Ladakh & W. Tibet. *Bull. Torr. Bot. Club* ***43***: 571-590.

Strachey, R. 1852. *Catalogue of the Plants of Kumaon and of the Adjacent Portions of Garhwal and Tibet*. Periodical Experts, New Delhi, India.

Subramani, S.P. & Goraya, G.S. 2003. Some folklores medicinal plants of Kolli hills: record of a Natti Vaidyas Sammelan: 665-678. *In:* Singh, V. & Jain, A.P. (*eds.*) *Ethnobotany and Medicinal Plants of India and Nepal.* Vol. II. Sci. Publ., Jodhpur, India.

Sudhakar, A. & Vedavathy, S. 2000. Wild edible plants used by the tribals of Chotoor district (Andhra Pradesh), India: 321-329. *In*: Maheshwari, J.K. (*eds.*) *Ethnobotany and Medicinal Plants of Indian Subcontinent.* Sci. Publ., Jodhpur, India.

Taylor, P.M. 1990. *The Folk Biology of the Tobelo People - A Study in Folk Classification.* Smith Inst. Press, Washington.

Temple Wright, R. 2007. *Flowers and Gardens in India-A Manual for Beginners*. Trollope Press, U.K.

Thakur, P., Dhiman, S.R., Gupta, Y.C., Kashyap, B., Chauhan, N.S & Dhall, S.P. 2010. *Wild Ornamentals of Himachal Pradesh*. DEE, UHF, Nauni (Solan) H.P.

Thakur, P., Kashyap, B., & Gupta, Y.C. 2009. Wild ornamental bulbous plants of Himachal Pradesh for landscaping. *Intl. J. USUFRUCTS Mgmt.* ***10***(1): 67-74.

Thakur, P., Sharma, Y.D., Dhiman. S.R., Gupta, Y.C. & Thakur, A. 2003. HIMFLORDAT: a computerized database of ornamental plants for Himachal Pradesh. *J. Ornam. Hort.* ***6***(2): 156-158.

Thakur, R.S., Puri, H.S, & Husain, A. 1989. *Major Medicinal Plants of India.* C.I.M.A.P., Lucknow, India.

Thakur, S. 2011. Medicinal plants used by tribal inhabitants of Sirmaur district, H.P. *Indian J. Sci. Res.* ***2***(4): 125-127.

Thakur, V. 2005. *Ethnic Indian Plants in Religious Ceremonies*, M.Phil. Thesis, H.P. Univ. Shimla.

Thomas, G.S. 1990. *Perennial Garden Plants.* J.M. Dent & Sons, London.

Thomas, G.S. 1992. *Ornamental Shrubs, Climbers and Bamboos.* Murray Publ. House, London.

Thomas, J. & Britto, A. John De. 2003. Ethnobotanical study in Wynad district in Kerela: 815-824. *In:* Singh, V. & Jain, A.P. (*eds.*) *Ethnobotany and Medicinal Plants of India and Nepal.* Vol. II. Sci. Publ., Jodhpur, India.

Tigga, M., Lakra, G, S. & Rao, P.S.N. 2003. Medicinal value of latex yielding plants from Andaman and Nicobar Islands: 765-772. *In:* Singh, V. & Jain, A.P. (*eds.*) *Ethnobotany and Medicinal Plants of India and Nepal,* Vol. II. Sci. Publ., Jodhpur, India.

Tirkey, A., Khan, F. & Khan, S.S. 2003a. Medicinal plants used in the treatment of hair in Raijarh district of Madhya Pradesh (India): 104-110. *In*: *Vistas in Ethnobotany,* Vol. I, *Indian J. Pure & Appl. Biol.* Naval Kishore Press, Lucknow.

Tirkey, A., Khan, F. & Khan, S.S. 2003b. Medicinal plants used by the tribals of Raijarh district (M.P.) for Jaundice: 163-170. *In*: *Vistas in Ethnobotany*, Vol. I, *Indian J. Pure & Appl. Biol.* Naval Kishore Press, Lucknow.

Tiwari, K.C. & Tiwari, V.P. 1996. Some important medicinal plants of the tropical, sub tropical and temperate region of Siang, Subansiri and Tirap district of Arunachal Pradesh. *J. Econ. Taxon. Bot.* Addl. Ser. ***12***: 359-363.

Trapp, L.A. 2006. *Ornamental Plants and Flowers of Tropical Mexico.* Abbott Publ., Mexico.

Trivedi, P.C. 2004. *Medicinal Plants Conservation and Utilization.* Avishkar Publ., Jaipur.

Trivedi, P.C. 2009. *Medicinal Plants Conservation and Utilization* (IInd edition). Avishkar Publ., Jaipur.

Trivedi, P.C. 2010. *Ethnic Tribes and Medicinal Plants.* Pointer Publ., Jaipur.

Trivedi, P.P. 1990. *Beautiful Shrubs.* I.C.A.R., New Delhi, India.

Uniyal, B. 2003. Utilization of medicinal plants by the rural women of Kullu, Himachal Pradesh. *Indian J. Trad. Know.* ***2***(4): 366-370.

Uniyal, M.R. 1968. Medicinal plants of Bhagirathi valley lying in Uttarkashi forest division. *Indian For.* ***94***: 407-420.

Uniyal, M.R., Bhatt, A.B. & Chaturvedi, P.N. 1973. Preliminary observations on medicinal plants of Lahoul Spiti forest division in Himachal Pradesh. *Bull. Med. Ethnobot. Res.* ***3***: 1-26.

Uniyal, M.R. & Chauhan, N.S. 1971. Medicinal plants of Uhal valley in Kangra forest division, Himachal Pradesh. *J. Res. Indian Med.* ***6***: 287-299.

Uniyal, M.R. & Chauhan, N.S. 1972. Commercially important medicinal plants of Kullu forest division, H.P. *Nagarjun* ***4***(2): 28-32.

Uniyal, M.R. & Chauhan, N.S. 1973a. Traditionally important medicinal plants of Kullu forest Division, H.P. *Jour. Res. Ind. Med. 7* (1): 76-85.

Uniyal, M.R. & Chauhan, N.S. 1973b. Traditionally important medicinal plants of Kangra Valley in Dharamshala Forest Circle, Himachal Pradesh. *J. Res. Indian Med.* ***8***(1): 76-85.

Uniyal, S.K., Singh, K.N., Jamwal, P. & Lal, B. 2006. Traditional use of medicinal plants among the tribal communities of Chhota Bhangal, Western Himalaya. *J. Ethnobiol. & Ethnomed.* ***2***(14): 1-8.

Upadhye, A.S., Vartak, V.D. & Kumbhojkar, M.S. 1994. Ethno-medicobotanical studies in Western Maharastra, India. *Ethnobotany* ***6***: 25-31.

Uphof, J.C. Th. 1968. *Dictionary of Economic Plants*. H.R. Engelmann (J. Cramer), Weinheim.

Usher, G. 1974. *A Dictionary of Plants Used by Man*. Constable & Comp. Ltd., London.

Valtonen, A., Jantunen, J., & Saarinen, K. (2006). Flora and Lepidoptera fauna adversely affected by invasive *Lupinus polyphyllus* along road verges. *Biol. Cons.* ***133***: 389–396.

Van Rheede, H. 1667. *Hortus Indicus Malabaricus*, Vols. I-XII. Amsterdam.

Vanker, P.S. 2007. *Handbook on Natural Dyes for Industrial Applications*. Natl. Inst. of Industr. Res. Kamla Nagar, New Delhi.

Ved, D.K., Kinchal, G.A., Ravikumar, K., Pravharan, V., Uttakarsh, G., Vijaya Sankar, R. & Indresha, J.H. 2003. *Conservation Assessment and Management Prioritisation for the Medicinal Plants of Jammu and Kashmir, Himachal Pradesh and Uttranchal.* FRLHT, Banglore, India.

Ved, D.K. & Tandon, V. (*eds.*) 1998. *CAMP Report for High Altitudinal Medicinal Plants of Jammu Kashmir and Himachal Pradesh*. FRLHT, Banglore.

Verma, P., Khan, A.A. & Singh, K.K. 1995. Traditional phyotherapy among the Baiga tribe of Shahdol district of Madhya Pradesh, India. *Ethnobotany 7*: 69-73.

Verma, R., Parkash, V. & Kumar, D. 2012. Ethno-medicinal uses of some plants of Kanag Hill in Shimla, Himachal Pradesh, India. *Intl. J. Res. Ayurv. Pharma.* ***3***(2): 319-322.

Verma, S. & Chauhan, N.S. 2006. Studies on ethno-medico-botany of Kunihar forest division, district Solan (H.P.). *Ethnobotany* ***18***: 160-165.

Verma, S. & Chauhan, N.S. 2007. Indigenous medicinal plants knowledge of Kunihar Forest Division, district Solan. *Indian J. Trad. Know.* ***6***(3): 494-497.

Vickery, A.R. 1994. Traditional herbal remedies in British Isles: some recently collected examples. *Ethnobotany* ***6***: 1-4.

Vihari, V. 1995. Ethnobotany of cosmetics of Indo-Nepal border. *Ethnobotany 7*: 89-94.

Virjee, G., Dar, H., Kachroo, P. & Bhat, G.M. 1984. Taxo-ethnobotanical studies of the rural areas in district Rajouri (Jammu). *J. Econ. Taxon. Bot.* ***5*** (4): 831-838.

Viswanathan, M.B., Harrison, E., Kumar, H.P., Ramesh, N. 2006. *Ethnobotany of the Kanis (Kalakkad-Mundanthurai Tiger Reserve in Tirunelveli, Tamilnadu, India).* B.S.M.P.S., Dehradun.

Vishwanathan, M.V. & Singh, H.B. 1996. Plants used as household remedies in India: 105-107. *In:* Jain, S.K. (*ed.*) *Ethnobiology in Human Welfare*. Deep Publ., New Delhi.

Vitousek, P. M., Antonio, C.M.D., Loope, L.L., & R. Westbrooks 1996. Biological invasions as global environmental change. *Am. Scient.* ***84***: 468-478.

Wangchuk, P., Samten, Ugyen, Thinley, J., & Afaq, S.H. 2008. High altitude plants used in Bhutanese traditional medicine. *Ethnobotany* ***20***: 54-64.

Warda, H.D. 1985. *Gerd Krussmann-Manual of Cultivated Conifers*. B.T. Batsford Ltd., London.

Warman, C.K.1999. *Trees of India (Medicinal, Commericial, Religious and Ornamental (A Colour Atlas).* CBS Publ. & Distr., New Delhi, India.

Wasson, E. (*ed.*) 2003. *The Complete Encyclopedia of Trees and Shrubs: Descriptions, Cultivation Requirements, Pruning, Planting*. Thunder Bay Press, San Diego, California.

Watt, G. 1889-96. *A Dictionary of the Economic Products of India*, Vols. I-VI. Periodical Experts, New Delhi (India).

Wight, R. 1838-1853. *Icones Plantarum Indiae Orientalis*, Madras (Reprint edition: 1963, Wheldon & Wesley, New York).

Wight, R. & Walker-Arnolt, G.A. 1834. *Prodromus Florae Peninsulae Indiae Orientalis*. Panbury Allen & Co., London (Reprint edition: 1976, B.S.M.P.S., Dehra Dun).

Wu, S.H., Hsieh, C.F., Chaw, S. M., & Rejmanek, M. 2004b. Plant invasions in Taiwan: insights from the flora of casual and naturalized alien species. *Diversity and Distribution* ***10***: 349–362.

Wu, S.H., Hsieh, C.F., & Rejmánek. M. 2004a. Catalogue of the naturalized flora of Taiwan. *Taiwania* ***49***: 16-31.

Wu, S.H., Sun, H.T., Teng, Y.C., Rejmanek, M., Chaw, S.M., Yang, T.Y.A., & Hsieh, C. F. 2010. Patterns of plant invasions in China: taxonomic, biogeographic, climatic approaches and anthropogenic effects. *Biol. Invas.* ***12***: 2179–2206.

Yang, M.H. & Blunden, G. 1993. Compounds with antiviral activity from Chinese medicinal plants. *Ethnobotany* ***5***: 67-72.

Yoganarisimhan, S.N., 1996. *Medicinal Plants of India* Vol. I, *Karnataka*. Interline Publ., Banglore.

Yonzone, G.S., Yonzone, A. & Bhujel, R.B. 1996. Contribution to the ethnobotany of Darjeeling district, India: 388-389. *In*: Jain, S.K. (*ed.*) *Ethnobiology in Human Welfare.* Deep Publ., New Delhi.

Zeven, A.C. & de Wet, J.M.J. 1982. *Dictionary of Cultivated Plants and their Regions of Diversity*. Wageningen.

Appendix I

INDEX TO TOTAL NUMBER OF GENERA/ SPECIES OF PLANTS OF HERITAGE SITES UNDER VARIOUS DIVISIONS OF PLANT KINGDOM

Division	Total No. of Genera/ Species	Generic Name alongwith Number of Plant Species
Gymnosperms	17/26	*Abies* (1), *Afrocarpus* (1), *Cedrus* (1), *Araucaria* (3), *Cryptomeria* (1), *Cunninghamia* (1), *Cupressus* (2), *Cycas* (1), *Ginkgo* (1), *Juniperus* (5), *Picea* (1), *Pinus* (2), *Taxodium* (1), *Taxus* (1), *Thuja* (2), *Thujopsis* (1), *Tsuga* (1)
Dicots	103/173	*Abutilon* (2), *Acacia* (1), *Acer* (3), *Actinidia* (1), *Aesculus* (1), *Aloysia* (1), *Amygdalus* (1), *Berberis* (2), *Bougainvillea* (1), *Buddleja* (3), *Buxus* (1), *Callistemon* (1), *Camellia* (1), *Campsis* (1), *Celtis* (1), *Cerasus* (1), *Cestrum* (3), *Citrus* (2), *Clematis* (2), *Colquhounia* (1), *Coriaria* (1), *Cornus* (1), *Cotoneaster* (1), *Daphne* (1), *Datura* (1), *Debregeasia* (1), *Desmodium* (2), *Diospyros* (2), *Duranta* (1), *Elaeagnus* (1), *Eriobotrya* (1), *Erythrina* (1), *Eucalyptus* (1), *Euonymus* (4), *Fagus* (1), *Ficus* (8), *Forsythia* (2), *Fuchsia* (2), *Girardinia* (1), *Grevillea* (1), *Hedera* (4), *Hibiscus* (2), *Himalrandia* (1), *Hydrangea* (1), *Hypericum* (3), *Ilex* (1), *Indigofera* (1), *Inula* (1), *Ipomoea* (1), *Jacaranda* (1), *Jasminum* (4), *Juglans* (1), *Kerria* (1), *Lagerstroemia* (1), *Leea* (1), *Ligustrum* (2), *Liriodendron* (1), *Lonicera* (2), *Lyonia* (1), *Magnolia* (3), *Malus* (1), *Malvaviscus* (1), *Morus* (3), *Myrica* (1), *Myrsine* (1), *Myrtus* (1), *Nerium* (1), *Opuntia* (1), *Padus* (1), *Parthenocissus* (1), *Paulownia* (1), *Pistacia* (1), *Platanus* (1), *Populus* (2), *Prinsepia* (1), *Prunus* (6), *Punica* (1), *Pyracantha* (1), *Pyrus* (3), *Quercus* (4), *Rhamnus* (2), *Rhododendron* (2), *Ricinus* (1), *Robinia* (1), *Rosa* (7), *Rubia* (1), *Rubus* (5), *Salix* (3), *Salvia* (2), *Sambucus* (1), *Sarcococca* (1), *Schefflera* (1), *Skimmia* (1), *Solanum* (3), *Spartium* (1), *Spiraea* (2), *Trachelospermum* (1), *Tropaeolum* (1), *Ulmus* (1), *Viburnum* (2), *Wisteria* (1), *Xanthium* (1), *Zanthoxylum* (1)
Monocots	9/13	*Agave* (1), *Arundo* (1), *Asparagus* (5), *Chamaerops* (1), *Drepanostachyum* (1), *Musa* (1), *Rhapis* (1), *Smilax* (1), *Yucca* (1)

Appendix II

INDEX TO PLANT FAMILIES OF HERITAGE SITES

Name of Family	Genus/Species	Total No of Genera and Species
Aceraceae	1/3	*Acer* (3)
Actinidiaceae	1/1	*Actinidia* (1)
Agavaceae	2/2	*Agave* (1), *Yucca* (1)
Anacardiaceae	1/1	*Pistacia* (1)
Apocyanaceae	2/2	*Nerium* (1), *Trachelospermum* (1)
Aquifoliaceae	1/1	*Ilex* (1)
Araliaceae	2/5	*Hedera* (4), *Schefflera* (1)
Araucariaceae	1/3	*Araucaria* (3)
Arecaceae	2/2	*Chamaerops* (1), *Rhapis* (1)
Asparagaceae	1/5	*Asparagus* (5)
Asteraceae	2/2	*Inula* (1), *Xanthium* (1)
Berberidaceae	1/2	*Berberis* (2)
Bignoniaceae	2/2	*Campsis* (1), *Jacaranda* (1)
Buddlejaceae	1/3	*Buddleja* (3)
Buxaceae	2/2	*Buxus* (1), *Sarcococca* (1)
Cactaceae	1/1	*Opuntia* (1)
Caprifoliaceae	3/5	*Lonicera* (2), *Sambucus*(1), *Viburnum* (2)
Celastraceae	1/4	*Euonymus* (4)
Clusiaceae	1/3	*Hypericum* (3)
Convolvulaceae	1/1	*Ipomoea* (1)
Coriariaceae	1/1	*Coriaria* (1)
Cornaceae	1/2	*Cornus* (2)
Cupressaceae	4/10	*Cupressus* (2), *Juniperus* (5), *Thuja* (2), *Thujopsis* (1)
Cycadaceae	1/1	*Cycas* (1)
Ebenaceae	1/2	*Diospyros* (2)
Elaeagnaceae	1/1	*Elaeagnus* (1)
Ericaceae	2/3	*Lyonia* (1), *Rhododendron* (2)
Euphorbiaceae	1/1	*Ricinus* (1)

Name of Family	Genus/Species	Total No of Genera and Species
Fabaceae	7/8	*Acacia* (1), *Desmodium* (2), *Erythrina* (1), *Indigofera* (1), *Robinia* (1), *Spartium* (1), *Wisteria* (1)
Fagaceae	2/5	*Fagus* (1), *Quercus* (4)
Ginkgoaceae	1/1	*Ginkgo* (1)
Hippocastanaceae	1/1	*Aesculus* (1)
Hydrangeaceae	1/1	*Hydrangea* (1)
Juglandaceae	1/1	*Juglans* (1)
Lamiaceae	2/3	*Colquhounia* (1), *Salvia* (2)
Leeaceae	1/1	*Leea* (1)
Lythraceae	2/2	*Lagerstroemia* (1), *Punica* (1)
Magnoliaceae	2/4	*Liriodendron* (1), *Magnolia* (3)
Malvaceae	3/5	*Abutilon* (2), *Hibiscus* (2), *Malvaviscus* (1)
Moraceae	2/11	*Ficus* (8), *Morus* (3)
Musaceae	1/1	*Musa* (1)
Myricaceae	1/1	*Myrica* (1)
Myrsinaceae	1/1	*Myrsine* (1)
Myrtaceae	3/3	*Callistemon* (1), *Eucalyptus* (1), *Myrtus* (1)
Nyctaginaceae	1/1	*Bougainvillea* (1)
Oleaceae	3/8	*Forsythia* (2), *Jasminum* (4), *Ligustrum* (2)
Onagraceae	1/2	*Fuchsia* (2)
Pinaceae	5/6	*Abies* (1), *Cedrus* (1), *Picea* (1), *Pinus* (2), *Tsuga* (1)
Platanaceae	1/1	*Platanus* (1)
Poaceae	2/2	*Arundo* (1), *Drepanostachyum* (1)
Podocarpaceae	1/1	*Afrocarpus* (1)
Proteaceae	1/1	*Grevillea* (1)
Ranunculaceae	1/2	*Clematis* (2)
Rhamnaceae	1/2	*Rhamnus* (2)
Rosaceae	14/32	*Amygdalus* (1), *Cerasus* (1), *Cotoneaster* (1), *Eriobotrya* (1), *Kerria* (1), *Malus* (1), *Padus* (1), *Prinsepia* (1), *Prunus* (6), *Pyracantha* (1), *Pyrus* (3), *Rosa* (7), *Rubus* (5), *Spiraea* (2)
Rubiaceae	2/2	*Himalrandia* (1), *Rubia* (1)
Rutaceae	3/4	*Citrus* (2), *Skimmia* (1), *Zanthoxylum* (1)
Salicaceae	2/5	*Populus* (2), *Salix* (3)
Scrophulariaceae	1/1	*Paulownia* (1)
Smilacaceae	1/1	*Smilax* (1)
Solanaceae	3/7	*Cestrum* (3), *Datura* (1), *Solanum* (3)

Name of Family	Genus/Species	Total No of Genera and Species
Taxaceae	1/1	*Taxus* (1)
Taxodiaceae	3/3	*Cryptomeria* (1), *Cunninghamia* (1), *Taxodium* (1)
Theaceae	1/1	*Camellia* (1)
Thymelaeaceae	1/1	*Daphne* (1)
Tropaeolaceae	1/1	*Tropaeolum* (1)
Ulmaceae	2/2	*Celtis* (1), *Ulmus* (1)
Urticaceae	2/2	*Debregeasia* (1), *Girardinia* (1)
Verbenaceae	2/2	*Aloysia* (1), *Duranta* (1)
Vitaceae	1/1	*Parthenocissus* (1)

Appendix III

INDEX TO ETHNOBOTANICAL USES OF PLANTS OF HERITAGE SITES

Uses	Botanical Names
Anthelmintic	*Aesculus indica, Myrsine africana, Ricinus communis, Rubia cordifolia*
Antidote	*Datura stramonium, Ficus religiosa, Rubus niveus*
Aphrodisiac	*Prunus amygdalus*
Arthritis	*Aesculus indica*
Asthma	*Datura stramonium, Opuntia dillenii, Pistacia khinjuk*
Astringent	*Pyrus communis, Rhododendron arboreum, Rosa moschata*
Avenue	*Acacia dealbata, Eucalyptus globulus, Grevillea robusta, Jacaranda acutifolia, Populus ciliata, Populus nigra, Robinia pseudoacacia*
Beverages	*Taxus baccata*
Blood Purifier	*Myrica esculenta, Myrsine africana, Populus ciliata, Zanthoxylum armatum*
Boils	*Datura stramonium, Ficus auriculata, Hypericum oblongifolium, Hypericum patulum*
Bone Fracture	*Agave americana, Cerasus cerasoides*
Breast Cancer	*Taxus baccata*
Brooms	*Arundo donax, Drepanostachyum falcatum, Myrsine africana, Pinus roxburghii*
Burns	*Cunninghamia lanceolata, Prunus amygdalus, Sambucus nigra*
Cardiovascular Disorders	*Ginkgo biloba*
Carminative	*Myrica esculenta*
Catarrh	*Sambucus nigra*
Chest Infection	*Solanum virginianum*
Cold	*Taxus baccata, Zanthoxylum armatum*
Constipation	*Euonymus tingens, Ficus palmata, Prunus amygdalus, Ricinus communis, Rosa damascena*
Coolant	*Pyrus pashia, Xanthium strumarium*

Uses	Botanical Names
Cough	*Ficus palmata*, *Ficus religiosa*, *Pistacia khinjuk*, *Punica granatum*, *Solanum virginianum*, *Taxus baccata*, *Zanthoxylum armatum*
Cuts	*Hypericum patulum*, *Lagerstroemia indica*, *Leea asiatica*
Culinary Preparation	*Morus alba*, *Rhododendron arboreum*
Dental Problems	*Cerasus cerasoides*, *Desmodium multiflorum*, *Ficus religiosa*, *Jasminum grandiflorum*, *Jasminum humile*, *Juglans regia*, *Ricinus communis*, *Robinia pseudoacacia*, *Salix pycnostachya* var. *oxycarpa*, *Zanthoxylum armatum*
Diabetes	*Thuja orientalis*
Diarrhoea	*Eriobotrya japonica*,
Digestion (Vet.)	*Myrsine africana*
Diuretic	*Xanthium strumarium*
Dye	*Juglans regia*, *Rubia cordifolia*
Dysentery	*Asparagus racemosus*, *Pyracantha crenulata*
Edible	*Abutilon* X *hybridum*, *Acacia dealbata*, *Actinidia chinensis*, *Agave americana*, *Aloysia citriodora*, *Amygdalus mira*, *Berberis aristata*, *Berberis lycium*, *Buddleja asiatica*, *Cerasus cerasoides*, *Chamaerops humilis*, *Citrus limon*, *Citrus maxima*, *Coriaria nepalensis*, *Cornus capitata*, *Daphne papyracea*, *Debregeasia saeneb*, *Diospyros kaki*, *Diospyros lotus*, *Drepanostachyum falcatum*, *Elaeagnus umbellata*, *Eriobotrya japonica*, *Ficus auriculata*, *Ficus palmata*, *Ficus religiosa*, *Ficus sarmentosa*, *Ficus subincisa*, *Girardinia diversifolia*, *Hibiscus rosa-sinensis*, *Hibiscus syriacus*, *Indigofera dosua*, *Juglans regia*, *Leea asiatica*, *Lonicera japonica*, *Malus pumila*, *Morus alba*, *Morus nigra*, *Morus serrata*, *Musa paradisiaca*, *Myrica esculenta*, *Myrsine africana*, *Opuntia dillenii*, *Padus cornuta*, *Parthenocissus semicordata*, *Pinus roxburghii*, *Pistacia khinjuk*, *Prinsepia utilis*, *Prunus amygdalus*, *Prunus armeniaca*, *Prunus domestica*, *Prunus persica*, *Prunus salicina*, *Punica granatum*, *Pyracantha crenulata*, *Pyrus communis*, *Pyrus pashia*, *Pyrus pyrifolia*, *Rhamnus persica*, *Rhododendron arboreum*, *Rosa chinensis*, *Rosa damascena*, *Rosa moschata*, *Rosa multiflora*, *Rubus biflorus*, *Rubus ellipticus*, *Rubus macilentus*, *Rubus niveus*, *Rubus paniculatus*, *Skimmia laureola*, *Smilax aspera*, *Solanum virginianum*, *Spiraea canescens*, *Taxus baccata*, *Viburnum cotinifolium*, *Viburnum mullaha*, *Yucca aloifolia*, *Zanthoxylum armatum*
Emollient	*Salix babylonica*
Eye Diseases	*Berberis aristata*, *Euonymus tingens*

Uses	Botanical Names
Fever	*Indigofera dosua, Smilax aspera, Solanum virginianum*
Fibre	*Agave americana, Chamaerops humilis, Clematis buchananiana, Debregeasia saeneb, Girardinia diversifolia, Spartium junceum, Taxodium distichum,Yucca aloifolia*
Flavourant	*Aloysia citriodora, Debregeasia saeneb, Hypericum oblongifolium, Spiraea canescens, Zanthoxylum armatum*
Fodder	*Aesculus indica, Berberis aristata, Berberis lycium, Buddleja crispa, Buxus wallichiana, Celtis australis, Cerasus cerasoides, Colquhounia coccinea, Cornus macrophylla, Cotoneaster bacillaris, Debregeasia saeneb, Desmodium elegans, Desmodium multiflorum, Drepanostachyum falcatum, Ficus auriculata, Ficus palmata, Ficus sarmentosa, Ficus subincisa, Girardinia diversifolia, Hedera helix, Himalrandia tetrasperma, Hypericum oblongifolium, Ilex dipyrena, Indigofera dosua, Leea asiatica, Lonicera quinquelocularis, Morus alba, Morus nigra, Morus serrata, Myrica esculenta, Myrsine africana, Padus cornuta, Parthenocissus semicordata, Pistacia khinjuk, Populus ciliata, Populus nigra, Pyrus pashia, Quercus floribunda, Quercus incana, Quercus oblongata, Rhamnus persica, Rhamnus virgatus, Robinia pseudoacacia, Rosa moschata, Rubus ellipticus, Rubus niveus, Salix babylonica, Ulmus wallichiana, Viburnum cotinifolium, Viburnum mullaha*
Fuel	*Abies pindrow, Acacia dealbata, Acer oblongum, Berberis aristata, Berberis lycium, Cerasus cerasoides, Colquhounia coccinea, Coriaria nepalensis, Cornus capitata, Debregeasia saeneb, Desmodium multiflorum, Duranta erecta, Elaeagnus umbellata, Erythrina arborescens, Eucalyptus globulus, Euonymus tingens, Juniperus squamata, Lonicera quinquelocularis, Morus alba, Morus nigra, Morus serrata, Myrsine africana, Padus cornuta, Picea smithiana, Pinus roxburghii, Populus nigra, Prunus domestica, Pyrus pashia, Quercus floribunda, Quercus incana, Quercus oblongata, Rhamnus persica, Rhododendron arboreum, Robinia pseudoacacia, Ulmus wallichiana, Viburnum cotinifolium*
Gargle	*Sambucus nigra*
Gastro-Intestinal Disorder	*Inula cappa, Smilax aspera*
Gonorrohea	*Sarcococca pruniformis*
Green Manure	*Grevillea robusta*
Hair Care (Baldness, Dandruff, Hair Lice)	*Datura stramonium, Hedera nepalensis, Hibiscus rosa-sinensis, Pyrus pashia, Ricinus communis*

Uses	Botanical Names
Headache	*Girardinia diversifolia, Nerium oleander, Salix pycnostachya* var. *oxycarpa, Solanum virginianum*
Heart Complaints	*Rhododendron arboreum*
Hedge	*Abutilon* X *hybridum, Afrocarpus gracilior, Agave americana, Arundo donax, Bougainvillea glabra, Buddleja asiatica, Buxus wallichiana, Clematis montana, Cupressus torulosa, Drepanostachyum falcatum, Duranta erecta, Elaeagnus umbellata, Erythrina arborescens, Euonymus japonicus, Ficus neriifolia, Forsythia* X *intermedia, Hedera canariensis, Hedera colchica, Hedera helix, Hibiscus syriacus, Hydrangea macrophylla, Hypericum hookerianum, Juniperus chinensis, Ligustrum lucidum, Ligustrum* X *vicaryi, Lonicera japonica, Myrsine africana, Myrtus communis, Nerium oleander, Opuntia dillenii, Populus ciliata, Populus nigra, Prinsepia utilis, Pyracantha crenulata, Rosa banksiae, Rosa moschata, Rosa multiflora, Rubus niveus, Sambucus nigra, Solanum laxum, Spartium junceum, Thuja occidentalis, Yucca aloifolia*
House-Hold Articles	*Cryptomeria japonica, Drepanostachyum falcatum*
Incense	*Cryptomeria japonica, Cupressus torulosa, Juniperus horizontalis, Juniperus squamata, Skimmia laureola, Taxus baccata*
Indigestion	*Viburnum mullaha*
Inflammation	*Nerium oleander, Xanthium strumarium*
Injuries	*Opuntia dillenii*
Lactation	*Ficus neriifolia, Quercus oblongata*
Laxative	*Myrsine africana, Pyrus communis, Rubia cordifolia*
Leucorrhoea	*Rosa damascena*
Magico-Religious	*Cerasus cerasoides, Cestrum nocturnum, Citrus limon, Datura stramonium, Ficus religiosa, Hibiscus rosa-sinensis, Jasminum grandiflorum, Jasminum humile, Jasminum multiflorum, Malvaviscus arboreus* var. *mexicanus, Musa paradisiaca, Punica granatum, Prinsepia utilis, Rhamnus virgatus, Ricinus communis, Rosa alba, Rosa chinensis, Rosa damascena, Tropaeolum majus, Zanthoxylum armatum*
Menstruation Complaints	*Asparagus racemosus, Rubia cordifolia*
Miscarriage	*Asparagus racemosus*
Mouth Odour	*Zanthoxylum armatum*
Mouth Sores/ Ulcers	*Eucalyptus globulus, Ficus religiosa, Jasminum grandiflorum, Pyrus pashia, Solanum virginianum, Zanthoxylum armatum*
Mouth Wash	*Sambucus nigra*

Uses	Botanical Names
Nail Disorder	*Ilex dipyrena*
Nose Bleeding	*Pyrus pashia, Rhododendron arboreum*
Oil	*Prunus amygdalus, Prunus armeniaca, Prunus persica*
Ornamental	*Abutilon* X *hybridum, Abutilon megapotamicum, Acer negundo, Acer palmatum, Afrocarpus gracilior, Agave americana, Aloysia citriodora, Araucaria bidwillii, Araucaria columnaris, Araucaria cunninghamii, Asparagus aethiopicus, Asparagus densiflorus, Asparagus setaceus, Asparagus virgatus, Bougainvillea glabra, Buddleja asiatica, Buddleja davidii, Buxus wallichiana, Callistemon lanceolatus, Camellia japonica, Campsis radicans, Cestrum elegans, Cestrum nocturnum, Cestrum parqui, Chamaerops humilis, Clematis montana, Colquhounia coccinea, Cryptomeria japonica, Cunninghamia lanceolata, Cupressus cashmeriana, Cupressus torulosa, Cycas revoluta, Diospyros lotus, Duranta erecta, Erythrina arborescens, Euonymus japonicus, Euonymus lucidus, Fagus sylvatica, Ficus benjamina, Ficus elastica, Ficus religiosa, Forsythia* X *intermedia, Forsythia ovata, Fuchsia* X *hybrida, Fuchsia magellanica, Ginkgo biloba, Grevillea robusta, Hedera canariensis, Hedera colchica, Hedera helix, Hibiscus rosa-sinensis, Hibiscus syriacus, Hydrangea macrophylla, Hypericum hookerianum, Hypericum oblongifolium, Ipomoea purpurea, Jacaranda acutifolia, Jasminum grandiflorum, Jasminum humile, Jasminum mesnyi, Jasminum multiflorum, Juniperus chinensis, Juniperus horizontalis, Juniperus procera, Juniperus pseudosabina, Juniperus squamata, Kerria japonica, Lagerstroemia indica, Ligustrum lucidum, Ligustrum* X *vicaryi, Liriodendron tulipifera, Lonicera japonica, Magnolia grandiflora, Magnolia obovata, Magnolia* X *soulangeana, Malvaviscus arboreus* var. *mexicanus, Myrtus communis, Nerium oleander, Opuntia dillenii, Paulownia tomentosa, Platanus orientalis, Prunus laurocerasus, Quercus robur, Rhapis excelsa, Rhododendron simsii* var. *vittatum, Rosa alba, Rosa banksiae, Rosa chinensis, Rosa damascena, Rosa jacquemontii, Rosa multiflora, Salix alba, Salix babylonica, Salvia leucantha, Salvia splendens, Sambucus nigra, Schefflera arboricola, Solanum laxum, Solanum pseudocapsicum, Spartium junceum, Spiraea cantoniensis,*
	Taxodium distichum, Taxus baccata, Thuja occidentalis, Thuja orientalis, Thujopsis dolabrata, Trachelospermum jasminoides, Tropaeolum majus, Tsuga dumosa, Wisteria sinensis, Yucca aloifolia

Uses	Botanical Names
Pain and Aches	*Opuntia dillenii, Pistacia khinjuk, Prunus amygdalus, Sarcococca pruniformis, Smilax aspera*
Piles	*Jasminum grandiflorum, Punica granatum*
Poison (Vet.)	*Buxus wallichiana, Clematis buchananiana, Lyonia ovalifolia, Prunus laurocerasus, Solanum laxum*
Pulp	*Picea smithiana*
Refrigerant	*Citrus maxima, Malvaviscus arboreus* var. *mexicanus, Punica granatum, Rhododendron arboreum, Rubus niveus*
Respiratory Problems	*Eucalyptus globulus, Pistacia khinjuk, Taxus baccata*
Rheumatic Pain	*Citrus limon, Girardinia diversifolia, Prinsepia utilis, Prunus persica, Ricinus communis*
Root Stock	*Cerasus cerasoides, Prunus persica*
Scalds	*Cunninghamia lanceolata, Sambucus nigra*
Skin Diseases	*Buddleja asiatica, Girardinia diversifolia, Hedera nepalensis, Jasminum grandiflorum, Jasminum humile, Juniperus squamata, Lyonia ovalifolia, Nerium oleander, Prunus amygdalus, Rhododendron arboreum, Ricinus communis, Rubia cordifolia, Salix alba, Salix babylonica, Xanthium strumarium*
Smallpox	*Xanthium strumarium*
Sores	*Datura stramonium*
Spasm	*Taxus baccata*
Sprain	*Clematis buchananiana*
Suppuration of Boils	*Daphne papyracea, Hibiscus rosa-sinensis, Jasminum multiflorum, Juniperus squamata, Prinsepia utilis, Ricinus communis, Sarcococca pruniformis*
'Sur' (Local Drink)	*Buddleja crispa, Myrica esculenta, Rhododendron arboreum, Skimmia laureola, Zanthoxylum armatum*
Swelling	*Girardinia diversifolia, Rubia cordifolia, Xanthium strumarium*
Thatching	*Tsuga dumosa*
Tonic	*Asparagus racemosus, Prunus amygdalus, Prunus armeniaca, Rosa damascena, Rubia cordifolia*
Tonsil	*Desmodium elegans*
Ulcer	*Cunninghamia lanceolata, Jasminum humile, Xanthium strumarium*
Vomiting	*Citrus maxima*

Uses	Botanical Names
Walking Sticks	*Arundo donax, Buddleja asiatica, Buddleja crispa, Cerasus cerasoides, Cornus macrophylla, Cotoneaster bacillaris, Drepanostachyum falcatum, Himalrandia tetrasperma, Lonicera quinquelocularis, Pyracantha crenulata, Pyrus pashia, Quercus floribunda, Spiraea canescens, Viburnum mullaha, Zanthoxylum armatum*
Welcome Arches	*Cerasus cerasoides, Drepanostachyum falcatum, Rhododendron arboreum, Sarcococca pruniformis*
Wood	*Abies pindrow, Acer oblongum, Aesculus indica, Araucaria cunninghamii, Buxus wallichiana, Cedrus deodara, Celtis australis, Cornus macrophylla, Cotoneaster bacillaris, Cryptomeria japonica, Cunninghamia lanceolata, Cupressus cashmeriana, Cupressus torulosa, Diospyros kaki, Drepanostachyum falcatum, Erythrina arborescens, Eucalyptus globulus, Euonymus fimbriatus, Euonymus tingens, Fagus sylvatica, Grevillea robusta, Ilex dipyrena, Jacaranda acutifolia, Juglans regia, Lonicera quinquelocularis, Morus alba, Morus nigra, Morus serrata, Picea smithiana, Pinus roxburghii, Pinus wallichiana, Pistacia khinjuk, Populus ciliata, Populus nigra, Pyrus pashia, Quercus floribunda, Quercus incana, Quercus oblongata, Rhamnus persica, Rhamnus virgatus, Rhododendron arboreum, Robinia pseudoacacia, Salix babylonica, Skimmia laureola, Taxodium distichum, Taxus baccata, Ulmus wallichiana*
Wounds	*Cunninghamia lanceolata, Hypericum oblongifolium, Hypericum patulum, Jacaranda acutifolia, Jasminum humile, Lagerstroemia indica, Leea asiatica, Rosa multiflora, Thuja orientalis*
Miscellaneous	*Agave americana, Aloysia citriodora, Araucaria bidwillii, Arundo donax, Berberis aristata, Cerasus cerasoides, Cycas revoluta, Euonymus fimbriatus, Ficus palmata, Grevillea robusta, Himalrandia tetrasperma, Inula cappa, Picea smithiana, Pinus roxburghii, Prunus amygdalus, Salix alba, Salix babylonica, Spartium junceum, Yucca aloifolia*

Appendix IV
INDEX TO ADDITIONAL USES OF PLANTS

Agricultural implements: *Cornus macrophylla, Rhamnus persica, Rhamnus virgat*us
Astringent: *Rhododendron arboreum, Rosa moschata*
Baldness: *Hibiscus rosa-sinensis*
Basket making: *Salix alba*
Beads: *Euonymous fimbriatus*
Beverages: *Aloysia citriodora*
Blood purifier: *Zanthoxylum armatum*
Boil: *Ficus auriculata*
Bone fracture: *Agave americana*
Broom: *Arundo donax, Myrsine africana, Pinus roxburghii*
Burning sensation: *Rhododendron arboreum*
Constipation: *Rosa damascena, Ricinus communis*
Coolant: *Pyrus pashia*
Cough: *Ficus reliogiosa, Ficus palmata*
Cut: *Hypericum patulum, Lagerstroemia indica*
Dandruff: *Datura stramonium, Pyrus pashia*
Decoration: *Cycas revoluta, Inula cappa, Picea smithiana, Pinus roxburghii*
Dental problems: *Citrus limon, Desmodium multiflorum, Ricinus communis*
Diabetes: *Thuja orientalis*
Diarrhoea: *Eriobotrya japonica*
Dysentery: *Pyracantha crenulata*
Emollient: *Salix babylonica*
Evil spirit: *Prinsepia utilis*
Fever: *Indigofera dosua, Smilax aspera*
Flavourant: *Aloysia citriodora*
Fodder: *Cerasus cerasoides, Cotoneaster bacillaris, Desmodium multiflorum, Hypericum oblongifolium, Leea asiatica, Myrica esculenta, Padus cornuta, Robinia pseudoacacia, Viburnum cotinifolium, Viburnum mullaha*
Fuel: *Cornus capitata, Desmodium multiflorum, Duranta erecta, Erythrina arborescens, Padus cornuta, Prunus domestica, Rhamnus persica, Viburnum cotinifolium*
Gastric ailment: *Smilax aspera*
Gonorrhoea: *Sarcococca pruniformis*
Hair lice: *Hedera nepalensis*

Hairfall: *Datura stramonium*

Headache: *Nerium oleander*, *Salix pycnostachya* var. *oxycarpa*, *Solanum virginianum*

Hedge: *Abutilon* X *hybridum*, *Afrocarpus gracilior*, *Arundo donax*, *Buddleja asiatica*, *Juniperus chinensis*, *Myrsine africana Nerium oleander*, *Prinsepia utilis*, *Pyracantha crenulata*, *Rosa moschata*, *Rosa multiflora*, *Rubus niveus*, *Sambucus nigra*, *Solanum laxum*

Indigestion: *Viburnum mullaha*

Laxative: *Pyrus communis*, *Rubus ellipticus*

Leucorrhoea: *Rosa damascena*

Local wine, Sur: *Hedera helix*, *Skimmia laureola*

Magico religious: *Zanthoxylum armatum*

Minor construction: *Erythrina arborescens*

Mouth odour: *Citrus limon*

Mouth Sore: *Solanum virginianum*

Nail disorder: *Ilex dipyrena*

Nose bleeding: *Pyrus pashia*

Ornamental: *Acer negundo*, *Agave americana*, *Aloysia citriodora*, *Colquhounia coccinea*, *Diospyros lotus*, *Erythrina arborescens*, *Euonymus lucidus*, *Fagus sylvatica*, *Hypericum oblongifolium*, *Opuntia dillenii*, *Taxodium distichum*, *Taxus baccata*, *Yucca aloifolia*

Packaging: *Pinus roxburghii*

Pen: *Arundo donax*, *Cerasus cerasoides*

Pile: *Punica granatum*

Refrigerant: *Malvaviscus arboreus* var. *mexicanus*, *Rubus niveus*

Rootstock: *Cerasus cerasoides*

Sacred: *Cerasus cerasoides*, *Citrus limon*, *Jasminum grandiflorum*, *Jasminum humile*, *Malvaviscus arboreus* var. *mexicanus*, *Musa paradisiaca*, *Punica granatum*, *Rhamnus virgatus*, *Ricinus communis*, *Rosa alba*, *Rosa chinensis*, *Rosa damascena*, *Tropaeolum majus*

Scouring teeth: *Desmodium multiflorum*

Skin diseases: *Cunninghamia lanceolata*, *Rubia cordifolia*

Skin eruption: *Salix alba*

Soil erosion: *Arundo donax*

Sprain: *Clematis buchananiana*

Suppuration of boils: *Daphne papyracea*, *Sarcococca pruniformis*

Swelling: *Xanthium strumarium*

Tonsil: *Desmodium elegans*

Toothbrush: *Cerasus cerasoides*, *Robinia pseudoacacia*

Ulcer: *Cunninghamia lanceolata*

Utensil: *Cryptomeria japonica*

Walking sticks: *Arundo donax, Buddleja crispa, Cornus macrophylla, Drepanostachyum falcatum, Lonicera quinquelocularis, Pyracantha crenulata, Zanthoxylum armatum*

Washing cloth: *Himalrandia tetrasperma*

Welcome gate: *Cerasus cerasoides, Drepanostachyum falcatum, Sarcococca pruniformis*

Wounds: *Hypericum patulum, Jacaranda acutifolia, Lagerstroemia indica, Thuja orientalis*

Appendix V

INDEX TO PLANT PARTS USED FOR VARIOUS MEDICINAL PURPOSES

Aerial Parts	**Daphne papyracea**
Bark	*Cerasus cerasoides, Cunninghamia lanceolata, Euonymus tingens, Ficus religiosa, Populus ciliata, Rubia cordifolia, Salix alba, Taxus baccata, Zanthoxylum armatum*
Flower	*Buddleja asiatica, Hibiscus rosa-sinensis, Hypericum oblongifolium, Indigofera dosua, Jasminum grandiflorum, Lagerstroemia indica, Lyonia ovalifolia, Opuntia dillenii, Rhododendron arboreum, Rosa chinensis, Rosa damascena, Rosa moschata, Sambucus nigra*
Fruit	*Aesculus indica, Citrus maxima, Datura stramonium, Ficus palmata, Myrsine africana, Punica granatum, Pyracantha crenulata, Pyrus communis, Pyrus pashia, Rosa multiflora, Rubus ellipticus, Sambucus nigra, Solanum virginianum, Viburnum mullaha, Xanthium strumarium, Zanthoxylum armatum*
Gall	*Pistacia khinjuk*
Latex	*Ficus palmata, Ficus religiosa*
Leaves	*Aloysia citriodora, Citrus limon, Clematis buchananiana, Datura stramonium, Desmodium elegans, Eucalyptus globulus, Ginkgo biloba, Girardinia diversifolia, Hypericum oblongifolium, Hypericum patulum, Ilex dipyrena, Jacaranda acutifolia, Jasminum grandiflorum, Jasminum humile, Jasminum multiflorum, Leea asiatica, Myrica esculenta, Myrsine africana, Nerium oleander, Ricinus communis, Rubia cordifolia, Salix pycnostachya* var. *oxycarpa, Sarcococca pruniformis, Taxus baccata, Thuja orientalis, Xanthium strumarium, Zanthoxylum armatum*
Rhizome	*Agave americana*
Root	*Berberis aristata, Hypericum patulum, Inula cappa, Jasminum humile, Prinsepia utilis, Rubia cordifolia, Salix babylonica, Sarcococca pruniformis, Smilax aspera, Solanum virginianum, Xanthium strumarium*
Root-bark	*Salix pycnostachya* var. *oxycarpa*
Seed	*Aesculus indica, Prinsepia utilis, Prunus amygdalus, Prunus persica, Ricinus communis, Solanum virginianum, Thuja orientalis,* Xanthium strumarium

Aerial Parts	Daphne papyracea
Stem	*Solanum virginianum*
Twig	*Pistacia khinjuk, Robinia pesudoacacia, Rubus niveus*
Whole Plant	*Girardinia diversifolia, Hedera helix, Hedera nepalensis, Juniperus squamata*
Wood	*Cunninghamia lanceolata, Desmodium multiflorum, Pistacia khinjuk*

Appendix VI

INDEX TO LOCAL NAMES OF PLANTS OF HERITAGE SITES

S.No.	Local Name/s	Botanical Name	Page No.
30.	Bhekuyi	*Colquhounia coccinea*	143
31.	Bhelar	*Lyonia ovalifolia*	209
32.	Bicchu Buti	*Girardinia diversifolia*	178
33.	Boganvillea	*Bougainvillea glabra*	126
34.	Bottle Brush	*Callistemon lanceolatus*	130
35.	Braah, Brass	*Rhododendron arboreum*	250
36.	Bragh Bel	*Smilax aspera*	272
37.	Brahm Phool	*Hibiscus syriacus*	184
38.	Brera	*Ulmus wallichiana*	285
39.	Buddleya	*Buddleja davidii*	128
40.	Buddliya, Bana	*Buddleja asiatica*	127
41.	Bunya Araucaria	*Araucaria bidwillii*	116
42.	Camellia	*Camellia japonica*	131
43.	Chakhu	*Spiraea canescens*	276
44.	Chakotra	*Citrus maxima*	140
45.	Chamar-Chumb, Raabal	*Daphne papyracea*	151
46.	Chameli	*Jasminum grandiflorum*	193
47.	Chameli	*Jasminum mesnyi*	195
48.	Chameli	*Jasminum multiflorum*	196
49.	Chamla	*Desmodium elegans*	154
50.	Changma Beeuins	*Salix pycnostachya* var. *oxycarpa*	267
51.	Charkha, Holly Tree	*Ilex dipyrena*	189
52.	Charoti Mirch	*Solanum pseudocapsicum*	274
53.	Chaunsh	*Lonicera quinquelocularis*	208
54.	Cherry-Laurel	*Prunus laurocerasus*	237
55.	Chil	*Pinus roxburghii*	227
56.	China Rose	*Hibiscus rosa-sinensis*	183
57.	Chinar	*Platanus orientalis*	230
58.	Chinese Juniperus	*Juniperus chinensis*	198
59.	Chinese Lantern	*Abutilon* X *hybridum*	106
60.	Chuli, Khumani	*Prunus armeniaca*	235
61.	Cryptomeria	*Cryptomeria japonica*	146
62.	Cycas	*Cycas revoluta*	150
63.	Cypress	*Taxodium distichum*	277
64.	Dakari	*Hedera helix*	181
65.	Deaar, Devdar	*Cedrus deodara*	132
66.	Dhaab	*Viburnum mullaha*	286
67.	Dhatura	*Datura stramonium*	152

S.No.	Local Name/s	Botanical Name	Page No.
106.	Junglee Dakh, Ricchnabu	*Viburnum cotinifolium*	285
107.	Juniper	*Juniperus horizontalis*	199
108.	Juniper	*Juniperus pseudosabina*	200
109.	Kabra	*Ficus benjamina*	168
110.	Kachog	*Coriaria nepalensis*	143
111.	Kail	*Pinus wallichiana*	228
112.	Kainth	*Pyrus pashia*	243
113.	Kakad	*Pistacia khinjuk*	229
114.	Kaner	*Nerium oleander*	221
115.	Kanhor, Khanor	*Aesculus indica*	111
116.	Kant	*Solanum virginianum*	274
117.	Kaphal	*Myrica esculenta*	218
118.	Kashmal	*Berberis aristata*	123
119.	Kashmal	*Berberis lycium*	125
120.	Kashmiri Cypress	*Cupressus cashmeriana*	148
121.	Kathee	*Indigofera dosua*	190
122.	Kaunia-Bali	*Clematis montana*	142
123.	Kela	*Musa paradisiaca*	216
124.	Keora	*Agave americana*	113
125.	Khair	*Acacia dealbata*	107
126.	Khirak	*Celtis australis*	134
127.	Kimu	*Morus serrata*	215
128.	Kiraya	*Hedera colchica*	180
129.	Kiwi	*Actinidia chinensis*	110
130.	Klenchu	*Rubus niveus*	263
131.	Kuja	*Rosa moschata*	258
132.	Lady Palm	*Rhapis excelsa*	249
133.	Lantern Flower	*Abutilon megapotamicum*	107
134.	Lokat	*Eriobotrya japonica*	161
135.	Lushai	*Lonicera japonica*	207
136.	Lyoon	*Hypericum oblongifolium*	187
137.	Magnolia	*Magnolia grandiflora*	210
138.	Majnu	*Salix babylonica*	266
139.	Manjith	*Rubia cordifolia*	260
140.	Maple	*Acer negundo*	108
141.	Marmakoul	*Euonymus tingens*	166
142.	Martoi	*Desmodium multiflorum*	155
143.	Morpankh	*Thuja orientalis*	280

S.No.	Local Name/s	Botanical Name	Page No.
182.	Shakarpara	*Prunus salicina*	239
183.	Shamshad	*Buxus wallichiana*	129
184.	Shehnai Phool	*Campsis radicans*	132
185.	Silver Fir	*Grevillea robusta*	179
186.	Simbli	*Cotoneaster bacillaris*	145
187.	Spirea	*Spiraea cantoniensis*	277
188.	Sprengeri Asparagus	*Asparagus aethiopicus*	119
189.	Star Jasmine	*Trachelospermum jasminoides*	282
190.	Syaru	*Debregeasia saeneb*	153
191.	Teelahan	*Cornus capitata*	144
192.	Thalta	*Rhamnus persica*	248
193.	Thanara	*Juniperus squamata*	201
194.	Thingia	*Tsuga dumosa*	284
195.	Thuja	*Thuja occidentalis*	279
196.	Thujopsis	*Thujopsis dolabrata*	282
197.	Thuno	*Taxus baccata*	278
198.	Timbal	*Ficus auriculata*	167
199.	Tindu Phool	*Tropaeolum majus*	283
200.	Tirmir	*Zanthoxylum armatum*	290
201.	Toot	*Morus alba*	213
202.	Toot	*Morus nigra*	215
203.	Tulip Tree	*Liriodendron tulipifera*	206
204.	Umbrella Plant	*Schefflera arboricola*	271
205.	Vaimi	*Amygdalus mira*	115
206.	Velvet Salvia	*Salvia leucantha*	267
207.	Verbena	*Aloysia citriodora*	114
208.	Videshi Phool	*Paulownia tomentosa*	225
209.	Vilayati Suri	*Opuntia dillenii*	222
210.	Weavers Broom	*Spartium junceum*	275
211.	Wisteria	*Wisteria sinensis*	287
212.	Zarainth	*Pyrus pyrifolia*	244